Handbook of Marine Mammals

Volume 6 The Second Book of Dolphins and the Porpoises

Bottlenose dolphin, *Tursiops truncatus*, mother and 5-month-old calf. The calf represents the second generation in human care. Both parents of the calf came from the northern Gulf of Mexico. (Official US Navy Photograph.)

Dedication

In memory of our collaborators in this series,

Michael Bigg
Nigel Bonner
Francis Fay
Stephen Leatherwood
Masaharu Nishiwaki
Howard Winn

Handbook of Marine Mammals

Volume 6 The Second Book of Dolphins and the Porpoises

Edited by

SAM H. RIDGWAY

Naval Ocean Systems Center
San Diego, California, USA

and

SIR RICHARD HARRISON, F.R.S.

Downing College
Cambridge, UK

ACADEMIC PRESS

San Diego London Boston
New York Sydney Tokyo Toronto

This book is printed on acid-free paper

ACADEMIC PRESS
24–28 Oval Road, London NW1 7DX, UK
http://www.hbuk.co.uk/ap/

ACADEMIC PRESS
525 B Street, Suite 1900, San Diego, California 92101-4495, USA
http://www.apnet.com

ISBN 0-12-588506-7

A catalogue record for this book is available from the British Library

Typeset by Keyword Typesetting Services, Wallington, Surrey
Printed in Great Britain by The University Press, Cambridge
99 00 01 02 03 04 CU 9 8 7 6 5 4 3 2 1

Contents

Contributors

Numbers in parentheses indicate the pages on which the authors' contributions begin.

H. J. BERNARD (245), 4175 Hana Highway, Haiku, HI 96708, USA

R. L. BROWNELL, JR (1, 31, 57, 85, 105, 121, 357, 379, 393), Southwest Fisheries Science Center, National Marine Fisheries Service, P.O. Box 271, La Jolla, CA 92038-0271, USA

D. K. CALDWELL (deceased) (183), Marineland of Florida, Rt 1, Box 122, St Augustine, FL 32086, USA

M. C. CALDWELL (deceased) (183), Marineland of Florida, Rt 1, Box 122, St Augustine, FL 32086, USA

F. CIPRIANO (85) Kewalo Marine Laboratory, 41 Ahui Street, Honolulu, HI 96813, USA

P. J. CLAPHAM (379, 393), Smithsonian Institution, NHB 390, MRC 108, Washington, DC 20560, USA

E. A. CRESPO (105), Centro Nacional Patagónico, CONICET, Boulevard Brown s/n, 9120 Puerto Madryn, Chubut, Argentina

M. E. DAHLHEIM (281), National Marine Mammal Laboratory, National Marine Fisheries Services, 7600 Sand Point Way NE, Seattle, WA 98115, USA

M. A. DONAHUE (105, 121), Southwest Fisheries Science Center, National Marine Fisheries Service, P.O. Box 271, La Jolla, CA 92038-0271, USA

L. T. FINDLEY (357), Carretera al Varadero Nacional km. 6.6, Apartado Postal 284, CP 85400, Guaymas, Sonora, Mexico

K. A. FORNEY (57), Southwest Fisheries Science Center, National Marine Fisheries Service, P.O. Box 271, La Jolla, CA 92038-0271, USA

J. E. HEYNING (281), Section of Mammals, Natural History Museum of Los Angeles County, 900 Expo Blvd, Los Angeles, CA 90007, USA

W. J. HOUCK (443), P.O. Box 2190, McKinleyville, CA 95521, USA

T. A. JEFFERSON (443), Ocean Park Conservation Foundation, Ocean Park Corporation, Aberdeen, Hong Kong

T. KASUYA (411), Faculty of Bioresources, Mie University, Kamihama-cho, Tsu, Mie, 514 Japan

C. C. KINZE (1, 31), Zoologisk Museum, Universitetsparken 15, DK-2100 Copenhagen, Denmark

S. KRUSE (183), Marine Mammals Mgt, 1011 East Tudor, Anchorage, AK 99503, USA

J. LIEN (1), Whales Research Group, Memorial University of Newfoundland, 230 Mt Scio Road, St John's, Newfoundland A1C 5S7, Canada

K. M. McCLUNE (213), Hubbs Research Institute, 2595 Ingraham Street, San Diego, CA 92109, USA

D. K. ODELL (213), Sea World Inc., 7007 Sea World Drive, Orlando, FL 32821-8097, USA

J. READ (323), Duke University, Nicholas School of the Environment, Marine Laboratory, 135 Duke Marine Lab Road, Beaufort, NC 28516, USA

R. R. REEVES (1, 31), 27 Chandler Lane, Hudson, Québec, J0P 1H0, Canada

S. B. REILLY (245), Southwest Fisheries Science Center, National Marine Fisheries Service, P.O. Box 271, La Jolla, CA 92038-0271, USA

M. D. SCOTT (137), Inter-American Tropical Tuna Commission, c/o Scripps Institute of Oceanography, La Jolla, CA 92037, USA

C. SMEENK (1, 31), Mammalia, National Museum of Natural History, P.O. Box 9517, 2300 RA Leiden, The Netherlands

O. VIDAL (357), Water Branch UNEP, P.O. Box 47074, Nairobi, Kenya

W. A. WALKER (57), P.O. Box 1346, Vashon Island, WA 98070, USA

R. S. WELLS (137), Mote Marine Laboratory, 1600 Thompson Parkway, Sarasota, FL 34236, USA

Preface to the Series

The idea of producing a Handbook of Marine Mammals of this type was the result of many discussions between the authors and the late Mr John Cruise of Academic Press during his visits to Cambridge over ten years ago. It would be, it was hoped, a comprehensive account of all marine mammal species with each chapter written by someone who had actually worked on a particular form. We felt that this would give a more personal flavour to each chapter, provide opportunity to include original observations, and increase accuracy. Some species of marine mammal are worldwide in their distribution and known to many investigators, others are restricted to certain rivers, lakes, seas and even to relatively short ranges along a coast. We know of no one person who claims to have seen every extant species alive in its environment. We hoped that such a series of chapters written by experts on each species would naturally draw attention to subtle differences in form, coloration, behaviour and many other characteristics, many of which might not be known to an ordinary reviewer.

There are obvious difficulties and problems to vex the editors of multi-author works and they all affected this one. We knew it would take time; active marine mammalogists spend long periods away from home on expeditions and at conferences. The majority of contributors have remained loyal to the project but a few have had to withdraw with inevitable difficulties over replacement and with continual postponement of completion dates. New information has become available about many species over the last ten years, due mainly to efforts by countries to find out more about their local species but also to enterprising attempts to display little-known species in captivity. The last ten years have also seen a much increased interest in conservation of marine mammals and an awareness by the public of their importance as interesting animals from many points of view. There has been much debate about whether a moratorium should be introduced by international agreement over the taking of large, or even any, cetaceans. The arguments have affected many of the classical views held on harvesting, culling, controlling or managing stocks of all marine mammal species. They have exposed our ignorance on many basic facts concerning reproduction, school composition and behaviour, and even what is really meant by many of the terms and concepts used so glibly in marine mammalogy in the past.

The last decade has seen much burgeoning in scientific investigation of marine mammals, though not always in the direction leading to important results. For example, the delightfully satisfying construction of mathematical models has occupied many folk. The results give joy to the constructor, the administrator, the politician and this whole game is not only fun but really quite cheap. What the computer tells us is only as good as the data put into it which are often incomplete or inadequate for detailed analysis. At the very best the results are always based on past events. What is happening now in the seas is still mysterious as it always was.

Many advances have been made in our knowledge of how to keep marine mammals in captivity, about their diseases and the associated pathological changes, and on treatment. The life styles, life history, expectation of life, growth rates, vulnerability and responses to adverse effects of disease and pollution are becoming better known and understood. Formulae have been constructed to provide the optimum size of pool for holding marine mammals of varying number and size. Codes of practice have been issued to ensure the welfare of marine mammals during transport and in captivity. This concern has increased life expectation of captive animals, and has improved their lot from being mere objects of curiosity and participants in circus acts. The list of different species displayed to the public has steadily lengthened; to an extent that many countries demand permits to be issued for taking of rare and even not so rare animals from local stocks, and records must be kept about captive animals.

Film and television companies have realized the increasing popularity of productions involving marine mammals, and this has spread knowledge more than any other factor, and unfortunately in some cases misinformation, about the various species. Conservationists have also promoted public awareness of which nations still hunt whales and the purposes to which the carcasses are put. These groups have put pressure on national and international bodies to provide more accurate information about endangered species and the state of stocks generally. Again and again it has to be confessed that many species are still little known, their habits obscure, and assessment of their numbers mere guesswork.

Problems of nomenclature are bound to arise in presenting any group as diverse and widely dispersed as marine animals. We have avoided strict rules and have left decisions as to the use of particular names to each expert author. Authors were asked to summarize the scientific and common nomenclature of the genera and species about which they have written. As a guide to the various genera and species Dale W. Rice's "A List of Marine Mammals of the World" (NOAA Technical Report NMFS SSRF-711, US Department of Commerce, Washington, D.C., 1977) is especially useful.

This Handbook is intended as a guide to marine mammal types and is all about them as animal forms. It is meant for use in the field and laboratory as a practical aid to identification and to provide useful basic information. It is not concerned with management, husbandry or treatment of disease, stock levels or whether there should be a moratorium on the taking of certain marine mammals. We have both worked for many years on various aspects of the biology of what we have found an entirely fascinating group of mammals. We know that all contributors have an interest equal to ours and far more expertise, so we have tried to keep editorial intervention to a minimum. We dare to hope that only marine mammals will know about themselves what is not in this Handbook, or that it is the editors' fault that useful information has been overlooked.

We thank Academic Press for their patience and guidance.

Sam H. Ridgway
Richard J. Harrison
1981

Preface to Volume 6

In the first two volumes of this handbook series, we wrote about the amphibious marine mammals — pinnipeds and sea otter. Although many regard the polar bear as a marine mammal, we have left it out of this series with apology to those who find this a defect. In the third volume we turned to the wholly aquatic marine mammals, covering manatees, dugongs and baleen whales. In the fourth volume we started our account of the large suborder Odontoceti with a book on the river dolphins and larger toothed whales. In the 17 chapters of the fifth volume, we covered 20 species representing only a portion of the largest marine mammal family in terms of number of species, the dolphin family, Delphinidae. In this second book of dolphins and the porpoises we provide another 17 chapters to comprise the sixth and final volume of the handbook series. We complete our accounts of the Delphinidae and also cover the porpoises — the family Phocoenidae.

The larger members of the family Delphinidae such as killer whale and bottlenose dolphin discussed in this volume, are among the best known of marine mammals. These are the animals that have been kept for display and research at oceanaria. Since these animals have been observed closely, a great many biological details have been worked out through close observation of their lives from the time of conception, through birth, on into old age and death.

Each chapter has been reviewed by scientific peers. We thank all who were so generous with their time in providing these assessments. We thank Jeanette Ridgway and Michelle Reddy for additional help on this volume. This final volume has been long in the preparation process and we trust that the reader will find the data-rich material worth the wait.

Naval Ocean Systems Center,
San Diego, USA — Sam H. Ridgway

Downing College
Cambridge, UK — Sir Richard J. Harrison

Contents of Other Volumes

Volume 3
THE SIRENIANS AND BALEEN WHALES

Volume 4
RIVER DOLPHINS AND THE LARGER TOOTHED WHALES

Volume 5
THE FIRST BOOK OF DOLPHINS

1

White-beaked Dolphin
Lagenorhynchus albirostris Gray, 1846

Randall R. Reeves, Chris Smeenk, Carl C. Kinze,
Robert L. Brownell, Jr and Jon Lien

Genus and Species

Taxonomy

A dolphin caught off Great Yarmouth, England, and reported by Brightwell (1846) as belonging to *Delphinus tursio* Fabricius, 1780, was described by Gray (1846a) as a new genus and species, *Lagenorhynchus albirostris*. The skeleton of this specimen is now in the British Museum (Natural History), London. After Gray's publication, Reichenbach (1846) proposed the name *Delphinus pseudotursio* for Brightwell's specimen. Another animal, from the west coast of Jutland, Denmark, was described as *Delphinus ibsenii* by Eschricht (1846). The genus *Lagenorhynchus* was more formally diagnosed by Gray (1846b) and included five putative species. Based on Gray's earlier publication, however, *L. albirostris* is the type species for the genus by monotypy.

Common names

The white-beaked dolphin is called belomordyi del'fin in Russian, kvitnos (or hvitnos) in Norwegian, hvidnæse in Danish, vitnosdelfin in Swedish, hnýðingur in Icelandic, kjafthvítur springari in Faroese, Weißschnauzendelphin in

Handbook of Marine Mammals, Volume 6
ISBN 0-12-588506-7

German, witsnuitdolfijn in Dutch, delfín de pico blanco in Spanish, lagénorhynque à bec blanc in French and aarluarsuk in Greenlandic. Some Norwegian coastal residents and mariners call both the white-beaked dolphin and the Atlantic white-sided dolphin (*Lagenorhynchus acutus*) "springer" (Jonsgård and Nordli, 1952). Sergeant and Fisher (1957) gave "squidhound" as the local name in Newfoundland and Labrador. Although these authors gave "jumper" as the name used in Newfoundland and Labrador for *L. acutus*, Alling and Whitehead (1987) reported that white-beaked dolphins were called "jumpers" in Labrador. The social science literature from Labrador is muddled by the use of vernacular names. For example, according to Brice-Bennett (1977) white-beaked dolphins are called "grampus" or "herring hogs", but the same terms are sometimes used for minke whales (*Balaenoptera acutorostrata*). The name "scoulter" was at least formerly applied to the white-beaked dolphin on the Norfolk coast of England (Southwell, 1885).

Distribution

The white-beaked dolphin is endemic to the cold temperate and subarctic North Atlantic, with a wide distribution across the northern rim of this ocean basin (Fig. 1). It is common in the eastern Greenland and Barents seas, from Finnmark and the Murman coast to western Svalbard and Novaya Zemlya and along portions of the Atlantic coast of Norway (Collett, 1912; Jonsgård, 1962; Christensen, 1972; Benjaminsen *et al.*, 1976; Øien, 1996).

The main concentrations around the British Isles are off northern Scotland (including the Outer and Inner Hebrides, Orkney and Shetland islands) and along portions of the Atlantic coast of Ireland (Evans, 1991; also see Evans, 1980, 1990; McBrearty *et al.*, 1986; Northridge *et al.*, 1997). British seabird surveys produced the highest sighting rates for white-beaked dolphins off the Outer Hebrides and in a strip down the western edge of the North Sea from the Orkneys to Flamborough Head in Yorkshire, between about 50 km and 200 km offshore (Northridge *et al.*, 1995). The white-beaked dolphin's occurrence off northern Scotland in August may be linked with spawning concentrations of herring (*Clupea harengus*) (Evans, 1980). It is one of the most frequently stranded cetaceans along the North Sea coast of the UK (Klinowska, 1985). Strandings occur infrequently on the Irish coasts (Berrow and Rogan, 1997), and occasional sightings are made in the Irish Sea (Jones, 1984). Strandings have also been recorded from the German North Sea coast (van Bree, 1970; Schultz, 1970; Goethe, 1983; Kinze *et al.*, 1997).

White-beaked dolphins are common in the northern and central North Sea and in the Skagerrak between Jutland (Denmark) and Norway (Kinze *et al.*, 1987, 1997). They occur frequently in the Kattegat between Jutland and Sweden and occasionally in inner Danish waters (Lepiksaar, 1966; Amundin and Amundin,

1975; Kinze, 1995; Kinze *et al.*, 1997) and the southern approaches to the Baltic Sea (Aguayo L., 1978; Schulze, 1991). They rarely enter the Baltic Sea proper (Schultz, 1970; Schulze, 1991; Kinze *et al.*, 1997).

The allopatric distribution of strandings of white-beaked dolphins and bottlenose dolphins (*Tursiops truncatus*) around the British Isles was interpreted as suggesting that the two species had almost "complementary" distributions (Harmer, 1927). Fraser (1974) likened the distribution of *L. albirostris* to that of the long-finned pilot whale (*Globicephala melas*) and suggested that these two species were associated with the North Atlantic Current.

On the basis of stranding frequencies, the white-beaked dolphin is believed to have become more common in Dutch waters since about 1960; it is now the most common delphinid stranded and sighted in Dutch waters (Smeenk, 1986, 1989, 1992, 1995; Bakker and Smeenk, 1987, 1990; Kinze *et al.*, 1997; see also van Bree and Nijssen, 1964; van Bree, 1970, 1977; Baptist, 1987; Baptist *et al.*, 1990). The great majority of records are from the northern and northwestern parts of the Dutch sector of the North Sea; most strandings have occurred in the

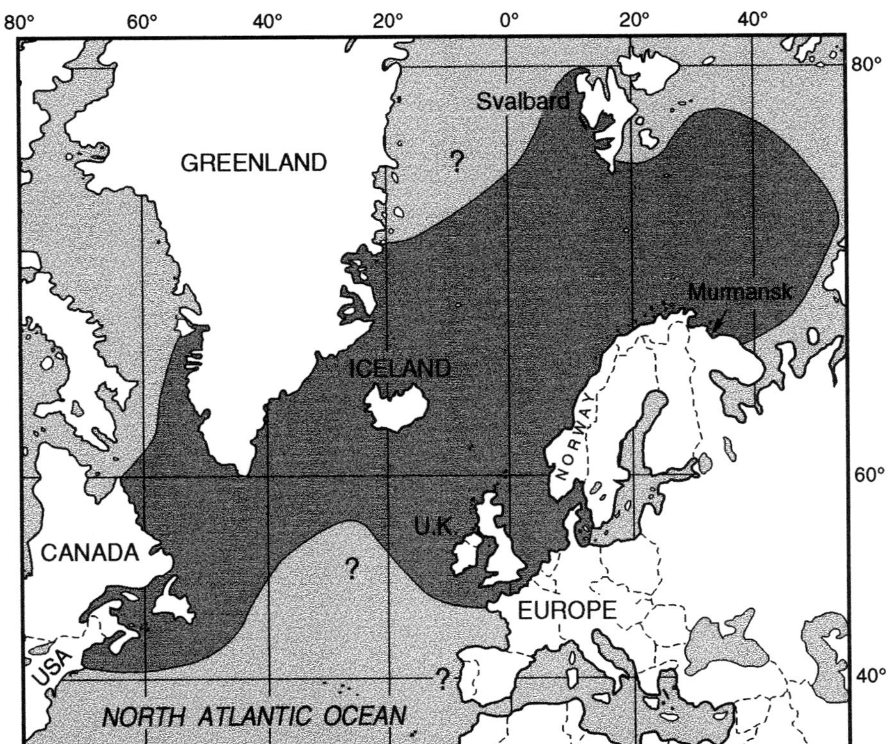

FIG. 1 Approximate distribution of the white-beaked dolphin. Question marks indicate that there is uncertainty about the species' regular occurrence in the area.

northern part of the Netherlands. Belgian records were reported by de Smet (1974, 1981) and Van Gompel (1991, 1996), and several white-beaked dolphins have stranded in winter months on the English Channel coast of France (Collet and Duguy, 1981; Duguy, 1984, 1987, 1988). They have been reported as far south as 47°N along the coast of France (Bierman and Slijper, 1948), and there is a single record for the southern Bay of Biscay in Spain (Castells and Mayo, 1992). Two sightings were reported off Portugal — a single animal with a group of common dolphins (*Delphinus delphis*) and another group of five white-beaked dolphins at 37°N, 12°30′W (Bierman and Slijper, 1948, but see Aguayo L., 1978; McBrearty *et al.*, 1986). Observations have also been reported in the Strait of Gibraltar (Hashmi and Adloff, 1991, 1992). A published record from Turkish waters (Hennipman *et al.*, 1961; Kumerloeve, 1975) is very doubtful.

The white-beaked dolphin has been described as common around the Faroe Islands (Degerbøl, 1940), and Evans (1991) mapped the Faroes as a "concentration area" for the species. It is also considered the most common dolphin off southeastern Greenland (Jonsgård and Christensen, 1968) and in the Denmark Strait and the seas around Iceland (Martin *et al.*, 1984; Sigurjónsson, 1983, 1985; Gunnlaugsson *et al.*, 1988). Born (1983) reported a probable record near Cape Swainson at about 70°N on the east coast of Greenland. Sightings off Iceland in June–July 1986 were mainly along or seaward of the 100 m depth contour, off the southwest and northeast coasts (Gunnlaugsson *et al.*, 1988). Previous surveys had found concentrations west and northwest of Iceland in June–July 1981 (Sigurjónsson, 1983) and 1982 (Martin *et al.*, 1984) and to the northeast in August 1983 (Sigurjónsson, 1985).

Phenotypical differences have been demonstrated between white-beaked dolphins from the eastern and western Atlantic (Mikkelsen and Lund, 1994). Their distribution in the west extends from approximately Cape Cod in the south (Katona *et al.*, 1983) to central Davis Strait in the north (Winge, 1902; Heide-Jørgensen and Leatherwood, 1987; Mikkelsen and Lund, 1994). Specimens from Nuuk (Godthåb) and Aasiaat (Egedesminde), West Greenland, are in the Zoological Museum, Copenhagen. A dolphin taken in Uummannaq Fjord at 70°45′N in early September was identified as a *Lagenorhynchus* sp. from a photograph but could not be identified to species (Anonymous, 1981; F.O. Kapel, personal communication). The white-beaked dolphin is much less common than the Atlantic white-sided dolphin in the Gulf of Maine (Katona *et al.*, 1983; Kenney *et al.*, 1996). Both are commonly seen in parts of the Gulf of St. Lawrence, the former primarily along the north shore near the Strait of Belle Isle during the summer and autumn (Sears *et al.*, 1981; Sears and Williamson, 1982; Kingsley and Reeves, 1998). Sightings are common around the island of Newfoundland during the winter and spring (J. Lien, unpublished data). Fishermen along the Labrador coast claim that white-beaked dolphins approach the coast in late June and remain until October (Alling and Whitehead, 1987; also see Boles, 1980). Densities on the Southeast Shoal of the Grand Banks decreased from mid June to mid July (Whitehead and Glass, 1985).

External Characteristics

Colour pattern

The coloration of this species shows considerable variability (Harmer, 1927; Mercer, 1973; Camphuysen, 1991; Ree, 1994). Mitchell (1970) made detailed comparisons between the pigmentation patterns of *L. albirostris* and other species of *Lagenorhynchus*. The beak of most white-beaked dolphins is white, often mottled with light grey or with greyish or blackish spots or fields (Fraser, 1946, 1953; Mercer, 1973; van Bree and Smeenk, 1982, their Fig. 3; Ree, 1994). In others it is partly or almost entirely greyish (though paler than the head) (Jonsgård, 1962; Smeenk, 1986, his Fig. 3; 1989, his Fig. 7; 1995, his Fig. 1). The dark dorsal field anterior to the dorsal fin is sometimes separated from the dark melon by a transverse light grey stripe, a brownish-grey patch (Mercer, 1973) or a bold whitish "chevron" around and behind the blowhole (Mitchell, 1970; Camphuysen, 1991; Smeenk, 1992, his Fig. 3; Ree, 1994). It may extend downwards from the melon to encircle the eye. The eye patch often is light grey and surrounded by a narrow, whitish-grey circle (van Bree and Smeenk, 1982, their Fig. 3; Camphuysen, 1991; Smeenk, 1992, his Fig. 4). In front of the dark grey zones on the sides there is a paler grey, rather ill-defined thoracic patch (Smeenk, 1989, his Fig. 7; Camphuysen, 1991; Ree, 1994). Above and behind this patch, between the dark grey dorsal and lateral fields, the body is varying shades of lighter grey to nearly white. The whitish or light grey flank pigmentation extends dorsally onto the back behind the dorsal fin. This lightly pigmented area (saddle) provides a useful way of distinguishing the white-beaked dolphin from the Atlantic white-sided dolphin at sea (it is also very striking from the air), as the latter has no such intrusion of light pigmentation across the dorsal ridge (Baptist, 1987, his Fig. 4; Camphuysen, 1991; Ree, 1994). The underside is white, with the white central part of the abdomen forming a narrow band between two pale grey patches (van Bree and Smeenk, 1982, their Fig. 3; Smeenk, 1986, his Fig. 3). The flukes and tail stock immediately in front of the flukes, and also the flippers, are generally dark, but often spotted or marbled with white near the insertions of the flippers and on the undersides of the flukes (van Bree and Smeenk, 1982, their Figs 3 and 4; Smeenk, 1992, his Fig. 4). Another field mark for distinguishing *L. albirostris* from *L. acutus* is that in the former the dark colour of the flippers is continuous with the slightly paler thoracic field, whereas in the latter it is separated from the dark dorsal area by the white ventrum (Smeenk, 1986, his Figs 3 and 4; 1989, his Figs 5 and 9; Camphuysen, 1991; Ree, 1994).

Four to six hair follicles are present on each side of the upper jaw (Guldberg and Nansen, 1894; Fraser, 1946). Hairs are present on the upper lip of young individuals (Harmer, 1927).

Size and shape

The largest measured specimen was a 310 cm male that stranded in the British Isles (Fraser, 1974). Few individuals reach 300 cm in length (Guldberg and Nansen, 1894; Harmer, 1927). Males grow longer than females (Dong *et al.*, 1996).

This species has a robust body, with a short, thick beak about 5–8 cm long in adults (Table 1; Figs 2 and 3). The beak is distinctly set off from the melon. The dorsal fin is at mid body. It is proportionally large (to 15% of body length), often rounded at the peak, and strongly recurved. Both the dorsal fin and the flukes apparently increase in size relative to other body dimensions as the dolphin ages (Tomilin, 1967). The pointed flippers can be up to 19% of the total adult length. The thickened tail stock tapers gradually, in marked contrast to that of the Atlantic white-sided dolphin.

FIG. 2 White-beaked dolphin breaching in St. Lawrence River Estuary, eastern Canada. (Photo by Richard Sears)

FIG. 3 Female white-beaked dolphin that stranded on the isle of Texel, the Netherlands, 24 January 1990. (Photo courtesy *Texelse Courant*, Texel)

TABLE 1 Selected measurements (cm) for a sample of *L. albirostris* from Danish waters[a]

Sex	TL	BKL	DORF	AFLP	DFBS	DFHT	FLLA	FLLP	FLMW	NAV	VENT	ANUS	SPFL
F	182.5	4.5	110	37	40	19	35	24	11.5	83.5	121	128	41
F	242	5	153	43.5	50	—	—	—	—	112	167	175	—
F	243	3.5	150	44	57	34	34	32	15	106	168	170	69
F	245	4.5	154	45	48	30	45	32	16.5	110	168	175	60
F	247	6.0	155	41	56	31	47.5	33.5	16	107	168	175	62
F	252	5.5	157	48	50	31.5	46.5	33.5	16	123	167	176	65
F	253	—	—	47	—	—	45	34	—	—	165	175	65
F	255	—	138	45	40	30	49	26	—	111	158	—	—
F	255	—	—	—	—	—	—	—	—	—	—	—	63
F	258	—	—	—	—	34	—	—	—	—	—	184	—
F	260	6	—	48	—	30	45	—	17	110	—	—	65
F	265	8	174	52	50	36	47	33	18	115	176	182	70
F	282.5	—	157	—	47	31.5	—	—	—	—	—	—	83.5
M	118	3	71	24	20	16	25	21	10	54	73	82	32
M	247	5.5	133.5	—	46	32.5	—	—	—	—	150.5	182	62.5
M	248	5.5	125	50	42	27	46.5	34.5	17	—	144	180	—
M	272	6.5	159	44	50	40	49.5	35	18.5	116.5	156	188	74

[a]TL, total length; BKL, beak length; DORF, tip of snout to top of dorsal fin; AFLP, tip of snout to anterior insertion of dorsal fin; DFBS, base of dorsal fin; DFHT, height of dorsal fin; FLLA, flipper length from anterior insertion to tip; FLLP, flipper length from posterior insertion to tip; FLMW, maximum width of flipper; NAV, tip of snout to navel; VENT, tip of snout to vent; ANUS, tip of snout to anus; and SPFL, fluke span.

TABLE 2 Lengths and body weights of 10 male and 25 female *Lagenorhynchus albirostris*[a]

Males		Females	
209 cm	95 kg	154 cm	55 kg
225.5 cm	200 kg	179 cm	67.5 kg
239 cm	134 kg	182.5 cm	80 kg
248 cm	250 kg	210 cm	175 kg
250.5 cm	354 kg	218 cm	186.5 kg
257 cm	220 kg	223 cm	248.5 kg
264 cm	272 kg	236 cm	160 kg
269 cm	309 kg	240 cm	287 kg (pregnant)
275 cm	275 kg	241 cm	179 kg
277 cm	255 kg	242 cm	100 kg
		246 cm	132.5 kg
		251 cm	254 kg
		252 cm	255 kg
		255 cm	240.5 kg
		255 cm	249 kg
		256 cm	191 kg
		259 cm	242 kg
		262 cm	c. 265 kg
		262.5 cm	306 kg
		263 cm	206 kg
		264 cm	210 kg
		264 cm	219 kg
		264 cm	242 kg
		269 cm	280 kg
		278 cm	232 kg (pregnant)

[a]Sources: van Bree and Nijssen (1964), Duguy (1987), Goethe (1983), C.C. Kinze (unpublished data), Kinze *et al.* (1997), Mathiasson (1990, 1992), van Oort (1918), Jóhann Sigurjónsson (personal communication), C. Smeenk (unpublished data), de Smet *et al.* (1985), Van Utrecht (1981), Weber (1896).

Weight

Few weights have been published (Table 2). The largest reported male weighed 354 kg, the largest female 306 kg.

Internal Anatomy

Skull

The rostrum of the white-beaked dolphin (Fig. 4) is relatively shorter and wider at the base than that of the Atlantic white-sided dolphin. Skull measurements ($\bar{x}$, range in mm, and *n*) from Tomilin (1967) are: condylobasal length, 443, 422–460, 8; rostral length, 214, 202–229, 8; basal width of rostrum, 144, 140–

152, 7; orbital width, 230, 219–236, 8; length of upper tooth row, 196, 182–223, 4. The white-beaked dolphin's skull can be more than 18% of the total length of the skeleton (Lütken, 1888; Tomilin, 1967). The distance from the tip of the cleaned lower jaw to the posterior end of the tooth row in most specimens larger than 265 cm is 188–194 mm (Harmer, 1927). White-beaked dolphins have fewer teeth than Atlantic white-sided dolphins, but their teeth are larger. The numbers of upper and lower teeth are 23–28 and 22–28, respectively; the first three in each row are often concealed within the gum (van Bree and Nijssen,

(A)

(B)

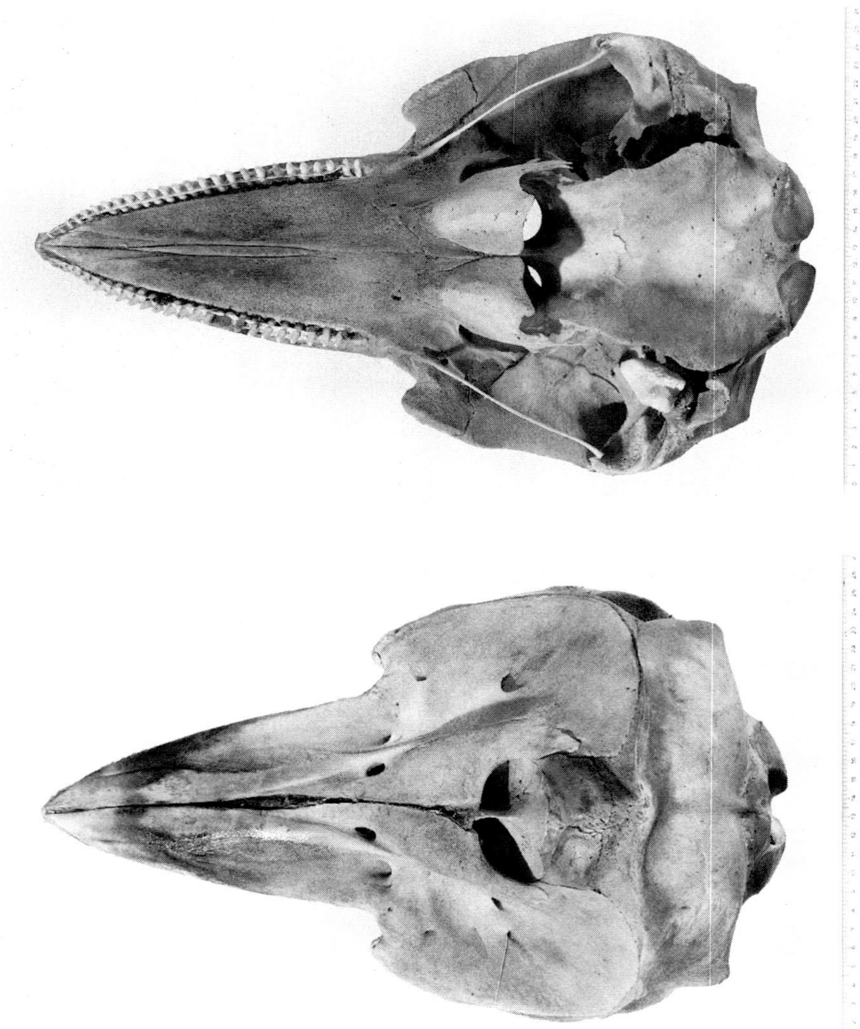

FIG. 4 (A) Ventral and (B) dorsal views of cranium of the white-beaked dolphin.

(C)

(D)

FIG. 4 *Continued* (C) Left lateral view of skull; and (D) dorsal view of mandibles. (Courtesy Zoological Museum, Copenhagen; photos by Geert Brouad)

1964). Generally only 22–26 teeth per row are visible within the mouth. The teeth of white-beaked dolphins are up to 19 mm long and 7 mm in diameter (Lütken, 1888; Harmer, 1927).

Postcranial skeleton

The vertebral skeleton is C_7, T_{15-16}, L_{23-24}, Ca_{43-45}, giving a total of 88–92 vertebrae (Lütken, 1888; True, 1889), or C_7, T_{14}, L_{27}, Ca_{45}, total 93 (Tomilin, 1967). Van Beneden (1862) gave the range as 90–94. The atlas and axis are fused; the other five cervicals are free and have thin, overlapping transverse processes (Cunningham, 1876). There are 14 (Clark, 1876) or 15 (Tomilin, 1967) pairs of ribs, of which 5 are true ribs. The transition between double-headed and single-headed ribs occurs in the area of the 6th or 7th thoracic vertebra. In a study of 22 skeletons, de Smet (1975) found that in 73% of the cases the transverse process of the 7th thoracic vertebra had a protuberance. The phalangeal formula is I_2, II_6, III_4, IV_1, V_0 (Tomilin, 1967).

Organs, Tissues and Physiology

Organ systems are described in the following references: integument (Scholander and Schevill, 1955), muscular system (Murie, 1873), nervous system (brain) (Weber, 1897), nervous system (spinal) (Cunningham, 1877), respiratory system (lungs) (Cleland, 1884; Weber, 1887), digestive system (stomach) (Clark, 1876; Cleland, 1884; Weber, 1887; Turner, 1889), digestive system (small and large intestines) (Van Beneden, 1860; Cleland, 1884), circulatory system (blood vessels) (Slijper, 1936), circulatory system (heart) (de Smet *et al.*, 1985; van Nie, 1985, 1986), female reproductive system (vagina) (Van Beneden, 1860), male reproductive system (Cleland, 1884; Meek, 1918) and sensory system (auditory region) (Fraser, 1952).

Mead (1975) found a lateral diverticulum appended to the right premaxillary nasal sac in a single juvenile specimen. This feature, which is also found in at least juvenile *Lagenodelphis hosei*, was not observed in the nine *L. albirostris* dissected by Schenkkan (1973) and thus may be a juvenile character that disappears with age (Mead, 1975). A distinctive feature of the nasal complex in *L. albirostris* is the hypertrophy of the nasofrontal sacs (Schenkkan, 1973; Mead, 1975). Reidenberg and Laitman (1988) illustrated laryngeal folds and proposed that these may have a vocal function.

Organ weights of two female white-beaked dolphins are given in Table 3.

Life History and Population Dynamics

In a sample of four males, the smallest sexually mature individual was 251 cm, and the largest sexually immature specimen was 257 cm (Van Utrecht, 1981; see also Jonsgård, 1962). The smallest sexually mature female was a pregnant individual 174 cm long (Dong *et al.*, 1996), and the largest sexually immature female was 246 cm (Van Utrecht, 1981). A 230 cm female from Aasiaat, West Greenland, was lactating (C.C. Kinze, unpublished data). Physical maturity, defined by complete fusion of the phalanges in the flippers, was attained in a sample of 4 males at 281 cm (SD = 2.9) and in a sample of 4 females at 261 cm (SD = 6.2) (Dong *et al.*, 1996). These lengths corresponded to estimated ages of 13 and 16 dentinal growth-layer-groups, or GLGs (years?), respectively.

Fraser (1974) reported a stranded 119 cm neonate in late September and estimated the calving season in British waters to span at least three months (July–

TABLE 3 Organ weights (g) of two white-beaked dolphins from the North Sea. Female 1 was taken as a by-catch in fishing gear; Female 2 was found stranded[a]

Organ	Female 1, 2.23 m (ZMA 9478)		Female 2, 2.59 m (ZMA 6815)	
Heart	4 000	(1.613)	4 630	(1.913)
Lungs				
left	4 800	(1.935)	4 235	(1.750)
right	4 900	(1.976)	5 059	(2.090)
Stomach	4 300	(1.734)	—	—
Liver	8 500	(3.427)	6 333	(2.617)
Kidneys				
left	703.5	(0.248)	1 059	(0.438)
right	634	(0.256)	1 375.5	(0.568)
Pancreas	423	(0.171)	—	—
Spleen	183.5	(0.074)	128.5	(0.053)
Adrenals				
left	8.5	(0.003)	19	(0.008)
right	9	(0.004)	18	(0.007)
Intestine	9 800	(3.951)	—	—
Uterus	—	—	1 246	(0.515)
Ovary	450	(0.181)	—	—
Blubber	30 050	(12.11)	57 000	(23.55)
Total body weight	248 000	(100.00)	242 000	(100.00)

[a]Unpublished data from W.L. Van Utrecht; courtesy P.J.H. van Bree.

September). Evans (1991) gave May–August as the birth season and speculated that births take place mainly offshore in the northern North Sea and in the Atlantic north and west of the British Isles. Young dolphins 118–134 cm long have been reported from June to September (Southwell, 1885; Harmer, 1927; Fraser, 1946, 1974; Sheldrick, 1989; Smeenk, 1989; Kinze *et al.*, 1997; Berrow and Rogan, 1997).

The birth length is 110–120 cm (Tomilin, 1967; Collet and Duguy, 1981). The largest foetus was 122 cm (Millais, 1906) and the smallest free-swimming individual 118 cm (Fraser, 1974). Foetuses 79, 92 and 94 cm long were found in dolphins that were ice-entrapped off Newfoundland in late March (J. Lien, unpublished data). Foetal lengths summarized in Table 4 suggest either a multiphasic or prolonged birth season.

Mortality

Fatal entrapment by ice can occur along the coasts of Newfoundland in years with heavy pack ice (Sergeant and Fisher, 1957; Dong *et al.*, 1996; Fig. 5). In March 1982 at least 40 white-beaked dolphins were trapped by wind-driven pack ice in Grand Bay, Port-aux-Basques, and in March 1983 at least 200 were

TABLE 4 Foetal lengths (cm) of *Lagenorhynchus albirostris*[a]

January	56, c. 60	Refs 9, 14
February	67, 75	Refs 2, 13
March	45, 79, 92, 94	Refs 6, 11
April	70, 76.5	Ref. 8
May	72, 75, 106, 113	Refs 3, 4, 12
June	—	
July	79, 91, 91, $\geq$112	Refs 7, 8, 13
August	—	
September	122	Ref. 10
October	—	
November	35.5, c. 45	Refs 1, 5
December	—	

[a]Sources: 1, van Bree and Nijssen (1964); 2, Collet and Duguy (1981); 3, Grieg (1888); 4, Japha (1908); 5, Kinze *et al.* (1997); 6, J. Lien (unpublished data); 7, Lilljeborg (1866); 8, Lütken (1888); 9, Mathiasson (1990); 10, Millais (1906); 11, Nilsson (1847); 12, Smeenk (1989); 13, C. Smeenk (unpublished data); and 14, Van Utrecht (1981).

trapped in Placentia Bay. Dong *et al.* (1996) estimated that more than half of ice-entrapped dolphins die either from suffocation, crushing or forced stranding.

In contrast to the Atlantic white-sided dolphin, which sometimes strands in large numbers ("mass strands"), the white-beaked dolphin usually strands singly

FIG. 5 Part of a group of white-beaked dolphins ice-entrapped along the Burin Peninsula between Lord's Cove and Lamaline, Newfoundland, on 8 April 1987. (Photo by Jon Lien)

or in small groups. It is not always possible to distinguish stranded animals from those that have been taken in fishing gear and discarded. Fraser (1974) reported only two multiple strandings: one of five white-beaked dolphins (alive on mud-flats) and one of a mother and large calf. In the Netherlands van Bree and Nijssen (1964) reported multiple strandings of two animals in January 1941 and three in May 1964, Husson and van Bree (1976) reported a stranding of two females in November 1974, and Smeenk (1986) reported a stranding of a male and female in June 1981. In Germany U. Siebert (personal communication) recorded a stranding of three females in May 1993. Smeenk and Addink (1990) and Smeenk (1995) reported a stranding of four individuals (3 adult females and one subadult male) within four days; a fifth animal (adult female) was found a few days later. Sheldrick (1989) listed four strandings involving two or three animals. Five of nine strandings of white-beaked dolphins in Newfoundland between 1979 and 1990 involved groups of two to seven animals. On three occasions live dolphins came ashore, including groups of three and four (Dong *et al.*, 1996). Live strandings are not unusual in the Netherlands (e.g. four in April 1992; Smeenk, 1995).

Predation by killer whales (*Orcinus orca*) almost certainly occurs but direct evidence is lacking. Reeves observed killer whales and white-beaked dolphins swimming together east of Greenland, but it was impossible to tell in this instance whether it was a predatory attack by the killer whales, interaction similar to avian mobbing (cf. Wilson, 1975) with the dolphins swimming quickly and erratically among the killer whales, or simultaneous predation by the whales and dolphins on the same school of fish, as has been described from the herring fishing grounds off Iceland (Sigurjónsson *et al.*, 1988).

Abundance

Published estimates indicate a population of at least several thousand white-beaked dolphins in portions of the northwestern Atlantic: shoreward of the 200 m contour between St. Anthony, Newfoundland, and Nain, Labrador (Alling and Whitehead, 1987) and in coastal and offshore waters east of Newfoundland and southeast of Labrador (Hay, 1981, 1982). Gunnlaugsson *et al.* (1988) recorded 942 white-beaked dolphins in primary sightings during 58 hr of non-overlapping aerial survey around Iceland. Since their coverage was 20% or less in all the survey blocks, it is reasonable to assume that at least a few thousand white-beaked dolphins inhabit Icelandic waters. Ship surveys designed primarily to estimate abundance of minke whales in the northeastern Atlantic allowed rough estimates of dolphin abundance. Øien (1996) estimated that there were 132 000 small delphinids (CV 0.266, 95% CI 79 000–220 000) in the areas surveyed in 1989, including the Barents Sea, the eastern part of the Norwegian Sea and the North Sea north of 56°N. He reasoned that most of them (about 90%) were white-beaked dolphins based on information from Jonsgård and Nordli (1952) and Jonsgård (1962) and on the species ratio for those opportunistic and survey sightings that were identified to species. A survey of the North Sea and adjacent waters in 1994, designed for harbour porpoise (*Phocoena phocoena*) abundance estimation, provided an estimate of 7856 (CV = 0.30) white-beaked dolphins, or 11 760 (CV = 0.26) *Lagenorhynchus* spp., primarily *L. albirostris* (Hammond *et al.*, 1995). The total number of white-beaked dolphins throughout the North Atlantic thus may be in the high tens or low hundreds of thousands.

Evans (1990) suggested that the number of white-beaked dolphins in the North Sea, Irish Sea and English Channel and off western Britain had increased since the mid 1960s, with a possible decrease during the 1980s off western Britain and in the northern North Sea. Bakker and Smeenk (1987) referred to an apparent increase in the southern North Sea (Dutch waters) since 1960. A pronounced decline in abundance has occurred in the Gulf of Maine since the early 1970s (Kenney *et al.*, 1996).

Behaviour

Social organization

Tomilin (1967), Fraser (1974) and Katona *et al.* (1983) referred to herds of up to 1500 of these dolphins. However, such large groups seem exceptional and would probably consist of many smaller subgroups. The mean group size in 131 observations made from aircraft off Iceland was 7.2 animals (Gunnlaugsson *et al.*, 1988). Øien (1996) reported group sizes of mainly one to five individuals, with occasional herds of several hundred. Groups of 10–35, or occasionally as many as 100, were sighted along the north shore of the Gulf of St. Lawrence (Sears and Williamson, 1982). Two groups of dolphins observed in early August consisted entirely of adults accompanied by calves. Sears and Williamson (1982) interpreted these to be "nursery groups" and noted that they were more difficult to approach than other groups of white-beaked dolphins. Based on sightings in the eastern North Atlantic, Evans (1991) stated that most groups were of less than ten individuals but that herds of 30–50 were not unusual. He also noted that there had been a few sightings of herds of 100–500 off northwestern Scotland.

In two separate ice entrapments off Newfoundland an attempt was made to determine the composition of herds. One group (at Port-aux-Basques) consisted mainly of young animals 189–241 cm long ($n = 31$, $x = 213$ cm, SD 11), with estimated ages of 2–7 GLGs ($n = 34$, $x = 2.9$ GLGs, SD 1.1) (Dong *et al.*, 1996). The other group (at Point Verde) included animals ranging in length from 168 cm to 305 cm ($n = 37$, $x = 229$ cm, SD 39). Of the 28 dolphins from Point Verde whose ages were estimated, 13 had only 1 GLG and the rest more than 6 ($n = 15$, $x = 11.9$ GLGs, SD 2.6) (Dong *et al.*, 1996). At least 24 of the Point Verde group were females, and four of them were pregnant. Because both groups of ice-entrapped dolphins were incompletely sampled, conclusions about group composition need to be made cautiously. However, there is a suggestion that groups of juveniles travel apart from mixed groups of adults and calves. A group of five that live-stranded in the UK consisted of three males (229, 267 and 274 cm) and two females (254 cm and 269 cm) (Fraser, 1974). Four multiple strandings on the Dutch and German coasts consisted of: (1) an adult male and subadult female (about 260 cm and 215 cm, respectively; van Bree and Nijssen, 1964), (2) all adult females (260, 264 and 259 cm stranded and a fourth, 264 cm, found floating at sea in the same area; van Bree and Nijssen, 1964); (3) all adult females (three together; U. Siebert, personal communication); and (4) adult females (261, 253, 251 and 239 cm) with a subadult male (212 cm; Smeenk, 1995).

Swimming and diving

White-beaked dolphins frequently ride the bow and stern waves of vessels (Jonsgård, 1962; de Boer, 1989; Evans, 1991). They are acrobatic at the surface, often leaping vertically from the water (Sears and Williamson, 1982; Katona *et al.*, 1983; Baptist, 1987; Fig. 2). They seem to be attracted to small boats, and this can bias the results of vessel surveys.

Feeding

In all areas where stomach contents have been examined, clupeids (e.g. herring), gadids (e.g. Atlantic cod, *Gadus morhua*, haddock, *Melanogrammus aeglefinus*, poor-cod, *Trisopterus minutus*, bib, *T. luscus*, and whiting, *Merlangius merlangus*) and hake (*Merluccius merluccius*) have been found to be the principal prey of white-beaked dolphins (Southwell, 1885; Harmer, 1927; Fraser, 1946, 1964; Sergeant and Fisher, 1957; van Bree and Nijssen, 1964; Evans, 1980; Smeenk and Gaemers, 1987; Dong *et al.*, 1996; Kinze *et al.*, 1997). In addition to the above, Evans (1991; also see Tomilin, 1967) listed the fishes *Scomber*, *Pleuronectes*, *Limanda*, *Eleginus* and *Hyperoplus* as well as squid, octopus and benthic crustaceans as prey. The octopus *Eledone cirrhosa* was found in dolphins from Scottish waters (Santos *et al.*, 1993). In a sample of 20 dolphins with food in their stomachs examined off southwestern Newfoundland in March, 90% contained otoliths and bones from Atlantic cod (Dong *et al.*, 1996). White-beaked dolphins have been seen off southeastern Greenland feeding on capelin (*Mallotus villosus*) in association with fin whales (*Balaenoptera physalus*) (Jonsgård and Christensen, 1968).

Katona *et al.* (1983) cited the fact that the name "squidhound" is applied in Newfoundland to mean that white-beaked dolphins eat more squid than do Atlantic white-sided dolphins. The suggestion has been made that *L. albirostris* are benthoichthyophagous and prefer cod, whiting and capelin, whereas *L. acutus* are nektonichthyophagous and prefer salmonids, mackerel and herring (Gaskin, 1982). There is considerable overlap in diet, however, so this generalization requires rigorous validation.

Cooperative feeding has been described. Dolphins herd the fish into a tight cluster and trap them against the surface (Evans, 1991; also observed by C. Smeenk, unpublished).

Association with other species

Jonsgård and Christensen (1968) reported a close association between fin whales and white-beaked dolphins off southeastern Greenland: "In areas where fin whales were plentiful, white-beaked dolphins in great numbers were almost always present". These authors believed that the dolphins were exploiting opportunities to catch capelin "out of the mouth" of the surfacing fin whales. Similar observations of white-beaked dolphins associating with feeding fin whales and humpback whales (*Megaptera novaeangliae*) have been made along the north shore of the Gulf of St. Lawrence (Sears and Williamson, 1982; R. R. Reeves, unpublished observations) and around Newfoundland (J. Lien, unpublished observations). Small numbers of white-beaked dolphins are sometimes seen in close association with herds of pilot whales (Sears and Williamson, 1982), and with sei whales (*Balaenoptera borealis*) and killer whales (R. R. Reeves, unpublished observations). Off southwestern Ireland they have been seen with bottlenose dolphins (Evans, 1980), and in the North Sea with Atlantic white-sided dolphins (Haase, 1987; de Boer, 1989). White-beaked and common dolphins have been seen together in the Irish Sea (Jones, 1984).

Sound production

The "squeals" of white-beaked dolphins recorded in the Gulf of Maine in February were in the frequency range of 6.5–15 kHz (Schevill and Watkins, 1962). Burst-pulse sounds, but not whistles, were recorded on a ship's sonar from a group of white-beaked dolphins in the southern North Sea (Mitson and Morris, 1988; Mitson, 1990). These sounds, presumably used for echolocation, had significant energy to at least 325 kHz. Pulse rates varied from 100 pulses to 750 pulses per second, and crude estimates of source levels were between 123 dB and >207 dB re 1 micropascal at 1 m. A click with a dominant frequency of 65 kHz, recorded by Kamminga, was cited by Hartmann *et al.* (1992).

Parasites and Diseases

A high incidence of diseased jaws and teeth has been reported in stranded specimens (Harmer, 1927; Fraser, 1946; van Bree and Nijssen, 1964). Many teeth were missing, their alveoli were inflamed, and large abscesses had formed and invaded the bone of the jaw. A specimen described by Baker (1992) had such a pathologically deformed lower jaw that it had led to sepsis and death. Harmer

(1927) suggested that diseased jaws are a sign of old age and approaching death. One white-beaked dolphin caught at sea was found to have a severely deformed spine; and one stranded specimen, in addition to having badly diseased jaws, had vertebrae that were enlarged and roughened by periostosis (Slijper, 1936; Fraser, 1946). Van Assen (1975) reported a stranded animal with a deformed spine in the Netherlands (kyphosis) and another was found in the Netherlands in 1993 (J. S. van der Kamp, personal communication). A high incidence of spondylosis deformans has been reported for white-beaked dolphins (van Bree and Nijssen, 1964; Kompanje, 1995).

The nematodes *Anisakis simplex* and *Halocercus lagenorhynchi* have been found in the digestive tract and bronchi, respectively (Gibson and Harris, 1979). White-beaked dolphins examined off Newfoundland had infestations of *Anisakis* sp. in their stomachs (Sergeant and Fisher, 1957; Brattey and Clark, 1992; Dong *et al.*, 1996). *Pseudoterranova* sp. also have been found in the stomach (Arvy, 1982).

Two species of whale lice (*Isocyamus delphinii* and *Scutocyamus parvus*) have been reported from white-beaked dolphins (Lincoln and Hurley, 1974; Husson and van Bree, 1976; Stock, 1977; Fransen and Smeenk, 1991). *Scutocyamus parvus* have thus far been reported only from white-beaked dolphins.

Bacteria found in stranded white-beaked dolphins were listed by Buck *et al.* (1991) and Kinze *et al.* (1997). A rhabdovirus was found in a specimen that live-stranded in the Netherlands (Osterhaus *et al.*, 1993). Another Dutch specimen showed a fibrinous pleuro-pneumonia (morbilli IF positive; see Kinze *et al.*, 1997).

Live Maintenance

At least three white-beaked dolphins were captured with seine nets off Iceland and delivered to the Hafnarfjörður marine zoo in 1977 (Andrésson, 1978). Six ice-entrapped female white-beaked dolphins were removed alive from a bay in Newfoundland and transported to an aquarium in Connecticut (Buck and Spotte, 1986a). One lived for 101 days. Pathogens implicated in the deaths of the five dolphins that reached the aquarium alive included staphylococci, *Erysipelothrix rhusiopathiae* and Gram-negative bacteria (Buck and Spotte, 1986a,b). Stranded specimens have been brought into captivity for short periods on at least two occasions in the United Kingdom (Klinowska and Brown, 1986). Seven dolphins that stranded in the Netherlands, Belgium and France have been brought into captivity (Smeenk, 1986, 1995, unpublished data); one of them was released after five months.

Human Effects

Directed fisheries

The white-beaked dolphin has been hunted at least opportunistically along the coasts of Norway, particularly in Kjollefjord, western Finnmark (Jonsgård, 1962), Iceland (Sæmundsson, 1939; Mitchell, 1975), Newfoundland (Sergeant and Fisher, 1957) and possibly Nova Scotia (Gilpin, 1878) and Greenland (Heide-Jørgensen, 1990). The white-beaked dolphin "fishery" in Davis Strait mentioned by Tomilin (1967) is probably a reference to the occasional capture of dolphins (*L. albirostris* and/or *L. acutus*) in southern Greenland. During the early 1980s an estimated 366 white-beaked dolphins were taken annually by the residents of 12 Labrador harbours (Alling and Whitehead, 1987; also see Brice-Bennett, 1977; Mackey and Orr, 1987). It was estimated that about 25–50% of the dolphins killed were not secured (Alling and Whitehead, 1987). White-beaked dolphins may have been taken in small numbers by indigenous porpoise hunters in the Bay of Fundy during the 19th century, although the specimens described by Gilpin (1878) could have been Atlantic white-sided dolphins. At least some of the dolphins taken at the Faroe Islands in early years (Degerbøl, 1940) and more recently (e.g. Bloch and Joensen, 1985) may have been white-beaked dolphins.

By-catches

White-beaked dolphins have been taken in fishing gear in many areas (Southwell, 1879, 1886; Van Utrecht and Husson, 1968; Kerkhoff *et al.*, 1981; Northridge, 1987, 1991; Anonymous, 1989; Van Gompel, 1991, 1996; Kinze *et al.*, 1997; Fig. 6). There are many by-caught specimens in the zoological museums in Amsterdam and Leiden. Of three by-caught animals reported off Newfoundland during 1987–1988, one died in a groundfish gill net, one in a herring gill net and one in a cod trap (Dong *et al.*, 1996). The Newfoundland/Labrador by-catch is substantially under-reported in published accounts. One specimen found dead off eastern Newfoundland was thought to have died from a collision with a vessel (Mercer, 1973).

Pollution

Like other North Atlantic marine mammals, white-beaked dolphins are contaminated by organochlorines, other anthropogenic compounds and heavy metals (Andersen and Rebsdorff, 1976; Kerkhoff *et al.*, 1981; Muir *et al.*, 1988; Law *et al.*, 1991, 1992).

FIG. 6 This male white-beaked dolphin, 277 cm long and weighing 255 kg, died in a fishing net southwest of Iceland in April 1992. (Photo by Selvin, courtesy Jóhann Sigurjónsson)

Acknowledgments

The authors wish to thank Marjan Addink and Manuel García Hartmann for advice and assistance, and Richard Sears, Jóhann Sigurjónsson and the *Texelse Courant* for providing photographs.

References

Aguayo L., A. (1978). Smaller cetaceans in the Baltic Sea. *Rep. Int. Whal. Commn* **28**, 131–146.

Alling, A. K. and Whitehead, H. P. (1987). A preliminary study of the status of white-beaked dolphins, *Lagenorhynchus albirostris*, and other small cetaceans off the coast of Labrador. *Can. Field-Naturalist* **101**, 131–135.

Amundin, B. and Amundin, M. (1975). Sightings of whitenose dolphins, *Lagenorhynchus albirostris* (Gray) in Danish waters. *Z. Säugetierk.* **40**, 58–59.

Andersen, S. H. and Rebsdorff, A. (1976). Polychlorinated hydrocarbons and heavy metals in harbour porpoise (*Phocoena phocoena*) and whitebeaked dolphin (*Lagenorhynchus albirostris*) from Danish waters. *Aquat. Mammals* **4**, 14–20.

Andrésson, G. (1978). Chasing the elusive dolphin. *Atlantica Iceland Rev.* **16**(2), 28–31.

Anonymous (1981). Denmark (Greenland) progress report on cetacean research 1978–79. *Rep. Int. Whal. Commn* **31**, 185–186.

Anonymous (1989). United Kingdom. Progress report on cetacean research, May 1987 to April 1988. *Rep. Int. Whal. Commn* **39**, 191–192.

Arvy, L. (1982). Phoresies and parasitism in cetaceans: a review. *Invest. Cetacea* **14**, 233–335.

van Assen, R. (1975). Over een merkwaardige witsnuitdolfijn van het strand van Terschelling. *De Levende Natuur* **78**, 63–64.

Baker, J. R. (1992). Causes of mortality and parasites and incidental lesions in dolphins and whales from British waters. *Vet. Rec.* **130**, 569–572.

Bakker, J. and Smeenk, C. (1987). Time-series analysis of *Tursiops truncatus*, *Delphinus delphis*, and *Lagenorhynchus albirostris* strandings on the Dutch coast. *In* "The European Cetacean Society Report of the 1987 Meeting, Hirtshals, Denmark" (Eds J.W. Broekema and C. Smeenk) pp. 14–19.

Bakker, J. and Smeenk, C. (1990). Dolphins on the Dutch coast: an analysis of stranding records. *Lutra* **33**, 190–191.

Baptist, H. J. M. (1987). Waarnemingen van zeezoogdieren in de Nederlandse sector van de Noordzee. *Lutra* **30**, 93–104.

Baptist, H., Camphuysen, C. and Leopold, M. (1990). Sea mammal sightings in the Netherlands. *In* "European Research on Cetaceans — 4, Proceedings of the Fourth Annual Conference of the European Cetacean Society, Palma de Mallorca, Spain, 2–4 March 1990" (Eds P.G.H. Evans, A. Aguilar and C. Smeenk) pp. 31–33.

Beneden, P. J. Van (1860). Recherches sur la faune littorale de Belgique. Cétacés. *Mém. Acad. R. Belg.* **32**, 1–38.

Beneden, P. J. Van (1862). Recherches sur les Cétacés. *Nouv. Mém. Acad. R. Bruxelles* **32**.

Benjaminsen, T., Berlund, J., Christensen, D., Christensen, I., Huse, I. and Sandnes, O. (1976). Merking, observasjoner og adferdsstudier av hval i Barentshavet og ved Svalbard i 1974 og 1975. *Fisken Hav.* **76**(2), 9–23.

Berrow, S. D. and Rogan, E. (1997). Review of cetaceans stranded on the Irish coast, 1901–95. *Mammal Rev.* **27**, 51–76.

Bierman, W. H. and Slijper, E. J. (1948). Remarks upon the species of the genus *Lagenorhynchus*, II. *Proc. Konink. Ned. Akad. Wet.* **51**, 127–133.

Bloch, D. and Joensen, J. S. (1985). Part 2 [of Denmark progress report]. Faroe Islands. *Rep. Int. Whal. Commn* **35**, 165–166.

Boer, R. de (1989). Waarneming van een gemengde groep witflankdolfijnen *Lagenorhynchus acutus* en witsnuitdolfijnen *L. albirostris* in de centrale Noordzee. *Lutra* **32**, 181–184.

Boles, B. K. (1980). Data report listing sightings of cetaceans in the Labrador Sea. Offshore Labrador Biological Studies, 1979. Report by Atlantic Biological Services Ltd, St John's, Newfoundland, for Total Eastcan Exploration Ltd, Calgary, Alberta.

Born, E. W. (1983). Havpattedyr og havfugle i Scoresby Sund: fangst og forekomst. Rapport til Råstofforvaltningen for Grønland og Grønlands Fiskeri- og Miljøundersøgelser, Danbiu ApS, Hellerup, Denmark.

Brattey, J. and Clark, K. J. (1992). Effect of temperature on egg hatching and survival of larvae of *Anisakis simplex* B (Nematoda: Ascaridoidea). *Can. J. Zool.* **70**, 274–279.

van Bree, P. J. H. (1970). Über Weißschnauzdelphine (*Lagenorhynchus albirostris*) von den deutschen Nordseeküsten. *Natur Museum* **100**, 264–268.

van Bree, P. J. H. (1977). On former and recent strandings of cetaceans on the coast of the Netherlands. *Z. Säugetierk.* **42**, 101–107.

van Bree, P. J. H. and Husson, A. M. (1974). Strandingen van Cetacea op de Nederlandse kust in 1972 en 1973. *Lutra* **16**, 1–10.

van Bree, P. J. H. and Nijssen, H. (1964). On three specimens of *Lagenorhynchus albirostris* (Gray, 1846) (Mammalia, Cetacea). *Beaufortia* **11**, 85–93.

van Bree, P. J. H. and Smeenk, C. (1982). Strandingen van Cetacea op de Nederlandse kust in 1978, 1979 en 1980. *Lutra* **25**, 19–29.

Brice-Bennett, C. (1977). Land use in the Nain and Hopedale regions. *In* "Our Footprints Are Everywhere. Inuit Land Use and Occupancy in Labrador" (Eds C. Brice-Bennett, A. Cooke and N. Davis) pp. 97–203. Labrador Inuit Association, Nain, Labrador.

Brightwell, T. (1846). Observations on a specimen of the bottle-nosed dolphin, *Delphinus Tursio*, Fabr., taken at Great Yarmouth, October 1845. *Ann. Mag. Nat. Hist.* **17**, 21–22, pl. II.

Buck, J. D., Overstrom, N. A., Patton, G. W., Anderson, H. F. and Gorzelany, J. F. (1991). Bacteria associated with stranded cetaceans from the northeast USA and southwest Florida Gulf coasts. *Dis. Aquat. Organisms* **10**, 147–152.

Buck, J. D. and Spotte, S. (1986a). Microbiology of captive white-beaked dolphins (*Lagenorhynchus albirostris*) with comments on epizootics. *Zoo Biol.* **5**, 321–329.

Buck, J. D. and Spotte, S. (1986b). The occurrence of potentially pathogenic vibrios in marine mammals. *Mar. Mamm. Sci.* **2**, 319–324.

Camphuysen, C. J. (1991). Het herkennen van de witflankdolfijn *Lagenorhynchus acutus* en de witsnuitdolfijn *Lagenorhynchus albirostris* op zee. *Huid en Haar* **10**, 20–30.

Castells, A. and Mayo, M. (1992). Relación de citas de cetáceos (Mammalia, Cetacea): España y Portugal, No. I, Vol. II. Centro de Estudios de Mamíferos Marinos, Madrid.

Christensen, E. (1972). Vaagehvalundersøkelser ved Spitsbergen og i Barentshavet i mai og juni 1972 [Minke whale investigations in the area Spitsbergen–Barents Sea in May–June 1972]. *Fiskets Gang* **48**, 961–965.

Clark, J. W. (1876). Notes on a dolphin taken off the coast of Norfolk. *Proc. Zool. Soc. Lond.* **1876**, 686–691.

Cleland, J. (1884). Notes on the viscera of the porpoise (*Phocoena communis*) and the white-beaked dolphin (*Delphinus albirostris*). *J. Anat. Physiol.* **18**, 327–336.
Collet, A. and Duguy, R. (1981). *Lagenorhynchus albirostris* (Cetacea, Odontoceti): espèce nouvelle pour la faune de France. *Mammalia* **45**, 387–388.
Collett, R. (1912). "Norges Pattedyr". H. Aschehoug and Co. (W. Nygaard), Kristiania, Norway.
Cunningham, D. J. (1876). Description of a young specimen of the *Delphinus albirostris*. *Proc. Zool. Soc. Lond.* **1876**, 679–686.
Cunningham, D. J. (1877). The spinal nervous system of the porpoise and dolphin. *J. Anat. Physiol.* **11**, 209–228.
Degerbøl, M. (1940). Mammalia. "Zoology of the Faroes", LXV, 132 pp. Andr. Fred. Høst and Søn, Copenhagen.
Dong, J. H., Lien, J., Nelson, D. and Curren, K. (1996). A contribution to the biology of the white-beaked dolphin, *Lagenorhynchus albirostris*, in waters off Newfoundland. *Can. Field-Naturalist* **110**, 278–287.
Duguy, R. (1984). Rapport annuel sur les cétacés et pinnipèdes trouvés sur les côtes de France XIII. Année 1983. *Ann. Soc. Sci. Nat. Charente-Maritime* **7**, 189–205.
Duguy, R. (1987). Rapport annuel sur les cétacés et pinnipèdes trouvés sur les côtes de France XVI. Année 1986. *Ann. Soc. Sci. Nat. Charente-Maritime* **7**, 617–639.
Duguy, R. (1988). Rapport annuel sur les cétacés et pinnipèdes trouvés sur les côtes de France XVII. Année 1987. *Ann. Soc. Sci. Nat. Charente-Maritime* **7**, 753–769.
Eschricht, D. F. (1846). Undersögelser over hvaldyrene 5. Finhvalernes osteologie og arts-adskillelse. *Kongel. Danske Vidensk. Selsk. Naturv. Math. Afh.* **12**, 227–396.
Evans, P. G. H. (1980). Cetaceans in British waters. *Mammal Rev.* **10**, 1–52.
Evans, P. G. H. (1990). European cetaceans and seabirds in an oceanographic context. *Lutra* **33**, 95–125.
Evans, P. G. H. (1991). Whales, dolphins and porpoises: order Cetacea. *In* "The Handbook of British Mammals", 3rd Edn (Eds G. B. Corbet and S. Harris) pp. 299–350. Mammal Society, Blackwell, Oxford.
Fabricius, O. (1780). "Fauna Groenlandica". Ioannis Gottlob Rothe, Copenhagen and Leipzig.
Fransen, C. H. J. M. and Smeenk, C. (1991). Whale-lice (Amphipoda: Cyamidae) recorded from The Netherlands. *Zool. Med. Leiden* **65**, 393–405.
Fraser, F. C. (1946). "Report on Cetacea stranded on the British coasts from 1933 to 1937", No. 12, British Museum (Natural History), London.
Fraser, F. C. (1952). "Handbook of R. H. Burne's Cetacean Dissections". British Museum (Natural History), London.
Fraser, F. C. (1953). "Report on Cetacea stranded on the British coasts from 1938 to 1947", No. 13, British Museum (Natural History), London.
Fraser, F. C. (1964). Order Cetacea. *In* "The Handbook of British Mammals" (Ed. H.N. Southern) pp. 316–346. Blackwell, Oxford.
Fraser, F. C. (1974). "Report on Cetacea stranded on the British coasts from 1948 to 1966", No. 14, British Museum (Natural History), London.
Gaskin, D. E. (1982). "The Ecology of Whales and Dolphins". Heinemann, London.
Gibson, D. I. and Harris, E. A. (1979). The helminth-parasites of cetaceans in the collection of the British Museum (Natural History). *Invest. Cetacea* **10**, 309–324.
Gilpin, J. B. (1878). On the smaller cetaceans inhabiting the Bay of Fundy and shores of Nova Scotia. *Proc. Trans. Nova Scotian Inst. Nat. Sci.* **14**, 21–34.
Goethe, F. (1983). Wale und Delphine in niedersächsischen Küstengewässern und Flüssen. *Drosera* **1983**, 49–68.
Gray, J. E. (1846a). On the British Cetacea. *Ann. Mag. Nat. Hist.* **17**, 82–85.

Gray, J. E. (1846b). On the cetaceous animals. *In* "The Zoology of the Voyage of H.M.S. Erebus and Terror under the command of Sir James Clark Ross, R.N., F.R.S., during the years 1839 to 1843", Vol. 1, "Mammalia, Birds" (Eds J. Richardson and J. E. Gray) pp. 13–53, 37 pls.

Grieg, J. A. (1888). Hvalstænget ved bildøen. *Bergens Mus. Aarsberetning* **1888**(3), 1–17.

Guldberg, G. and Nansen, F. (1894). On the development and structure of the whale, Part I, On the development of the dolphin. *Bergens Mus. Skr.* **5**, 1–70, pls. I–VII.

Gunnlaugsson, T., Sigurjónsson, J. and Donovan, G. P. (1988). Aerial survey of cetaceans in the coastal waters off Iceland, June–July 1986. *Rep. Int. Whal. Commn* **38**, 489–500.

Haase, B. J. M. (1987). A mixed herd of white-beaked dolphins *Lagenorhynchus albirostris* and white-sided dolphins *L. acutus* in the southern North Sea. *Lutra* **30**, 105–106.

Hammond, P. S., Benke, H., Berggren, P., Borchers, D. L., Buckland, S. T., Collet, A., Heide-Jørgensen, M. P., Heimlich-Boran, S., Hiby, A. R., Leopold, M. F. and Øien, N. (1995). Distribution & abundance of the harbour porpoise & other small cetaceans in the North Sea & adjacent waters. Contract LIFE 92-2/UK/027, Final Report to the European Commission.

Harmer, S. F. (1927). "Report on Cetacea stranded on the British coasts from 1913 to 1926", No. 10, British Museum (Natural History), London.

Hartmann, M. G., Kamminga, C., Klooswijk, A. I. J. and Fleischer, G. (1992). The use of computer tomography in odontocete morphology—preliminary results. *In* "Whales: Biology—Threats—Conservation", Symposium held in Brussels, 5–7 June 1991 (Ed. J. J. Symoens) pp. 157–166. Royal Academy of Overseas Sciences, Brussels.

Hashmi, D. and Adloff, B. (1991). Surface frequency of cetaceans in the Strait of Gibraltar. *In* "European Research on Cetaceans—5". Proceedings of the Fifth Annual Conference of the European Cetacean Society, Sandefjord, Norway, 21–23 February 1991 (Ed. P. G. H. Evans) pp. 16–17.

Hashmi, D. and Adloff, B. (1992). Cetacean research in the Strait of Gibraltar. *In* "Whales: Biology—Threats—Conservation", Symposium held in Brussels, 5–7 June 1991 (Ed. J. J. Symoens) pp. 59–62. Royal Academy of Overseas Sciences, Brussels.

Hay, K. (1981). Line-transect Estimates of Whale Abundance in the Newfoundland–Labrador Area, 1980. Report to the Canadian Atlantic Fisheries Scientific Advisory Committee, 36 pp. (cited in Alling and Whitehead, 1987).

Hay, K. (1982). Aerial line-transect estimates of abundance of humpback, fin, and long-finned pilot whales in the Newfoundland–Labrador area. *Rep. Int. Whal. Commn* **32**, 475–486.

Heide-Jørgensen, M.-P. and Leatherwood, S. (1987). "Cetaceans of North and West Greenland. Report of a Reconnaissance Expedition August–September 1987". SWRI/HMRC Tech. Rep. 87-206, Sea World Research Institute, Hubbs Marine Research Center, San Diego, CA.

Hennipman, E., Nijhoff, P., Swennen, C., Vader, W. J. M., de Wilde, W. J. J. O. and Tulp, A. S. (1961). Verslag van de Nederlandse biologische expeditie Turkije 1959. *De Levende Natuur* **64**(5), bijlage, 3–27.

Husson, A. M. and van Bree, P. J. H. (1976). Strandingen van Cetacea op de Nederlandse kust in 1974 en 1975. *Lutra* **18**, 25–32.

Japha, A. (1908). Zusammenstellung der in der Ostsee bisher beobachteten Wale. *Schrift. Phys.-Ökon. Ges. (Königsberg)* **49**, 119–189.

Jones, P. H. (1984). Cetaceans seen in the Irish Sea and approaches, late summer 1983. *Nature in Wales* **3**, 62–64.

Jonsgård, Å. (1962). On the species of dolphins found on the coast of northern Norway and in adjacent waters. *Norsk Hvalfangst-tid.* **51**, 1–13.

Jonsgård, Å. and Christensen, I. (1968). A preliminary report on the 'Harøybuen' cruise in 1968. *Norsk Hvalfangst-tid.* **57**, 174–175.

Jonsgård, Å. and Nordli, O. (1952). Concerning a catch of white-sided dolphins (*Lagenorhynchus acutus*) on the west coast of Norway, winter 1952. *Norsk Hvalfangst-tid.* **41**, 229–232.

Katona, S. K., Rough, V. and Richardson, D. T. (1983). "A Field Guide to the Whales, Porpoises, and Seals of the Gulf of Maine and Eastern Canada — Cape Cod to Newfoundland". Charles Scribner's, New York.

Kenney, R. D., Payne, P. M., Heinemann, D. W. and Winn, H. E. (1996). Shifts in northeast shelf cetacean distributions relative to trends in Gulf of Maine/Georges Bank finfish abundance. *In* "The Northeast Shelf Ecosystem: Assessment, Sustainability, and Management" (Eds K. Sherman, N. A. Jaworski and T. J. Smayda) pp. 169–196. Blackwell, Cambridge, MA.

Kerkhoff, M., de Boer, J. and Geerdes, J. (1981). Heptachlor epoxide in marine mammals. *Sci. Total Environ.* **19**, 41–50.

Kingsley, M. C. S. and Reeves, R. R. (1998). Aerial surveys of cetaceans in the Gulf of St. Lawrence in 1995 and 1996. *Can. J. Zool.* (in press).

Kinze, C. C. (1995). Danish whale records (Mammalia, Cetacea). Review of whale specimens stranded, directly or incidentally caught along the Danish coasts. *Steenstrupia* **21**, 155–196.

Kinze, C. C., Baagøe, H. J. and Jensen, B. (1987). Danish strandings and sightings 1977–1985. *In* "The European Cetacean Society Report of the 1987 Meeting, Hirtshals, Denmark" (Eds J. W. Broekema and C. Smeenk) pp. 47–51.

Kinze, C. C., Addink, M., Smeenk, C., García Hartmann, M., Richards, H. W., Sonntag, R. P. and Benke, H. (1997). The white-beaked dolphin (*Lagenorhynchus albirostris*) and the white-sided dolphin (*Lagenorhynchus acutus*) in the North and Baltic seas: Review of available information. *Rep. Int. Whal. Commn* **47**, 675–681.

Klinowska, M. (1985). Interpretation of the UK cetacean strandings records. *Rep. Int. Whal. Commn* **35**, 459–467.

Klinowska, M. and Brown, S. (1986). "A Review of Dolphinaria". Department of the Environment, London.

Kompanje, E. J. O. (1995). On the occurrence of spondylosis deformans in white-beaked dolphins *Lagenorhynchus albirostris* (Gray, 1846) stranded on the Dutch coast. *Zool. Med. Leiden* **69**, 231–250.

Kumerloeve, H. (1975). Die Säugetiere (Mammalia) der Türkei. *Veröff. Zool. Staatssamml. München* **18**, 69–158.

Law, R. J., Fileman, C. F., Hopkins, A. D., Baker, J. R., Harwood, J., Jackson, D. B., Kennedy, S., Martin, A. R. and Morris, R. J. (1991). Concentrations of trace metals in the livers of marine mammals (seals, porpoises and dolphins) from waters around the British Isles. *Mar. Poll. Bull.* **22**, 183–191.

Lepiksaar, J. (1966). Zahnwalfunde in Schweden. *Bijdr. Dierk.* **36**, 3–16.

Lilljeborg, W. (1866). Synopsis of the cetaceous Mammalia of Scandinavia (Sweden and Norway). *In* "Recent Memoirs on the Cetacea" (Ed. W.H. Flower) pp. 219–309. Ray Society, Robert Hardwicke, London.

Lincoln, R. J. and Hurley, D. E. (1974). *Scutocyamus parvus*, a new genus and species of whale-louse (Amphipoda: Cyamidae) ectoparasitic on the North Atlantic white-beaked dolphin. *Bull. Brit. Mus. (Nat. Hist.) Zool.* **27**, 59–64.

Lütken, M. C. (1888). Critical studies upon some Odontoceti of the genera *Tursiops, Orca,* and *Lagenorhynchus. Ann. Mag. Nat. Hist.*, 6th Ser., **2**, 179–186.

Mackey, M. G. A. and Orr, R. D. (1987). An evaluation of household country food use in Makkovik, Labrador, July 1980–June 1981. *Arctic* **40**, 60–65.

Martin, A. R., Hembree, D., Waters, T. D. and Sigurjónsson, J. (1984). IDCR cruise/aerial survey in the north eastern Atlantic 1982: cruise report. *Rep. Int. Whal. Commn* **34**, 645–653.

Mathiasson, S. (1990). Faunistiskt nytt från Göteborgs Naturhistoriska Museum 1989. Göteborgs Naturhistoriska Museum Årstryck, 1990, pp. 13–17.

Mathiasson, S. (1992). Faunistiskt nytt från Göteborgs Naturhistoriska Museum 1991. Göteborgs Naturhistoriska Museum Årstryck, 1992, pp. 15–18.

McBrearty, D. A., Message, M. A. and King, G. A. (1986). Observations on small cetaceans in the north-east Atlantic Ocean and the Mediterranean Sea: 1978–1982. *In* "Research on Dolphins" (Eds M.M. Bryden and R. Harrison) pp. 225–249. Clarendon Press, Oxford.

Mead, J. G. (1975). Anatomy of the external nasal passages and facial complex in the Delphinidae (Mammalia: Cetacea). *Smithsonian Contrib. Zool.* **207**, 1–72.

Meek, A. (1918). The reproductive organs of Cetacea. *J. Anat.* **52**, 186–210.

Mercer, M. C. (1973). Observations on distribution and intraspecific variation in pigmentation patterns of odontocete Cetacea in the western North Atlantic. *J. Fish. Res. Bd Can.* **30**, 1111–1130.

Mikkelsen, A. M. H. and Lund, A. (1994). Intraspecific variation in the dolphins *Lagenorhynchus albirostris* and *L. acutus* (Mammalia: Cetacea) in metrical and non-metrical skull characters, with remarks on occurrence. *J. Zool. Lond.* **234**, 289–299.

Millais, J. G. (1906). "The Mammals of Great Britain and Ireland", Vol. III. Longmans, London.

Mitchell, E. (1970). Pigmentation pattern evolution in delphinid cetaceans: an essay in adaptive coloration. *Can. J. Zool.* **48**, 717–740.

Mitchell, E. (1975). "Porpoise, Dolphin and Small Whale Fisheries of the World. Status and Problems". IUCN Monograph 3, International Union for the Conservation of Nature and Natural Resources, Morges, Switzerland.

Mitson, R. B. (1990). Very-high-frequency acoustic emissions from the white-beaked dolphin (*Lagenorhynchus albirostris*). *In* "Sensory Abilities of Cetaceans: Laboratory and Field Evidence" (Eds J. A. Thomas and R. A. Kastelein) pp. 283–294. Plenum Press, New York.

Mitson, R. B. and Morris, R. J. (1988). Evidence of high-frequency acoustic emissions from the white-beaked dolphin (*Lagenorhynchus albirostris*). *J. Acoust. Soc. Am.* **83**, 825–826.

Morris, R. J. (1992). Trace metals in the livers of marine mammals from the Welsh coast and the Irish Sea. *Mar. Poll. Bull.* **24**, 296–304.

Muir, D. C. G., Wagemann, R., Grift, N. P., Norstrom, R. J., Simon, M. and Lien, J. (1988). Organochlorine chemical and heavy metal contaminants in white-beaked dolphins (*Lagenorhynchus albirostris*) and pilot whales (*Globicephala melaena*) from the coast of Newfoundland, Canada. *Arch. Environ. Contam. Toxicol.* **17**, 613–629.

Murie, J. (1873). Notes on the whitebeaked bottlenose, *Lagenorhynchus albirostris* Gray. *J. Linn. Soc.* **11**, 141–153.

van Nie, C. J. (1985). The conducting system of the heart of the white-snouted dolphin *Lagenorhynchus albirostris* (Gray, 1846). *Lutra* **28**, 106–112.

van Nie, C. J. (1986). The Purkinje fibres in the heart of the common seal *Phoca vitulina vitulina* (Linnaeus, 1758), the Baikal seal *Pusa sibirica* (Gmelin, 1788), the fur seal *Arctocephalus australis* (Zimmermann, 1782), the harbour porpoise *Phocoena phocoena* (Linnaeus, 1758) and the white-beaked dolphin *Lagenorhynchus albirostris* (Gray, 1846). *Aquat. Mamm.* **12**, 61–64.

Nilsson, S. (1847). Cetacea hvaldjur. *In* "Skandinavisk Fauna", I, "Däggdjuren", 2nd Edn, pp. 583–645. C. W. K. Gleerup Förlag, Lund.

Northridge, S. (1987). By-catch working group. *European Cetacean Society (ECS) Newsletter* **2**, 4–5.

Northridge, S. P. (1991). An Updated World Review of Interactions Between Marine Mammals and Fisheries. FAO Fish. Tech. Pap. 251, Suppl. 1, Food and Agriculture Organization of the United Nations, Rome.

Northridge, S. P., Tasker, M. L., Webb, A. and Williams, J. M. (1995). Distribution and relative abundance of harbour porpoises (*Phocoena phocoena* L.), white-beaked dolphins (*Lagenorhynchus albirostris* Gray), and minke whales (*Balaenoptera acutorostrata* Lacépède) around the British Isles. *ICES J. Mar. Sci.* **52**, 55–66.

Northridge, S., Tasker, M., Webb, A., Camphuysen, K. and Leopold, M. (1997). White-beaked *Lagenorhynchus albirostris* and Atlantic white-sided dolphin *L. acutus* distributions in northwest European and US North Atlantic waters. *Rep. Int. Whal. Commn* **47**, 797–805.

van Oort, E. D. (1918). Over een te Noordwijk aan Zee aangespoelden *Lagenorhynchus albirostris*, benevens een lijst van de cetaceen-soorten, die tot heden aan de Nederlandsche kust zijn waargenomen. *Zool. Meded. Leiden* **4**, 54–62.

Osterhaus, A. D. M. E., Broeders, H. W. G., Teppema, K. S., Kuiken, T., House, J. A., Vos, H. W. and Visser, I. K. G. (1993). Isolation of a virus with Rhabdovirus morphology from a white beaked dolphin *Lagenorhynchus albirostris*. *Arch. Vir.* **133**, 189–193.

Ree, V. (1994). Feltbestemmelse av kvitnos og kvitskjeving. En oppsummering av artenes utseende i eldre og nyere litteratur og en presentasjon av materiale fra norske farvann de senere år. *Fauna (Oslo)* **47**, 132–165.

Reichenbach, H. G. L. (1846). "Die Cetaceen oder Walthiere". Naturgeschichte der Säugethiere des In- und Auslandes, No. I(i).

Reidenberg, J. S. and Laitman, J. T. (1988). Existence of vocal folds in the larynx of Odontoceti (toothed whales). *Anat. Rec.* **221**, 884–891.

Santos, M. B., Pierce, G. J., Ross, H. M. and Reid, R. J. (1993). Fish and cephalopod prey of small cetaceans from the Scottish coast. Abstr. 10th Bienn. Conf. Biol. Mar. Mamm., Galveston, TX, USA, p. 95.

Sæmundsson, B. (1939). Mammalia. *In* "The Zoology of Iceland", Vol. 4(76). Munksgaard, Copenhagen.

Schenkkan, E. J. (1973). On the comparative anatomy and function of the nasal tract in odontocetes (Mammalia, Cetacea). *Bijdr. Dierk.* **43**, 127–159.

Schevill, W. E. and Watkins, W. A. (1962). "Whale and Porpoise Voices. A Phonograph Record". Woods Hole Oceanographic Institution, Woods Hole, MA, USA.

Scholander, P. F. and Schevill, W. E. (1955). Countercurrent vascular heat exchange in the fins of whales. *J. Appl. Physiol.* **8**, 279–282.

Schultz, W. (1970). Über das Vorkommen von Walen in der Nord- und Ostsee (Ordn. Cetacea). *Zool. Anz.* **185**, 172–264.

Schulze, G. (1991). Wale an der Küste von Mecklenburg-Vorpommern. *Meer und Museum* **7**, 22–52.

Sears, R., Wenzel, F. W. and Williamson, J. M. (1981). Aerial survey of marine mammals along the Quebec lower north shore and its principal river estuaries (Mingan Islands to Strait of Belle Isle) Fall 1981. Report for Hydro-Québec, Direction Environnement. Mingan Island Cetacean Study, East Falmouth, MA, USA, and Sept-Iles, Québec, Canada.

Sears, R. and Williamson, J. M. (1982) Distribution and relative abundance of marine mammals along the Quebec lower north shore and its principal river estuaries (Mingan Island to Strait of Belle Isle) July to October 1982. Report for Hydro-Québec, Vice Présidence Environnement. Mingan Island Cetacean Study, East Falmouth, MA, USA, and Sept-Iles, Québec, Canada.

Sergeant, D. E. and Fisher, H. D. (1957). The smaller Cetacea of eastern Canadian waters. *J. Fish. Res. Bd Can.* **14**, 83–115.
Sheldrick, M. C. (1989). Stranded whale records for the entire British coastline, 1967–1986. *Invest. Cetacea* **22**, 298–329.
Sigurjónsson, J. (1983). The cruise of the Ljósfari in the Denmark Strait (June–July 1981) and recent marking and sightings off Iceland. *Rep. Int. Whal. Commn* **33**, 667–682.
Sigurjónsson, J. (1985). Sightings survey in the Irminger Sea and off Iceland in 1983. *Rep. Int. Whal. Commn* **35**, 499–503.
Sigurjónsson, J., Lyrholm, T., Leatherwood, S., Jónsson, E. and Víkingsson, G. (1988). Photoidentification of killer whales, *Orcinus orca*, off Iceland, 1981 through 1986. *Rit Fisk.* **11**, 99–114.
Slijper, E. J. (1936). "Die Cetaceen, Verleichend-anatomisch und Systematisch". Martinus Nijhoff, 's-Gravenhage.
Smeenk, C. (1986). Strandingen van Cetacea op de Nederlandse kust in 1981, 1982 en 1983. *Lutra* **29**, 268–281.
Smeenk, C. (1989). Strandingen van Cetacea op de Nederlandse kust in 1984, 1985 en 1986. *Lutra* **32**, 164–180.
Smeenk, C. (1992). Strandingen van Cetacea op de Nederlandse kust in 1987, 1988 en 1989. *Lutra* **35**, 51–66.
Smeenk, C. (1995). Strandingen van Cetacea op de Nederlandse kust in 1990, 1991 en 1992. *Lutra* **38**, 90–104.
Smeenk, C. and Addink, M. (1990). "Massastranding" op de Nederlandse kust—Witsnuitdolfijnen in storm en branding. *Zoogdier* **1**(1), 5–9.
Smeenk, C. and Gaemers, P. A. M. (1987). Fish otoliths in the stomachs of white-beaked dolphins *Lagenorhynchus albirostris*. *In* "The European Cetacean Society Report of the 1987 Meeting, Hirtshals, Denmark" (Eds J. W. Broekema and C. Smeenk) pp. 12–13.
Smet, W. M. A. de (1974). Inventaris van de walvisachtigen (Cetacea) van de Vlaamse kust en de Schelde. *Bull. Inst. R. Sci. Nat. Belg.* **50**(1), 1–156.
Smet, W. M. A. de (1975). Les rapports entre les côtes et les vertèbres thoraciques chez le dauphin à bec blanc, *Lagenorhynchus albirostris* (Gray, 1846). *Lutra* **17**, 38–48.
Smet, W. M. A. de (1981). Gegevens over de walvisachtigen (Cetacea) van de Vlaamse kust en de Schelde uit de periode 1969–1975. *Bull. Inst. R. Sci. Nat. Belg.* **53**(4), 1–34.
Smet, W. M. A. de, Asselberg, R. and Duyndam, R. (1985). Gegevens over een witsnuitdolfijn, *Lagenorhynchus albirostris*, uit de zuidelijke Noordzee. *Marswin* **6**, 3–19.
Southwell, T. (1879). White-beaked dolphin at Yarmouth. *Zoologist*, 3rd Ser., **3**, 421–422.
Southwell, T. (1885). Notes on the white-beaked dolphin (*Lagenorhynchus albirostris*). *Trans. Norfolk Norwich Nat. Soc.* **4**, 120–124.
Southwell, T. (1886). White-beaked dolphin at Lowestoft. *Zoologist*, 3rd Ser., **10**, 364.
Stock, J. H. (1977). Whale-lice (Amphipoda, Cyamidae) on *Lagenorhynchus albirostris* in Dutch waters. *Crustaceana* **32**, 206.
Tomilin, A. G. (1967). "Mammals of the U.S.S.R. and Adjacent Countries", Vol. IX, "Cetacea" (Ed. V. G. Heptner). Nauk USSR, Moscow (English translation of 1957 Russian version, Israel Program for Scientific Translations, Jerusalem).
True, F. W. (1889). Contributions to the natural history of the cetaceans. A review of the family Delphinidae. *Bull. U.S. Nat. Mus.* **36**, 1–191, 47 pls.
Van Gompel, J. (1991). Cetacea aan de Belgische kust, 1975–1989. *Lutra* **34**, 27–36.
Van Gompel, J. (1996). Cetacea aan de Belgische kust, 1990–1994. *Lutra* **39**, 45–51.
Van Utrecht, W. L. (1981). Comparison of accumulation patterns in layered dentinal tissue of some Odontoceti and corresponding patterns in baleen plates and ear plugs of Balaenopteridae. *Beaufortia* **31**, 111–122.

Winn, H. E. and Edel, R. K. (1982). "A Characterization of Marine Mammals and Turtles in the Mid- and North-Atlantic Areas of the U.S. Outer Continental Shelf", CETAP Annual Report for 1980. Contract No. AA551-CT8-48, Cetacean and Turtle Assessment Program, University of Rhode Island, sponsored by U.S. Bureau of Land Management.

Øien, N. (1990). Sightings surveys in the northeast Atlantic in July 1988: distribution and abundance of cetaceans. *Rep. Int. Whal. Commn* **40**, 499–511.

Øien, N. (1996). *Lagenorhynchus* species in Norwegian waters as revealed from incidental observations and recent sighting surveys. Document SC/48/SM15, International Whaling Commission, Scientific Committee, Cambridge.

2

Atlantic White-sided Dolphin *Lagenorhynchus acutus* (Gray, 1828)

Randall R. Reeves, Chris Smeenk, Robert L. Brownell, Jr and Carl C. Kinze

Genus and Species

Taxonomy

A skull in the Museum D. Brookes, London, was described by Gray (1828) as *Delphinus* (*Grampus*) *acutus*. This specimen is now in the National Museum of Natural History, Leiden, the Netherlands (see Flower, 1884). The type locality is unknown. Gray (1846) placed the species in the genus *Lagenorhynchus*. Synonyms include *Delphinus leucopleurus* Rasch, 1843, *D. eschrichtii* Schlegel, 1841, *Lagenorhynchus perspicillatus* Cope, 1876, and *L. gubernator* Cope, 1876 (see Hershkovitz, 1966, for further details).

Common names

This species is called belobokii del'fin in Russian and kvitskjeving (or hvitskjeving) in Norwegian, although some Norwegians refer to coastal dolphins generally as "springere" (Guldberg and Nansen, 1894; Jonsgård and Nordli, 1952). It is called vitskjäving (or vitsiding) in Swedish, hvidskæving in Danish, hvítskjórutur springari in Faroese, leiftur in Icelandic, Weißseitendelphin in German, witflankdolfijn in Dutch, lagénorhynque à flancs blancs in French,

Handbook of Marine Mammals, Volume 6
ISBN 0-12-588506-7

delfin de flancos blancos in Spanish and aarluarsuk in Greenlandic. The term "jumper" is applied to Atlantic white-sided dolphins (and to some extent white-beaked dolphins, *Lagenorhynchus albirostris*) in Newfoundland and Labrador (Sergeant and Fisher, 1957).

Evolution

Mitchell (1970) suggested, on the basis of pigmentation comparisons, that *L. acutus* may be the most specialized species in the genus. There is no published fossil record for *L. acutus*.

Distribution

The range of the Atlantic white-sided dolphin is restricted to temperate and sub-arctic portions of the North Atlantic, including continental shelves and slopes as well as deep submarine canyons (Fig. 1). Mikkelsen and Lund (1994) found no evidence of separate populations based on a study of metrical and non-metrical skull characters of 123 dolphins from much of the species' range.

Although most early references define the Atlantic white-sided dolphin's southern limit in the west as Cape Cod near 41°N (Schevill, 1956), recent evidence shows that it occurs at least as far south as 38°N (Testaverde and Mead, 1980). Winn and Edel (1982) plotted a distribution off the northeastern United States centred in shelf waters of the Gulf of Maine year-round (see also Kenney *et al.*, 1996). There are large concentrations in the southwestern Gulf of Maine (40–42°N) and along the continental slope at approximately 39°N in spring (Selzer and Payne, 1988). The distribution is more scattered throughout the Gulf of Maine in autumn, with most sightings between 42°N and 44°N. Mean surface water temperature for a sample of 86 sightings off the northeastern United States was 7.0 ± 2.89 °C. Selzer and Payne (1988) suggested that Atlantic white-sided dolphins move south along the continental shelf edge in winter and spring, in association with the relatively cold, less saline Gulf of Maine water flowing southwards through Northeast Channel during these seasons. The same authors compared the distribution of Atlantic white-sided dolphins with that of the short-beaked common dolphin (*Delphinus delphis*). They concluded that, while both species were sighted most frequently in areas with high sea floor relief, the former inhabited slightly cooler, less saline water than the latter.

Occurrence of Atlantic white-sided dolphins off Newfoundland has been said to be seasonal, mainly from July to October (Sergeant and Fisher, 1957). Several sightings were reported in October along and seaward of the 200 m contour off Labrador at 55–56°N (Boles, 1980). Gaskin (1992) judged the observation of one specimen in the northern part of Flemish Cap at 52° 30′N, 43° 31′W in early April (Sergeant *et al.*, 1980) to be the most easterly record of the species in the western North Atlantic.

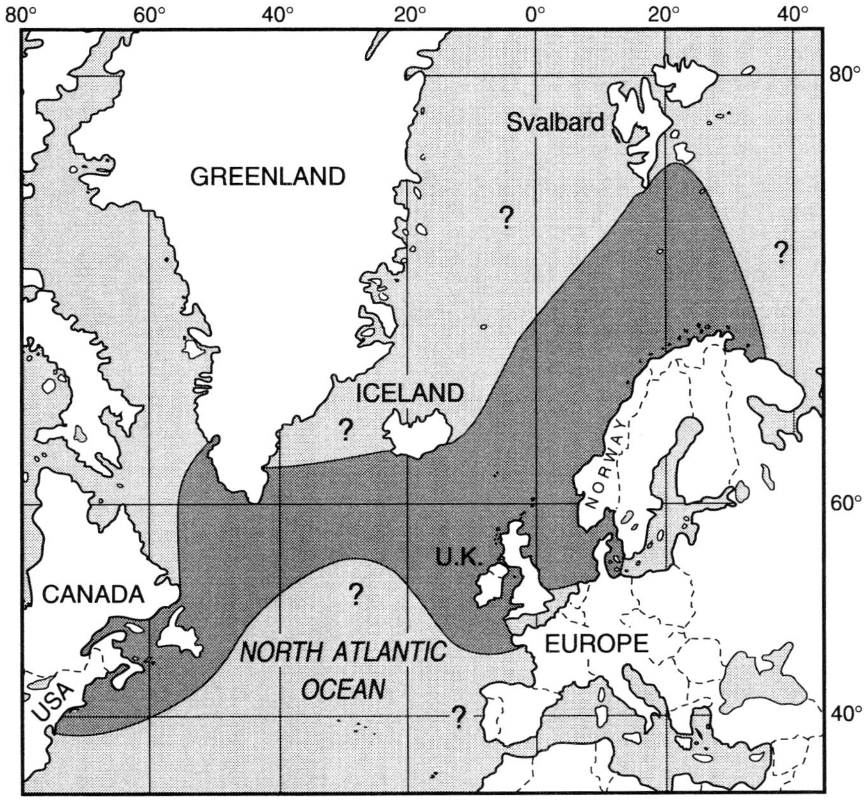

FIG. 1 Approximate normal distribution of the Atlantic white-sided dolphin. Question marks indicate that there is uncertainty about the species' regular occurrence in the area.

In some years Atlantic white-sided dolphins move as far up the St. Lawrence River estuary as the Saguenay River confluence (Sergeant *et al.*, 1980). They are common near the Mingan Islands and elsewhere in the Gulf of St. Lawrence although they are largely replaced by the white-beaked dolphin in the northeastern Gulf and the Strait of Belle Isle (Sears *et al.*, 1981; Sears and Williamson, 1982; Katona *et al.*, 1983; Kingsley and Reeves, 1998). Atlantic white-sided dolphins commonly enter the Bay of Fundy (Gaskin and Smith, 1979; Gaskin, 1983), and they are present but not common on the Scotian Shelf in late summer and early autumn (Gaskin, 1992). Strandings have occurred on Sable Island (Sergeant *et al.*, 1970). These dolphins are regularly seen over the Gully, a large submarine canyon about 2000 m deep located 40 km southeast of Sable Island (Gowans and Whitehead, 1995).

The northern limit in the western and central North Atlantic is not well known. References to Atlantic white-sided dolphins in Davis Strait are poorly documented (Gaskin, 1992). There is a record of a small group near Nuuk (Godthåb), Greenland, in September 1858 (Winge, 1902). Specimens possibly of this species have been taken also as far north as Uummannaq (70° 45′N) (Anonymous, 1981a) and Maniitsoq (Sukkertoppen; about 65°N) (Kapel, 1985). Other catch records for Greenland (e.g. Mitchell, 1975; Heide-Jørgensen, 1990) provide no details. According to F.O. Kapel (personal communication) catch statistics from Greenland should be interpreted cautiously because the same local name is applied to both *L. acutus* and *L. albirostris*.

Atlantic white-sided dolphins are present in the Irminger Sea and common in waters south of Iceland between approximately 50°N and 60°N during summer (Sigurjónsson, 1985; Gunnlaugsson *et al.*, 1988; Sigurjónsson *et al.*, 1989; Sigurjónsson and Gunnlaugsson, 1988, 1990). They have been considered abundant around the Faroe Islands and in the seas between the Faroe and Shetland Islands, especially in late summer and autumn (Degerbøl, 1940). Hundreds are taken in drive hunts at the Faroes in some years (see below).

Despite sightings reported in the early literature, Jonsgård (1962) concluded that there was no reliable evidence of the Atlantic white-sided dolphin's presence off northern Norway. However, on the basis of data from systematic whale surveys and recent opportunistic sightings, Øien (1996) concluded that *L. acutus* occurs more regularly in the Barents Sea, from Finnmark and the Murman coast to southern Svalbard, than was suggested by Jonsgård and Nordli (1952) and Jonsgård (1962). Tomilin's (1967) statements concerning the presence of *L. acutus* in the southwestern Barents Sea and off Kanin Peninsula are difficult to evaluate because the species identifications for the records in question were not confirmed.

Degerbøl (1940) described Atlantic white-sided dolphins as common off southwestern Norway, but relatively few sightings have been reported in recent years (Northridge *et al.*, 1997). They sometimes stray into the Kattegat and inner Danish waters (Aguayo L., 1978). Strandings have occurred on the North Sea and Skagerrak coasts of Denmark and the Swedish Bohuslän coast (Lepiksaar, 1966; Kinze, 1995). At least 50 animals of a larger school were caught in the Isefjord and Roskilde Fjord inlets of Sjælland, Denmark, in October/November 1942 (Degerbøl, 1943; Kinze, 1995). White-sided dolphins have not been recorded from the Baltic proper.

Fraser (1974) concluded, on the basis of strandings, that concentrations of *L. acutus* around the British Isles were somewhat farther north than those of *L. albirostris*. Stranding patterns also suggested to Fraser (1974) that *L. acutus* were most common off the north coasts of the UK during summer but also fairly common there in February and March (cf. Waller and Tyler, 1979). Evans (1980) interpreted sightings along the edge of the continental shelf west of Ireland to mean that the Atlantic white-sided dolphin has an offshore distribution and therefore is under-represented in the stranding records. He regarded the Atlantic white-sided dolphin as having a more pelagic distribution than the white-beaked dolphin (Evans, 1980), and this conclusion was shared by Mikkelsen and Lund (1994; see also Northridge *et al.*, 1997). Atlantic white-sided dolphins strand occasionally on the west and south coasts of Ireland, with a considerable increase in frequency since the 1980s (Berrow and Rogan, 1997; Rogan *et al.*, 1997). Records of by-catches by Dutch freezer trawlers show that the species is common on the edge of the continental shelf west and south of Ireland in February and March (Couperus, 1994; Leopold and Couperus, 1995). Despite the fact that the same trawlers are fishing west of Scotland in January and February, no by-catches of Atlantic white-sided dolphins have been reported from north of 54°N. Information from stomach contents suggests that white-sided dolphins migrate from deeper oceanic waters to the shelf area (approx. 200 m depth) when the mackerel schools arrive on their southward migration (Couperus, 1997b). Observations of large groups (hundreds of animals) have been made in deeper water (close to or seaward of the 200 m depth contour) southwest of Ireland and west of Brittany during March and May (Leopold and Couperus, 1995). McBrearty *et al.* (1986) listed sighting records around the British Isles for all months except January and October, with the highest frequencies in May–September. Evans (1991) stated that these dolphins occur in peak numbers in British waters between August and October, and he suspected a possible offshore movement during winter. Evans (1980) also suggested that there is a northward movement from the North Sea east of Shetland and Orkney in late autumn.

Strandings on the Dutch coast are reported infrequently (Van Utrecht and Husson, 1968; van Bree and Husson, 1974; Smeenk, 1986, 1989). Baptist (1987) mentioned a few sightings of white-sided dolphins in Dutch waters, and Haase (1987) and de Boer (1989) referred to observations in and just outside Dutch waters, respectively. Nearly all records are from the northern part of the Dutch sector of the North Sea. Strandings have occurred in winter and spring on the Finistère and Charente-Maritime coasts of France (Duguy, 1978, 1980, 1987) and the Atlantic coast of Spain (Castells and Mayo, 1992; López Fernández, 1997). Several dolphins identified as this species were sighted in the Strait of Gibraltar (Hashmi and Adloff, 1991, 1992).

External Characteristics

Colour pattern

The most striking aspect of the colour pattern is the elongated white patch on either side, beginning below the dorsal fin and continuing back to near the origin of the flukes (see Katona *et al.*, 1983; Fig. 2). An equally well defined longitudinal patch of yellow-brown or tan ("mustard" by some accounts, "ochreous yellow" according to Gaskin, 1992) extends caudally from near the front end of the white flank patch. This yellowish patch, which begins as a very narrow line or stripe cranially and widens posteriorly, extends onto the upper part of the tail stock but does not cross over the dorsal ridge. The Atlantic white-sided dolphin

FIG. 2 An Atlantic white-sided dolphin surfacing in Cape Cod Bay, MA, showing the well-demarcated flank pigmentation. (Photo by William Rossiter)

is the only living species in its genus to have evolved such distinctive white and brownish-yellow patches on the sides (Mitchell, 1970). It has been likened to the common dolphin because of the white and brownish-yellow patches and the sharpness of the demarcation between the dark spinal field (including the yellowish stripe) and the white abdominal field. The Atlantic white-sided dolphin is dark grey to black on the back, including the upper jaw, melon, dorsal fin, flippers and flukes. The belly and lower jaw are white. There is a black patch around the genital slit. Typically a black eye-ring is connected with the upper jaw by a black line, and a very thin black line connects the eye-ring with the external ear opening (Fig. 3). A grey stripe connects the eye and flipper. The sides of the body are light grey, apart from the two patches (one white, one yellowish) described above. According to Haase (1987) and de Boer (1989) there is less contrast between the zones of different colours on calves than on adults.

Size and shape

Maximum body length is 282 cm (Harmer, 1927). There is sexual dimorphism in size. In one sample of 111 dolphins, the maximum sizes of males were 267 cm and 234 kg and of females were 243 cm and 182 kg (Sergeant *et al.*, 1980). The

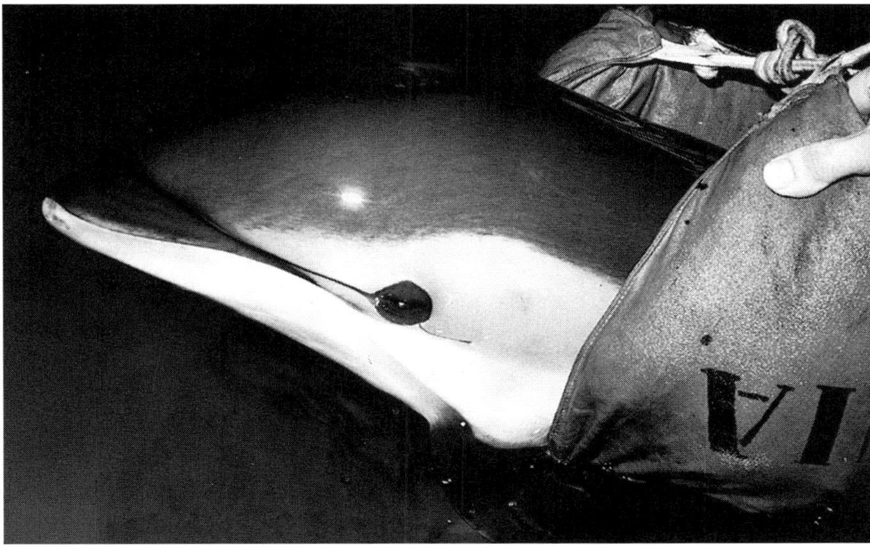

FIG. 3 A female Atlantic white-sided dolphin (150 cm) that stranded alive on the isle of Texel, the Netherlands, 24 December 1985. The thin black line from eye-ring to ear is clearly shown. (Photo courtesy Texelse Courant, Texel)

maximum girth can be as much as 60% of the body length (Testaverde and Mead, 1980).

The beak is short (5 cm or less) and not sharply demarcated from the melon (Fig. 4). The dorsal fin, at mid-body, is tall (up to 33 cm, or more than 12% of body length), pointed and recurved. The pointed flippers can be up to 36 cm long, or about 15% of body length. The caudal peduncle is laterally compressed and deepened dorso-ventrally. In lateral view it is parallel-sided anteriorly, then abruptly narrowed just in front of the tail flukes. In adult males it is higher and more abruptly narrowed caudally than in females (C. Smeenk, unpublished data).

Schevill (1956) described 8–12 small dermal tubercles, reminiscent of those often found on the harbour porpoise (*Phocoena phocoena*), on the distal leading edges of the flippers of one young female Atlantic white-sided dolphin.

Internal Anatomy

Skull

The skull of this species (Fig. 5) generally is longer and narrower than that of *L. albirostris* but shorter than that of the Pacific white-sided dolphin *L. obliquidens*. Skull measurements for ten specimens (*x*, range in mm) from True (1889; also

FIG. 4 This female and calf were part of a mass stranding of Atlantic white-sided dolphins in Cape Cod Bay, MA, 27 August 1992. Note the robustness of the body and the short, broad beak. The light-coloured splashes on the side of the mother are not typical of the species. (Photo by Daniel Morast/International Wildlife Coalition)

see Harmer, 1927) are: condylobasal length, 406, 388–425; rostral length, 207, 196–216; basal width of rostrum, 109, 102–117; orbital width, 209, 198–224; and length of upper tooth row, 181, 165–191. The dental formula is 29–40/31–38. The teeth are 4 mm or less in diameter.

Postcranial skeleton

The flipper is narrow and pointed, approximately 1/7 the length of the entire skeleton. The phalangeal formula is: I_{1-2}, II_{10}, III_6, IV_{2-3} and V_2 (Tomilin, 1967).

The vertebral formula is: C_7, T_{14-15}, L_{18-22}, Ca_{38-41}, for a total of 77–82 (Lütken, 1888). The first two cervicals are fused together by both the centra and the spinous processes; the third and sometimes fourth are fused by the spinous processes only (Tomilin, 1967).

Organs, Tissues and Physiology

Organ weights of *L. acutus* are shown in Table 1. The anatomy and possible function of laryngeal folds were discussed by Reidenberg and Laitman (1988).

TABLE 1 Organ weights (absolute in g and relative as % of total body weight) of Atlantic white-sided dolphins, from a sample of animals stranded at Lingley Cove, ME, USA[a]

Organ	*N*	Sex	Range	*X* ± SD	Relative (*X* ± SD)
Brain	11	M	680–1 330	1 070 ± 240	1.48 ± 0.82
	30	F	680–1 300	1 140 ± 140	0.98 ± 0.48
Heart	11	M	220–1 400	720 ± 450	0.75 ± 0.14
	30	F	250–1 800	1 000 ± 330	0.76 ± 0.13
Lungs	10	M	800–4 420	2 430 ± 1 480	2.63 ± 0.78
	28	F	1 220–5 720	4 290 ± 940	3.31 ± 1.51
Liver	11	M	560–5 780	2 290 ± 1 760	2.26 ± 0.34
	27	F	620–5 600	3 520 ± 1 210	2.52 ± 0.41
Pancreas	10	M	35.1–450	180.0 ± 134.8	0.001 66 ± 0.000 60
	24	F	40.0–520	255.4 ± 119.7	0.001 80 ± 0.000 60
Spleen	10	M	16.6–146	54.8 ± 36.1	0.000 58 ± 0.000 27
	28	F	17.0–72.0	33.8 ± 16.9	0.000 28 ± 0.000 23
Pituitary	5	M	0.11–0.60	0.32 ± 0.18	0.000 005 ± 0.000 003
	23	F	0.19–1.57	0.83 ± 0.42	0.000 006 ± 0.000 002
Thyroid	4	M	8.4–13.9	12.0 ± 2.5	0.000 31 ± 0.000 16
	18	F	8.3–50.1	27.2 ± 12.8	0.000 21 ± 0.000 08
Thymus	5	M	60.3–129.6	93.8 ± 30.7	0.002 00 ± 0.000 40
	5	F	27.8–180.0	99.1 ± 65.1	0.001 81 ± 0.000 62
Adrenal[b]	11	M	1.1–11.9	5.0 ± 3.6	0.000 05 ± 0.000 01
	28	F	1.3–22.2	7.8 ± 4.8	0.000 056 ± 0.000 028
Mature ovaries	23	F			
right			2.6–21.5	9.1 ± 5.7	0.000 058 ± 0.000 036
left			2.5–19.7	7.6 ± 5.5	0.000 049 ± 0.000 035
Mature testes	4	M	160–370	287.5 ± 84.9	0.001 38 ± 0.000 32
Mature penis	4	M	330–400	376.2 ± 33.0	0.001 83 ± 0.000 19
Mature uterus	24	F	140–650	357.7 ± 257.8	0.002 30 ± 0.001 50
Mature mammary glands[b]	13	F	130–1150	630.2 ± 239.6	0.004 00 ± 0.001 40
Viscera	9	M	3 600–26 100	13 300 ± 8 000	12.2 ± 2.6
	26	F	4 300–23 300	17 600 ± 4 400	12.3 ± 2.5
Body blubber (incl. peduncle)	9	M	7 100–40 300	20 900 ± 13 700	21.7 ± 3.7
	29	F	7 500–45 900	27 200 ± 8 900	20.3 ± 3.6
Body muscle	8	M	6 800–105 700	46 000 ± 41 300	38.9 ± 8.1
	30	F	10 100–75 900	57 400 ± 19 000	41.2 ± 3.7
Total weight	11	M	24 700–233 500	110 000 ± 78 600	
	31	F	29 200–182 000	138 000 ± 41 500	

[a]From Geraci *et al.* (1978a).

[b]Individual organ weight, obtained from average of paired weight.

(A)

(B)

FIG. 5 (A) Dorsal and (B) ventral views of the cranium of the Atlantic white-sided dolphin.

Life History and Population Dynamics

Growth and maturation

Length at birth is 108–122 cm (Sergeant *et al.*, 1980). The largest reported foetus was 114 cm long and the smallest free-swimming individual was 108 cm (Perrin and Reilly, 1984). A 126 cm calf weighed 24 kg (Sergeant *et al.*, 1980), and a 150 cm female weighed 33 kg (in December; C. Smeenk, unpublished data).

(C)

(D)

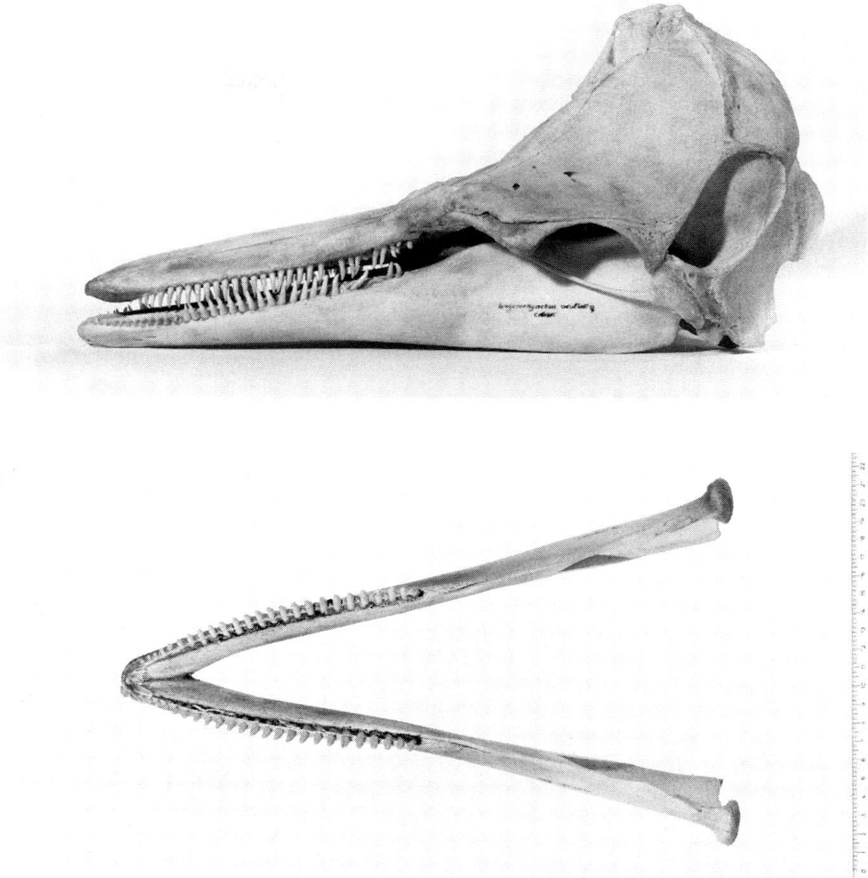

FIG. 5 *Continued* (C) Left lateral view of skull and (D) dorsal view of mandibles. Specimen No. CN 21X, condylobasal length 39.7 cm. (Courtesy Zoological Museum, Copenhagen; photos by Geert Brouad)

Fraser (1934) reported a calf 109 cm long whose teeth had not yet erupted through the gums.

Length at sexual maturity was estimated as 201–222 cm from a sample of 36 females; the range in ages (in dentinal GLGs, considered annual) for the specimens 201 cm or longer was 6–12 (Sergeant *et al.*, 1980). The smallest reported sexually mature female was 194 cm long (Perrin and Reilly, 1984). Sexual maturity may be reached in males at lengths of 230–250 cm (Sergeant *et al.*, 1980). Samples of males examined by Addink *et al.* (1997), however, indicated attainment of sexual maturity at ages of 7–11 GLGs, corresponding to lengths of about 215–230 cm.

Maximum age is at least 22 GLGs for males (a 246 cm specimen) and 27 for females (a 224 cm specimen with 15 corpora albicantia on the ovaries) (Sergeant *et al.*, 1980).

Geraci *et al.* (1978a) calculated a growth rate of 34.3 cm per year for the first 2.5 years of life. Males apparently grow faster than females and thus generally are larger than females of equivalent age. Physical maturity is reached at an age of close to 8 GLGs (Geraci *et al.*, 1978a; Addink *et al.*, 1997).

Reproduction

Parturition in the western North Atlantic usually takes place between May and August, with a peak in June and July, following an estimated 11 month gestation period (St. Aubin and Geraci, 1979; Sergeant *et al.*, 1980). The timing of parturition is apparently similar in the eastern North Atlantic, where sightings have been interpreted to suggest "breeding areas" offshore in the North Sea and in the Atlantic to the north and west of there (Evans, 1991). Other data, however, indicate a more extended reproductive period. Jonsgård and Nordli (1952) found foetuses ranging from 35 cm to 70 cm in length in a school caught in Norway in March. Addink *et al.* (1997), in animals by-caught south of Ireland, found a 53.5 cm, 2890 g foetus in late February, a 44 cm, 1340 g foetus in late March, as well as a female in late February which had recently given birth. The presence of foetuses 32 mm and 40 mm long in females that stranded in Ireland in the second half of September was interpreted as suggesting that conception occurs in late August and early September (Rogan *et al.*, 1997; see also Palka *et al.*, 1997). A stranding on the Irish coast of a female with a partly born neonate (109 cm long) in July (Smiddy, 1992) was interpreted as suggesting a gestation period of 11 months (Rogan *et al.*, 1997). Perrin and Reilly (1984) gave estimates of gestation period ranging from 10 months to 12 months. Lactation may last about 18 months (St. Aubin and Geraci, 1979; Sergeant *et al.*, 1980). In a sample of 22 pregnant or lactating dolphins from a stranded herd, four individuals were both pregnant and lactating (St. Aubin and Geraci, 1979); this was the case in one out of six mature females in the group examined by Rogan *et al.* (1997). Twin foetuses have been found once (Jonsgård and Nordli, 1952).

Abundance

The number of Atlantic white-sided dolphins in the western North Atlantic south of Nova Scotia was estimated to be of the order of 30 000 based on results of systematic aerial surveys in the late 1970s and early 1980s (Winn and Edel, 1982). More recent estimates for July–September 1995 are about 27 000 in the area from the southern Gulf of Maine and northeastwards on the continental shelf and slope to Cabot Strait (Palka *et al.*, 1997) and at least 12 000 in the Gulf of St. Lawrence

(Kingsley and Reeves, 1998). Evans (1987) suggested a total population throughout the North Atlantic of tens of thousands to low hundreds of thousands.

Behaviour

Social organization

Lagenorhynchus acutus sometimes occurs in large herds numbering in the high hundreds (Evans, 1980; Katona *et al.*, 1983; Leopold and Couperus, 1995); groups of 50–60 dolphins were considered typical in Newfoundland inshore waters (Sergeant and Fisher, 1957). Near the British Isles groups usually contain less than ten individuals, although herds of 50–100 are sometimes sighted offshore (Evans, 1991). Sightings from aerial surveys near Iceland yielded a mean group size of 6.4 (Gunnlaugsson *et al.*, 1988). In year-round aerial surveys of the North American continental shelf between Nova Scotia and Cape Cod, the range in estimated group size was 2–500, with a mean of 42 and a mode of 6 ($n = 215$) (Winn and Edel, 1982). While observing a herd of more than 400 Atlantic white-sided dolphins off Cape Cod in late April 1985, Rossiter (1985) noted that it consisted of subgroups of 4–12 animals. Haase (1987), de Boer (1989) and Leopold and Couperus (1995) referred to subgroups of 2–15, 2–5 and 5–15 animals, respectively. Gaskin (1992) hypothesized that Atlantic white-sided dolphins split into small groups for feeding and that such small groups merge into large aggregations "while migrating".

Segregation on the basis of age and sex has been inferred from the composition of stranded groups. Two mass strandings in the northeastern United States contained mostly adults and very young animals; a high percentage of Atlantic white-sided dolphins that stranded alone were 170–200 cm long, with 2–5 GLGs. This was interpreted as evidence that "older immatures" remain apart from the "reproductive herds" (Sergeant *et al.*, 1980). A mass stranding of 19 dolphins on the west coast of Ireland in September 1994 consisted of ten males and nine females, including two male calves and two female calves 1–2 GLGs of age (Rogan *et al.*, 1997). No larger juveniles were present in this group. Addink *et al.* (1997), on the other hand, found juveniles 3–6 GLGs of age by-caught in the same hauls with adults. Large herds have been observed to include small calves as well as "immatures" and adults (Rossiter, 1985; Haase, 1987; de Boer, 1989).

A herd of 52 Atlantic white-sided dolphins was driven ashore easily by boats off Norway in early March 1952. The animals remained together throughout the drive. Approximately one-third of the herd consisted of pregnant females, with foetuses 35–70 cm long (Jonsgård and Nordli, 1952). At Hvalba, Faroe Islands, on 1 April 1988 a large school of Atlantic white-sided dolphins "interfered with a drive of pilot whales and beached in front of the pod of pilot whales". The reported catch of dolphins in this drive was 544 (Bloch and Hoydal, 1990). Jóhann Sigurjónsson (personal communication, 1992) described an incident at sea in which one dolphin was observed to circle and "lift" its companion that had recently died.

Feeding

Herring (*Clupea harengus*) and small mackerel (*Scomber scombrus*) are major food sources in some areas (Jonsgård and Nordli, 1952; Berrow and Smiddy, 1989; Berrow and Stark, 1990). The stomach contents of 17 animals in a mass stranding on the Irish coast consisted mainly of otoliths from gadid fishes, of which only *Trisopterus* sp. could be identified, herring, horse mackerel (*Trachurus trachurus*), one argentine (*Argentina sphyraena*) and a squid beak (Rogan *et al.*, 1997). Couperus (1997a, 1997b) examined the stomach contents of 49 animals (2911 prey items) by-caught by Dutch freezer trawlers in 1993–1995, from January to April, southwest of Ireland. Here 87% of the prey weight consisted of fish, mainly mackerel (67%). Silvery pout (*Gadiculus argenteus*) was frequently eaten (71% by frequency of occurrence, 44% as proportion of the total number of prey), but was much less important as a proportion of the total prey weight (9%). Cephalopods made up 13% of total prey weight and consisted of a variety of species, of which Ommastrephidae were most important (9% of total prey weight). A stomach of a white-sided dolphin caught in autumn southwest of Ireland contained 1118 prey items, mainly Gadidae (45% in weight, most of which were blue whiting *Micromesistius poutassou* and silvery pout) and Ommastrephidae (51%; Couperus, 1997a).

In addition to herring and various other fishes (e.g. smelt, *Osmerus mordax*, and silver hake, *Merluccius bilinearis*), short-finned squid (*Illex illecebrosus*) are consumed off New England and Newfoundland (Schevill, 1956; Sergeant and Fisher, 1957; Geraci *et al.*, 1978a; Katona *et al.*, 1978; St. Aubin and Geraci, 1979; Sergeant *et al.*, 1980). The distribution of Atlantic white-sided dolphins in the Gulf of Maine during spring and autumn was said to coincide with that of American sand lance (*Ammodytes americanus*), probably "a major prey species" (Selzer and Payne, 1988). Smith *et al.* (1990) analysed a data set that apparently included the data used by Selzer and Payne (1988), and reported catches of northern sand lance (*Ammodytes dubius*) in 40–60% of 130 otter-trawl stations taken within 50 km of sightings of Atlantic white-sided dolphins during spring. However, during autumn no sand lance were caught within 50 km of white-sided dolphins. Gaskin (1992) noted that American sand lance occur mainly in waters $\leq$ 20 m deep and that northern sand lance are present in the deeper waters (73–90 m) where Atlantic white-sided dolphins are usually found. He questioned the inference that American sand lance are an important prey, and suggested that northern sand lance could be the more important prey species (see also Kenney *et al.*, 1996). Smeenk and Gaemers (1987) found whiting (*Merlangius merlangus*), Atlantic cod (*Gadus morhua*), poor-cod (*Trisopterus minutus*) and unidentified gadids in the stomach of a dolphin stranded in the Netherlands. In addition to the species mentioned above, the fish *Micromesistius*, the squid *Loligo* and unspecified gammarid crustaceans were listed as prey of white-sided dolphins in the eastern North Atlantic (Evans, 1991).

Atlantic white-sided dolphins apparently cooperate in their efforts to contain and attack schools of fish (Rossiter, 1985). Such behaviour is in many ways similar to that described in detail for dusky dolphins (*Lagenorhynchus obscurus*) off Argentina (Würsig, 1986).

Association with other species

Groups of Atlantic white-sided dolphins often associate and probably feed with fin whales (*Balaenoptera physalus*) and humpback whales (*Megaptera novaeangliae*) (Katona *et al.*, 1978; Sears *et al.*, 1981; Sears and Williamson, 1982; Winn and Edel, 1982; Evans, 1982) and with long-finned pilot whales (*Globicephala melas*) (Sergeant and Fisher, 1957; Perkins *et al.*, 1982; Bloch and Hoydal, 1990). Pods of 5–15 dolphins swim alongside the head of a fin whale or humpback whale that is feeding near the surface. Mixed herds of Atlantic white-sided dolphins and white-beaked dolphins have been observed in the North Sea (Haase, 1987; de Boer, 1989; M.J. Addink, personal communication). Evans (1976) referred to mixed schools of bottlenose dolphins (*Tursiops truncatus*), white-beaked dolphins and Atlantic white-sided dolphins seen off southwestern Ireland in August and September, but he gave few details. Atlantic white-sided dolphins and short-beaked common dolphins have been seen in a mixed school near the edge of the continental shelf southwest of Ireland (Couperus, 1994).

Swimming and diving

Atlantic white-sided dolphins are active at the surface and often breach (Fig. 6) or perform tail slaps. Also they sometimes "ride" bow waves and swim in the stern wakes of fast-moving vessels. Baptist (1987) noted that *L. acutus* leaps more "horizontally" over the tops of waves and appears to travel "lower" and faster than *L. albirostris*. Data from one satellite-monitored dolphin indicated an ability to travel long distances at a speed of at least 14 km per hr (Mate *et al.*, 1994). None of the dives recorded for this individual lasted longer than 4 min, and most dives lasted for less than a minute (Mate *et al.*, 1994).

Sound production

"Squeals" in the range of 1–24 kHz, and a few clicks, have been recorded from wild *L. acutus* (Schevill and Watkins, 1962). Particularly considering that it is a common and frequently observed species, the Atlantic white-sided dolphin's acoustic behaviour has been surprisingly little studied.

Parasites and Diseases

The gut helminths of 37 Atlantic white-sided dolphins that stranded in Maine, USA, included representatives of six species: two digeneans, *Pholeter gastrophilus*

FIG. 6 A high breach by an Atlantic white-sided dolphin, Provincetown harbour, Cape Cod, MA. (Photo by William Rossiter)

in the stomach and duodenum and *Oschmarinella laevicaecum* in the bile ducts and hepatopancreatic ducts; the cestode *Tetrabothrius forsteri* mainly in the first 8 m of the intestine; two nematodes, *Anisakis* sp. in the stomach of one female dolphin and *Stenurus globicephalae* in the stomach; and the acanthocephalan *Bolbosoma* sp. in an unspecified location in a male dolphin (Beverley-Burton, 1978). The "lung-worms" *S. globicephalae* were thought to have been transitory in the stomachs, having passed up the trachea and then been swallowed. Although *Bolbosoma* sp. was poorly represented in this sample of dolphins, it was found in the intestinal tracts of five of six specimens examined from a previous mass stranding on Cape Cod (J.R. Geraci *in* Beverley-Burton, 1978).

Thirty females from a mass stranding of approximately 150 Atlantic white-sided dolphins were examined for parasites, and the mammary glands of 47% were found to be infected with *Crassicauda grampicola*, a spirurid nematode (Geraci *et al.*, 1978b). The severity of the lesions and destruction of the mammary tissue could have compromised the reproductive success of some of the infected females.

Evans (1991) listed the internal parasites found in white-sided dolphins from the eastern North Atlantic as follows: *Anisakis simplex* (stomach), *Monorygma grimaldii* (abdomen, peritoneum and testes), *Phyllobothrium* sp. (abdominal blubber) and *Pseudalius inflexus* (bronchi and lungs). In addition to the above, Gibson and Harris (1979) mentioned finding the nematode *Torynurus convolutus*, a lungworm, in the head sinuses. The cestodes *Phyllobothrium delphini* and *Strobilocephalus triangularis* were reported by Baer (1932) as occurring in Atlantic white-sided dolphins. Rogan *et al.* (1997) found the following parasites in 18 animals in a mass stranding on the Irish coast: *Stenurus globicephalae* (cranial sinuses), *Pholeter gastrophilus* (stomach wall), *Crassicauda* sp. (subcutaneous tissue, mammary glands), *Phyllobothrium delphini* (blubber), *Monorygma grimaldii* (retro-peritoneally in the abdomen) and nematodes (gastric lumina). Rupture of an aortic valve with consequent decompensating heart failure was found in the oldest male of this group. Other diseases in these animals included benign intestinal leiomyomas, arthritis of the shoulder joint and genital herpes (see Rogan *et al.*, 1997, for further details).

Sarcocysts were found just beneath the blubber in one of four Atlantic white-sided dolphins examined from the St. Lawrence estuary, Canada (De Guise *et al.*, 1993). No gross lesions or inflammatory response of any kind was observed to be associated with the sarcocysts. Geraci *et al.* (1987) reported some tumours from a stranded group including papillomas of the tongue, pharynx, and penis as well as leiomyomas of the small intestine.

A variety of microorganisms, including several bacteria, were identified from swab samples of the anus and blowhole, and from sampling of the internal organs of three Atlantic white-sided dolphins that stranded alive (Buck *et al.*, 1988; also see Tangredi and Medway, 1980; Buck and Spotte, 1986; Buck *et al.*, 1991).

Slijper (1931) reported a synostosis of the chevron bones in a specimen at the Leiden Museum.

Live Maintenance

Atlantic white-sided dolphins commonly strand singly or in groups of as many as 150 animals (e.g. Fraser, 1934; Geraci *et al.*, 1978a; Béland *et al.*, 1987; Sheldrick, 1989; Rogan *et al.*, 1997). Efforts are made in some areas to push live-stranded dolphins back to sea or to rehabilitate and release them. Although the species has not been specifically captured for maintenance in aquaria (Klinowska, 1991), some stranded individuals have been taken to aquaria for rehabilitation. Smeenk (1989) mentioned a young female (about 150 cm) that lived in captivity for six days after stranding. Three Atlantic white-sided dolphins were taken to Mystic Marinelife Aquarium in Connecticut, and one of them was released 126 days after it had stranded (Buck *et al.*, 1988). The other two died of pneumonia less than four months after the stranding. A male that had stranded in Massachusetts was released after eight months in captivity. It was monitored for six days with a satellite-linked radio tag, providing an opportunity to study movements, travelling speed and diving behaviour (Mate *et al.*, 1994). Studies of behaviour (Nelson and Lien, 1994) and of swimming kinematics based on observations in an aquarium (Curren *et al.*, 1994) have also been published.

Human Effects

Directed fisheries

Atlantic white-sided dolphins have been taken, at times in large numbers, along the Norwegian coast (Guldberg and Nansen, 1894; Jonsgård and Nordli, 1952). Nets and harpoons were used to trap and kill groups of dolphins after they entered small fjords or inlets. Small numbers of these dolphins have been taken off southwestern Greenland (Anonymous, 1979, 1981b; Kapel, 1985; Larsen, 1990; Heide-Jørgensen, 1990) and in the Faroe Islands (Bloch and Joensen, 1985). In the latter area, they have been taken in large numbers (hundreds) by driving in some years, often but not always in drives also involving long-finned pilot whales (Degerbøl, 1940; International Whaling Commission, 1978; Bloch and Hoydal, 1989, 1990; Bloch and Desportes, 1993; Bloch, 1995). They also were mixed occasionally in the catch of long-finned pilot whales in a drive fishery in Trinity Bay, Newfoundland (Sergeant and Fisher, 1957), and they have been taken deliberately by shooting elsewhere in Canada (Brice-Bennett, 1977; Sergeant *et al.*, 1980; Mackey and Orr, 1987).

By-catches

Incidental mortality in fishing gear has been documented off Canada (Mercer, 1973; Gaskin, 1984, 1992; Béland *et al.*, 1987; Palka *et al.*, 1997), the United States (Katona *et al.*, 1978; Gilbert and Wynne, 1987; Waring *et al.*, 1990; Bisack, 1993; Anonymous, 1993; Blaylock *et al.*, 1995; Weinrich, 1996; Palka *et al.*, 1997), the United Kingdom (Anonymous, 1986) and Ireland (Berrow and Smiddy, 1989; Berrow and Stark, 1990; Addink *et al.*, 1997; see also Northridge, 1991). Gaskin (1992) judged Atlantic white-sided dolphins to be less vulnerable to capture in pelagic near-surface drift nets and fixed groundfish gill nets than are many other small cetaceans. They may, however, be especially susceptible to capture in midwater trawl nets (Addink *et al.*, 1997). Substantial numbers have been by-caught in pelagic trawl fisheries for horse mackerel and mackerel southwest of Ireland (Couperus, 1994, 1997a, 1997b).

Pollution

A juvenile dolphin from the northwest coast of Ireland was found to have a relatively high concentration of mercury in its liver (44 ng per g wet weight) (Law *et al.*, 1991). An adult male from Nova Scotia had moderately high levels of organochlorines in its blubber (Taruski *et al.*, 1975). Three Atlantic white-sided dolphins (203 cm male, 209 cm female, and 228 cm male) were collected after having been killed incidental to fishing operations. The highest contaminant concentrations (ng per g lipid) of the most frequently detected pesticides in the blubber were: 3290 dieldrin, 145 HCB, 73 mirex, 63 lindane, 23 100 *p*, *p*′-DDE, 401 heptachlor epoxide, 767 oxychlordane, 1230 *cis*-chlordane, and 7020 *trans*-nonachlor (Kuehl *et al.*, 1991). The higher values were always in one of the two males sampled, except in the case of HCB. Additional data on contaminants are presented in Borrell (1993), Kuehl *et al.* (1994), Palka *et al.* (1997) and McKenzie *et al.* (1998).

Acknowledgments

The authors wish to thank Marjan Addink, Bram Couperus and Manuel García Hartmann for advice and assistance, and Bill Rossiter, Dan Morast and the *Texelse Courant* for providing photographs.

References

Addink, M., García Hartmann, M. and Couperus, B. (1997). A note on life-history parameters of the Atlantic white-sided dolphin (*Lagenorhynchus acutus*) from animals bycaught in the northeastern Atlantic. *Rep. Int. Whal. Commn* **47**, 637–639.

Aguayo L., A. (1978). Smaller cetaceans in the Baltic Sea. *Rep. Int. Whal. Commn* **28**, 131–146.

Anonymous (1979). Denmark progress report on cetacean research June 1977–May 1978. *Rep. Int. Whal. Commn* **29**, 115.

Anonymous (1981a). Denmark (Greenland) progress report on cetacean research 1978–79. *Rep. Int. Whal. Commn* **31**, 185–186.

Anonymous (1981b). Denmark (Greenland) progress report on cetacean research 1979–80. *Rep. Int. Whal. Commn* **31**, 186–188.

Anonymous (1986). United Kingdom. Progress report on cetacean research, June 1984 to May 1985. *Rep. Int. Whal. Commn* **36**, 170.

Anonymous (1993). United States. Progress report on cetacean research, May 1991 to May 1992. *Rep. Int. Whal. Commn* **43**, 295–307.

Baer, J. G. (1932). Contribution à l'étude des cestodes de cétacés. *Rev. Suisse Zool.* **39**, 195–228.

Baptist, H. J. M. (1987). Waarnemingen van zeezoogdieren in de Nederlandse sector van de Noordzee. *Lutra* **30**, 93–104.

Béland, P., Martineau, D., Robichaud, P., Plante, R. and Greendale, R. (1987). Échouages de mammifères marins sur les côtes du Québec dans l'estuaire et le golfe du Saint-Laurent de 1982 à 1985. *Rapp. tech. Can. Sci. halieut. aquat.* **1506**, 1–44.

Berrow, S. D. and Rogan, E. (1997). Review of cetaceans stranded on the Irish coast, 1901–95. *Mammal Rev.* **27**, 51–76.

Berrow, S. and Smiddy, P. (1989). White-sided dolphin *Lagenorhynchus acutus*. *Irish Nat. J.* **23**, 158–159.

Berrow, S. and Stark, D. (1990). White-sided dolphin *Lagenorhynchus acutus*. *Irish Nat. J.* **23**, 334.

Beverley-Burton, M. (1978). Helminths of the alimentary tract from a stranded herd of Atlantic white-sided dolphin, *Lagenorhynchus acutus*. *J. Fish. Res. Bd Can.* **35**, 1356–1359.

Bisack, K. (1993). Estimates of total bycatch in the Gulf of Maine sink gillnet fishery. National Oceanic and Atmospheric Administration, National Marine Fisheries Service, NEFSC Ref. Doc. 93-11, pp. 1–23.

Blaylock, R. A., Hain, J. W., Hansen, L. J., Palka, D. L. and Waring, G. T. (1995). U.S. Atlantic and Gulf of Mexico marine mammal stock assessments. NOAA Tech. Memo. NMFS-SEFSC-363.

Bloch, D. (1995). Faroe Islands progress report on marine mammal research 1994. North Atlantic Marine Mammal Commn Ann. Rep., pp. 129–134.

Bloch, D. and Desportes, G. (1993). Faroe Islands progress report on cetacean research 1993. North Atlantic Marine Mammal Commission, Doc. SC/2/Prog. Rep. — Faroes 1993. Tromsø, Norway.

Bloch, D. and Hoydal, K. (1989). Part 2 [of Denmark progress report]. Faroe Islands (Denmark). *Rep. Int. Whal. Commn* **39**, 177–178.

Bloch, D. and Hoydal, K. (1990). Part 2 [of Denmark progress report]. Faroe Islands. *Rep. Int. Whal. Commn* **40**, 192–194.

Bloch, D. and Joensen, J. S. (1985). Part 2 [of Denmark progress report]. Faroe Islands. *Rep. Int. Whal. Commn* **35**, 165–166.

de Boer, R. (1989). Waarneming van een gemengde groep witflankdolfijnen *Lagenorhynchus acutus* en witsnuitdolfijnen *L. albirostris* in de centrale Noordzee. *Lutra* **32**, 181–184.

Boles, B. K. (1980). Data report listing sightings of cetaceans in the Labrador Sea. Offshore Labrador Biological Studies, 1979. Report by Atlantic Biological Services Ltd, St. John's, Newfoundland, for Total Eastcan Exploration Ltd, Calgary, Alberta.

Borrell, A. (1993). PCB and DDTs in blubber of cetaceans from the northeastern North Atlantic. *Mar. Poll. Bull.* **26**, 146–151.

van Bree, P. J. H. and Husson, A. M. (1974). Strandingen van Cetacea op de Nederlandse kust in 1972 en 1973. *Lutra* **16**, 1–10.

Brice-Bennett, C. (1977). Land use in the Nain and Hopedale regions. *In* "Our Footprints Are Everywhere. Inuit Land Use and Occupancy in Labrador" (Eds C. Brice-Bennett, A. Cooke and N. Davis) pp. 97–203. Labrador Inuit Association, Nain, Labrador.

Buck, J. D., Bubucis, P. M. and Spotte, S. (1988). Microbiological characterization of three Atlantic whiteside dolphins (*Lagenorhynchus acutus*) from stranding through captivity with subsequent rehabilitation and release of one animal. *Zoo Biol.* **7**, 133–138.

Buck, J. D., Overstrom, N. A., Patton, G. W., Anderson, H. F. and Gorzelany, J. F. (1991). Bacteria associated with stranded cetaceans from the northeast USA and southwest Florida Gulf coasts. *Dis. Aquat. Organisms* **10**, 147–152.

Buck, J. D. and Spotte, S. (1986). The occurrence of potentially pathogenic vibrios in marine mammals. *Mar. Mamm. Sci.* **2**, 319–324.

Castells, A. and Mayo, M. (1992). "Relación de Citas de Cetáceos (Mammalia, Cetacea): España y Portugal", No. I, Vol. II. Centro de Estudios de Mamíferos Marinos, Madrid.

Cope, E. D. (1876). Fourth contribution to the history of the existing Cetacea. *Proc. Acad. Nat. Sci. Phila.* 1876, 129–139, pls. III–IV.

Couperus, A. S. (1994). Een Oriënterende Studie van de Bijvangsten van Zeezoogdieren door Nederlandse Vriestrawlers. Internal Report, Netherlands Institute for Fisheries Research (RIVO-DLO), IJmuiden. CO. 33/94.

Couperus, A. S. (1997a). By-catch of Marine Mammals and Discards in Pelagic Fisheries. Report by the Netherlands Institute for Fisheries Research (RIVO-DLO) to the European Commission, CFP 94/018.

Couperus, A. S. (1997b). Interactions between Dutch midwater trawl and Atlantic white-sided dolphins (*Lagenorhynchus acutus*) southwest of Ireland. *J. Northwest Atlantic Fish. Sci.* **22**, 209–218.

Curren, K., Bose, N. and Lien, J. (1994). Swimming kinematics of a harbor porpoise (*Phocoena phocoena*) and an Atlantic white-sided dolphin (*Lagenorhynchus acutus*). *Mar. Mamm. Sci.* **10**, 485–492.

Degerbøl, M. (1940). Mammalia. In "Zoology of the Faroes", LXV, 132pp. Andr. Fred. Høst & Søn, Copenhagen.

Degerbøl, M. (1943). Hvidsiden eller hvidskævingen (*Lagenorhynchus acutus* Gray) i danske farvande. *Naturens Verden* **27**, 164–174.

De Guise, S., Lagacé, A., Girard, C. and Béland, P. (1993). Intramuscular *Sarcocystis* in two beluga whales and an Atlantic white-sided dolphin from the St. Lawrence estuary, Quebec, Canada. *J. Vet. Diagn. Invest.* **5**, 296–300.

Duguy, R. (1978). Un dauphin rare trouvé sur les côtes du Finistère: *Lagenorhynchus acutus* (Gray, 1828). *Penn ar Bed* **11**, 365–366.

Duguy, R. (1980). Rapport annuel sur les cétacés et pinnipèdes trouvés sur les côtes de France. IX. Année 1979. *Ann. Soc. Sci. Nat. Charente-Maritime* **6**, 615–632.

Duguy, R. (1987). Rapport annuel sur les cétacés et pinnipèdes trouvés sur les côtes de France. XVI. Année 1986. *Ann. Soc. Sci. Nat. Charente-Maritime* **7**, 617–639.

Evans, P. G. H. (1976). An analysis of sightings of Cetacea in British waters. *Mamm. Rev.* **6**, 5–14.

Evans, P. G. H. (1980). Cetaceans in British waters. *Mamm. Rev.* **10**, 1–52.

Evans, P. G. H. (1982). Associations between seabirds and cetaceans: a review. *Mamm. Rev.* **12**, 187–206.

Evans, P. G. H. (1987). "The Natural History of Whales and Dolphins". Christopher Helm, London.

Evans, P. G. H. (1990). European cetaceans and seabirds in an oceanographic context. *Lutra* **33**, 95–125.

Evans, P. G. H. (1991). Whales, dolphins and porpoises: Order Cetacea. *In* "The Handbook of British Mammals", 3rd Edn (Eds G. B. Corbet and S. Harris) pp. 299–350. Mammal Society and Blackwell, Oxford.

Flower, W. H. (1884). On the characters and divisions of the family Delphinidae. *Proc. Zool. Soc. Lond.* **1883**, 466–513.

Fraser, F. C. (1934). "Report on Cetacea stranded on British coasts from 1927 to 1932", No. 11, British Museum (Natural History), London.

Fraser, F. C. (1974). "Report on Cetacea stranded on the British coasts from 1948 to 1966", No. 14, British Museum (Natural History), London.

Gaskin, D. E. (1983). The marine mammal community. *In* "Marine and Coastal Systems of the Quoddy Region, New Brunswick" (Ed. M. L. H. Thomas) pp. 245–268. Can. Spec. Publ. Fish. Aquat. Sci. No. 64.

Gaskin, D. E. (1984). The harbour porpoise *Phocoena phocoena* (L.): regional populations, status, and information on direct and incidental catches. *Rep. Int. Whal. Commn* **34**, 569–586.

Gaskin, D. E. (1992). Status of the Atlantic white-sided dolphin, *Lagenorhynchus acutus*, in Canada. *Can. Field-Naturalist* **106**, 64–72.

Gaskin, D. E. and Smith, G. J. D. (1979). Observations on marine mammals, birds and environmental conditions in the Head Harbour region of the Bay of Fundy. *In* "Evaluation of Recent Data Relative to Potential Oil Spills in the Passamaquoddy Area" (Ed. D. J. Scarratt) pp. 69–86. Can. Fish. Mar. Serv. Tech. Rep. No. 901.

Geraci, J. R., Testaverde, S. A., St. Aubin, D. J. and Loop, T. H. (1978a). "A Mass Stranding of the Atlantic White-sided Dolphin, *Lagenorhynchus acutus*: A Study into Pathobiology and Life History". Contract report to U.S. Marine Mammal Commission. Available from National Technical Information Service, Springfield, Virginia, PB-289 361.

Geraci, J. R., Dailey, M. D. and St. Aubin, D. J. (1978b). Parasitic mastitis in the Atlantic white-sided dolphin, *Lagenorhynchus acutus*, as a probable factor in herd productivity. *J. Fish. Res. Bd Can.* **35**, 1350–1355.

Geraci, J. R., Palmer, N. C. and St. Aubin, D. J. (1987). Tumors in cetaceans: analysis and new findings. *Can. J. Fish. Aquat. Sci.* **44**, 1289–1300.

Gibson, D. I. and Harris, E. A. (1979). The helminth-parasites of cetaceans in the collection of the British Museum (Natural History). *Invest. Cetacea* **10**, 309–324.

Gilbert, J. R. and Wynne, K. M. (1987). "Marine Mammal Interactions with New England Gillnet Fisheries". Report for U.S. National Marine Fisheries Service, Northeast Fisheries Science Center, Woods Hole, Massachusetts, Contract No. NA-84-EA-C-00070.

Gowans, S. and Whitehead, H. (1995). Distribution and habitat partitioning by small odontocetes in the Gully, a submarine canyon on the Scotian Shelf. *Can. J. Zool.* **73**, 1599–1608.

Gray, J. E. (1828). "Spicilegia Zoologica; or Original Figures and Short Systematic Descriptions of New and Unfigured Animals". Treüttel, Würtz and Co. & W. Wood, London.

Gray, J. E. (1846). On the cetaceous animals. *In* "The Zoology of the Voyage of H.M.S. Erebus and Terror under the Command of Captain Sir James Clark Ross, R.N., F.R.S., during the Years 1839 to 1843", Vol. 1, "Mammalia, Birds" (Eds J. Richardson and J. E. Gray) pp. 13–53, 37 pls.

Guldberg, G. and Nansen, F. (1894). On the development and structure of the whale. Part I. On the development of the dolphin. *Bergens Mus. Skr.* **5**, 1–70, pls. I–VII.

Gunnlaugsson, T., Sigurjónsson, J. and Donovan, G. P. (1988). Aerial survey of cetaceans in the coastal waters off Iceland, June–July 1986. *Rep. Int. Whal. Commn* **38**, 489–500.

Haase, B. J. M. (1987). A mixed herd of white-beaked dolphins *Lagenorhynchus albirostris* and white-sided dolphins *L. acutus* in the southern North Sea. *Lutra* **30**, 105–106.

Harmer, S. F. (1927). "Report on Cetacea stranded on the British coasts from 1913 to 1926", No. 10, British Museum (Natural History), London.

Hashmi, D. D. K. and Adloff, B. B. (1991). Surface frequency of cetaceans in the Strait of Gibraltar. *In* "European Research on Cetaceans — 5". Proceedings of the Fifth Annual Conference of the European Cetacean Society, Sandefjord, Norway, 21–23 February 1991 (Ed. P. G. H. Evans) pp. 16–17.

Hashmi, D. and Adloff, B. (1992). Cetacean research in the Strait of Gibraltar. *In* "Whales: Biology — Threats — Conservation", Proceedings of a Symposium held in Brussels, 5–7 June 1991 (Ed. J. J. Symoens) pp. 59–62. Royal Academy of Overseas Sciences, Brussels.

Heide-Jørgensen, M.-P. (1990). Small cetaceans in Greenland: hunting and biology. *In* "Whaling Communities" (Ed. E. Vestergaard) pp. 55–58, North Atlantic Studies No. 2, Aarhus University, Aarhus, Denmark.

Hershkovitz, P. (1966). Catalog of living whales. *Bull. U.S. Nat. Mus.* **246**, 1–259.

International Whaling Commission (1978). Report of the standing Sub-committee on Small Cetaceans. *Rep. Int. Whal. Commn* **28**, 79–82.

Jonsgård, Å. (1962). On the species of dolphins found on the coast of northern Norway and in adjacent waters. *Norsk Hvalfangst-tid.* **51**, 1–13.

Jonsgård, Å. and Nordli, O. (1952). Concerning a catch of white-sided dolphins (*Lagenorhynchus acutus*) on the west coast of Norway, winter 1952. *Norsk Hvalfangst-tid.* **41**, 229–232.

Kapel, F. O. (1985). Denmark. Progress report on cetacean research, June 1983 to May 1984, Part 1, Greenland. *Rep. Int. Whal. Commn* **35**, 162–165.

Katona, S. K., Rough, V. and Richardson, D. T. (1983). "A Field Guide to the Whales, Porpoises, and Seals of the Gulf of Maine and Eastern Canada — Cape Cod to Newfoundland". Charles Scribner's, New York.

Katona, S. K., Testaverde, S. A. and Barr, B. (1978). Observations on a white-sided dolphin, *Lagenorhynchus acutus*, probably killed in gill nets in the Gulf of Maine. *Fish. Bull.* **76**, 475–476.

Kenney, R. D., Payne, P. M., Heinemann, D. W. and Winn, H. E. (1996). Shifts in northeast shelf cetacean distributions relative to trends in Gulf of Maine/Georges Bank finfish abundance. *In* "The Northeast Shelf Ecosystem: Assessment, Sustainability, and Management" (Eds K. Sherman, N. A. Jaworski and T. J. Smayda) pp. 169–196. Blackwell, Cambridge, MA.

Kingsley, M. C. S. and Reeves, R. R. (1998). Aerial surveys of cetaceans in the Gulf of St. Lawrence in 1995 and 1996. *Can J. Zool.*, in press.

Kinze, C. C. (1995). Danish whale records 1575–1991 (Mammalia, Cetacea). Review of whale specimens stranded, directly or incidentally caught along the Danish coasts. *Steenstrupia* **21**, 155–196.

Kinze, C. C., Addink, M., Smeenk, C., García Hartmann, M., Richards, H. W., Sonntag, R. P. and Benke, H. (1997). The white-beaked dolphin (*Lagenorhynchus albirostris*) and the white-sided dolphin (*Lagenorhynchus acutus*) in the North and Baltic seas: review of available information. *Rep. Int. Whal. Commn* **47**, 675–681.

Klinowska, M. (1991). "Dolphins, Porpoises and Whales of the World", The IUCN Red Data Book. IUCN, Gland, Switzerland and Cambridge, UK.

Kuehl, D. W., Haebler, R. and Potter, C. (1991). Chemical residues in dolphins from the U.S. Atlantic coast including Atlantic bottlenose obtained during the 1987/88 mass mortality. *Chemosphere* **22**, 1071–1084.

Kuehl, D. W., Haebler, R. and Potter, C. (1994). Coplanar PCB and metal residues in dolphins from the United States Atlantic coast including Atlantic bottlenose obtained during the 1987/88 mass mortality. *Chemosphere* **28**, 1245–1253.

Law, R. J., Fileman, C. F., Hopkins, A. D., Baker, J. R., Harwood, J., Jackson, D. B., Kennedy, S., Martin, A. R. and Morris, R. J. (1991). Concentrations of trace metals in the livers of marine mammals (seals, porpoises and dolphins) from waters around the British Isles. *Mar. Poll. Bull.* **22**, 183–191.

Larsen, F. (1990). Denmark. Progress report on cetacean research, June 1988 to May 1989, Part 1, Greenland and Denmark. *Rep. Int. Whal. Commn* **40**, 190–192.

Leopold, M. F. and Couperus, A. S. (1995). Sightings of Atlantic white-sided dolphins *Lagenorhynchus acutus* near the south-eastern limit of the known range in the North-East Atlantic. *Lutra* **38**, 77–80.

Lepiksaar, J. (1966). Zahnwalfunde in Schweden. *Bijdr. Dierk.* **36**, 3–16.

López Fernández, A. (1997). Some unusual cetacean strandings and sightings on the Galician coast, north-west Spain. *In* "European Research on Cetaceans — 10", Proc. 10th Ann. Conf. European Cetacean Society, Lisbon, Portugal, 11–13 March 1996 (Ed. P. G. H. Evans) pp. 131–133.

Lütken, M. C. (1888). Critical studies upon some Odontoceti of the genera *Tursiops, Orea,* and *Lagenorhynchus*. *Ann. Mag. Nat. Hist.*, 6th Ser., **2**, 179–186.

Mackey, M. G. A. and Orr, R. D. (1987). An evaluation of household country food use in Makkovik, Labrador, July 1980–June 1981. *Arctic* **40**, 60–65.

Mate, B. R., Stafford, K. M., Nawojchik, R. and Dunn, J. L. (1994). Movements and dive behavior of a satellite-monitored Atlantic white-sided dolphin (*Lagenorhynchus acutus*) in the Gulf of Maine. *Mar. Mamm. Sci.* **10**, 116–121.

McBrearty, D. A., Message, M. A. and King, G. A. (1986). Observations on small cetaceans in the north-east Atlantic Ocean and the Mediterranean Sea: 1978–1982. *In* "Research on Dolphins" (Eds M. M. Bryden and R. Harrison) pp. 225–249. Clarendon Press, Oxford.

McKenzie, C., Rogan, E., Reid, R. J. and Wells, D. E. (1998). Concentrations and patterns of organic contaminants in Atlantic white-sided dolphins (*Lagenorhynchus acutus*) from Irish and Scottish coastal waters. *Environ. Poll.* **98**, 15–27.

Mercer, M. C. (1973). Observations on distribution and intraspecific variation in pigmentation patterns of odontocete Cetacea in the western North Atlantic. *J. Fish. Res. Bd Can.* **30**, 1111–1130.

Mikkelsen, A. M. H. and Lund, A. (1994). Intraspecific variation in the dolphins *Lagenorhynchus albirostris* and *L. acutus* (Mammalia: Cetacea) in metrical and non-metrical skull characters, with remarks on occurrence. *J. Zool. Lond.* **234**, 289–299.

Mitchell, E. (1970). Pigmentation pattern evolution in delphinid cetaceans: an essay in adaptive coloration. *Can. J. Zool.* **48**, 717–740.

Mitchell, E. (1975). "Porpoise, Dolphin and Small Whale Fisheries of the World. Status and Problems". IUCN Monograph 3, International Union for the Conservation of Nature and Natural Resources, Morges, Switzerland.

Nelson, D. L. and Lien, J. (1994). Behaviour patterns of two captive Atlantic white-sided dolphins, *Lagenorhynchus acutus*. *Aquat. Mamm.* **20**, 1–10.

Northridge, S. P. (1991). "An Updated World Review of Interactions between Marine Mammals and Fisheries", FAO Fisheries Tech. Pap. 251, Suppl. 1. Food and Agriculture Organization of the United Nations, Rome.

Northridge, S., Tasker, M., Webb, A., Camphuysen, K. and Leopold, M. (1997). White-beaked *Lagenorhynchus albirostris* and Atlantic white-sided dolphin *L. acutus* distributions in northwest European and US North Atlantic waters. *Rep. Int. Whal. Commn* **47**, 797–805.

Palka, D., Read, A. and Potter, C. (1997). Summary of knowledge of white-sided dolphins (*Lagenorhynchus acutus*) from US and Canadian Atlantic waters. *Rep. Int. Whal. Commn* **47**, 729–734.

Perkins, J. S., Bryant, P. J., Nichols, G. and Patten, D. R. (1982). Humpback whales (*Megaptera novaeangliae*) off the west coast of Greenland. *Can. J. Zool.* **60**, 2921–2930.

Perrin, W. F. and Reilly, S. B. (1984). Reproductive parameters of dolphins and small whales of the family Delphinidae. *Rep. Int. Whal. Commn*, Spec. Issue No. 6, pp. 97–133.

Rasch, H. (1843). *Delphinus leucopleurus*. Nova species. *Nyt Mag. Naturvidenskaberne* **4**, 100–103, Tab. II, III.

Reidenberg, J. S. and Laitman, J. T. (1988). Existence of vocal folds in the larynx of Odontoceti (toothed whales). *Anat. Rec.* **221**, 884–891.

Rogan, E., Baker, J. R., Jepson, P. D., Berrow, S. and Kiely, O. (1997). A mass stranding of white-sided dolphins (*Lagenorhynchus acutus*) in Ireland: biological and pathological studies. *J. Zool., Lond.* **242**, 217–227.

Rossiter, W. (1985). Observations of Atlantic white-sided dolphins during feeding and transit. Unpublished manuscript (available from the author: 21 Laurel Hill Road, Ridgefield, CT 06877, USA), 4 typescript pp. + photographs + sighting form.

Schevill, W. E. (1956). *Lagenorhynchus acutus* off Cape Cod. *J. Mamm.* **37**, 128–129.

Schevill, W. E. and Watkins, W. A. (1962). "Whale and Porpoise Voices. A Phonograph Record". Woods Hole Oceanographic Institution, Woods Hole, MA, USA.

Schlegel, H. (1841). Beiträge zur Charakteristik der Cetaceen. *Abh. Geb. Zool. Vergl. Anat.* **1**, 1–44, Taf. I–VI.

Sears, R., Wenzel, F. W. and Williamson, J. M. (1981). Aerial survey of marine mammals along the Quebec lower north shore and its principal river estuaries (Mingan Islands to Strait of Belle Isle) fall 1981. Report for Hydro-Quebec, Direction Environnement, Mingan Island Cetacean Study, East Falmouth, MA, USA and Sept-Iles, Québec, Canada.

Sears, R. and Williamson, J. M. (1982). Distribution and relative abundance of marine mammals along the Quebec lower north shore and its principal river estuaries (Mingan Island and Strait of Belle Isle) July to October 1982. Report for Hydro-Québec, Vice-Présidence Environnement, Mingan Island Cetacean Study, East Falmouth, MA, USA and Sept-Iles, Québec, Canada.

Selzer, L. A. and Payne, P. M. (1988). The distribution of white-sided (*Lagenorhynchus acutus*) and common dolphins (*Delphinus delphis*) vs. environmental features of the continental shelf of the northeastern United States. *Mar. Mamm. Sci.* **4**, 141–153.

Sergeant, D. E. and Fisher, H. D. (1957). The smaller Cetacea of eastern Canadian waters. *J. Fish. Res. Bd Can.* **14**, 83–115.

Sergeant, D. E., Mansfield, A. W. and Beck, B. (1970). Inshore records of Cetacea for eastern Canada, 1949–68. *J. Fish. Res. Bd Can.* **27**, 1903–1915.

Sergeant, D. E., St. Aubin, D. J. and Geraci, J. R. (1980). Life history and northwest Atlantic status of the Atlantic white-sided dolphin, *Lagenorhynchus acutus*. *Cetology* **37**, 1–12.

Sheldrick, M. C. (1989). Stranded whale records for the entire British coastline, 1967–1986. *Invest. Cetacea* **22**, 298–329.

Sigurjónsson, J. (1985). Sightings survey in the Irminger Sea and off Iceland in 1983. *Rep. Int. Whal. Commn* **35**, 499–503.

Sigurjónsson, J. and Gunnlaugsson, T. (1988). Distribution and abundance of cetaceans in Icelandic and adjacent waters from sightings surveys conducted in June–July 1987. ICES Doc. C.M. 1988/N:14, International Council for the Exploration of the Sea, Copenhagen (cited in Gaskin, 1992).

Sigurjónsson, J. and Gunnlaugsson, T. (1990). Distribution and abundance of cetaceans in Icelandic and adjacent waters during sightings surveys July–August 1989. Marine Mammals Committee Doc. C.M. 1990/N:5, International Council for the Exploration of the Sea, Copenhagen.

Sigurjónsson, J., Gunnlaugsson, T. and Payne, M. (1989). NASS-87: Shipboard sightings surveys in Icelandic and adjacent waters June–July 1987. *Rep. Int. Whal. Commn* **39**, 395–408.

Slijper, E. J. (1931). Über Verletzungen und Erkrankungen der Wirbelsäule und Rippen bei den Cetaceen. *Anat. Anz.* **71**, 156–185.

Smeenk, C. (1986). Strandingen van Cetacea op de Nederlandse kust in 1981, 1982 en 1983. *Lutra* **29**, 268–281.

Smeenk, C. (1989). Strandingen van Cetacea op de Nederlandse kust in 1984, 1985 en 1986. *Lutra* **32**, 164–180.

Smeenk, C. and Gaemers, P. A. M. (1987). Fish otoliths in the stomachs of white-beaked dolphins *Lagenorhynchus albirostris*. *In* "The European Cetacean Society Report of the 1987 Meeting, Hirtshals, Denmark" (Eds J.W. Broekema and C. Smeenk) pp. 12–13.

Smiddy, P. (1992). White-sided dolphin *Lagenorhynchus acutus* (Gray). *Irish Nat. J.* **24**, 164.

Smith, T. D., Payne, P. M., Heinemann, D., Waring, G. and Lange, A. (1990). Simultaneous fishery resource and seabird and cetacean sighting surveys. *In*

"Whaling Communities" (Ed. E. Vestergaard) pp. 90–101. North Atlantic Studies No. 2, Aarhus University, Aarhus, Denmark.

St. Aubin, D. J. and Geraci, J. R. (1979). Strandings: a rare look into the biology of the Atlantic white-sided dolphin, *Lagenorhynchus acutus*. *In* "Biology of Marine Mammals: Insights through Strandings" (Eds J. R. Geraci and D. J. St. Aubin) pp. 190–206. PB-293 890, US Marine Mammal Commission, National Technical Information Service, Springfield, VA.

Tangredi, B. P. and Medway, W. (1980). Post-mortem isolation of *Vibrio alginolyticus* from an Atlantic white-sided dolphin (*Lagenorhynchus acutus*). *J. Wildl. Dis.* **16**, 329–331.

Taruski, A. G., Olney, C. E. and Winn, H. E. (1975). Chlorinated hydrocarbons in cetaceans. *J. Fish Res. Bd Can.* **32**, 2205–2209.

Testaverde, S. A. and Mead, J. G. (1980). Southern distribution of the Atlantic white-sided dolphin, *Lagenorhynchus acutus*, in the western North Atlantic. *Fish. Bull.* **78**, 167–169.

Tomilin, A. G. (1967). "Mammals of the U.S.S.R. and Adjacent Countries", Vol. IX, "Cetacea" (Ed. V. G. Heptner). Nauk USSR, Moscow (English translation of 1957 Russian version, Israel Program for Scientific Translations, Jerusalem).

True, F. W. (1889). Contributions to the natural history of the cetaceans. A review of the family Delphinidae. *Bull. U.S. Nat. Mus.* **36**, 1–191, 47 pls.

Van Utrecht, W. L. and Husson, A. M. (1968). Strandingen van Cetacea in het voorjaar van 1967 op de Nederlandse kusten. *Lutra* **10**, 7–17.

Waller, G. H. and Tyler, N. J. C. (1979). Observations on *Lagenorhynchus acutus* stranded on the Yorkshire coast. *Naturalist* **104**, 61–64.

Waring, G. T., Gerrior, P., Payne, P. M., Parry, B. L. and Nicholas, J. R. (1990). Incidental take of marine mammals in foreign fishery activities off the northeast United States, 1977–88. *Fish. Bull.* **88**, 347–360.

Weinrich, M. (1996). Abandonment of an entangled conspecific by Atlantic white-sided dolphins (*Lagenorhynchus acutus*). *Mar. Mamm. Sci.* **12**, 293–296.

Winge, H. (1902). Grønlands pattedyr. *Meddr Grønland* **21**, 317–521.

Winn, H. E. and Edel, R. K. (1982). "A Characterization of Marine Mammals and Turtles in the Mid- and North-Atlantic Areas of the U.S. Outer Continental Shelf". CETAP Annual Report for 1980. Contract No. AA551-CT8-48, Cetacean and Turtle Assessment Program, University of Rhode Island, sponsored by US Bureau of Land Management.

Würsig, B. (1986). Delphinid foraging strategies. *In* "Dolphin Cognition and Behavior: A Comparative Approach" (Eds R. J. Schusterman, J. A. Thomas and F. G. Wood) pp. 347–359. Lawrence Erlbaum, Hillsdale, NJ.

Øien, N. (1996). *Lagenorhynchus* species in Norwegian waters as revealed from incidental observations and recent sighting surveys. Document SC/48/SM15, International Whaling Commission, Scientific Committee, Cambridge.

3

Pacific White-sided Dolphin *Lagenorhynchus obliquidens* Gill, 1865

Robert L. Brownell, Jr, William A. Walker and Karin A. Forney

Genus and Species

Taxonomy

Lagenorhynchus obliquidens was described by Gill (1865), based on three skulls that he examined at the National Museum of Natural History, Smithsonian Institution, Washington, DC. The type and co-type specimens (USNM 1961, 1962, and 1963) were collected near San Francisco, California by W.P. Trowbridge (Gill, 1865; True, 1889). Members of the genus *Lagenorhynchus* are considered, by most researchers, to belong to the subfamily Delphininae (Slijper, 1936).

Several other species have been described and placed in the synonymy of *L. obliquidens*. A dolphin skull collected, from an unknown locality, by the US Exploring Expedition (1838–1842) was described by Cope (1866) as *Delphinus longidens*. True (1889) placed this specimen in the genus *Lagenorhynchus* and considered it to be conspecific with *L. obliquidens*. Sleptsov (1955) described four specimens of *Lagenorhynchus* from the Nemoro Sea, 15–20 miles east of Kunashir Island, South Kurils, Russia as *L. ognevi*. Tomilin (1967) considered Sleptsov's specimens to be the same as *L. obliquidens*. Hershkovitz (1966) also treated *L. ognevi* and *L. longidens* as junior synonyms of *L. obliquidens*. Some authors

Handbook of Marine Mammals, Volume 6
ISBN 0-12-588506-7

have considered *L. obliquidens* a junior synonym of *L. obscurus* (Honacki *et al.*, 1982), but this has not been adopted by any subsequent authors.

Common names

The most common English name for *L. obliquidens* is the Pacific white-sided dolphin. This species is sometimes called the Pacific white-striped dolphin. Foreign names include dauphin a flancs blancs du Pacifique (French), and delfin de costados blancos del Pacifico (Spanish). Hershkovitz (1966) also listed tikhookeanskii belobokii delfin (Russian) and kama-iruka (Japanese) as additional vernacular names for this species.

Evolution

For the subfamily Delphininae, the fossil record extends back at least 5–7 million years to the late Miocene. The oldest known occurrence of the Delphinidae is based on a fossil skull from approximately 10–11 million years ago (Barnes, 1976). A Late Pleistocene skull belonging to the genus *Lagenorhynchus* was collected from the Palos Verdes Sand at Newport Bay in southern California (Barnes, 1976).

Distribution

Range

In the western Pacific Ocean, the Pacific white-sided dolphin is found from the South China Sea northward (Wang, 1993), throughout Japanese waters, and around the Kuril Islands, extending north to the Commander Islands (Fig. 1). Pacific white-sided dolphins also occur in the Sea of Japan (Miyazaki and Shikano, 1989), the southwestern Okhotsk Sea (Miyashita and Berzin, 1991), and infrequently, in the southern Bering Sea (Leatherwood *et al.*, 1984; Kajimura and Loughlin, 1988). In the eastern Pacific, the species occurs primarily in shelf and slope waters from the southern Gulf of California, Mexico along the western coast of North America (Stacey and Baird, 1991; Green *et al.*, 1992; Carretta and Forney, 1993; Mangels and Gerrodette, 1994; Barlow, 1995) north to the Gulf of Alaska and as far west as Amchitka in the Aleutian Islands (Leatherwood *et al.*, 1984; Walker *et al.*, 1986). Across the North Pacific, the species is generally found to have a relatively narrow distribution between 38°N and 47°N (Hobbs and Jones, 1993).

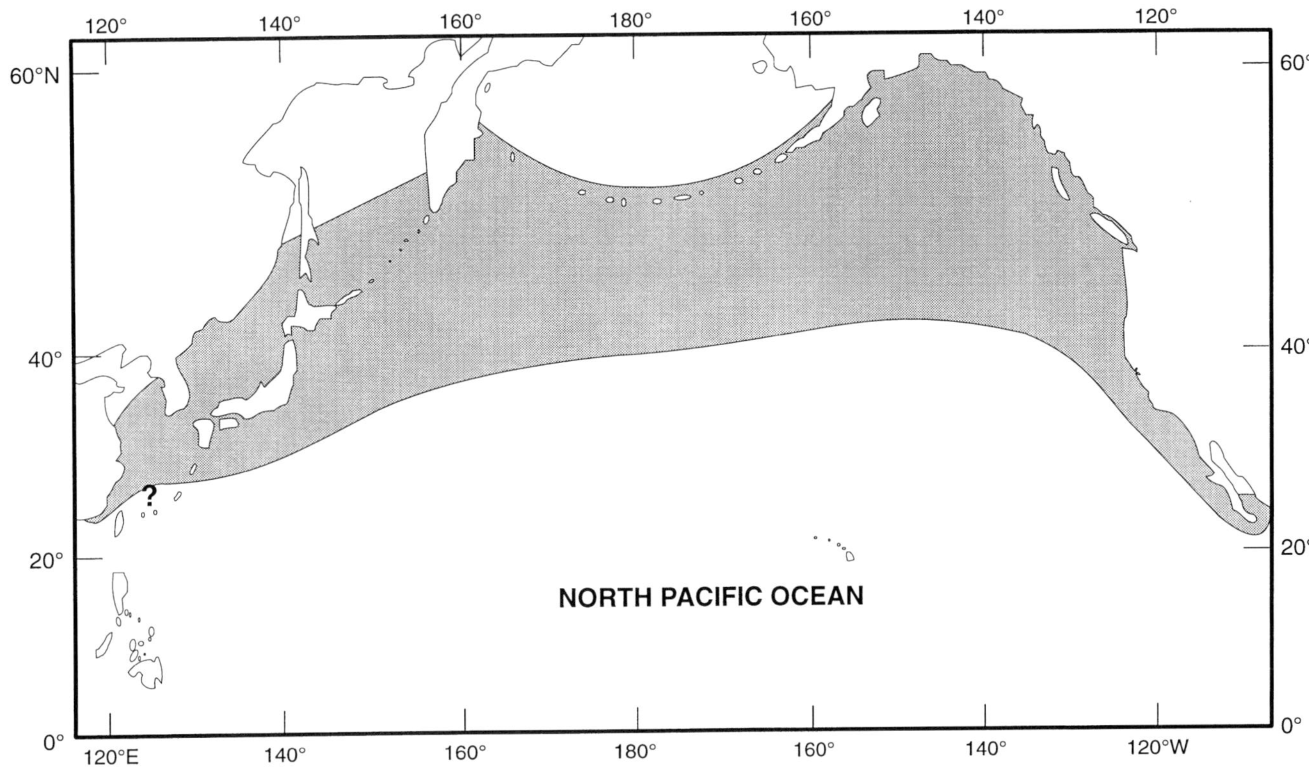

FIG. 1 Distribution of the Pacific white-sided dolphin, *Lagenorhynchus obliquidens* (modified from Stacey and Baird, 1991). The southern range of this species in the western North Pacific is poorly known.

Habitat

The Pacific white-sided dolphin is found in cold–temperate waters of the North Pacific. The species is found in deep waters of the North Pacific Ocean between approximately 38°N and 47°N latitudes. Along the continental margins the species is most common in shelf and slope waters, and infrequently is observed beyond the slope (Green *et al.*, 1992; Carretta and Forney, 1993; Barlow, 1995). Pacific white-sided dolphins are known to occur close to shore in regions such as the inshore passes of Alaska (Dahlheim and Towell, 1994), British Columbia, and Washington, and seasonally off southern California (Pike, 1960; Leatherwood *et al.*, 1984).

External Characteristics

Colour pattern

Pacific white-sided dolphins have a white belly, and dark grey to black coloration on the sides and back. A black line separates the white ventrum from darker lateral areas. The dark coloration of the upper body is interrupted by a light grey thoracic patch extending from below the dorsal fin onto the facial area. A light blowhole chevron (Mitchell, 1970), commonly referred to as "suspender stripes", begins above the eye on each side and runs high on the dorsum, widening behind the dorsal fin to form a lateral blaze (flank patch of Mitchell, 1970). The width of these stripes is variable between individuals. The upper beak, chin, and lips are black to dark grey, and the eye is encircled in black. Light patches often are found on the flippers. The dorsal fin is darker anteriorly, and the posterior two-thirds is a variably light grey patch (Fig. 2). The flukes are completely dark.

Fraser (1966) noted that among the members of the genus *Lagenorhynchus* the general distribution of pigmentation in *L. obliquidens* is most similar to that of *L. australis*. The primary differences in coloration patterns of these two species are in the extent of pigmentation on the chin, the pattern of the flank patch at the flipper insertion, and the relatively light coloured dorsal fin of *L. obliquidens* (Fraser, 1966). Detailed descriptions of the colour pattern of the Pacific white-sided dolphin can be found in Fraser (1966) and Mitchell (1970).

Variations from the normal colour pattern have been recorded for individuals in the eastern North Pacific Ocean. These include all-black individuals and largely all-white individuals (Stacey and Baird, 1991; Black, 1994). Brownell (1965) illustrated an animal with a black thoracic patch surrounded by an extensive white stripe that started from above the eye. This anomalously coloured form has been reported also by Walker *et al.* (1986) and Black (1994). In Monterey Bay, CA, individuals have been observed with shades of orange or tan dorsally where a normally coloured Pacific white-sided dolphin would be grey (Black, 1994).

FIG. 2 Lateral view of Pacific white-sided dolphin, *Lagenorhynchus obliquidens* (44° 49′N, 176° 29′E) on 9 August 1991. (Photograph by S. Mizroch)

Size and shape

The largest reported male and female specimens from the eastern North Pacific were 250 cm and 236 cm, respectively ($n = 243$; Walker *et al.*, 1986). The largest male and female from the western North Pacific off Japan were 238 cm and 229 cm, respectively ($n = 174$; Kasuya, 1985). Iwasaki (1991) reports a maximum length of 230 cm for both sexes based on 114 male and 127 female specimens from the central North Pacific between 34°N and 46°N. A separate sample of 341 specimens, taken from the central North Pacific between 38°N and 46°N, resulted in maximum lengths of 237 cm for males and 210 cm for females (Ferrero and Walker, 1996). Miyazaki and Shikano (1989) found that specimens from the Sea of Japan were longer (and had greater condylobasal length) than those from the northwestern North Pacific. Two forms have also been distinguished from the eastern North Pacific, with southern animals larger than northern ones (Walker *et al.*, 1986). Selected external morphometrics can be found in Scheffer and Slipp (1948), Scheffer (1950), Sleptsov (1955), and Norris and Prescott (1961). Ranges for selected external measurements (previously unpublished) from specimens collected north of 35°N in the eastern North Pacific are given in Table 1. Insufficient data are available to provide similar measurements for the larger southern animals.

TABLE 1 Ranges of selected external measurements (cm) for Pacific white-sided dolphins, *Lagenorhynchus obliquidens*, collected north of 35°N off California, Oregon and Washington. Both sexes are combined, and the sample includes only specimens greater than or equal to 175 cm in length and presumed to be mature

	N	Range	Mean	SD	% of mean total length
Total length	27	175–206	181.9	6.5	—
Tip of jaw to:					
apex of melon	19	3–5	4.4	0.6	2.4%
end of gape	26	20–25	21.9	1.5	12.0%
umbilicus	21	72–88	82.8	3.8	45.5%
tip of dorsal fin	26	103–128	109.8	5.6	60.4%
Girth at axilla	22	101–122	105.9	5.0	58.2%
Anterior length of flipper	27	29–40	34.1	2.4	18.8%
Width of flipper	25	10–13	11.6	0.8	6.4%
Span of flukes	25	38–58	47.6	4.4	26.2%
Width of flukes	23	11–17	13.9	1.4	7.6%
Height of dorsal fin	25	14–28	20.4	2.7	11.2%

All six dolphin species of the genus *Lagenorhynchus* have abbreviated, ploughshare-like beaks (True, 1889). Considerable variation in dorsal fin shape has been observed in *L. obliquidens*. Dorsal fin shape can range from sharply falcate to heavily lobate in both western and eastern North Pacific populations (Kasuya, 1981; Walker *et al.*, 1986). The lobate condition may be age-related and become more pronounced at the onset of physical maturity (Walker *et al.*, 1986; Kasuya and Yoshida, 1990).

Weight

The heaviest recorded female and male Pacific white-sided dolphins, both from Mexican waters, were 145 kg (212 cm) and 198 kg (228 cm), respectively (Walker *et al.*, 1986). Although sample sizes are small, the largest reported Mexican specimens appear to have greater weights than specimens of similar size from southern Californian waters (Walker *et al.*, 1986).

Internal Anatomy

Skull

Lagenorhynchus obliquidens has a broad skull with a moderately long, wide rostrum. Nishiwaki (1963) noted that the rostrum is approximately twice as long as it is

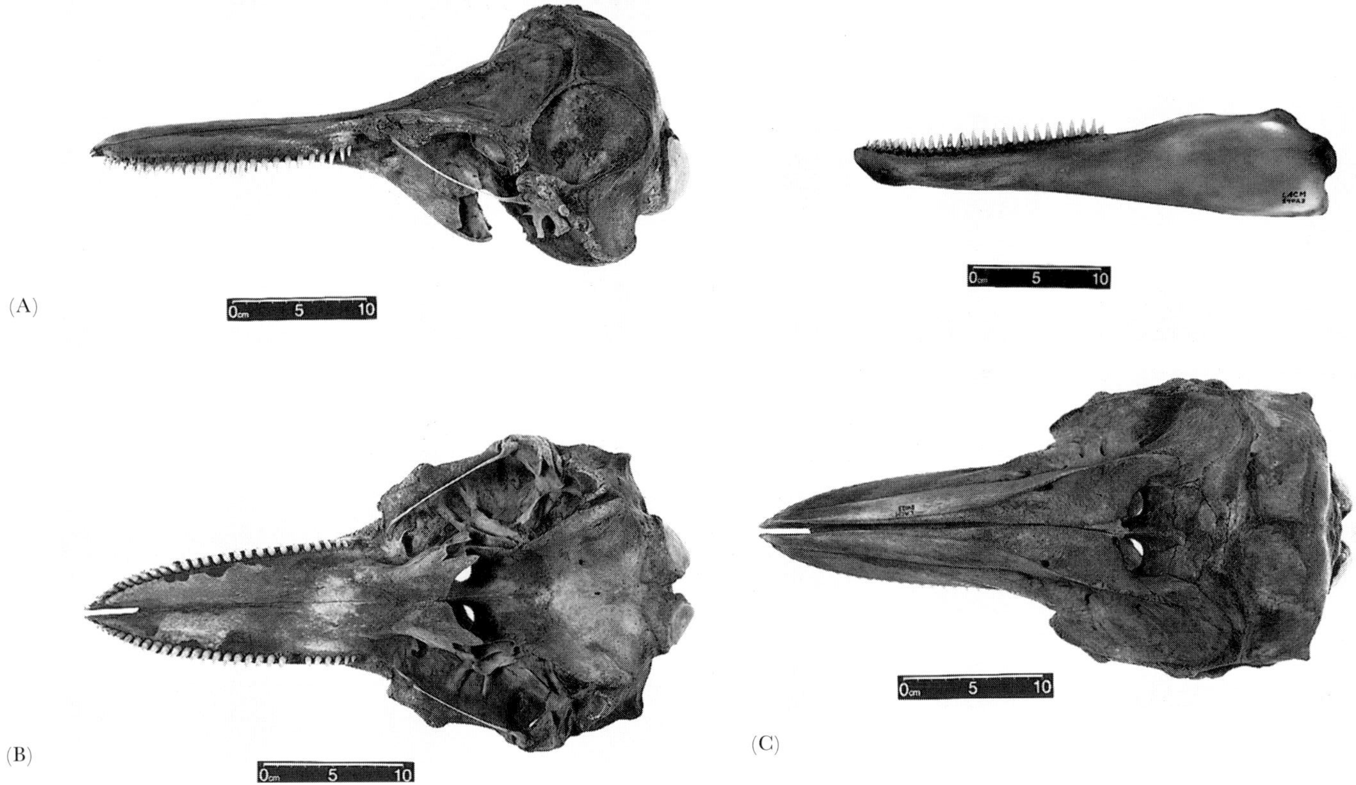

FIG. 3 Lateral (A) (with mandible), ventral (B) and dorsal (C) views of a skull (LACM 84023) from a 225 cm Pacific white-sided dolphin, *Lagenorhynchus obliquidens*, in the Natural History Museum of Los Angeles County. (Photographs courtesy of the Natural History Museum of Los Angeles County)

wide. In the proximal portion of the rostrum, the tooth rows are approximately parallel to one another, converging gradually towards the tip of the distal portion. The premaxillae are slightly raised in relation to the adjacent maxillae. Distal rostral fusion does not occur in adults of this species as in some other delphinids (Walker *et al.*, 1986). There are no grooves present in the palate. Sexual dimorphism in skull morphometrics (males with wider cranium and rostrum than females) was observed in specimens from the Sea of Japan (Miyazaki and Shikano, 1989).

Webber (1987) examined interspecific differences between *L. obliquidens* and *L. obscurus* (dusky dolphin), and found that the cranium and rostrum in *L. obliquidens* were larger and proportionally broader than in *L. obscurus*. There was, however, overlap between the two species for each of the characters measured. Walker *et al.* (1986) used skull morphometrics to demonstrate stock differences in the eastern North Pacific. The mean condylobasal length of 144 sexually mature specimens of *L. obliquidens* was 388 mm, with a range of 350–446 mm (Walker *et al.*, 1986). The skull of this species is illustrated in Fig. 3. Selected skull measurements and tooth count ranges are given in Table 2. The tympanoperiotic bones, illustrated in Kasuya (1973), are intermediate in length between those of *Tursiops* and *Delphinus*.

Postcranial skeleton

Scheffer (1950) described the complete skeleton of an adult male. The vertebral count is: C_7, T_{13-14}, L_{20-24}, Ca_{30-33} with a total of 73–78 (True, 1889; Tomilin, 1967; Nishiwaki, 1972). Unpublished data on 40 specimens from the eastern North Pacific are presented in Table 3. These results, C_7, T_{12-14}, L_{20-25} and Ca_{30-35} with a total of 72–76, generally agree with previously reported counts.

Organs, tissues and physiology

No complete organ weight sets have been published for Pacific white-sided dolphins. However, specific organ weights have been recorded for some systems. An average brain weight of 1515 g ($n = 4$; mean total body weight 90 kg) was reported by Gihr and Pilleri (1979). Ridgway and Brownson (1984) examined relative brain surface areas in odontocetes, and reported a mean brain weight for *L. obliquidens* of 1256.5 g ($n = 2$; mean total body length 207.5 cm; mean total body weight 99.5 kg). A brain/spinal cord ratio of 34.5:1 was reported for this species by Ridgway *et al.* (1966). Flanigan (1966) examined the spinal cord anatomy of a single specimen of *L. obliquidens* and found 44 pairs of spinal nerves.

The mean heart weight of five Pacific white-sided dolphins was 0.85% of the body weight, with a range of 0.70–0.92%. The mean blood volume for four specimens was 108 ml per kg body weight (Ridgway and Johnston, 1966). Total blood-oxygen content may be used as a general indication of physiological capabilities of the species, and this was estimated as 23% by volume for a 100 kg

Pacific white-sided dolphin (Ridgway and Johnston, 1966). These physiological characteristics may be associated with such ecological factors as fast swimming, deep diving, and an oceanic habitat.

Gonad weights for some specimens of Pacific white-sided dolphins were reported by Harrison *et al.* (1972) and Walker *et al.* (1986), and the weights of some endocrine organs have been published (Harrison, 1969). The appearance of the ovaries was reported by Harrison and McBrearty (1977). The adrenal weight to body weight (g/kg) in two specimens was 0.11 and 0.20 (Harrison, 1969). Harrison (1969) reported also that the thyroid in a single specimen of *L. obliquidens* appeared to be heavier relative to body weight than in other delphinids of any age. Anatomical and physiological data on the thyroids of seven Pacific white-sided dolphins also indicated that the gland is relatively large, but its role in accommodating the high metabolic requirements of this species is not fully understood (Ridgway and Patton, 1971).

Several other physiological and anatomical studies of the Pacific white-sided dolphin have been published. Nasal and facial anatomy were described by Schenkkan (1973) and Heyning (1989). Observations of the digestive system were considered in a description of the delphinid small intestine (Harrison *et al.*, 1977). Wever *et al.* (1972) described the cochlear structure in the Pacific white-sided dolphin, with reference to its function for hearing high frequency sounds. Coulombe *et al.* (1965) and Simpson and Gardner (1972) described the microscopic anatomy of various tissues in this species. Clinical data on haematology were presented and discussed by Ridgway *et al.* (1970).

Genetics

Recent genetic analyses of mitochondrial DNA have identified significant geographic variation in Pacific white-sided dolphins from the eastern and central North Pacific (Lux *et al.*, 1997). Four geographic strata (Baja California; California, Oregon and Washington; British Columbia and Southeast Alaska; and the central North Pacific) were included in a molecular analysis of variance. Consistent with the morphological evidence presented by Walker *et al.* (1986), Pacific white-sided dolphins off Baja California, Mexico were found to be significantly different from animals farther north and offshore, although some Baja California haplotypes were shared with animals from southern California. In addition, Pacific white-sided dolphins from British Columbia and Southeast Alaska were determined to be significantly different from those in all other areas. Further genetic work may reveal additional population structure in other regions of the North Pacific.

TABLE 2 Skull measurements (mm) and meristics of Pacific white-sided dolphins, *Lagenorhynchus obliquidens*, from the Eastern North Pacific (from Walker *et al.*, 1986)

Variable		Sample	Range (mm)	Mean	SD
1	Condylobasal length	144	350–446	388.0	16.2
2	Rostrum length	144	171–228	195.6	10.3
3	Rostrum width at base	146	91–123	105.6	6.6
4	Rostrum width at 60 mm	145	66–95	78.6	6.2
5	Rostrum width at mid-length	142	60–86	70.9	5.5
6	Premaxillary width at rostrum midlength	144	36–56	43.8	3.9
7	Rostrum width at 3/4 length	142	43–70	54.6	5.2
8	Rostrum tip to external nares	144	208–273	237.9	11.3
9	Rostrum tip to internal nares	123	206–277	243.8	12.9
10	Preorbital width	139	164–198	178.0	7.6
11	Postorbital width	141	181–214	194.5	7.2
13	External nares width	146	44–90	56.9	4.8
14	Zygomatic width	144	178–218	195.1	8.3

15	Greatest width of premaxillaries	147	80–103	89.9	3.6
16	Parietal width	144	148–184	168.8	6.2
19	Temporal fossa length	147	65–93	79.7	5.8
20	Temporal fossa width	148	50–78	60.8	4.9
25	Orbital length	141	47–65	53.9	3.0
26	Antorbital process length	142	25–47	36.1	3.6
27	Internal nares width	144	48–75	61.7	4.7
28	Pterygoid length	142	53–74	62.8	4.6
30	Bulla length	89	29–38	35.2	1.6
31	Periotic length	81	28–37	30.5	1.4
32	Upper tooth row length	142	154–213	179.1	9.9
33–36	Teeth (No.)	140/138	24–36/25–35	30.3/30.5	2.1/2.2
		145/144	23–34/25–35	30.1/30.1	2.0/2.0
37	Lower tooth row length	140	165–214	180.6	9.0
38	Ramus length	137	231–366	320.2	15.6
39	Ramus height	139	60–81	68.6	3.9
40	Tooth width	127	3.2–6.5	4.6	0.5

TABLE 3 Regional survey of stomach contents of Pacific white-sided dolphins, *Lagenorhynchus obliquidens*, by prey family, summarized from the literature[a]

Taxon	Number of species reported	Pacific coast of North America and Mexico	Pacific coast of Japan	High seas between 38°N and 46°N
Fish				
Petromyzonidae, Lampreys	1	12		
Clupeidae, Herrings	2	1, 10, 13, 16		
Engraulididae, Anchovy	1	4, 6, 7, 8, 10, 11, 12, 13, 14	9	
Salmonidae, Salmon	2	12		
Osmeridae, Smelts	2	10, 11		
Argentinidae, Argentines	2			15
Bathylagidae, Deepsea smelt	4	13		15
Gonostomatidae, Lightfishes	2			15
Melanostomiatidae, Dragonfishes	2			15
Myctophidae, Lanternfishes	19	12, 13	3	15
Paralepididae, Barracudinas	2			15
Scopelarchidae, Pearleyes	1			15
Scopelosauridae, Wearyfish	1			15
Batrachoididae, Toadfishes	2	10, 11, 13, 14		
Ophidiidae, Cusk-eels	2	10, 13, 14		
Merlucciidae, Whiting	1	7, 8, 10, 11, 12, 13, 14		
Scomberesocidae, Saury	1	5, 11, 12, 16		15
Exocoetidae, Flyingfishes	1	14		
Atherinidae, Silversides	1	10		
Sphyraenidae, Barracudas	1	14		
Serranidae, Sea basses	1	14		
Kyphosidae, Sea chubs	1	14		
Labridae, Wrasses	1	14		
Pomacentridae, Damselfishes	1	14		
Trachipteridae, Ribbonfishes	1	12		15

Scorpaenidae, Scorpionfishes	2+	10, 11, 12, 13		
Anoplopomatidae, Sablefish	1	10, 13		
Carangidae, Jacks	1	5, 6, 11, 12, 13, 14, 16		
Sciaenidae, Croakers	4	11, 13, 14, 16		
Pentacerotidae, Armorhead	1			15
Embiotocidae, Surfperches	1	11, 14		
Gempylidae, Snake mackerels	1			15
Scombridae, Mackerels	1	13, 14	3	
Centrolophidae, Medusafishes	1	12		15
Tetragonuridae, Squaretail	1			15
Stromateidae, Butterfish	1	13		
Gobiidae, Gobies	1	14		
Cynoglossidae, Tonguefish	1	13		
Pleuronectidae, Righteye flounders	3	13, 14		
Bothidae, Lefteye flounders	1	11, 12, 14, 16		
Cephalopoda				
Sepiolidae	1	10		
Loliginidae	1	4, 10, 12, 13, 14		
Enoploteuthidae	3	10, 12, 16	3	15
Octopoteuthidae	2	12, 13, 16		15
Onychoteuthidae	1	10, 12, 13		15
Gonatidae	5+	7, 10, 12, 13, 14, 16		15
Histioteuthidae	1	16		
Ommastrephidae	1		9	15

Continued

TABLE 3 Regional survey of stomach contents of Pacific white-sided dolphins, *Lagenorhynchus obliquidens*, by prey family, summarized from the literature[a] (*contd.*)

Taxon	Number of species reported	Pacific coast of North America and Mexico	Pacific coast of Japan	High seas between 38°N and 46°N
Cephalopoda (continued)				
Chiroteuthidae	1	12, 16		15
Mastigoteuthidae	1			15
Cranchiidae	2+	12		15
Octopodidae	2	10, 13, 14		15
Ocythoidae	1	12		
Unidentified Cephalopoda	—	11		
Unidentified Octopoda	—	11		

[a]Literature sources: 1, Higgins (1919); 2, Scheffer (1953); 3, Wilke *et al.* (1953); 4, Brown and Norris (1956); 5, Houck (1961); 6, Norris and Prescott (1961); 7, Fiscus and Niggol (1965); 8, Fitch and Brownell (1968); 9, Hotta *et al.* (1969); 10, Morejohn *et al.* (1978); 11, Jones (1981); 12, Stroud *et al.* (1981); 13, Walker *et al.* (1986); 14, Schwartz *et al.* (1992); 15, Walker and Jones (1993); and 16, Black (1994).

Life History and Population Dynamics

Growth and reproduction

The average length at attainment of sexual maturity for Pacific white-sided dolphins in the eastern North Pacific has been estimated as 170–180 cm for males (Harrison *et al.*, 1972). There are no estimates of the average age or length at attainment of sexual maturity for females from this region (but see Harrison *et al.*, 1969, 1972). Ovulation rates appear to be higher in this species than in other delphinids. The gestation period has been estimated to be 10 months (Harrison, 1969; Harrison *et al.*, 1972); however, Dalton *et al.* (1995) gave a gestation of "approximately 12 months" for a calf born at Sea World of Texas.

Ferrero and Walker (1996) examined 341 specimens of *L. obliquidens* taken from squid fisheries in the central North Pacific and produced information on age, growth and reproduction in these animals. Due to insufficient samples of indeterminate age classes for both sexes, age at onset of sexual maturity was not calculated. The youngest sexually mature female was 8.3 years of age and the oldest sexually immature female 11 years. The youngest sexually mature male was 10 years of age and the oldest sexually immature male 11 years. The average length at sexual maturation was 177.5 cm for females and 173.7 cm for males. Males and females reached an asymptotic length of 184 cm and 177 cm, respectively. The estimated gestation period was 11–12 months with an average length at birth of 91.8 cm for both sexes combined. Early postnatal growth was calculated to be about 4.5 cm per month for the first 6 months of life.

Based on a different set of samples from the central North Pacific, Iwasaki (1991) estimated length at attainment of sexual maturity to be 170–210 cm for males, and 170–220 cm for females. Age at attainment of sexual maturity was estimated at 7–9 years of age for both males and females (Iwasaki, 1991).

Miyazaki and Shikano (1989) estimated that skulls of *L. obliquidens* in Japanese waters reach adult size at approximately 4.5 years of age.

Mortality

The trematode *Nasitrema* sp. and the nematode *Crassicauda* sp. have been implicated as natural mortality factors in single strandings in southern Californian waters (Dailey and Walker, 1978; Cowan *et al.*, 1986; Walker *et al.*, 1986).

Dahlheim and Towell (1994) reported three separate incidents of pursuit and subsequent predation on Pacific white-sided dolphins by killer whales (*Orcinus orca*), in southeastern Alaska.

Abundance

Buckland *et al.* (1993) estimated the abundance of Pacific white-sided dolphins in the North Pacific to be 931 000 (95% C.I. = 206 000–4 216 000) animals. This is in close agreement with the estimate of 989 000 (95% C.I. = 287 000–3 401 000) made by Miyashita (1993). However, precision is low for both studies, and vessel attraction probably resulted in overestimation of population size (Buckland, 1993; Hobbs and Jones, 1993; Miyashita, 1993).

In the eastern North Pacific, separate abundance estimates have been made for different regions and seasons. Off Oregon and Washington, Green *et al.* (1993) estimated a peak abundance of 23 400 animals (95% C.I. = 12 200–45 100) during the month of May based on 1992 surveys. In February–April of 1991 and 1992, aerial surveys conducted along the continental shelf and slope of California resulted in a population estimate of 122 000 (95% C.I. = 51 000–290 000; Forney *et al.*, 1995). This contrasts with a ship-based estimate of 5 900 (95% C.I. = 2 530–13 800) in August–November 1991 for the same study area (Forney and Barlow, 1998). This discrepancy may be explained by seasonal migrations (see Movements below).

Behaviour

Social organization

Pacific white-sided dolphins are considered to be gregarious and generally are seen in groups of tens to thousands of animals. In the eastern and central North Pacific, groups of less than one hundred are most common (Leatherwood *et al.*, 1984; Hill and Barlow, 1992; Carretta and Forney, 1993; Buckland *et al.*, 1993). Age and sex segregation has been reported for schools collected off Iki Island (Takemura, 1986).

Movements

In coastal areas off Southern California, peak abundance of Pacific white-sided dolphins appears to occur from November to April (Leatherwood *et al.*, 1984). Recent seasonal abundance estimates off the entire coast of California are an order of magnitude higher in February–April (Forney *et al.*, 1995) than in August–November (Forney and Barlow, 1998). Few sightings have been made beyond the continental shelf in summer and fall (Barlow, 1995). Peak abundances are observed off Oregon and Washington during May (Green *et al.*, 1993). Overall, this pattern strongly suggests seasonal north–south movements of Pacific white-sided dolphins in the eastern North Pacific.

The larger southern form of the species, found primarily off the west coast of Baja California, also has been reported off southern California (Walker *et al.*,

1986). However, most of the recent specimens from southern Californian waters were found to belong to the northern form, based on genetic studies (Lux *et al.*, 1997). Thus the occurrence of the southern form off Southern California appears to be variable, possibly relating to changes in oceanographic conditions on seasonal or inter-annual time scales (i.e. El Niño events).

The species is found seasonally, in spring and summer, in the southwestern Gulf of California (Aurioles *et al.*, 1988). In Alaskan waters, published sighting records are sparse, but the occurrence of Pacific white-sided dolphins may be related to periods of warmer waters (Dahlheim and Towell, 1994). Off Japan, Pacific white-sided dolphins occupy the Korean Strait and waters of western Japan in the winter, and appear to move to the east from March to July (Kasuya, 1971). Nothing is known about the movements of the two forms described from Japanese coastal waters (Miyazaki and Shikano, 1989).

Feeding

Pacific white-sided dolphins consume a wide variety of fishes and cephalopods (Table 3). Considerable differences in feeding preference are evident between animals from coastal and offshore regions. Reports of feeding habits have been confined largely to the coastal waters of the eastern North Pacific. In this region, Pacific white-sided dolphins feed primarily on epipelagic fishes and cephalopods. The fishes most commonly consumed were: the northern anchovy, *Engraulis mordax*; Pacific hake, *Merluccius productus*; Pacific saury, *Cololabis saira*; juvenile rock fish, *Sebastes* spp.; and horse mackerel, *Trachurus symmetricus*. The market squid, *Loligo opalescens*, also is frequently ingested. Over 35 species representing an additional 23 families of nearshore and epipelagic fishes and cephalopods also have been observed in stomach contents of Pacific white-sided dolphins.

Walker and Jones (1993) reported on the stomach contents of 33 Pacific white-sided dolphins collected between 38°N and 46°N in the central North Pacific. In this oceanic environment, these dolphins feed heavily on mesopelagic fish and cephalopods. Mesopelagic fish species of the families Argentinidae, Bathylagidae, Gonostomatidae, Melanostomiatidae, Myctophidae, Paralepididae, Scopelarchidae, and Scopelosauridae made up 81.6% of the total species ingested. Seventeen species of lanternfish (Myctophidae) made up 78.6% of the total. Epipelagic fish and squid were consumed also, but were of lesser importance.

In the coastal waters of northern Japan *L. obliquidens* feed on both mesopelagic and epipelagic fishes and cephalopods (Wilke *et al.*, 1953; Hotta *et al.*, 1969). Fishes of the family Myctophidae comprised 77% (by volume) in 13 stomachs (Wilke *et al.*, 1953). The epipelagic fish species *Engraulis japonica*; *Scomber japonicus* and the squids *Watasenia scintillans* and *Todarodes pacificus* also were consumed. Walker and Jones (1993) suggested that the similarity between the results from both high seas and coastal waters of northern Honshu are due to the narrow submarine shelf and nearshore deep water conditions off the Pacific coast of Honshu.

Association with other animals

Pacific white-sided dolphins are frequently encountered in mixed-species aggregations of cetaceans and pinnipeds, and are often accompanied by seabirds (Leatherwood *et al.*, 1982; Stacey and Baird, 1991; Black, 1994). Cetaceans recorded in association with Pacific white-sided dolphins include: the northern right whale dolphin, *Lissodelphis borealis*; Risso's dolphin, *Grampus griseus*; striped dolphin, *Stenella coeruleoalba*; short-beaked common dolphin, *Delphinus delphis*; long-beaked common dolphin, *Delphinus capensis*, short-finned pilot whale, *Globicephala macrorhynchus*; bottlenose dolphin, *Tursiops truncatus*; Dall's porpoise, *Phocoenoides dalli*; sperm whale, *Physeter catodon*; humpback whale, *Megaptera novaeangliae*, blue whale, *Balaenoptera musculus*, fin whale, *Balaenoptera physalus*, sei whale, *Balaenoptera borealis*; and gray whale, *Eschrichtius robustus* (Rice, 1989; Stacey and Baird, 1991; Hill and Barlow, 1992; Black, 1994; Mangels and Gerrodette, 1994; Carretta *et al.*, 1995).

The interspecific relationship with the northern right whale dolphin appears to be a unique association in which large groups of both species are frequently observed to form heterogeneous herds and subgroups (Walker, 1975; Leatherwood and Walker, 1979). The reason for this close association is not known, but it may be food related; particularly in the oceanic environment. Food habits analyses from individuals of both species in the central North Pacific revealed considerable overlap in preferred mesopelagic prey (Walker and Jones, 1993).

Swimming and diving

Ridgway and Johnston (1966) observed that Pacific white-sided dolphins could not overtake their boat at speeds faster than 26 km/hr, but noted that the animals could bowride for brief periods at 32 km/hr. Lang (1966) measured acceleration power and velocity in an individual Pacific white-sided dolphin. Tests were made as the animal swam through a series of hoops in a tank. Maximum velocity was measured to be about 28 km/hr (15 knots; Lang, 1966). Mean and maximum swim speeds for three wild Pacific white-sided dolphins tracked in Monterey Bay are reported as 5.0–8.9 km/hr, and 14.8–20.2 km/hr, respectively (Black, 1994). In this same study, mean dive times for the three tracked animals were 22.0–25.7 sec, and maximum dive times were 3.3–6.2 minutes.

Aerial and other behaviour

Pacific white-sided dolphins are well known for their aerial displays that include a variety of leaps, bellyflops and somersaults (Leatherwood *et al.*, 1982; Stacey and Baird, 1991). These behaviours have been observed repeated up to 20 times in rapid succession (Black, 1994).

Sound production

Caldwell and Caldwell (1967) recorded whistles emitted from captive Pacific white-sided dolphins, and described a burst-pulse "tin horn" sound that appeared to be elicited by any strong stimulus. It was suggested that this sound may be individually distinct, a type of "signature burst-pulse" sound (Caldwell and Caldwell, 1967). A similar sound was noted by Caldwell and Caldwell (1967) to have been described by Gales (1966). Schevill and Watkins (1971) noted whistlelike squeals from Pacific white-sided dolphins. In four captive Pacific white-sided dolphins, vocalization levels were highest at dawn, at dusk, and during feeding sessions (Goley, 1991).

Parasites and Disease

Eight genera of parasites have been reported from L. obliquidens (Table 4). Trematodes, *Nasitrema* spp. have been documented to cause brain pathology (Schroeder *et al.*, 1973; Dailey and Walker, 1978; Cowan *et al.*, 1986) and severe ulceration and necrosis of the air sinus and inner ear tissues (Dailey and Walker, 1978; Cowan *et al.*, 1986). Walker *et al.* (1986) reported air sinus infections of nematodes, *Crassicauda* sp., in four of 39 Pacific white-sided dolphins examined. Other interesting cases include: severe earbone erosion believed to be caused by parasites (Cowan *et al.*, 1986) and brain parasitism (Martin *et al.*, 1970; Cowan *et al.*, 1986). The presence of the commensal pseudo-stalked barnacle, *Xenobalanus globicipitus*, has been reported for this species (Norris and Prescott, 1961; Dailey and Walker, 1978).

In captivity, parasitic infection is largely curtailed through the feeding of fish which has been previously held in deep frozen storage. However, disease from a wide range of bacterial, viral and mycotic infections has been encountered by Pacific white-sided dolphins in the captive environment (Ridgway, 1972; Simpson and Gardner, 1972; Sweeney and Ridgway 1975a,b; Sweeney *et al.*, 1976; Migaki *et al.*, 1978; Migaki and Jones, 1983; Simpson and Cornell, 1983).

A high incidence of vertebral bone disease (osteonecrosis) has been reported in specimens of *L. obliquidens* collected in Californian waters. Walker *et al.* (1986) reported this bone disease in 34 of 63 postcranial skeletons.

Live Maintenance

The first live collection of Pacific white-sided dolphins was in 1954 (Brown and Norris, 1956) off southern California. Walker (1975) reported on 51 Pacific white-sided dolphins taken for live display in southern California during the

TABLE 4 Parasites from Pacific white-sided dolphins, *Lagenorhynchus obliquidens*

Parasite	Infection site and source	Reference[a]
Trematoda		
Nasitrema globicephalae	Air sinus	1
Nasitrema lagenorhynchus	Air sinus, larynx, nasal cavity and lungs	9, 10
Nasitrema sp.	Air sinus;	6, 7, 8
	Brain	4, 6, 7
Odhneriella gondo	Liver (bile ducts), intestine and stomach	10
Cestoda		
Phyllobothrium delphini	Blubber	5, 6
Phyllobothrium sp.	Blubber	3
Monorygma grimaldi	Peritoneum	3
Tetrabothrius sp. or *Tetrabothrium* sp. (Ref. 9)	Intestine (duodenum)	6
Strobilocephalus triangularis	Intestine (colon)	6
Nematoda		
Anisakis simplex	Stomach	6, 10
Anisakis sp.	Stomach	7
Crassicauda sp.	Air sinus;	6, 7, 8
	Brain[b]	2

[a]Literature sources: 1, Neiland *et al.* (1970); 2, Martin *et al.* (1970); 3, Dailey and Brownell (1972); 4, Schroeder *et al.* (1973); 5, Testa and Dailey (1977); 6, Dailey and Walker (1978); 7, Cowan *et al.* (1986); 8, Walker *et al.* (1986); 9, Kikuchi *et al.* (1987); and 10, Miyazaki *et al.* (1991).
[b]Reported as *Contracaecum* sp.

years 1966–1972, and Reeves and Leatherwood (1984) reported that a total of 28 animals were taken between 1973 and 1982. Since 1982, 12 Pacific white-sided dolphins have been captured off southern California, most recently in 1993 (Forney, 1994). In March 1997, there were 22 animals, 3 males and 19 females, in North American marine displays (L. Dalton, personal communication). Of this group, 2 females were born at Sea World of Texas in 1993 and 1996, respectively. Dalton *et al.* (1995) reported on the growth and development of the 1993 female.

Between 1973 and 1982 a total of 238 Pacific white-sided dolphins were collected from Japanese waters for aquaria. Some of these individuals were incidentally or directly taken by fishermen (Kasuya *et al.*, 1984).

Human Effects

Directed fisheries

Small numbers of *L. obliquidens* are killed in Japanese waters in drive fisheries and the hand harpoon fishery (Wilke *et al.*, 1953; Miyazaki, 1983; Anonymous,

1989). During the 1960s, fishermen from Iki Island in western Japan became convinced that small cetaceans were eating and damaging or scaring away yellowtail, *Seriola* spp. and squid (Kasuya, 1985). Government-supported cull programmes to control several small cetaceans, including Pacific white-sided dolphins, were initiated during the 1970s. Between 1976 and 1980, which were the peak years of this programme, at least 466 *L. obliquidens* are reported to have been killed (Kasuya, 1985).

By-catches

Pacific white-sided dolphins are one of the most commonly caught cetaceans in the Japanese and Korean high seas squid drift net fisheries (Hobbs and Jones, 1993). This species was also taken in the Japanese large-mesh and Taiwanese squid and large-mesh fisheries. In 1989, the estimated total by-catch of Pacific white-sided dolphins for only the Japanese squid drift net fishery was about 6100; in 1990, the total estimate for all drift net fisheries combined was 5759 animals (Hobbs and Jones, 1993). Effort for these fisheries was estimated to have increased during the late 1970s and early 1980s, and then remained relatively stable at least until 1990 (Hobbs and Jones, 1993). On 1 January 1993 a United Nations moratorium on these high seas drift net fisheries went into effect. Smaller catches (e.g. at least 194 in 1987) of these dolphins are reported from the Japanese land-based salmon drift net fishery (Anonymous, 1989; Miyazaki, 1983). Small numbers of Pacific white-sided dolphins are taken yearly in seines, set nets, and trap nets around Japan (Anonymous, 1989; Miyazaki *et al.*, 1991).

In the eastern North Pacific, these dolphins have been taken in a variety of fisheries (Barlow *et al.*, 1994). A total of 363 animals were estimated to have been killed in the shark and swordfish drift net fishery in California during the period from April 1988 to December 1995 (Perkins *et al.*, 1994; Julian and Beeson, 1998). Additional mortality has been documented for trammel and set nets in California coastal waters, for drift gill nets in British Columbia and Alaska, and for trawl fisheries in Alaska; however, no overall mortality estimates are available for these fisheries (Barlow *et al.*, 1994; Heyning *et al.*, 1994).

Pacific white-sided dolphins are rarely taken (e.g. 5 in 1988 and 45 in 1989) incidental to the tuna purse seine fishery in the eastern tropical Pacific (Chivers *et al.*, 1989; Hall and Boyer, 1991). Most of the fishing takes place south of the range of these dolphins.

Pollution

The maximum concentrations of DDT and PCBs reported in the blubber of Pacific white-sided dolphins in Japanese waters were 99 ppm and 71 ppm wet weight, respectively (Tanabe *et al.*, 1983, 1987). A single specimen that died in captivity had concentrations of 1023 ppm DDT and 147 ppm PCBs (Taruski *et al.*, 1975). Britt and Howard (1983) reported no detectable PCBs and 0.22 ppm total DDT in liver tissue from a single animal that stranded in southern California. Organochlorine levels in the blubber of two stranded Pacific white-sided dolphins from Californian waters were 2.08 ppm and 99.5 ppm DDT, and 0.23 ppm and 4.88 ppm PCBs (Schafer *et al.*, 1984). Overall, pollutant loads for this species appear to be variable.

Acknowledgments

The photograph of a wild Pacific White-sided dolphin was provided by Sally Mizroch. Skull photographs were provided by the Natural History Museum of Los Angeles County. John E. Heyning and Toshio Kasuya read an early draft of this manuscript and provided useful comments. We thank all of the above people for their help with this chapter.

References

Anonymous (1989). Japan. Progress report on cetacean research, June 1987 to April 1988. *Rep. Int. Whal. Commn* **39**, 201–204.

Aurioles, D., Leatherwood, S. and Munoz, E. (1988). Pacific white-sided dolphins (*Lagenorhynchus obliquidens*) in the Sea of Cortez. *Bull. South. Calif. Acad. Sci.* **87**, 44–45.

Barlow, J. (1995). The abundance of cetaceans in California waters. Part I: Ship surveys in summer and fall of 1991. *Fish. Bull.* **93**, 1–14.

Barlow, J., Baird, R. W., Heyning, J. E., Wynne, K., Manville II, A. M., Lowry, L. F., Hanan, D., Sease, J. and Burkanov, V. N. (1994). A review of cetacean and pinniped mortality in coastal fisheries along the west coast of the USA and Canada and the east coast of the Russian Federation. *Rep. Int. Whal. Commn*, Spec. Issue No. 15, pp. 405–426.

Barnes, L. G. (1976). Outline of eastern North Pacific fossil cetacean assemblages. *Systematic Zool.* **23**, 321–343.

Black, N. (1994). Behavior and ecology of Pacific white-sided dolphins (*Lagenorhynchus obliquidens*) in the Monterey Bay, California. M. S. thesis, San Francisco State University, CA.

Britt Jr., J. O. and Howard, E. B. (1983). Tissue residues of selected environmental contaminants in marine mammals. *In* "Pathobiology of Marine Mammal Diseases", Vol. II (Ed. E. B. Howard) pp. 79–94. CRC Press, Boca Raton, FL.

Brown, D. H. and Norris, K. S. (1956). Observations of captive and wild cetaceans. *J. Mamm.* **37**, 311–326.

Brownell Jr., R. L. (1965). Observations of odontocetes in central California waters. *Norsk Hvalf.-Tid.* **53**, 60–66.

Buckland, S. T., Cattanach, K. L. and Hobbs, R. C. (1993). Abundance estimates of Pacific white-sided dolphin, northern right whale dolphin, Dall's porpoise and northern fur seal in the North Pacific, 1987–90. *Int. N. Pacif. Fish. Comm. Bull.* **53**, 387–407.

Caldwell, M. C. and Caldwell, D. K. (1967). Intraspecific transfer of information via the pulsed sound in captive odontocete cetaceans. *In* "Les Systèmes Sonars Animaux,

Biologie et Bionique" (Ed. R.-G. Busnel) pp. 879–936. Laboratoire de Physiologie Acoustique, Jouy-en-Josas, France.

Carretta, J. V. and Forney, K. A. (1993). Report of the two aerial surveys for marine mammals in California coastal waters utilizing a NOAA DeHavilland Twin Otter aircraft March 9–April 7, 1991 and February 8–April 6, 1992. NOAA Tech. Memo. NMFS, NOAA-TM-NMFS-SWFSC-185, 77 pp.

Carretta, J. V., Forney, K. A. and Barlow, J. (1995). Report of 1993–1994 marine mammal aerial surveys conducted within the US Navy Outer Sea Test Range off southern California. NOAA Tech. Memo. NMFS, NOAA-TM-NMFS-SWFSC-217, 90 pp.

Chivers, S. J., Hohn, A. A. and Miller, R. B. (1989). Composition of the 1987 incidental kill of small cetaceans in the U.S. purse-seine fishery for tuna in the eastern tropical Pacific. *Rep. Int. Whal. Commn* **39**, 315–319.

Cope, E. D. (1866). Third contribution to the history of the Delphinidae. *Proc. Acad. Nat. Sci. Philadelphia* **18**, 293–300.

Coulombe, H. N., Ridgway, S. H. and Evans, W. E. (1965). Respiratory water exchange in two species of porpoise. *Science* **149**, 86–88.

Cowan, D. R., Walker, W. A. and Brownell Jr., R. L. (1986). Pathology of small cetaceans stranded along southern California beaches. *In* "Research on Dolphins" (Eds M. M. Bryden and R. Harrison) pp. 323–367. Oxford University Press, Oxford.

Dahlheim, M. E. and Towell, R. G. (1994). Occurrence and distribution of Pacific white-sided dolphins (*Lagenorhynchus obliquidens*) in Southeastern Alaska, with notes on an attack by killer whales (*Orcinus orca*). *Mar. Mamm. Sci.* **10**, 458–464.

Dailey, M. D. and Brownell Jr., R. L. (1972). A checklist of marine mammal parasites. *In* "Mammals of the Sea: Biology and Medicine" (Ed\. S. H. Ridgway) pp. 528–589. Charles C. Thomas, Springfield, IL.

Dailey, M. D. and Walker, W. A. (1978). Parasitism as a factor (?) in single strandings of Southern California cetaceans. *J. Parasitol.* **64**, 593–596.

Dalton, L. M., Robeck, T. R. and Young, G. W. (1995). Growth and development of a Pacific white-sided dolphin (*Lagenorhynchus obliquidens*). *In* "Proceedings, International Association for Aquatic Animal Medicine" (Ed. J. McBain). Mystic Marinelife Aquarium, Mystic, CT, USA.

Ferrero, R. C. and Walker, W. A. (1996). Age, growth and reproductive patterns of the Pacific white-sided dolphin (*Lagenorhynchus obliquidens*) taken in high seas drift nets in the central North Pacific Ocean. *Can. J. Zool.* **74**, 1673–1687.

Fiscus, C. H. and Niggol, K. (1965). Observations of cetaceans off California, Oregon and Washington. US Fish and Wildlife Service Special Science Report on Fisheries No. 498.

Fitch, J. E. and Brownell Jr., R. L. (1968). Fish otoliths in cetacean stomachs and their importance in interpreting feeding habits. *J. Fish. Res. Bd Can.* **25**, 2561–2564.

Flanigan, N. J. (1966). The anatomy of the spinal cord of the Pacific striped dolphin, *Lagenorhynchus obliquidens*. *In* "Whales, Dolphins, and Porpoises" (Ed\. K. S. Norris) pp. 207–231. University of California Press, Berkeley, CA.

Forney, K. A. (1994). Recent information on the status of odontocetes in Californian waters. NOAA Tech. Memo. NMFS, NOAA-TM-NMFS-SWFSC-202, 87 pp.

Forney, K. A. and Barlow, J. (1998). Seasonal patterns in the abundance and distribution of California cetaceans, 1991–92. *Mar. Mamm. Sci.* **14**, 460–489.

Forney, K. A., Barlow, J. and Carretta, J. V. (1995). The abundance of cetaceans in California waters. Part II: Aerial surveys in winter and spring of 1991 and 1992. *Fish. Bull.* **93**, 15–26.

Fraser, F. C. (1966). Comments on the Delphinoidea. *In* "Whales, Dolphins, and Porpoises" (Ed. K. S. Norris) pp. 7–31. University of California Press, Berkeley, CA.

Gales, R. S. (1966). Pickup, analysis, and interpretation of underwater acoustic data. *In* "Whales, Dolphins, and Porpoises" (Ed. K. S. Norris) pp. 435–444. University of California Press, Berkeley, CA.
Gihr, M. and Pilleri, G. (1979). Interspecific body length–body weight ratio and body weight–brain weight ratio in Cetacea. *Invest. Cetacea* **10**, 245–253.
Gill, T. (1865). On two species of Delphinidae, from California, in the Smithsonian Institution. *Proc. Acad. Nat. Sci. Philadelphia* **17**, 177–178.
Goley, P. D. (1991). Dawn/dusk vocal choruses and nighttime resting behavior of captive Pacific white-sided dolphins. *In* Abstracts of the Ninth Biennial Conference on the Biology of Marine Mammals, Chicago, IL, 5–9 December 1991.
Green, G., Brueggeman, J. J., Grotefendt, R. A., Bowlby, C. E., Bonnell, M. L. and Balcomb III, K. C. (1992). Cetacean distribution and abundance off Oregon and Washington. *In* "Oregon and Washington Marine Mammal and Seabird Surveys". Final Report prepared for Pacific OCS Region, Minerals Management Service, US Department of the Interior, Los Angeles, CA, May 1992, Chap. 1.
Green, G., Grotefendt, R. A., Smultea, M. A., Bowlby, C. E. and Rowlett, R. A. (1993). Delphinid aerial surveys in Oregon and Washington offshore waters. Final Report prepared for National Marine Fisheries Service, National Marine Mammal Laboratory, Seattle, WA.
Hall, M. A. and Boyer, S. D. (1991). Incidental mortality of dolphins in the tuna purse-seine fishery in the eastern Pacific Ocean during 1989. *Rep. Int. Whal. Commn* **41**, 507–509.
Harrison, R. J. (1969). Endocrine organs: hypophysis, thyroid, and adrenal. *In* "The Biology of Marine Mammals" (Ed. H. T. Andersen) pp. 349–390. Academic Press, London.
Harrison, R. J. and McBrearty, D. A. (1977). Ovarian appearances in captive delphinids (*Tursiops* and *Lagenorhynchus*). *Aquat. Mamm.* 5, 57–66.
Harrison, R. J., Boice, R. C. and Brownell Jr., R. L. (1969). Reproduction in wild and captive dolphins. *Nature* **222**, 1143–1147.
Harrison, R. J., Brownell Jr., R. L. and Boice, R. C. (1972). Reproduction and gonadal appearance in some odontocetes. *In* "Functional Anatomy of Marine Mammals", Vol. 1 (Ed. R. J. Harrison) pp. 361–426. Academic Press, London.
Harrison, R. J., Johnson, F. R. and Young, B. A. (1977). The small intestine of dolphins (*Tursiops, Delphinus, Stenella*). *In* "Functional Anatomy of Marine Mammals", Vol. 3 (Ed. R. J. Harrison) pp. 297–332. Academic Press, London.
Hershkovitz, P. (1966). Catalog of living whales. *Bull. U.S. Nat. Mus.* **246**, 1–259.
Heyning, J. E. (1989). Comparative facial anatomy of beaked whales (Ziphiidae) and a systematic revision among the families of extant Odontoceti. *Contrib. Sci.* **405**, 1–64.
Heyning, J. E., Lewis, T. D. and Woodhouse, C. D. (1994). A note on odontocete mortality from fishing gear entanglements off Southern California. *Rep. Int. Whal. Commn*, Spec. Issue No. 15, pp. 439–442.
Higgins, E. (1919). Porpoise captured. *Calif. Fish and Game* **5**, 157.
Hill, P. S. and Barlow, J. (1992). Report of a marine mammal survey of the California coast aboard the research vessel McArthur July 28–November 5, 1991. NOAA Tech. Memo. NMFS, NOAA-TM-NMFS-SWFSC-169, 103 pp.
Hobbs, R. C. and Jones, L. L. (1993). Impacts of high seas driftnet fisheries on marine mammal populations in the North Pacific. *Int. N. Pacif. Fish. Comm. Bull.* **53**, 409–434.
Honacki, J. H., Kinman, K. E. and Koepple, J. W. (1982). "Mammal Species of the World: A Taxonomic and Geographic Reference". Allen Press, Inc. and The Association of Systematics Collections, Laurence, KS, 694 pp.

Hotta, H., Mako, K., Okada, K. and Yamada, U. (1969). On the stomach contents of dolphins and porpoises off Kyushu. *Bull. Seikai Reg. Fish. Res. Lab.* **37**, 71–85 (in Japanese with English summary).

Houck, W. J. (1961). Notes on the Pacific striped porpoise. *J. Mamm.* **42**, 107.

Iwasaki, T. (1991). Preliminary analysis of life history parameters of Pacific white-sided dolphins and northern right whale dolphins. Document submitted to the Scientific Review of the North Pacific high seas driftnet fisheries, Sidney, B. C. June 11–14, 1991, 13 pp.

Jones, R. E. (1981). Food habits of smaller marine mammals from northern California. *Proc. Calif. Acad. Sci.* **42**, 409–423.

Julian, F. and Beeson, M. (1998). Estimates of marine mammal, turtle, and seabird mortality for two California gillnet fisheries: 1990–1995. *Fish. Bull.* **96**, 271–284.

Kajimura, H. and Loughlin, T. R. (1988). Marine mammals in the oceanic food web of the eastern subarctic Pacific. *Bull. Ocean Res. Inst. Univ. Tokyo* **26**, 187–223.

Kasuya, T. (1971). Consideration of distribution and migration of toothed whales off the Pacific coast of Japan based upon aerial sighting record. *Sci. Rep. Whales Res. Inst.* **23**, 37–60.

Kasuya, T. (1973). Systematic consideration of recent toothed whales based on the morphology of tympano-periotic bone. *Sci. Rep. Whales Res. Inst.* **25**, 1–103.

Kasuya, T. (1981). "Identification Manual of Dolphins and Porpoises in the Japanese Waters". Research Division, Japan Fisheries Agency (in Japanese).

Kasuya, T. (1985). Fishery–dolphin conflict in the Iki Island area of Japan. *In* "Marine Mammals and Fisheries" (Eds J. R. Beddington, J. H. Beverton and D. M. Lavigne) pp. 253–275. George Allen and Unwin, London.

Kasuya, T. and Yoshida, S. (1990). Secondary sexual characteristics of the Pacific white-sided dolphin. *In* Abstracts of the Spring Meeting of the Japanese Society of Scientific Fisheries, Tokyo, Japan, 2–5 April 1990.

Kasuya, T., Tobayama, T. and Matsui, S. (1984). Review of the live-capture of small cetaceans in Japan. *Rep. Int. Whal. Commn* **34**, 597–602.

Kikuchi, S., Okuyama, Y. and Nakajima, M. (1987). *Nasitrema lagenorhynchus* from the larynx and lungs of a Pacific striped dolphin (Nasitrematidae, Trematoda). *Jap. J. Parasit.* **36**, 42–48.

Lang, T. G. (1966). Hydrodynamic analysis of cetacean performance. *In* "Whales, Dolphins, and Porpoises" (Ed. K. S. Norris) pp. 410–432. University of California Press, Berkeley, CA.

Leatherwood, S. and Walker, W. A. (1979). The northern right whale dolphin *Lissodelphis borealis* Peale in the eastern North Pacific Ocean. *In* "Behavior of Marine Animals", Vol. 3, "Cetaceans" (Eds H. E. Winn and B. L. Olla) pp. 85–141. Plenum Press, New York.

Leatherwood, S., Reeves, R. R., Perrin, W. F. and Evans, W. E. (1982). Whales, dolphins and porpoises of the eastern North Pacific and adjacent Arctic waters. A guide to their identification. NOAA Tech. Rep. NMFS Circular 444, 245 pp.

Leatherwood, S., Reeves, R. R., Bowles, A. E., Stewart, B. S. and Goodrich, K. R. (1984). Distribution, seasonal movements and abundance of Pacific white-sided dolphins in the eastern North Pacific. *Sci. Rep. Whales Res. Inst.* **35**, 129–157.

Lux, C. A., Costa, A. S. and Dizon, A. E. (1997). Mitochondrial DNA population structure of the Pacific white-sided dolphin. *Rep. Int. Whal. Commn* **47**, 645–652.

Mangels, K. F. and Gerrodette, T. (1994). Report of cetacean sightings during a marine mammal survey in the eastern Pacific Ocean and the Gulf of California aboard the NOAA ships *McArthur* and *David Starr Jordan* July 28–November 6, 1993. NOAA Tech. Memo. NMFS, NOAA-TM-NMFS-SWFSC-211, 88 pp.

Martin, W. E., Haun, C. K., Barrows, S. and Cravioto, H. (1970). Nematode damage to brain of striped dolphin, *Lagenorhynchus obliquidens*. *Trans. Am. Microsc.* **89**, 200–205.

Migaki, G., Font, R. L., Kaplan, W. and Asper, E. D. (1978). Sporotrichosis in a Pacific white-sided dolphin, *Lagenorhynchus obliquidens*. *Am. J. Vet. Res.* **39**, 1916–1919.

Migaki, G. and Jones, S. R. (1983). Mycotic diseases in marine mammals. *In* "Pathobiology of Marine Mammal Diseases", Vol. II (Ed. E. B. Howard) pp. 1–27. CRC Press, Boca Raton, FL.

Mitchell, E. (1970). Pigmentation pattern evolution in delphinid cetaceans: an essay on adaptive coloration. *Can. J. Zool.* **48**, 717–740.

Miyashita, T. (1993). Distribution and abundance estimates of some dolphins taken in the North Pacific driftnet fisheries. *Int. N. Pacif. Fish. Comm. Bull.* **53**, 435–449.

Miyashita, T. and Berzin, A. A. (1991). Report of the whale sighting survey in the Okhotsk Sea and adjacent waters in 1990. IWC Meeting Document SC/43/05.

Miyazaki, N. (1983). Catch statistics of small cetaceans taken in Japanese waters. IWC Meeting Document SC/34/SM10.

Miyazaki, N. and Shikano, C. (1989). Two forms of *Lagenorhynchus obliquidens* in Japanese waters. *In* Abstracts of the Fifth International Theriological Congress, Rome, 22–29 August, 1989, p. 491.

Miyazaki, N., Kuramochi, T. and Amano, M. (1991). Pacific white-sided dolphins (*Lagenorhynchus obliquidens*) off northern Hokkaido. *Mem. Nat. Sci. Mus. Tokyo* **24**, 131–139.

Morejohn, G. V., Harvey, J. T. and Krasnow, L. T. (1978). The importance of *Loligo opalescens* in the food web of marine vertebrates in Monterey Bay, California. *In* "Biological, Oceanographic and Acoustical Aspects of the Market Squid, *Loligo opalescens* Berry" (Eds C. W. Recksiek and H. W. Frey) pp. 67–97. California Department of Fish and Game, Fish Bulletin 169.

Neiland, K. A., Rice, D. W. and Holden, B. L. (1970). Helminths of marine mammals. I. The genus *Nasitrema*, air sinus flukes of delphinid cetacea. *J. Parasitol.* **56**, 304–316.

Nishiwaki, M. (1963). Taxonomical consideration on genera of Delphinidae. *Sci. Rep. Whales Res. Inst.* **17**, 93–103.

Nishiwaki, M. (1972). General Biology. *In* "Mammals of the Sea: Biology and Medicine" (Ed. S. H. Ridgway) pp. 1–204. Charles C. Thomas, Springfield, IL.

Norris, K. S. and Prescott, J. H. (1961). Observations on Pacific cetaceans of Californian and Mexican waters. *Univ. Calif. Publ. Zool.* **63**, 291–402.

Perkins, P., Barlow, J. and Beeson, M. (1994). Report on pinniped and cetacean mortality in California gillnet fisheries: 1988–1990. Southwest Fisheries Science Center Administrative Report LJ-94-11 Southwest Fisheries Science Center, La Jolla, CA.

Pike, G. C. (1960). Pacific striped dolphin, *Lagenorhynchus obliquidens*, off the coast of British Columbia. *J. Fish. Res. Bd Can.* **17**, 123–124.

Reeves, R. R. and Leatherwood, S. (1984). Live-capture fisheries for cetaceans in USA and Canadian waters, 1973–1982. *Rep. Int. Whal. Commn* **34**, 497–507.

Rice, D. W. (1989). Sperm whale *Physeter macrocephalus* Linnaeus, 1758. *In* "Handbook of Marine Mammals", Vol. IV (Eds S. H. Ridgway and R. Harrison) pp. 177–233. Academic Press, London.

Ridgway, S. H. (1972). Homeostasis in the aquatic environment. *In* "Mammals of the Sea: Biology and Medicine" (Ed. S. H. Ridgway) pp. 590–747. Charles C. Thomas, Springifield, IL.

Ridgway, S. H. and Brownson, R. H. (1984). Relative brain sizes and cortical surface areas in odontocetes. *Acta Zool. Fenn.* **172**, 149–152.

Ridgway, S. H. and Johnston, D. G. (1966). Blood oxygen and ecology of porpoises of three genera. *Science* **151**, 456–457.

Ridgway, S. H. and Patton, G. S. (1971). Dolphin thyroid: some anatomical and physiological findings. *Z. vergl. Physiol.* **71**, 129–141.

Ridgway, S. H., Flanigan, N. J. and McCormick, J. G. (1966). Brain–spinal cord ratios in porpoises: possible correlations with intelligence and ecology. *Psychon. Sci.* **6**, 491–492.

Ridgway, S. H., Simpson, J. G., Patton, G. S. and Gilmartin, W. G. (1970). Hematologic findings in certain small cetaceans. *J. Am. Vet. Med. Assn* **157**, 566–575.

Schafer, H. A., Gossett, R. W., Ward, C. F. and Westcott, A. M. (1984). Chlorinated hydrocarbons in marine mammals. *In* "Southern California Coastal Water Research Project Biennial Report, 1983–84" (Ed. W. Bascom). Southern California Coastal Water Research Project, Long Beach, CA.

Scheffer, V. B. (1950). The striped dolphin, *Lagenorhynchus obliquidens* Gill 1865, on the coast of North America. *Am. Midl. Nat.* **44**, 750–758.

Scheffer, V. B. (1953). Measurements and stomach contents of eleven delphinids from the northeast Pacific. *Murrelet* **32**, 27–30.

Scheffer, V. B. and Slipp, J. W. (1948). The whales and dolphins of Washington State. *Am. Midl. Nat.* **39**, 257–337.

Schenkkan, E. J. (1973). On the comparative anatomy and function of the nasal tract in odontocetes (Mammalia, Cetacea). *Bijdragen tot de Dierkunde* **43**, 127–159.

Schevill, W. E. and Watkins, W. A. (1971). Pulsed sounds of the porpoise *Lagenorhynchus australis*. *Breviora* **366**, 1–10.

Schroeder, R. J., Delli Quadri, C. A., McIntyre, R. W. and Walker, W. A. (1973). Marine mammal disease surveillance program in Los Angeles County. *J. Am. Vet. Med. Assn* **163**, 580–581.

Schwartz, M., Hohn, A., Bernard, H., Chivers, S. and Peltier, K. (1992). Stomach contents of beach cast cetaceans collected along the San Diego County coast of California, 1972–1991. Southwest Fisheries Science Center Administrative Report LJ-92-18, Southwest Fisheries Science Center, La Jolla, CA.

Simpson, J. G. and Cornell, L. H. (1983). Diseases associated with strandings and captivity. *In* "Pathobiology of Marine Mammal Diseases", Vol. II (Ed. E. B. Howard) pp. 30–63. CRC Press, Boca Raton, FL.

Simpson, J. G. and Gardner, M. B. (1972). Comparative microscopic anatomy of selected marine mammals. *In* "Mammals of the Sea: Biology and Medicine" (Ed. S. H. Ridgway) pp. 298–418. Charles C. Thomas, Springfield, IL.

Sleptsov, M. M. (1955). Noviyi vid del'fina dal'nevostochnykh morei *Lagenorhynchus ognevi* species nova (new species of dolphin from the far-eastern ocean . . .). *Trudy Inst. Okean* **18**, 60–68.

Slijper, E. J. (1936). Die Cetaceen, vergleichend-anatomisch und systematisch. Diss. Utrecht, 1936. *Capita Zool. (den Haag)* **7**, 1.

Stacey, P. J. and Baird, R. W. (1991). Status of the Pacific white-sided dolphin, *Lagenorhynchus obliquidens*, in Canada. *Can. Field-Naturalist* **105**, 219–232.

Stroud, R. K., Fiscus, C. H. and Kajimura, H. (1981). Food of the Pacific white-sided dolphin, *Lagenorhynchus obliquidens*, Dall's porpoise, *Phocoenoides dalli*, and northern fur seal, *Callorhinus ursinus*, off California and Washington. *Fish. Bull.* **78**, 951–959.

Sweeney, J. C. and Ridgway, S. H. (1975a). Common diseases of small cetaceans. *J. Am. Vet. Med. Assn* **167**, 533–540.

Sweeney, J. C. and Ridgway, S. H. (1975b). Procedures for the clinical management of small cetaceans. *J. Am. Vet. Med. Assn* **167**, 540–545.

Sweeney, J. C., Migaki, G., Vainik, P. M. and Conklin, R. H. (1976). Systemic mycosis in marine mammals. *J. Am. Vet. Med. Assn* **169**, 946–948.

Takemura, A. (1986). On the life history of the Pacific white-sided dolphin and bottlenose dolphin. *In* "Report of the Contract Study for Solution of Fishery Dolphin Conflict

(For years 56–50 of Showa)" (Eds T. Tamura, S. Ohsumi and S. Arai) pp. 161–177 (in Japanese).

Tanabe, S., Mori, T., Tatsukawa, R. and Miyazaki, N. (1983). Global pollution of marine mammals by PCBs, DDTs and HCHs (BHCs). *Chemosphere* **12**, 1269–1275.

Tanabe, S., Kannan, N., Subramanian, A., Watanabe, S. and Tatsukawa, R. (1987). Highly toxic coplanar PCBs: occurrence, source, persistency and toxic implications to wildlife and humans. *Environ. Pollut.* **47**, 147–163.

Taruski, A. G., Olney, C. E. and Winn, H. E. (1975). Chlorinated hydrocarbons in cetaceans. *J. Fish. Res. Bd Can.* **132**, 2205–2209.

Testa, J. and Dailey, M. D. (1977). Five new morphotypes of *Phyllobothrium* (Cestoda: Tetraphyllidea), their relationship to existing morphotypes, and their zoogeography. *Bull. So. Calif. Acad. Sci.* **76**, 99–110.

Tomilin, A. G. (1967). *In* "Mammals of the U.S.S.R. and Adjacent Countries", Vol. IX, "Cetacea" (Ed. V. G. Heptner) pp. 566–574 (English translation, 1967, Israel Program for Scientific Translation, Jerusalem).

True, F. W. (1889). Contributions to the natural history of cetaceans; a review of the family Delphinidae. *U.S. Nat. Mus. Bull.* **36**, 1–191.

Walker, W. A. (1975). Review of the live-capture fishery for smaller cetaceans taken in southern California waters for public display, 1966–73. *J. Fish. Res. Bd Can.* **32**, 1197–1211.

Walker, W. A. and Jones, L. L. (1993). Food habits of northern right whale dolphin, Pacific white-sided dolphin, and northern fur seal caught in the high seas driftnet fisheries of the northern North Pacific Ocean, 1990. *Int. North Pacif. Fish. Comm. Bull.* **53**, 285–295.

Walker, W. A., Leatherwood, S., Goodrich, K. R., Perrin, W. F. and Stroud, R. K. (1986). Geographical variation and biology of the Pacific white-sided dolphin, *Lagenorhynchus obliquidens*, in the north-eastern Pacific. *In* "Research on Dolphins" (Eds M. M. Bryden and R. Harrison) pp. 441–465. Clarendon Press, Oxford.

Wang Peilie (1993). Fauna of marine mammals in China. *Acta Oceanol. Sinica* **12**, 273–278.

Webber, M. A. (1987). A comparison of dusky and Pacific white-sided dolphins (genus *Lagenorhynchus*): morphology and distribution. MS thesis, San Francisco State University, CA.

Wever, E. G., McCormick, J. G., Palin, J. and Ridgway, S. H. (1972). Cochlear structure in the dolphin *Lagenorhynchus obliquidens*. *Proc. Nat. Acad. Sci. USA* **69**, 657–661.

Wilke, F., Taniwaki, T. and Kuroda, N. (1953). *Phocoenoides* and *Lagenorhynchus* in Japan, with notes on hunting. *J. Mamm.* **34**, 488–497.

4

Dusky Dolphin
Lagenorhynchus obscurus (Gray, 1828)

Robert L. Brownell, Jr and Frank Cipriano

Genus and Species

Taxonomy

Gray (1828) described a new dolphin species, *Delphinus* (*Grampus*) *obscurus*, from stuffed skins of adults and young (with skulls inside) that were originally lodged in the Royal College of Surgeons, London. The specimens were collected in South African waters off the Cape of Good Hope by Captain Haviside (commonly misspelled Heaviside, see Fraser, 1966). The adult skull (No. 354a) is now in the British Museum (Natural History) (Hershkovitz, 1966:64). The specimen number given by Hershkovitz (1966:64) is No. 41.1733, but this number refers to the type skin (True, 1889:174). Flower (1885) placed this species in the genus *Prodelphinus*. *Delphinus obscurus* was transferred to *Lagenorhynchus* by True (1889), but he also noted that *L. obscurus* was similar to *Prodelphinus* (= *Stenella*). Fraser and Purves (1960) reported that the extent of the development of the pterygoid sinus systems in *L. obscurus* and *Stenella* species was similar. Fraser (1966) believed, therefore, that the placement of this species in the genus *Lagenorhynchus* was still dubious. Mead (1975) suggested that *L. obscurus* may be the most generalized species within *Lagenorhynchus*.

Lesson and Garnot (1826) described a new species, *Delphinus supercilious*, based on a specimen collected off Castle-Forbes, 44°S, off Cape Diemen (= South Cape), Tasmania. The specimen was illustrated by Lesson and Garnot (1826)

Handbook of Marine Mammals, Volume 6
ISBN 0-12-588506-7

but was lost or not preserved. Schlegel (1841) referred to a dolphin collected from South Africa as *D. supercilious*. Based on the Schlegel specimen, True (1889) placed *D. supercilious* in the genus *Lagenorhynchus*. This specimen was incorrectly re-identified by Broekema (1983) as *L. cruciger*. The Schlegel specimen is a skull of *L. obscurus* (see discussion under *L. cruciger*, Chapter 6).

During the voyage of the *Beagle*, Charles Darwin collected a dolphin in the Golfo San José, Chubut, Agentina (42° 30′S) in April 1832. According to Darwin, "this porpoise, which was a female, was harpooned from the 'Beagle' out of several, in a large troop, which were sporting round the ship" (Flower, 1885:23). The rostrum and anterior portion of the lower jaw with all the teeth of this specimen were described as *Delphinus Fitzroyi* (Waterhouse, 1838). This specimen was named after the captain of the Beagle, Robert Fitzroy, who prepared the illustration of the type specimen used by Waterhouse (1838). *Delphinus Fitzroyi* is a junior synonym of *L. obscurus* (Kellogg, 1941).

Gray (1868) described a new species of dolphin from the Cape of Good Hope, South Africa as *Clymene similis*. The holotype is in the British Museum (Natural History), London and is catalogued as No. 1509.b. True (1889) placed this species in the genus *Lagenorhynchus*. Various other authors (Allen, 1939; Roberts, 1951; Ellerman *et al*., 1953; Hershkovitz, 1966) have treated *Clymene similis* as a junior synonym of *Stenella coeruleoalba*. True (1889) placed this species in the genus *Lagenorhynchus*. The tooth count of the type specimen is 32 upper and 33 lower (Gray, 1868), which agrees with the mean count of 32 for *L. obscurus* (Webber, 1987). We have examined photographs of this specimen and have confirmed the identification as *L. obscurus*.

A female dolphin 185 cm in total length was harpooned off the coast of Chile (48° 10′S, 77° 00′W) and named *Phocaena posidonia* (Philippi, 1893). The illustration of *P. posidonia* is very similar to *L. obscurus*, although Philippi remarked that it "was distinct from *P. Fitzroyi* because of the absence of the dark streak extending from the angle of the mouth to the base of the pectoral flipper and because of supposed differences of the head". Kellogg (1941) regarded *P. posidonia* as doubtfully distinct from *L. obscurus*. We believe that *P. posidonia* should be regarded as a junior synonym of *L. obscurus*.

Wagner (in Schreber and Wagner, 1846) described a dolphin collected from the Rio de la Plata "à 20 lieues de l'embouchure" during the expedition of the *Astrolabe* and the *Zélée* as a new *Delphinus breviceps*. Kellogg (1941) considered *D. breviceps* to be a colour variant of *L. fitzroyi*. The holotype of *D. breviceps* was re-examined and illustrated by Robineau (1990) and considered to be a synonym of *L. obscurus*. The holotype (specimen number JAC:1880-646) is in the Laboratoire d'Anatomie Comparée, Muséum National d'Histoire Naturelle, Paris (Robineau, 1990).

Rodolpho A. Philippi (1895, 1896) described a new species of dolphin as *Tursio? panope*, based on a skull in the Museo Nacional de Historia Natural, Santiago, Chile. True (1903) was unable to determine the generic affinity of *T.? panope* and suggested that it probably represented a new genus. Hershkovitz

(1966) and various authors listed *T.? panope* as a junior synonym of *Cephalorhynchus eutropia*. Brownell and Mead (1989) re-examined the holotype of *T.? panope* and concluded it was a junior synonym of *L. obscurus*.

All nominal Southern Hemisphere species of *Lagenorhynchus* were included in the synonymy of a single species *L. cruciger* by Bierman and Slijper (1948). This single species recognition was repeated by Scheffer and Rice (1963), Hershkovitz (1966), and Rice and Scheffer (1968). However, Fraser (1966) detailed how the Southern Hemisphere forms should be split into three species: *L. obscurus*, *L. cruciger*, and *L. australis*. These three species are now well recognized (Mitchell, 1970; Brownell, 1974; Rice, 1977).

Common names

The most common English name for this species is the dusky dolphin. In South America, the English name is used by translating it to delfin oscuro. This species is also known in South America as delfin listado (Andrade and Baez, 1980).

Distribution

Range

This species is found in coastal and continental shelf waters off parts of South America, southwest South Africa, most of New Zealand and around oceanic islands, namely Tristan da Cunha, Prince Edward, Amsterdam and St. Paul Islands (Van Waerebeek *et al.*, 1995). These populations are discontinuous (Fig. 1), although occasional interchange may be possible. There is potential for genetic divergence between such widely separated populations and some suggestion that dusky dolphins found in these areas may be distinct enough to deserve sub-specific or specific status. Consequently, life history, ecological, and behavioural attributes of the dolphins found in these different areas are discussed separately below.

In South American waters the northernmost west coast record is from Isla Mazorca, Peru (11°S, skull in USNM 270418), and on the east coast the northernmost record is Mar del Plata, Argentina (ca 38° 00′S) (Gallardo, 1912; Würsig and Bastida, 1986). This species is notably rare (one specimen) in the Tierra del Fuego region (Goodall, 1978, 1989). Historically, *L. obscurus* was assumed to be common in the Beagle Channel, Tierra del Fuego, because of the presence of remains in Indian kitchen middens from approximately 6000 years ago (Piana *et al.*, 1985). Records from around the Falkland Islands are reviewed by Van Waerebeek *et al.* (1995).

Dusky dolphins occur only off the west coast of South Africa, particularly in the Cape of Good Hope region (Webber, 1987). The northern limit is at least to

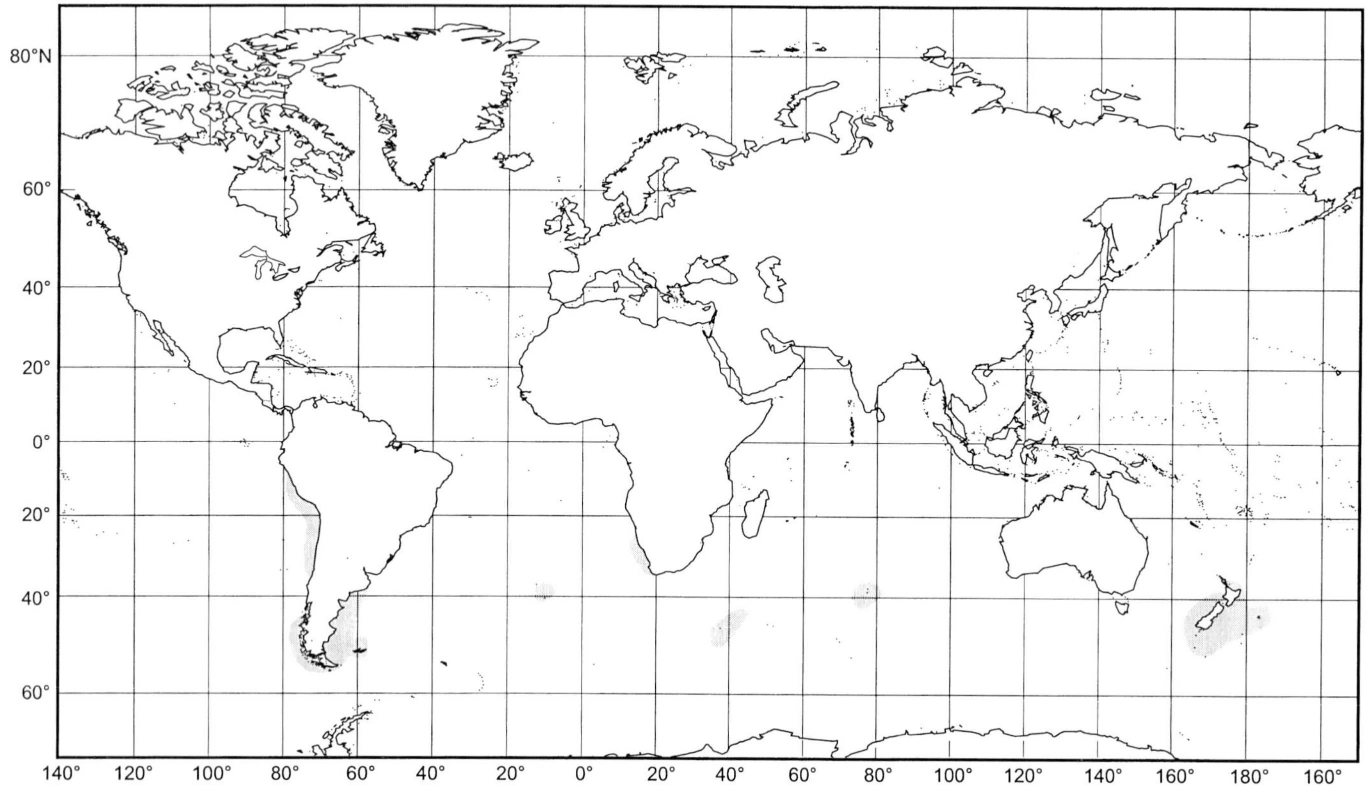

FIG. 1 Range of the dusky dolphin, *Lagenorhynchus obscurus*. (Modified from Van Waerebeek *et al.*, 1995)

Luderitz (26° 59′S, 14° 31′E) and probably north to Walvis Bay. The southern limit is to at least Hout Bay (34° 04′S, 18° 21′E) in the Cape region. Brownell (1974) erroneously mapped the distribution of this species eastward from the Cape to approximately 28° 30′E (based on a misunderstanding that captive dusky dolphins in the aquarium at Port Elizabeth were captured in local waters and an incorrect interpretation of information in Taylor and Saayman [1972]).

Webber (1987) noted that, in New Zealand waters, *L. obscurus* is common around South Island, except for the southern half of the west coast. Off North Island, these dolphins occur only around the southern third of the island, to around Hawke Bay on the east and to New Plymouth (39° 04′S) on the west side.

Dusky dolphins have been reported in the waters off southern Australia and Tasmania (Davies, 1963; Brownell, 1974; Baker, 1983). Occurrence in this region, at least in part, was based on the Lesson and Garnot (1826) unconfirmed report of *Delphinus supercilious* from off South Cape, Tasmania and the subsequent reference to this specimen as *L. obscurus*. In addition, Lillie (1915) reported several sightings of *L. obscurus* off southern Australia and one from the Bass Strait area between southeastern Australia and Tasmania. However, Webber (1987) considered that there is a great deal of uncertainty associated with the Lillie (1915) records. Nicol and Croome (1988) reported that a skull (No. A191) of unknown origin in the Tasmanian Museum had been referred to as *L. obscurus* by Pearson (1935) and *L. cruciger* by Guiler (1978). Van Waerebeek (1993a) reidentified this specimen as *Lissodelphis peronii*. The lack of any confirmed stranding or sighting makes the occurrence of *L. obscurus* in southern Australian waters very dubious. The southernmost records from New Zealand waters are three stranded dusky dolphins from Campbell Island (52° 30′S) (Baker, 1977).

Numerous authors have included the waters around Kerguelen Island as part of the range of *L. obscurus*. This resulted solely from Paulian's (1953) report of a skull from a young animal collected from Golfe du Morbihan, Kerguelen. This specimen was identified for Paulian by F.C. Fraser. Fraser (1964) made only minor comments on this specimen from Kerguelen. Robineau (1989) re-examined this specimen (MNHN-AC:n 1952–197) and concluded that it was a young specimen of *Cephalorhynchus commersonii*.

Habitat

The distribution of dusky dolphins along the west coast of South Africa and both coasts of South America is associated with the continental shelves and cool waters of the Benguela, Humboldt and Falkland Currents. Around New Zealand these dolphins are associated mainly with various cold water currents (Gaskin, 1972). Van Waerebeek *et al.* (1995) have suggested that normally this species may be limited to water shallower than 200 m.

Kasamatsu *et al.* (1990) reported sightings of dusky dolphins in waters with sea surface temperatures between 10 °C and 16 °C, except one anomalous sighting at 31° 46′S at a surface temperature of 25 °C. However, based on an examination of photographs from this sighting, Van Waerebeek *et al.* (1995) believe the identification as dusky dolphin to be doubtful.

Gaskin (1968a) reported sightings of this species offshore from New Zealand in the vicinity of the Subtropical Convergence at surface temperatures between 14 °C and 15 °C. Cipriano (1992) regularly observed *L. obscurus* off the Kaikoura coast of South Island, New Zealand, in waters that varied seasonally between 10 °C and 18 °C. Sightings of dusky dolphins in the Kaikoura region were concentrated over the flanks of the Kaikoura Submarine Canyon (Cipriano, 1992).

Würsig and Würsig (1980) observed dusky dolphins in Golfo San José, Argentina, in waters that varied seasonally between 11 °C and 17 °C.

External Characteristics

Colour pattern

The pigmentation of *L. obscurus* is similar to that of *L. obliquidens* (Figs 2 and 3). In lateral view, the pattern (Figs 2 and 3) can be divided into four parts: the spinal field, the flank patch, the thoracic patch, and the white abdominal field (termi-

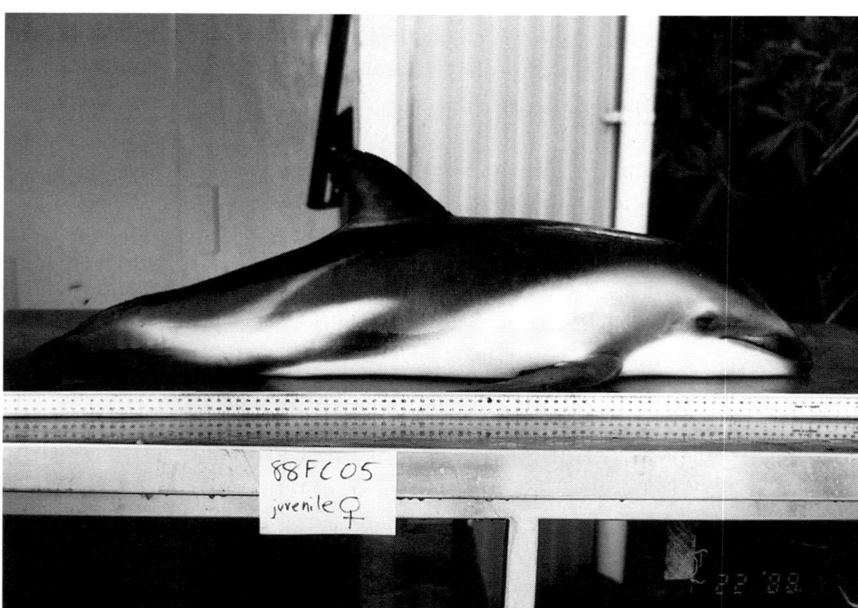

FIG. 2 External view of female dusky dolphin, *Lagenorhynchus obscurus* (88FCOS), 102 cm in total length from off Kaikoura, New Zealand. (Photo by F. Cipriano)

FIG. 3 Two dusky dolphins, *Lagenorhynchus obscurus*, off Kaikoura, New Zealand, 24 October 1987. (Photo by F. Cipriano)

nology from Mitchell, 1970). The most conspicuous features of the coloration are flank blazes, thoracic patch, and snout colour. Detailed descriptions of the pigmentation are found in Fraser (1966), Mitchell (1970), and Van Waerebeek (1993b). Also see Fig. 77 in Gaskin (1968b).

Size and shape

The largest reported dusky dolphin was a 240 cm female from South Africa (Best and Ross, 1984:617), but this specimen was actually a bottlenose dolphin (*Tursiops* sp.), which was misidentified because of a typographical error in the table (Peter B. Best, personal communication). The largest female from South Africa was 193 cm, and the largest male was 181 cm (Best and Ross, 1984). The largest female reported from Peru was 204.5 cm, while the largest males from Peru, 206 cm and 211 cm (Manzanilla, 1989), constitute the maximum size records for this species. The smallest adult female and male from Peru were 173 cm and 176 cm, respectively (Manzanilla, 1989). The largest of 19 adult dusky dolphins (four females, 15 males) collected in New Zealand was only 186 cm, somewhat smaller than the maximum size of specimens from South Africa and much smaller than those from Peru (Cipriano, 1992). Ranges for selected external measurements are given in Table 1.

TABLE 1 Ranges of selected external measurements (in cm) of dusky dolphins, *Lagenorhynchus obscurus*, from New Zealand waters (Cipriano, 1992). Sample size in parentheses. Age range of 18 individuals of known age *ca.* 1 month–36 years (Webber, 1987)

Measurement	Range	Mean	SD	Percentage of mean total length
Total length (4)	97.0–186.0	158.8	(23.60)	—
Tip of upper jaw to:				
angle of gape (25)	13.5–22.7	19.5	(2.27)	12.3
centre of blowhole (23)	13.0–29.0	24.1	(4.04)	15.2
centre of eye (27)	15.8–27.0	23.1	(2.73)	14.5
auditory meatus (22)	20.0–32.1	28.3	(3.19)	17.8
umbilical scar (23)	49.0–85.0	74.6	(8.99)	47.0
ant. insertion flipper (26)	26.0–44.0	38.4	(4.23)	24.2
tip of dorsal fin (20)	62.0–110.0	96.3	(13.20)	60.6
centre of genital slit (26)	69.0–125.0	103.3	(12.90)	65.1
centre of anus (25)	70.5–132.0	114.9	(15.10)	72.4
Ant. length of flipper (24)	23.5–39.0	32.4	(3.93)	20.4
Axilla length of flipper (20)	16.8–30.5	24.1	(3.81)	15.2
Max. width of flipper (25)	6.7–12.0	10.3	(1.41)	6.5
Height of dorsal fin (33)	10.5–28.0	20.5	(4.06)	12.9
Length of dorsal fin base (32)	17.0–34.0	26.9	(3.97)	16.9
Width of flukes (31)	22.5–56.0	40.8	(8.70)	25.7
Width of fluke, from notch to nearest point on anterior border (23)	8.7–17.0	12.8	(2.07)	8.1
Depth of notch (0)	—	—	—	—
Width of blowhole (21)	2.1–3.5	2.66	(0.337)	1.7
Girth, at axilla (30)	61.3–112.0	89.6	(13.50)	56.4
Girth, maximum (0)	—	—	—	—
Girth, at anus (20)	36.5–68.0	57.3	(8.80)	36.1

Like other species of *Lagenorhynchus*, this species has virtually no beak and the head slopes evenly down from the blowhole to the snout. The dorsal fin is high with a concave posterior border (Figs 2 and 3).

Weight

The fifteen adult dusky dolphins from New Zealand reported by Cipriano (1992) weighed 69–78 kg for females (total lengths 167–178 cm) and 70–85 kg for males (total lengths 165–175 cm). Best (1976) reported that two captive females from South Africa at death weighed 53.6 kg (191 cm) and 69.9 kg (184 cm). It appears both South African captive animals had below normal weights at the time of their death when compared with those of the New Zealand specimens.

Internal Anatomy

Skull

Distinctiveness of the skull of *L. obscurus* (Fig. 4) from either *L. cruciger* or *L. australis* was demonstrated by Fraser (1966). Skulls of *L. obscurus* have a longer, narrower rostrum than either *L. australis* or *L. cruciger* of comparable length and maturity. Also, the pre- and postorbital and zygomatic widths of *L. obscurus* skulls are smaller than in the other two species. Mean condylobasal length (CBL) in 16 specimens from Peru (all in the USNM) of *L. obscurus* is 373 mm (range 333–421 mm) (Brownell and Mead, 1989). Mean CBL of 25 specimens from New Zealand waters is 369.5 mm, with a range of 347–387 mm (Webber, 1987). Mean tooth count in New Zealand dolphins was 32 for each tooth row (Webber, 1987). The periotic and tympanic bulla are illustrated in Kasuya (1973).

Postcranial skeleton

The vertebral count given by Cárdenas *et al.* (1986) is C_7, T_{15}, L_{23}, Ca_{36}, giving a total of 81 vertebrae.

Organs and tissues

Gallardo (1913) provided some preliminary notes on the nasal tract complex of a dusky dolphin, and Schenkkan (1973) gave a more detailed description of this complex.

Hearts of six adult dusky dolphins collected by Cipriano (1992) in New Zealand weighed 595–657 g (2 females) and 571–773 g (4 males). Livers of these same specimens weighed 2068–2663 g (females) and 1347–2160 g (males); kidneys of the specimens weighed 339–425 g (females) and 191–417 g (males); right and left lungs of the two females weighed 646/663 g and 1054/1000 g, respectively; the lungs of a single male were weighed at 722/779 g (Cipriano, 1992).

Physiology and biochemistry

Manzanilla (1989) examined dentinal growth layers in the teeth of Peruvian dusky dolphins. She reported a pair of hypocalcified incremental layers about 75 μm in width within one of the growth layer groups. This anomalous dentine was deposited during 1983 at the time of the 1982–1983 El Niño event. Tetracycline marking of captive *L. obscurus* teeth (Best, 1976) supports formation of one growth layer group (GLG) per year.

(A)

(B)

(C)

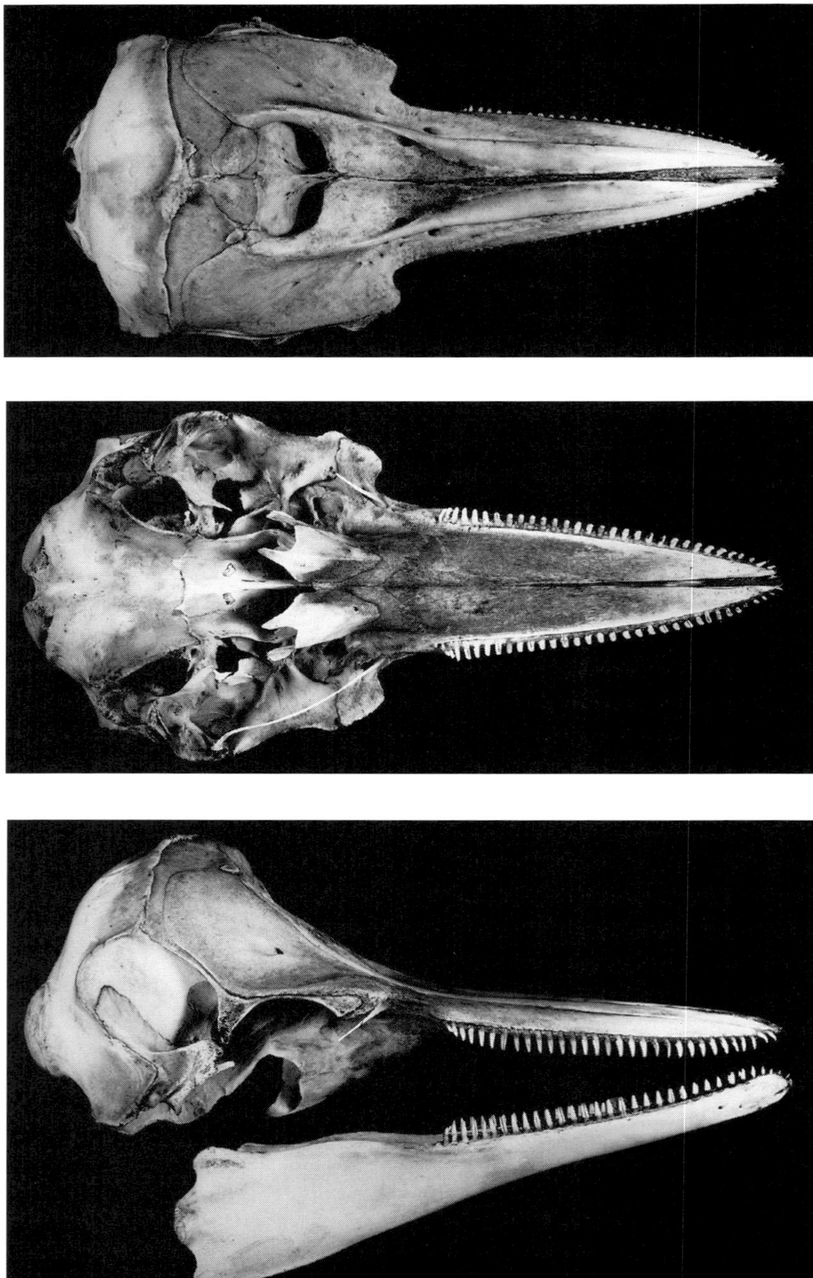

FIG. 4 Ventral (A) and dorsal (B) views of a dusky dolphin skull with lateral (C) view of skull and mandible in the US National Museum, Washington, DC (USNM 550739). (Photos courtesy of the National Museum of Natural History, Smithsonian Institution)

Life History and Population Dynamics

Growth and reproduction

Manzanilla (1989) reported on the reproductive status of dusky dolphins (26 females, 14 males) landed in the Peruvian ports of Pucusana and Cerro Azul between 1985 and 1986. Eight immature females ranged in total length from 127 cm to 179.5 cm. The three largest immature females (174, 177, 179.5 cm) had developing follicles in their ovaries. The remaining 18 females were all mature and ranged in total length from 173 cm to 193 cm. The age (based on dentine layers) for all adult females was 4.0–11.5 years (Manzanilla, 1989). Two immature males measured 141 cm and 172 cm. Twelve mature males, with seminal fluid in the epididymis, ranged in total length from 176 cm to 206 cm. The largest known male from Peru was 211 cm (Brownell, 1974).

Van Waerebeek and Read (1994) analysed growth and reproductive parameters of 852 dusky dolphins (522 females, 330 males) captured in a directed fishery off the central coast of Peru during 1985–1990. Another study of the same specimens indicated that female and male dolphins reached sexual maturity at 72.1 kg/174.9 cm and 75.0 kg/175.2 cm, respectively, at about 4–6 years of age (Van Waerebeek, 1992). The reproductive cycle of dusky dolphins from Peru totalled about 28.6 months, including 12.9 months gestation, 12.0 months lactation, and a resting period of 3.7 months. The main calving season in Peru extended from August to October, with some neonates collected in February; mean size at birth was 91.2 cm with a weight of 9.6 kg.

Webber (1987) and Cipriano (1992) analysed life-history characters of beach-cast and incidentally captured dusky dolphins from New Zealand. The largest of nineteen adult *L. obscurus* was 186 cm and weighed 77 kg (Cipriano, 1992). The maximum age of 32 New Zealand specimens was 35–36 years, while the second oldest was 26–30 years old (Webber, 1987). All the adult specimens examined by Webber (1987) had an age greater than 7 GLGs. Age of first reproduction of dusky dolphins from New Zealand was estimated as very approximately 7–8 years at 160–165 cm total length (Cipriano, 1992). A 66.5 cm foetus weighed 4 kg (Cipriano, 1992). Small calves were collected or sighted during boating surveys only between November and January, suggesting a limited calving period which also coincided with the months that dolphin groups were found closest to shore (Cipriano, 1992). In Golfo San José, Argentina, calves of *L. obscurus* also were born mainly during the summer (Würsig and Würsig, 1980).

Best (1976) reported that a 173 cm subadult male dusky dolphin from South African waters had a combined testes weight of only 134 g and that adult males have combined testes weights up to 1.6 kg. Five of the nine adult males collected by Cipriano (1992) in New Zealand had semen present in the epididymis and enlarged testes (995–2143 g each, including epididymis) and were collected in spring or summer months. The other four adult males collected in fall or winter (165–175 cm, minimum age 12–26 years) had smaller (252–479 g) and apparently inactive testes (Cipriano, 1992). Testis weight of juvenile and subadult males (130–160 cm) from New Zealand was 12.0–663 g (Cipriano, 1992). Testis weight (including epididymis) of 125 *L. obscurus* from Peru analysed by Van Waerebeek and Read (1994) was 53–5120 g (left) and 53–4930 g (right), with a maximum combined weight of 9730 g. Testis weight in these Peruvian dolphins increased in August, peaked in September–October, and then decreased again in November, in synchrony with the timing of ovulation and conception.

These relatively large testes in *L. obscurus* are similar in size to other species of *Lagenorhynchus* (see also Chapters 1, 2, 3, 5, 6).

Abundance

Little information is available on the abundance of *L. obscurus* throughout its range. Along the coast of Argentina, "several thousand (exact number unknown)" dusky dolphins are known to occur (Würsig and Bastida, 1986).

During the Southern Hemisphere minke whale assessment cruises (IWC/IDCR) between 1978/79 and 1987/88, a total of 2665 individuals in 27 schools of dusky dolphins were observed (Kasamatsu *et al.*, 1990). These observations were made while in transit between home ports and the Antarctic. No abundance estimates have been calculated for this species from the IWC/IDCR data base.

Behaviour

Social organization

School size of the 27 dusky dolphin schools encountered in transit by Kasamatsu *et al.* (1990) was fairly variable, with a range of 2–500 and a mean school size of 98.7 individuals.

Dusky dolphin school size in Golfo San José, Argentina, was in the range 6–300, with small groups of about 15 individuals that coalesced during bouts of cooperative feeding (Würsig and Würsig, 1980), although only 28% of the large feeding groups reported had more than 50 individuals. During winter months, groups with less than 20 individuals were more common than at other times of year (Würsig and Würsig, 1980). Stable subgroups were observed within a more fluid society of changing group size (Würsig and Würsig, 1980) and probably displayed a high degree of individual-to-individual fidelity (Würsig and Bastida, 1986).

School sizes of dusky dolphins off Kaikoura, New Zealand were 2–1000, and varied seasonally, with a mean school size of 50–100 in spring, summer and fall and 150–175 in winter months (Cipriano, 1992).

Movements

Dusky dolphins were found during most of the year in Golfo San José, Argentina, with a seasonal low in abundance during winter and a peak in summer (Würsig and Würsig, 1980). In summer, these dolphins were also found more often in deeper water near the mouth of the bay, at a time when a major prey species (southern anchovy, *Engraulis anchoita*) is probably moving into deeper water.

In the Kaikoura area off South Island, New Zealand, subgroups were usually closely associated within the larger school "envelope" (Würsig *et al.*, 1991). The school consisted of 50 to several hundred dolphins that usually travelled as a unit, sometimes covering several square kilometres. In summer, the dolphins moved from nearshore to offshore waters during the course of the day, apparently feeding on mesopelagic fishes in deep water during evening and night (Würsig *et al.*, 1991) but consistently remained closer to shore than during winter months (Cipriano, 1992).

Long range movements in this species were reported off Argentina (Würsig and Bastida, 1986). Two dolphins in the same school were tagged with spaghetti tags on 3 January 1975 in Golfo San José. One and five days after tagging, the two dolphins were sighted approximately 20 km and 35 km from the tagging site, respectively. On 1 December 1982, both dolphins were observed swimming side by side in a school of approximately 150 dusky dolphins about 10 km off Mar del Plata, approximately 780 km northeast of the original tagging location.

Feeding

In central Peruvian waters, the anchoveta, *Engraulis ringens*, comprised 88% by weight and 91% of the energy of the ingested prey of dusky dolphins (McKinnon, 1988). Other fish remains found in these dolphins included: jack mackerel, *Trachurus trachurus*, sardine, *Sardinops sagax*, and hake, *Merluccius gayi*. Remains of Patagonian squid, *Loligo gahi*, and jumbo flying squid, *Dosidicas gigas*, also were found in the stomachs of these dolphins. One adult male (TL 200 cm) collected on the beach at San Andrés, Peru, in February 1976 had 417 otoliths from sculpins, *Normanichthys crockeri* (Cottidae), in its stomach (R.L. Brownell, Jr, unpublished data). A few otoliths from *Trachurus* sp. and *Merluccius* sp., and some unidentified myctophids also were present, along with 144 unidentifiable otoliths that were small and badly worn. In Golfo San José, Chubut, Argentina, the major prey for these dolphins in non-winter months was the southern anchovy, *E. anchoita* (Würsig and Würsig, 1980).

In New Zealand waters, remains of small squid, *Nototodarus sloani*, were collected from stranded animals near Wellington (Gaskin, 1972). Stomachs from

24 dusky dolphins incidentally killed in fishing operations in New Zealand waters contained remains of mesopelagic fishes, mainly myctophids and Hoki (*Macruronus novaezelandiae*), and squids (*Nototodarus* spp., *Moroteuthopsis* spp. and *Teuthowenia* spp.) (Cipriano, 1989, 1992).

Association with other species

Dusky dolphins have been observed in mixed cetacean schools with southern right whale dolphins (*Lissodelphis peronii*) off Namibia (Cruickshank and Brown, 1981; Rose and Payne, 1991). In summer, *L. obscurus* groups off Kaikoura, New Zealand were occasionally accompanied by small groups of common dolphins (*Delphinus delphis*), which travelled as a cohesive subgroup within the larger dusky dolphin group (Cipriano, 1992). Dusky dolphins also were observed with pilot whales (*Globicephala* sp.) off Southwest Africa and the Prince Edward Islands (Cruickshank and Brown, 1981).

Swimming and diving

Mean dive time for 10 radiotagged dolphins off Argentina was only 21.0 sec, and the number of long dives (> 90 sec), probably associated with feeding, peaked in mid-day to afternoon in summer (Würsig, 1982).

Subgroups of dusky dolphins within larger schools off New Zealand sometimes were observed to dive synchronously, and occasionally almost the entire group would be underwater for several minutes (Würsig *et al.*, 1991; Cipriano, 1992). Most dives of nine radiotagged dolphins off New Zealand were 10–20 sec long, and maximum dive length rarely exceeded 150 sec (Cipriano, 1992). In winter, numbers of short, medium and long dives did not vary significantly over the course of the day, but in summer more long dives (60–180 sec) were made at night (Cipriano, 1992).

Dusky dolphins may be the most aerially acrobatic of any delphinid species; individuals may engage in bouts of 2–38 leaps, all of the same type, during one bout, especially after feeding (Würsig and Würsig, 1980).

Sound production

No sound analysis from this species has yet been published, but Schevill and Watkins (1971) noted that whistlelike squeals from *L. obscurus* were recorded in New Zealand waters. Sounds of dusky dolphins recorded in Argentina and New Zealand have been analysed recently, showing that their whistles and click trains are quite similar to those of *L. obliquidens* (Wang Ding and B. Würsig, personal communication).

Parasites and Disease

Three genera of endoparasites from stomachs of *L. obscurus* were reported by Gibson and Harris (1979): cestodes, *Trigonocotyle prudhoei* and *Phyllobothrium delphini*; and nematodes, *Anisakis typica* and *Anisakis* sp. Cipriano (1992) reported nematode, cestode and trematode parasites from dusky dolphins collected in New Zealand. Nematodes were found in *L. obscurus* intestines and stomachs and other types (*Stenurus* sp.) in upper respiratory tracts, ear capsules, pterygoid sinuses and lung tissue (Cipriano, 1992). Cestodes were found in stomachs and heart tissue, and plerocercoid cysts (probably *Phyllobothrium* sp.) were common in the blubber of most adult specimens (Cipriano, 1992). Trematodes were found in intestines, stomachs and liver tissue, and *Braunina cordiformis* was often attached to the lining of the second stomach (Cipriano, 1992). Nothing has been recorded about diseases in wild or captive dusky dolphins.

Live Maintenance

Live-capture of small cetaceans in South African waters started in 1961, and between 1961 and 1981, 37 of 106 dolphins obtained were *L. obscurus* (Best and Ross, 1977, 1984). Thirteen of these dolphins were released and five died at capture. Fifteen of the remaining 19 dolphins were maintained in South African oceanaria. Collet (1984) reported that the other four individuals were exported to Europe from South Africa prior to 1975. Best and Ross (1984) did not provide any details on the shipment. One of the South African dolphins lived at 2640 days (7.2 years). Gaskin (1972) and Baker (1983) noted that this species was kept at Marineland of New Zealand in Napier. Cawthorn and Gaskin (1984) provided details on 27 *L. obscurus* held at the Napier Marineland. All these dolphins were obtained from Hawke Bay, New Zealand. Three of the dolphins were released and the other 24 died. The greatest longevity was 2.5 years.

Human Effects

Directed fisheries

An expanded directed fishery for dolphins and porpoises may have started in Peru after the demise of the anchoveta fishery in 1972 (Read *et al.*, 1988). The commercial value of these small cetaceans has presumably stimulated increasing catches. Most of the small cetaceans are taken in the directed net fishery for dusky dolphins (Read *et al.*, 1988). Dusky dolphins are also taken by a harpoon fishery in Peru for human consumption (Van Waerebeek and Reyes, 1990).

By-catches

The incidental catch of dusky dolphins from Chile and Peru has been known for some time (Bini, 1951; Clarke, 1962; Grimwood, 1969; Mitchell, 1975). Early records indicated that small cetaceans were taken incidental to other fishing operations (Mitchell, 1975). Recently, the majority of dusky dolphins landed have been intentionally captured in a drift net fishery that targets this species (Read *et al.*, 1988). The commercial value of small cetaceans presumably stimulated deliberate catches of these animals for human consumption, particularly after the collapse of the anchoveta fishery in 1972 (Read *et al.*, 1988). Large catches (approximately 10 000) of small cetaceans were reported from the coastal waters of central Peru in 1985 (Read *et al.*, 1988). Large catches of this species continued through the 1980s (Van Waerebeek and Reyes, 1990). Guerra *et al.* (1987) reported the capture of a single specimen "en los aparejos de pesca de bacalao *Dissostichus eleginoides*". M.A. Webber and F. Cipriano (unpublished data) estimated that approximately 200 dusky dolphins were incidentally captured off Kaikoura, New Zealand, in 1984.

Pollution

The maximum concentration (ppm wet weight) of DDT reported in the blubber of this species in New Zealand waters was 175 (Koeman, 1973).

Acknowledgments

M.A. Webber, Bernd Würsig and Tom Jefferson read early drafts of this paper and provided useful comments. Meghan A. Donahue edited the manuscript.

References

Allen, G. M. (1939). A checklist of African mammals. *Bull. Mus. Comp. Zool.* **83**, 249–258.

Andrade, V. H. and Baez, R. P. (1980). Presencia del delfin listado: *Lagenorhynchus obscurus* (Gray, 1828) en la costa del Valparaiso (Mammalia, Cetacea, Delphinidae). *Notic. Mens. Mus. Hist. Nat. (Santiago)* **25**, 7–9.

Baker, A. N. (1977). Spectacled porpoise, *Phocoena dioptrica*, new to the subantarctic Pacific Ocean. *N. Z. J. Mar. Freshwater Res.* **11**, 401–406.

Baker, A. N. (1983). "Whales and Dolphins of New Zealand and Australia: an Identification Guide". Victoria University Press, Wellington.

Best, P. B. (1976). Tetracycline marking and the rate of growth layer formation in the teeth of a dolphin (*Lagenorhynchus obscurus*). *South African J. Sci.* **72**, 216–218.

Best, P. B. and Ross, G. J. B. (1977). Exploitation of small cetaceans off southern Africa. *Rep. Int. Whal. Commn* **27**, 494–497.

Best, P. B. and Ross, G. J. B. (1984). Live-capture fishery for small cetaceans in South African waters. *Rep. Int. Whal. Commn* **34**, 615–618.

Bierman, W. H. and Slijper, E. J. (1948). Remarks upon the species of the genus *Lagenorhynchus*, Pt. II. *Proc. Koninkl. Ned. Akad. Wet.* **51**, 127–133.

Bini, G. (1951). Osservazioni su alcuni mammiferi marini sulle coste del Cile e del Perù. *Bolletino di Pesca, Piscicoltura e Idrobiologia* (Nuova Serie), **6**, 5–19.

Broekema, J. W. (1983). Catalogue of cetacea in the collection of the Rijksmuseum van Natuurlijke, Leiden. *Zool. Mededelingen* **57**, 67–79.

Brownell Jr., R. L. (1974). "Small Odontocetes of the Antarctic", Antarctic Map Folio Series (Ed. V. C. Bushnell), Folio 18, "Antarctic Mammals". American Geographic Society, New York.

Brownell Jr., R. L. and Mead, J. G. (1989). Taxonomic status of the delphinid (Mammalia: Cetacea) *Tursio? panope* Philippi, 1895. *Proc. Biol. Soc. Wash.* **102**, 532–534.

Cárdenas, J. C., Stutzin, M. E., Oporto, J. A., Cabello, C. and Torres, D. (1986). "Manual de Identificacion de los Cetaceos Chilenos". Project WW-445, World Wildlife Fund, U.S., and Comite Nacional Pro Defensa de la Fauna y Flora (Chile), 102 pp.

Cawthorn, M. W. and Gaskin, D. E. (1984). Small cetaceans held in captivity in Australia and New Zealand. *Rep. Int. Whal. Commn* **34**, 613–614.

Cipriano, F. (1989). Prey types, prey distribution and foraging ecology of dusky dolphins (*Lagenorhynchus obscurus*) off Kaikoura, New Zealand. *In* Abstracts of the Eighth Biennial Conference on the Biology of Marine Mammals, Pacific Grove, CA.

Cipriano, F. (1992). Behavior and occurrence patterns, feeding ecology, and life history of Dusky dolphins (*Lagenorhynchus obscurus*) off Kaikoura, New Zealand. Ph.D dissertation, University of Arizona, Tucson.

Clarke, R. (1962). Whale observation and whale marking off the coast of Chile in 1958 and from Ecuador towards and beyond the Galapagos Islands in 1959. *Norsk Hvalfangst-tid.* **7**, 265–287.

Collet, A. (1984). Live-capture of cetaceans for European institutions. *Rep. Int. Whal. Commn* **34**, 603–607.

Cruickshank, R. A. and Brown, S. G. (1981). Recent observations and some historical records of southern right whale dolphins *Lissodelphis peronii*. *Fish. Bull. S. Afr.* **15**, 109–122.

Davies, J. L. (1963). The antitropical factor in cetacean speciation. *Evolution* **17**, 107–116.

Ellerman, J. R., Morrison-Scott, T. C. S. and Hayman, R. W. (1953). "Southern African Mammals 1758 to 1951: A Re-classification". British Museum, London.

Flower, W. H. (1885). List of the specimens of Cetacea in the Zoological Department of the British Museum, London.

Fraser, F. C. (1964). Whales and whaling. *In* "Antarctic Research: A Review of British Scientific Achievement in the Antarctic" (Eds R. Priestly, R. J. Adie, and G. de Q. Robin) pp. 191–205. Butterworths, London.

Fraser, F. C. (1966). Comments on the Delphinoidea. *In* "Whales, Dolphins, and Porpoises" (Ed. K. S. Norris) pp. 7–31. University of California Press, Berkeley, CA.

Fraser, F. C. and Purves, P. E. (1960). Hearing in cetaceans. *Bull. Brit. Mus. (Nat. Hist.)* **7**, 1–140.

Gallardo, A. (1912). El delfin *Lagenorhynchus fitzroyi* (Waterhouse) Flower capturado en Mar del Plata. *Anal. Mus. Nac. His. Nat. Buenos Aires* **23**, 391–398.

Gallardo, A. (1913). Notas sobre la anatomia del aparato espiracular, laringe y hioides de dos delfines: *Phocaena dioptrica* Lahille y *Lagenorhynchus fitzroyi* (Waterhouse) Flower. *Anal. Mus. Nac. His. Nat. Buenos Aires* **24**, 235–246.

Gaskin, D. E. (1968a). Distribution of Delphinidae (Cetacea) in relation to sea surface temperatures off eastern and southern New Zealand. *N.Z. J. Mar. Freshwater Res.* **2**, 527–534.

Gaskin, D. E. (1968b). The New Zealand Cetacea. *Fish. Res. Bull.* No. 1 (New Series), 1–92.

Gaskin, D. E. (1972). "Whales, Dolphins and Seals with Special Reference to the New Zealand Region". Heinemann Educational Books, London.

Gibson, D. I. and Harris, E. A. (1979). The helminth-parasites of cetaceans in the collection of the British Museum (Natural History). *Invest. Cetacea* **10**, 309–324.

Goodall, R. N. P. (1978). Report on the small cetaceans stranded on the coasts of Tierra del Fuego. *Sci. Rep. Whales Res. Inst. Tokyo* **30**, 197–230.

Goodall, R. N. P. (1989). The last whales of Tierra del Fuego. *Oceanus* **32**, 89–95.

Gray, J. E. (1828). "Spicilegia Zoologica; or Original Figures and Short Systematic Descriptions of New and Unfigured Animals". Trevittel, Wury and Co., London.

Gray, J. E. (1868). Notice of *Clymene similis*, a new dolphin sent from the Cape by Mr. Layard. *Proc. Zool. Soc. Lond.*, 146–149.

Grimwood, I. R. (1969). Notes on the distribution and status of some Peruvian mammals 1968. *New York Zoological Society Spec. Publication* No. 21, pp. 1–86.

Guerra Correa, C., Van Waerebeek, K., Portflitt Kandora, G. and Luna Jorquera, G. (1987). Presencia de cetaceos frente a la segunda region de Chile. *Estud. Oceanol.* **6**, 87–96.

Guiler, E. R. (1978). Whale strandings in Tasmania since 1945 with notes on some seal reports. *Papers Proc. Roy. Soc. Tasmania* **112**, 189–213.

Hershkovitz, P. (1966). Catalog of living whales. *Bull. U.S. Nat. Mus.* **246**, 1–259.

Kasamatsu, F., Joyce, G., Ensor, P. and Mermoz, J. (1990). Current occurrence of cetacea in the southern hemisphere minke whale assessment cruises, 1978/79–1987/88. Paper SC/42/015, presented to the International Whaling Commission Scientific Committee.

Kasuya, T. (1973). Systematic consideration of recent toothed whales based on the morphology of tympano-periotic bone. *Sci. Rep. Whales Res. Inst. Tokyo* **25**, 1–103 + 28 plts.

Kellogg, R. (1941). On the identity of the porpoise *Sagmatias amblodon*. *Field Mus. Nat. Hist. Zool. Ser.* **27**, 293–311 + 6 plts.

Koeman, J. H. (1973). PCB in mammals and birds in the Netherlands. *In* "PCB Conference II" (Eds E. Ahnland and A. Akerblom) pp. 35–43. National Swedish Environmental Protection Board, Publication 1973: 4E, Stockholm.

Lesson, R. P. and Garnot, P. (1826). Zoologie. *In* "Voyage autour du Monde, Exécuté par Ordre du Roi, sur la Corvette de Sa Majesté, *La Coquille*, pendant les Années 1822, 1823, 1824 et 1825", Vol. 1, Part 1. Arthus Bertrand, Paris.

Lillie, D. G. (1915). Cetacea. *British Antarctic "Terra Nova" Rep. Zool.* **1**, 85–124.

Manzanilla, S. R. (1989). The 1982–1983 El Niño event recorded in dentinal growth layers in teeth of Peruvian dusky dolphins (*Lagenorhynchus obscurus*). *Can. J. Zool.* **67**, 2120–2125.

McKinnon, J. (1988). Feeding habits of three species of small cetaceans from the coastal water of Peru. M.Sc. thesis, University of Guelph, Ontario.

Mead, J. G. (1975). Anatomy of the external nasal passages and facial complex in the Delphinidae (Mammalia: Cetacea). *Smithsonian Contrib. Zool.* **207**, 1–72.

Mitchell, E. (1970). Pigmentation pattern evolution in delphinid cetaceans: an essay in adaptive coloration. *Can. J. Zool.* **48**, 717–740.

Mitchell, E. (1975). Porpoise, dolphin and small whale fisheries of the world: Status and problems. International Union for Conservation of Nature and Natural Resources Monograph No. 3, pp. 1–129.

Nicol, D. J. and Croome, R. L. (1988). Trends in the Tasmanian cetacean stranding record. *In* "Marine Mammals of Australasia, Field Biology and Captive Management" (Ed. M. L. Augee) pp. 59–70. Royal Zoological Society of New South Wales, Sydney.

Paulian, P. (1953). Pinnepedes, cetaces, oiseaux des Iles Kerguelen et Amsterdam, Mission Kerguelen 1951. *Memn. Inst. Sci. Madagascar*, Ser. A., **8**, 111–234.

Pearson, J. (1935). The whales and dolphins of Tasmania. *Papers Proc. Roy. Soc. Tasmania* **1935**, 163–192.

Piana, E., Orquera, L., Goodall, R. N. P. , Galeazzi, A. R. and Sobral, A. P. (1985). Cetacean remains in Beagle Channel shell middens. Abstract, Sixth Biennial Conference on the Biology of Marine Mammals, Vancouver, British Columbia, Canada.

Philippi, R. A. (1893). Los delfines de la Punta Austral de la America del Sur. *Anal. Mus. Nac. Chile Sec. 1 Zool.*, entr. **6**, 1–18.

Philippi, R. A. (1895). Los delfines Chilenos. *Anal. Univ. Chile, Santiago*, 281–285.

Philippi, R. A. (1896). Los craneos de los delfines. *Anal. Mus. Nac. Chile Sec. 1 Zool.* **12**, 1–18 + 6 plts.

Read, A. J., Van Waerebeek, K., Reyes, J. C., McKinnon, J. S. and Lehman, L. C. (1988). The exploitation of small cetaceans in coastal Peru. *Biol. Conserv.* **46**, 53–70.

Rice, D. W. (1977). A list of the marine mammals of the world. U.S. Department of Commerce, NOAA Technical Report NMFS SSRF-711.

Rice, D. W. and Scheffer, V. B. (1968). A list of the marine mammals of the world. US Fish and Wildlife Service Special Science Report Fisheries No. 579, pp. 1–16.

Roberts, A. (1951). "The Mammals of South Africa". Trustees of the "Mammals of South Africa" Book Fund, Johannesburg, pp. 213–232.

Robineau, D. (1989). Les cétacés des Iles Kerguelen. *Mammalia* **53**, 265–278.

Robineau, D. (1990). Les types de cétacés actuels du Museum National d'Histoire Naturelle II. Delphinidae, Phocoenidae. *Bull. Mus. Natl. Hist. Nat., Paris* Ser. 4, **12**, Sect. A 1, 197–238.

Rose, B. and Payne, A. I. L. (1991). Occurrence and behavior of the southern right whale dolphin *Lissodelphis peronii* off Namibia. *Mar. Mamm. Sci.* **7**, 25–34.

Scheffer, V. B. and Rice, D. W. (1963). A list of the marine mammals of the world. US Fish and Wildlife Service Special Science Report Fisheries No. 431 [1–2].

Schenkkan, E. J. (1973). On the comparative anatomy and function of the nasal tract in odontocetes (Mammalia, Cetacea). *Bijdragen tot de Dierkunde* **43** 127–159.

Schevill, W. E. and Watkins, W. A. (1971). Pulsed sounds of the porpoise *Lagenorhynchus australis*. *Breviora* **366**, 1–10.

Schlegel, H. (1841). "Abhandlungen aus dem Gebiete der Zoologie und Vergleichenden Anatomie", 1 Heft. A. Arnz und Comp, Leiden.

von Schreber, J. C. D. and Wagner, J. A. (1846). "Die Saugethiere in Abbildungen", 7, Erlangen, 427 pp.

Taylor, C. K. and Saayman, G. S. (1972). A method for determining the composition, deployment and stability of groups of free-ranging dolphins. *Z. Saugtierkunde* **37**, 116–119.

True, F. W. (1889). Contributions to the natural history of the cetaceans: A review of the Family Delphinidae. *Bull. U.S. Nat. Mus.* **36**, 1–191 + 47 plts.

True, F. W. (1903). On species of South American Delphinidae described by Dr. R. A. Philippi in 1893 and 1896. *Proc. Biol. Soc. Washington* **16**, 133–144.

Van Waerebeek, K. (1992). Un analisis de la reproduccion del Delfin oscuro *Lagenorhynchus obscurus* en aguas peruanas. *In* Abstracts of the 5ta Reunión de Especialistas en Mamíferos Acuáticos de América del Sur, 28 September–2 October, 1992, Buenos Aires, Argentina.

Van Waerebeek, K. (1993a). Presumed *Lagenorhynchus* skull at Tasmanian Museum re-identified as *Lissodelphis peronii*. *Austr. Mamm.* **16**, 41–43.

Van Waerebeek, K. (1993b). External features of the dusky dolphin *Lagenorhynchus obscurus* (Gray, 1828) from Peruvian waters. *Estud. Oceanol.* **12**, 37–53.

Van Waerebeek, K. and Reyes, J. C. (1990). Catch of small cetaceans at Pucusana Port, Central Peru, during 1987. *Biol. Conserv.* **51**, 15–22.

Van Waerebeek, K. and Read, A. J. (1994). Reproduction of dusky dolphins, *Lagenorhynchus obscurus*, from coastal Peru. *J. Mammal.* **75**, 1054–1062.

Van Waerebeek, K., van Bree, P. J. H. and Best, P. B. (1995). On the identity of *Prodelphinus petersii* Lütken, 1889 and records of dusky dolphins *Lagenorhynchus obscurus* (Gray, 1828) from the southern Indian and Atlantic Oceans. *S. Afr. J. Mar. Sci.* **16**, 25–35.

Waterhouse, G. R. (1838). [A new species of *Delphinus*]. *Proc. Zool. Soc. Lond.*, 23–24.

Webber, M. A. (1987). A comparison of dusky and Pacific white-sided dolphins (Genus *Lagenorhynchus*): Morphology and Distribution. M.S. thesis, San Francisco State University, San Francisco, CA.

Würsig, B. (1982). Radio tracking dusky porpoises in the South Atlantic. *In* "Mammals in the Seas", FAO Fisheries Series No. 5, Volume IV. United Nations Food and Agriculture Organization, Rome, Italy.

Würsig, B. and Bastida, R. (1986). Long-range movement and individual association of two dusky dolphins (*Lagenorhynchus obscurus*) off Argentina. *J. Mamm.* **67**, 773–774.

Würsig, B., Cipriano, F. and Würsig, M. (1991). Dolphin movement patterns: Information from radio and theodolite tracking studies. *In* "Dolphin Societies: Discoveries and Puzzles" (Eds K. Pryor and K. S. Norris) pp. 79–111. University of California Press, Berkeley, CA.

Würsig, B. and Würsig, M. (1980). Behavior and ecology of the dusky dolphin, *Lagenorhynchus obscurus*, in the South Atlantic. *Fish. Bull.* **77**, 871–890.

5

Peale's Dolphin
Lagenorhynchus australis (Peale, 1848)

Robert L. Brownell, Jr, Enrique A. Crespo and Meghan A. Donahue

Genus and Species

Taxonomy

This species was described originally by Titian R. Peale as *Phocaena australis*, based solely on the external form and coloration of a single specimen. The type specimen was collected on 12 February 1839 off the coast of Patagonia, one day's sail north of the Straits of Le Maire between Staten Island and Cape San Diego, Tierra del Fuego, Argentina (Peale, 1848). The dolphin was harpooned from either the U.S.S. *Peacock* or the U.S.S. *Vincennes* of the United States Exploring Expedition but, unfortunately, "nothing is known about this specimen" (Poole and Schantz, 1942). No mention of any material collected from *Phocaena australis* is found in the list of specimens collected by the U.S. Exploring Expedition (Peale, 1848).

Cope (1866) described a new species of dolphin as *Sagmatias amblodon* based on a skull of an old adult animal in the Smithsonian Institution (USNM 3887). The type locality is unknown. This dolphin was collected from the U.S.S. *Vincennes*, U.S. Exploring Expedition, 1838–1842. Kellogg (1941) discussed all of the facts related to *S. amblodon* and stated that the skull may actually be from the type of *P. australis* taken off the coast of Patagonia by the U.S. Exploring

Handbook of Marine Mammals, Volume 6
ISBN 0-12-588506-7

Expedition. Cassin (1858) combined *P. australis* with *Delphinus obscurus*. Kellogg (1941) was the first to use the presently recognized combination of *Lagenorhynchus australis*.

Philippi (1900) described a skull of a dolphin from Ancud (41° 53′S, 14° 13′E), Chiloe, Chile as a new species, *Tursio chiloensis*. Additional details of this specimen were reported by Philippi (1901). Several authors (Hershkovitz, 1966; Donoso-Barros, 1975; Sielfeld, 1983) considered *T. chiloensis* to be a junior synonym of *Lissodelphis peronii*. Goodall (1986) and Van Waerebeek *et al.* (1991) have re-examined the type of *T. chiloensis* and placed it in the synonymy of *L. australis*. The holotype (number MNHN-S 586) is in Museo Nacional de Historia Natural in Santiago, Chile.

Bierman and Slijper (1948) proposed that all Southern Hemisphere species of *Lagenorhynchus* probably belonged to a single species, *L. cruciger*. This concept of a single species was followed by various authors until Fraser (1966) demonstrated clearly that three species should be recognized in the Southern Hemisphere: *L. cruciger*, *L. obscurus*, and *L. australis* (see additional discussion in Chapter 6 on *L. cruciger* and Chapter 4 on *L. obscurus*).

Common names

Peale's dolphin is known also as the black-chinned dolphin in English, or in South America as the delfin austral or llampa (Aguayo, 1975).

Distribution

Range

Peale's dolphins are found mainly in the coastal waters of southern South America (Fig. 1), especially in the central Strait of Magellan and the fjords of southern Chile (Mann, 1958; Cabrera, 1961; Gilmore, 1969, 1971; Cummings *et al.*, 1971; Sielfeld and Venegas, 1978; Oporto, 1984; Kasamatsu *et al.*, 1990). The most northern observations made in coastal waters along the western and eastern coasts of South America are Valparaiso (33°S), Chile (Aguayo, 1975) and Comodoro Rivadavia (45° 52′S, 67° 29′W) (Brownell, 1974), respectively.

However, one stranded specimen is from farther north at Miramar (38° 15′S, 57° 50′W), Pcia. Buenos Aires, Argentina (Brownell, 1974). In November and December of 1996, one of us (E.A.C.) surveyed the area of Cabo dos Bahias (44° 56′S, 55° 31′W) and sighted 12 schools of *L. australis* close to Isla Arce, Isla Rasa, Caleta Sara, Caleta Horno, Isla Leones and Isolote Moreno (E.A. Crespo and H. L. Capozzo, unpublished data).

These dolphins are observed regularly around the coast of the Falkland Islands (Hamilton, 1952) and over Burdwood Bank, south of the Falklands (Webber and Leatherwood, 1991).

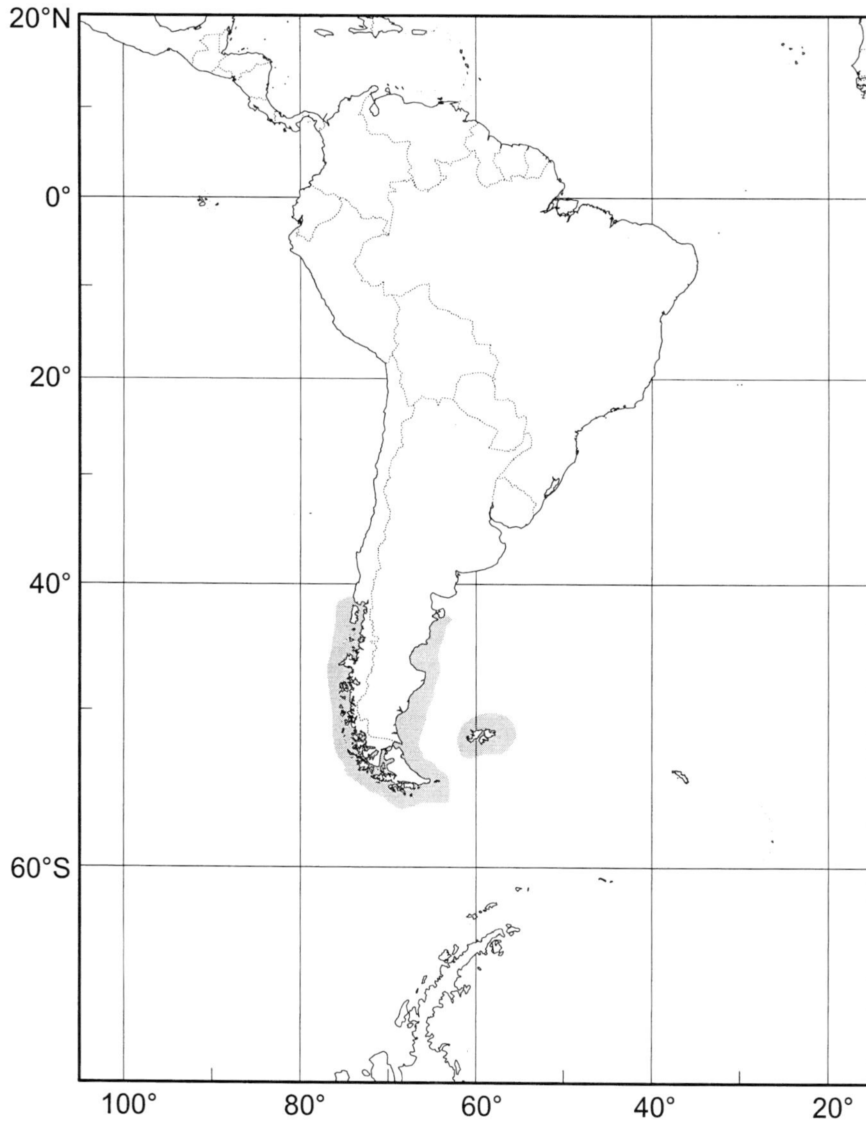

FIG. 1 Distribution of the Peale's dolphin, *Lagenorhynchus australis*.

One poorly documented offshore sighting of *L. australis* was recorded 220 nautical miles off the central Argentine coast at approximately 45° 21′S, 60° 16′W (Hanson and Erickson, 1985). The identity of this sighting remains uncertain.

Recent sightings and photos of a small group of dolphins in March 1988 near Palmerson Atoll (Cook Islands) in the South Pacific have been tentatively identified as *L. australis* (Webber and Leatherwood, 1991; Leatherwood *et al.*, 1991;

Van Waerebeek *et al.*, 1995). This observation extends the range of this species well outside its previous range and into much warmer waters. It is possible that this record simply represents an anomalous occurrence, perhaps related to a recent El Niño event or an undescribed species.

Peale's and dusky dolphins have been considered to be allopatric (Brownell, 1974; Mitchell, 1975). Between 42°S and 52°S on the Atlantic coast of Argentina both species have been observed; however, Peale's dolphins seem to be more common south of 47°S. Nevertheless, this species is seen frequently along the coast of southern Chubut, Argentina, which consists of many islands and peninsulas.

Habitat

This species inhabits waters very near to shore, commonly within or shoreward of kelp beds (*Macrocystis pyrifera*) (Leatherwood *et al.*, 1991; de Haro and Iñíguez, 1997; Lescrauwaet, 1997). In the waters of southern Chile and Tierra del Fuego they appear to prefer channels, fjords and deep bays (Goodall, 1978; Leatherwood and Reeves, 1983). Throughout the northern part of its range, this species inhabits the waters of the wide continental shelf off Argentina and the narrower shelf off Chile (Goodall *et al.*, 1997a). Although Peale's dolphins have been observed in waters at least 300 m deep, they appear to prefer shallower coastal waters (Iñíguez, 1991; Goodall *et al.*, 1997a).

Despite the wide overlap in distribution between *L. australis* and *L. obscurus*, the two species appear to differ in their habitat use, with *L. obscurus* having a wider offshore distribution.

External Characteristics

Colour pattern

Very few photographs have been published of *L. australis* (Kellogg, 1941; Aguayo, 1975; Webber and Leatherwood, 1991). The general colour pattern is illustrated in Fig. 2. Three main pigmentation areas distinguish *L. australis* (Fig. 2) from the other two southern hemisphere species of *Lagenorhynchus* (Fraser, 1966; Mitchell, 1970). These are the black lower lips, chin, and throat coloration (approximately posterior to the position of the eyes), simple flank patch without the dorsal and ventral flank blazes found in *L. obscurus* (whitish in colour), and the extension of the abdominal field (white in colour) pigmentation above the pectoral flippers entering at the axialla. The distinctive axillary marks are found also in *Cephalorhynchus eutropia* (Norris and Dohl, 1980). Other differences between *L. australis* and *L. obscurus* are (in *australis*): thoracic patch darker in colour; line of demarcation present (eye to anus stripe); posterior half of dorsal fin darker in colour.

De Haro and Iñíguez (1997) observed 58 *L. australis* calves, all of which had a lighter grey colour pattern than the adult animals.

Size and shape

Maximum size recorded for this species is a 216 cm male from Isla de Chiloe, Chile (FMNH 22248). However, Lichter (1992) refers to a 218 cm male specimen taken incidentally in Golfo San Jorge, Santa Cruz, Argentina, but no further details on this animal are provided. The largest female specimen is 210 cm long from 46° 25′S, 65° 50′W (Claver *et al.*, 1992). The general body shape resembles other members of *Lagenorhynchus* (Fig. 2). Goodall *et al.* (1997b) summarize lengths for 21 female and nine male specimens collected around Tierra del Fuego. The females ranged from 98 cm, an animal thought to be a foetus or neonate, to 210 cm. Male specimens ranged from 138 cm to 218 cm, the largest being the animal reported in Lichter (1992).

The snout is pointed, but the rostrum is not as clearly distinct as in other snouted delphinids, with the exception of *Cephalorhynchus* spp. The dorsal fin is well developed and falcate in shape.

Weight

A juvenile male Peale's dolphin measuring 138 cm and weighing approximately 39.5 kg and a 130 cm female weighing approximately 32 kg stranded in Tierra del Fuego (Goodall *et al.*, 1997b). Ovarian weights (in g) for three female specimens reported in Claver *et al.* (1992) appear to have been reported erroneously as body weights (in kg) in Iñíguez *et al.* (1990) and subsequent papers by other authors. No other information on body weight has been published.

FIG. 2 Lateral view of Peale's dolphin. (Drawing by P.A. Folkens)

Internal Anatomy

Skull

A skull and left mandible of *L. australis* is illustrated in Fig. 3. Fraser and Noble (1968) provided the following criteria for distinguishing skulls of *L. australis* from *L. cruciger*: "(1) The post-temporal fossa of *L. australis* is larger than that of *L. cruciger*, and its long axis is differently orientated; (2) the sinus of the preor-

(A)

(B)

FIG. 3 Dorsal (A) and ventral (B) views of a 370 mm skull from a *Lagenorhynchus australis* from Rio Inio, Chile in the Field Museum of Natural History, Chicago (FMNH 22248).

(C)

(D)

FIG. 3 *Continued* Lateral view of skull (C) and left mandible (D). (Photos courtesy of the National Museum of Natural History, Smithsonian Institution)

bital lobe of the pterygoid air sac of *L. australis* is larger than that of *L. cruciger*; (3) the zygomatic process of *L. cruciger* is more massive than that of *L. australis*; and (4) the guttering in the pre-narial triangle of *L. cruciger* extends nearly to the apex of the triangle, whereas in *L. australis* it is obscured anterior to the premaxillary foramen". The condylobasal length of five specimens reported in Kellogg (1941) and Fraser and Noble (1968) ranged from 334 to 371 mm. Tooth counts for four of these specimens were 29–33 in each tooth row. Condylobasal lengths for 11 specimens collected in Tierra del Fuego ranged from 335 mm to 372 mm ($\bar{x} = 361$) (Goodall *et al.*, 1997b). Maximum upper and lower tooth row counts for these specimens were 34 and 33, respectively (Goodall *et al.*, 1997b). The four cranial measurements summarized here are from the four largest specimens described by Kellogg (1941) and Fraser and Noble (1968). Measurements ($\bar{x}$, range in mm) are: condylobasal length, 365, 357–371; rostral length, 179, 174–183; basal width of rostrum, 97, 92–105; zygomatic width, 198, 191–207; and mandible length, 299, 292–308. The periotic and tympanic bulla are illustrated in Kasuya (1973).

Postcranial skeleton

No details have been published about the postcranial skeleton. The vertebral count reported by Cárdenas *et al.* (1986) is C_7, T_{15}, L_{24}, Ca_{35}, giving a total of 81 vertebrae, is incorrect. The vertebral counts from nine specimens described by Goodall *et al.* (1997b) are 7 cervical, 13–15 thoracic, 14–16 lumbar and 28–34 caudal with totals in the range 67–70. For eight of these specimens, the phalangeal formula was reported with the following ranges: I_{2-3}, II_{6-9}, III_{5-6}, IV_{2-4}, V_{0-2} (Goodall *et al.*, 1997b).

Organs, tissues and physiology

Histological details on the skin of *L. australis* were described by Iñíguez *et al.* (1990). The epidermis is 1–2 mm in thickness (Lacolla and Iñíguez, 1991). Blubber thickness from four specimens reported by Goodall *et al.* (1997b) was 18–35 mm in the mid-dorsal region, 19–25 on the lateral surface, and 15–23 mm on the ventral surface. Claver *et al.* (1992) described anatomical and histological features of gonads from three incidentally captured females. Various organ weights from one male and two female specimens are reported in Goodall *et al.* (1997b). No other studies have been made on any organ systems or any tissues.

McFarland (1971) reported on the different spectra of gross extracts of pigment from whole retinas of several marine mammal species, including *L. australis*.

Life History and Population Dynamics

Growth and reproduction

Nothing is known about growth and reproduction in this species. The ovaries from a 210 cm Peale's dolphin were mature and the ovaries in two others (185 cm and 195 cm) were immature, with the latter considered to be at the onset of puberty (Claver *et al.*, 1992).

Abundance

No substantial information is available about the abundance of *L. australis*. However, this species is reportedly the most common cetacean found around the coast of the Falkland Islands (Hamilton, 1952) and Chile (Oporto, 1984).

Behavior

Social organization

Peale's dolphins usually are seen in small groups of 2–30 (Aguayo, 1975; Olivos and Delgado, 1990; Oporto and Brieva, 1990a; Webber and Leatherwood, 1991). Reported average school size is 3–10 (Oporto and Brieva, 1990a). De Haro and Iñíguez (1997), Lescrauwaet (1997) and E. A. Crespo and H. L. Capozzo (unpublished data) collected the following information on mean school size (range, SD, *n*): 2.114 (1–13, 1.156, 1769); 4.33 (1–20, 1.05, 714); 4 (1–10, 3.14, 12). One offshore school of fifteen Peale's dolphins was observed on 23 December 1976 off Argentina at 54° 05′S, 62° 48′W (Mermoz and Goodall, 1980).

Movements

There is no published information about the movements of this species at this time.

Feeding

In the waters around Punta Arenas, Chile, female Peale's dolphins and their calves are found around kelp beds and are believed to be foraging there (Lescrauwaet, 1990). Three *L. australis* were incidentally killed in fishing gear off southern Patagonia, Argentina (Iñíguez and de Haro, 1994). The stomachs of these specimens contained four prey species: two species of molluscs, and one species each of crustacean and fish. The most frequently encountered prey were the kingklip fish, *Genypterus blacodes*, an Argentinian shrimp, *Pleoticus muelleri*, and a squid, *Loligo gahi*. Goodall and Galeazzi (1985) reported the food remains in two specimens from Tierra del Fuego. One dolphin had remains of fish, possibly *Eleginops*, and the other contained squid, probably *Loligo gahi*. The stomach of one dolphin collected around the Falkland Islands had "pieces of Octopod molluscs which were presumably being caught as they climbed about the Macrocystis plants" (Hamilton, 1952).

Schiavini *et al.* (1997) studied the stomach contents of nine specimens recovered from Tierra del Fuego which were thought to be incidentally captured in nets set for Patagonian blenny (*Eleginops maclovinus*). Fifteen taxa were represented in the stomach contents of these specimens including species of eight fish, three cephalopods, one bivalve mollusc, two crustaceans, and one species of salp. Of these the most important prey species were bottom fish, namely hagfish (*Myxine australis*), southern cod (*Salilota australis*) and Patagonian grenadier (*Macruronus magellanicus*), octopus (*Enteroctopus megalocyathus*) and squid (*Loligo gahi*). The feeding ecology of *L. australis* appears to be associated with demersal and bottom species taken in or near kelp beds (Schiavini *et al.*, 1997).

Association with other animals

Aguayo (1975) reported one sighting of a single Peale's dolphin with four Risso's dolphins, *Grampus griseus*, in coastal Chilean waters. Goodall *et al.* (1988) and de Haro and Iñíguez (1997) observed associations between Peale's and Commerson's dolphins (*Cephalorhynchus commersonii*). Also Peale's dolphins reportedly have interacted with southern right whales (*Eubalaena australis*) (de Haro and Iñíguez, 1997). Interspecific interactions between *L. australis* and great grebes (*Podiceps major*), Magellanic penguins (*Spheniscus magellanicus*), kelp gulls (*Larus dominicanus*), imperial shags (*Phalacrocorax atriceps*) and southern sea lions (*Otaria byronia*) also have been recorded (de Haro and Iñíguez, 1997; Lescrauwaet, 1997).

Swimming and diving

Hamilton (1952) reported dolphins swimming and diving slowly around beds of kelp (*Macrocystis* sp.) or floating at the surface with the blowholes open. When surfacing, Peale's dolphins expose the blowhole and a small portion of the back and dorsal fin, but seldom the flukes (de Haro and Iñíguez, 1997).

De Haro and Iñíguez (1997) observed *L. australis* surfacing three to four times per minute and then "disappearing" for 1–130 sec ($\bar{x} = 23.4$, SD = 23.2, $n = 3179$). Lescrauwaet (1997) observed an average dive period of 27.6 sec (range = 3–157, $n = 723$). The typical dive pattern consisted of three short dives, followed by a longer one (Lescrauwaet, 1997).

Aerial and other behaviour

Strange (1992) reported that this species can be "very aerobatic, leaping and tail slapping and will also ride waves". Peale's dolphins exhibit various types of behaviour common to many species of dolphins including tail-lobbing, head, tail and flipper slaps, leaps, breaching, and spyhopping (Lescrauwaet, 1990; de Haro and Iñíguez, 1997; Goodall *et al.*, 1997b).

Sound production

Pulsed sounds were recorded over five days from *L. australis* in southern Chilean water (Schevill and Watkins, 1971). Analyses showed two types of click sound: a broadband click, and a relatively restricted-bandwidth click (narrowband) at predominantly low frequencies. The broadband clicks had a 10 or 12-kHz bandwidth with the principal frequency in the lower 5 kHz. The narrowband clicks had a 2-kHz bandwidth centred near 1 kHz. No whistlelike squeals characteristic of other delphinids were recorded. It is likely that Peale's dolphins make sounds above 32 kHz, beyond the upper limit of the recording equipment used by Schevill and Watkins (1971), whose descriptions should not be taken as representing the entire range of sounds made by this species.

Parasites and Disease

No endoparasites have been reported from *L. australis* (Dailey and Brownell, 1972). However, Gibson and Harris (1979) reported on a cestode, *Trigonocotyle prudhoei*, that was probably collected from a Peale's dolphin from East Falkland Island.

Live Maintenance

No Peale's dolphins have been held in captivity.

Human Effects

Directed fisheries

Remains of *L. australis* have been found along the north coast of Beagle Channel in shell middens that date back about 6000 years (Piana *et al.*, 1985).

The direct hunting of dolphins has been prohibited in Chile since 1977 (Sielfeld *et al.*, 1977). However, crab traps for centolla [southern king crab], *Lithodes antarctica* and centollon [false king crab], *Paralomis granulosa*, are set with dolphin meat as bait in southern Chile. Sielfeld *et al.* (1977) estimated that 2350 dolphins would have been killed during the 1976/1977 crab fishing season (1 July to 31 January) to bait the estimated 3400 crab traps used in the Strait of Magellan and the Chilean part of the Beagle Channel. The two main species used for bait were *L. australis* and *C. commersonii*. In early 1984, a few fishermen at Isla Isabel, Porvenir and Punta Arenas were still using dolphins as bait for crabs (Leatherwood *et al.*, 1988). These authors were unable to estimate the total number taken. Goodall and Jordan (1986) believed that *L. australis* was the most likely dolphin harpooned in the fishery along with Commerson's dolphin, *Cephalorhynchus commersonii*, the Chilean dolphin, *C. eutropia*, and the dusky dolphin, *L. obscurus*. Fishermen who supply dolphins to crab fishermen claim that crabs prefer dolphins to other animals and birds (Cárdenas *et al.*, 1987). No recent estimates are available on the number of marine mammals killed for bait, and Perrin (1989) has recommended the collection of more definitive statistics on animals used for bait in the crab fishery.

By-catches

Local fishermen in southern Chile may incidentally catch Peale's dolphins (Mitchell, 1975). Small numbers of these dolphins were taken incidental to artesanal fishing operations out of Queule (39° 23′S) and Mehuin (39° 26′S), Chile (Oporto *et al.*, 1988; Oporto and Brieva, 1990b).

An unknown number of Peale's dolphins are taken incidental to mid-water fishing operations for kingklip fish [abadejo], *Genypterus blacodes*, off southern Patagonia, Argentina (Iñíguez and de Haro, 1994). Unknown numbers of these dolphins are also taken incidentally in a shore-based fishery primarily for Patagonian blenny [robalo], *Eleginops maclovinus*, using fixed gill nets set in the intertidal zones off northern and northeastern Tierra del Fuego and southern Santa Cruz, Argentina (Goodall and Cameron, 1980; Goodall *et al.*, 1988; Lichter and Goodall, 1989; Crespo *et al.*, 1994; Goodall *et al.*, 1994).

Pollution

Some residues of organochlorine contaminants were found in a single specimen of *L. australis* from Argentine waters. Dieldrin (0.620 ppm), Heptachlor (0.050 ppm), HCB (0.094 ppm), HCH (0.067 ppm) and DDT (0.405 ppm) were present in the blubber of this dolphin (Land *et al.*, 1990).

Acknowledgments

Steve Dawson and an anonymous reviewer read early drafts of this manuscript and provided useful comments. Natalie Goodall provided valuable information and suggestions, which added greatly to the manuscript. The Peale's dolphin drawing was prepared by Pieter A. Folkens. James G. Mead provided the skull photographs. Susan Wright typed and corrected various drafts of this paper.

References

Aguayo, L. A. (1975). Progress report on small cetacean research in Chile. *J. Fish. Res. Bd Can.* **32**, 1123–1143.

Bierman, W. H. and Slijper, E. J. (1948). Remarks upon the species of the genus *Lagenorhynchus*. Pt. II. *Proc. Koninkl. Ned. Akad. Wet.* **51**, 127–133.

Brownell Jr., R. L. (1974). Small odontocetes of the Antarctic. *In* "Antarctic Mammals", Antarctic Map Folio Series (Ed. V.C. Bushnell) pp. 13–19. American Geographical Society.

Cabrera, A. (1961). Catalogo de los mamiferos de America del Sur. *Rev. Mus. Argent. Cienc. Zool.* **4**, 603–625.

Cárdenas, J. C., Oporto, J., Stutzin, M. and Gibbons, J. (1987). Impacto de la pesqueria de centolla (*Lithodes antarctica*) y centollon (*Paralomis granulosa*) sobre las poblaciones de cetaceos y pinnipedos de Magallanes, Chile. Proposiciones para una politica de conservacion y manejo. *In* "Anais da 2a Reniao de Trabalho de Especialistas em Mamiferos Acuaticos de America do Sul" (Eds S. Siciliano and B. Fiori) pp. 32–35.

Cárdenas, J. C., Stutzin, M. E., Oporto, J. A., Cabello, C. and Torres, D. (1986). Manual de identificacion de los cetaceos chilenos. Proyecto WH-445 'Cetaceos-Chile', World Wildlife Fund and Comite Nacional Pro Defensa de la Fauna y Flora (Chile), 102 pp.

Cassin, J. (1858). United States Exploring Expedition, during the years 1838–1842, under the command of Charles Wilkes, U.S.N. Philadelphia, 8, "Mammalia and Ornithology".

Claver, J. A., Iñíguez, M. A., Lombardo, D. M. and von Lawzewitsch, I. (1992). Preliminary observations on ovarian activity and sexual maturity in female Peale's dolphin (*Lagenorhynchus australis*). *Aquat. Mamm.* **18**, 85–88.

Cope, E. D. (1866). Third contribution to the history of the Balaenidae and Delphinidae. *Proc. Acad. Nat. Sci. Philadelphia* **18**, 293–300.

Crespo, E. A., Corcuera, J. F. and Cazorla, A. L. (1994). Interactions between marine mammals and fisheries in some coastal fishing areas of Argentina. *Rep. Int. Whal. Commn* Special Issue **15**, 269–281.

Cummings, W. C., Fish, J. F., Thompson, P. O. and Jehl Jr., J. R. (1971). Bioacoustics of marine mammals off Argentina: Cruise 71-3 of R/V *Hero*. *Antarct. J. U.S.* **6**, 266–268.

Dailey, M. D. and Brownell Jr., R. L. (1972). A checklist of marine mammal parasites. *In* "Mammals of the Sea: Biology and Medicine" (Ed. S.H. Ridgway) pp. 528–589. Charles C. Thomas, Springfield, IL.

de Haro, J. C. and Iñíguez, M. A. (1997). Ecology and behavior of the Peale's dolphins *Lagenorhynchus australis* (Peale, 1848), at Cabo Virgenes (52° 30′S, 68° 28′W), Argentinian Patagonia. *Rep. Int. Whal. Commn* **47**, 723–727.

Donoso-Barros, R. (1975). Contribucion al conocimiento de los cetaceos vivientes y fosiles del territorio de Chile. *Gayana (Zoologia)* **36**, 1–127.

Fraser, F. C. (1966). Comments on the Delphinoidea. *In* "Whales, Dolphins, and Porpoises" (Ed. K.S. Norris) pp. 7–31. University of California Press, Berkeley, CA.

Fraser, F. C. and Noble, B. A. (1968). Skull of *Lagenorhynchus cruciger* from Livingston Island, South Shetland Islands. *Bull. Brit. Antarct. Surv.* **15**, 29–38.

Gibson, D. I. and Harris, E. A. (1979). The helminth-parasites of cetaceans in the collection of the British Museum (Natural History). *Invest. Cetacea* **10**, 309–324.

Gilmore, R. M. (1969). Populations, distribution, and behavior of whales in the western South Atlantic: Cruise 69-3 of R/V *Hero*. *Antarct. J. U.S.* **5**, 307–308.

Gilmore, R. M. (1971). Observations on marine mammals and birds off the coast of southern and central Chile, early winter 1970. *Antarct. J. U.S.* **6**, 10–11.

Goodall R. N. P. (1978). Report on small cetaceans stranded on Tierra del Fuego. *Sci. Rep. Whales Res. Inst. Tokyo* **30**, 197–230.

Goodall R. N. P. (1986). On the identity of *Tursio chiloensis* Philippi 1900. *Mar. Mamm. Sci.* **2**, 230–232.

Goodall R. N. P. and Cameron, I. S. (1980). Exploitation of small cetaceans off southern South America. *Rep. Int. Whal. Commn* **30**, 445–450.

Goodall R. N. P. and Galeazzi, A. R. (1985). A review of the food habits of the small cetaceans of the Antarctic and Sub-Antarctic. *In* "Antarctic Nutrient Cycles and Food Webs" (Eds W.R. Siegfried, P. R. Condy and R.M. Laws) pp. 566–572. Springer-Verlag, Berlin.

Goodall R. N. P., Galeazzi, A. R. and Lichter, A. A. (1988). Exploitation of small cetaceans off Argentina 1979–1986. *Rep. Int. Whal. Commn* **38**, 407–410.

Goodall R. N. P. and Jordan, J. (1986). Review of fisheries in the Magellan region of Chile. Paper SC/38/SM22 presented to the International Whaling Commission Scientific Committee.

Godall, R.N.P., de Haro, J. C., Iñíguez, M. A. and Norris, K. S. (1997a). Sightings and behavior of Peale's dolphins, *Lagenorhynchus australis*, with notes on dusky dolphins,

Lagenorhynchus obscurus, off southernmost South America. *Rep. Int. Whal. Commn* **47**, 757–775.

Goodall R. N. P., Norris, K. S., Schevill, W. E., Fraga, F., Paraderi, R., Iñiguez, M. A. and de Haro, J. C. (1997b). Review and update on the biology of Peale's dolphin, *Lagenorhynchus australis*. *Rep. Int. Whal. Commn* **47**, 777-796.

Goodall R. N. P., Schiavini A. M. C. and Fermani, C. (1994). Net fisheries and net mortality of small cetaceans off Tierra del Fuego, Argentina. *Rep. Int. Whal. Commn*, Spec. Issue No. 15, pp. 295–304.

Hamilton, J. E. (1952). Cetacea of the Falkland Islands. *Comun. Zool. Mus. Hist. Nat. Montevideo* **66**, 1–6.

Hanson, M. B. and Erickson, A. W. (1985). Sightings of toothed whales along the Antarctic coast and in the South Atlantic Ocean. *Antarct. J. U.S.* **20**, 16–19.

Hershkovitz, P. (1966). Catalog of living whales. *U.S. Nat. Mus. Bull.* **246**, 1–259.

Iñíguez, M. (1991). Delfin Austral, *Lagenorhynchus australis*. *In* "Estado de Conservación de los Mamíferos Marinos del Atlántico Sudoccidental" (eds H. L. Capozzo and M. Junín) pp. 48-51. Informes y Estudios del Programa de Mares Regionales del PNUMA No. 138.

Iñíguez, M. A. and de Haro, C. (1994). Preliminary report on the feeding habits of the Peale's dolphin *Lagenorhynchus australis* in southern Argentina. *Aquat. Mamm.* **20**, 35–37.

Iñiguez Bessega, M. A., Lacolla, D. V. and von Lawzewitsch, I. (1990). Aspectos histologicos de la piel del delfin austral (*Lagenorhynchus australis*). Presented at the Joint Meeting of the Argentine Mammal Society (SAREM) and the American Society of Mammalogists (ASM), Buenos Aires, Argentina, 17–20 June 1990, p. 47 abstract.

Kasamatsu, F., Joyce, G. G., Ensor, P. and Mermoz, J. (1990). Current occurrence of cetacea in the southern hemisphere; results from the IWC/IDCR southern hemisphere minke whale assessment cruises, 1978/79–1987/88. Paper SC/42/O 15 presented to the International Whaling Commission Scientific Committee.

Kasuya, T. (1973). Systematic consideration of recent toothed whales based on the morphology of tympano-periotic bone. *Sci. Rep. Whales Res. Inst. Tokyo* **25**, 1–103.

Kellogg, R. (1941). On the identity of the porpoise *Sagmatias amblodon*. *Field Mus. Nat. Hist.* Zool. Ser., **27**, 293–311 + 6 plts.

Lacolla, D. and Iñíguez, M. (1991). Tegumentary system of *Lagenorhynchus australis*. *In* Abstracts of the Ninth Biennial Conference on the Biology of Marine Mammals, 5–9 December 1991, Chicago, IL, p. 41.

Land, H. B., Aprile, N., Castello, H. P., Crespo, E. and Iñíguez, M. (1990). Contenido de pesticidas organoclorados en grasa de cachalote (*Physeter catodon*) y delfin austral (*Lagenorhynchus australis*) en aguas Patagonicas. *In* 4a Reunion de Trabajo de Especialistas en Mamiferos Acuaticos de America del Sur 12–15 Nov. 1990 Valdivia-Chile Programa y Resumenes, p. 55.

Leatherwood, S., Grove, J. S. and Zuckerman, A. E. (1991). Dolphins of the genus *Lagenorhynchus* in the tropical South Pacific. *Mar. Mamm. Sci.* **7**, 194–197.

Leatherwood, S., Kastelin, R. A. and Miller, K. W. (1988). Observations of Commerson's dolphin and other cetaceans in southern Chile, January–February 1984. *In* "Biology of the Genus *Cephalorhynchus*" (eds R.L. Brownell Jr. and G.P. Donovan) pp. 71–83. *Rep. Int. Whal. Commn* Special Issue 9.

Leatherwood, S. and Reeves, R. R. (1983). "The Sierra Club Handbook of Whales and Dolphins". Sierra Club Books, San Francisco.

Lescrauwaet, A.-K. (1990). Ecologia y comportamiento del delfin austral (*Lagenorhynchus australis*); un estudio preliminar—XIIA. Region Chile. *In* 4a Reunion de Trabajo de Especialistas en Mamiferos Acuaticos de America del Sur 12–15 Nov. 1990 Valdivia-Chile Programa y Resumenes, p. 36.

Lescrauwaet, A.-K. (1997). Notes on the behavior and ecology of the Peale's dolphin, *Lagenorhynchus australis*, in the Strait of Magellan, Chile. *Rep Int. Whal. Commn* **47**, 747–755.

Lichter, A. A. (1992). "Tracks in the Sand, Shadows on the Sea: Marine Mammals of Argentina and Antarctica". Ediciones Terra Nova, Buenos Aires, Argentina, 284 pp.

Lichter, A. A. and Goodall R. N. P. (1989). Argentina: Progress report on cetacean research, May 1987 to April 1988. *Rep. Int. Whal. Commn* **39**, 169–171.

Mann, F. G. (1958). Clave de determinacion para las especies de mamiferos silvestres de Chile. *Trab. Cent. Invest. Zool. Univ. Chile* **4**, 1–38.

McFarland, W. N. (1971). Cetacean visual pigments. *Vision Res.* **11**, 1065–1076.

Mermoz, J. F. and Goodall R. N. P. (1980). Argentina: Progress report on cetacean research 1975–79. *Rep. Int. Whal. Commn* **30**, 139–141.

Mitchell, E. (1970). Pigmentation pattern evolution in delphinid cetaceans: an essay in adaptive coloration. *Can. J. Zool.* **48**, 717–740.

Mitchell, E. (1975). "Porpoise, Dolphin and Small Whale Fisheries of the World: Status and Problems". IUCN Monograph No. 3, pp. 1–129.

Norris, K. S. and Dohl, T. P. (1980). The structure and function of the cetacean schools. *In* "Cetacean Behavior: Mechanisms and Processes" (Ed. L. Herman) pp. 211–262. Wiley, New York.

Olivos, J. and Delgado, C. (1990). Observaciones de delfin austral (*Lagenorhynchus australis*) en la playa de Santa Barbara, sur de Chile. Resultados proliminares. *In* 4a Reunion de Trabajo de Especialistas en Mamiferos Acuaticos de America del Sur 12–15 Nov. 1990 Valdivia, Chile Programa y Resumenes, p. 50.

Oporto, J. A. (1984). Observaciones de cetaceos en los canales del sur de Chile. p. 31. *In* Resumenes, Primera Reunion de Trabajo de Expertos en Mamiferos Acuaticos de America del Sud.

Oporto, J. and Brieva, L. (1990a). El delfin austral, *Lagenorhynchus australis* (Peale, 1848) en la costa de Chile. *In* 4a Reunion de Trabajo de Especialistas en Mamiferos Acuaticos de America del Sur 12–15 Nov. 1990 Valdivia, Chile, Programa y Resumenes, p. 51.

Oporto, J. and Brieva, L. (1990b). Mortalidad de pequenos cetaceos por la actividad pesquera artesanal en la localidad de Queule, sur de Chile. *In* 4a Reunion de Trabajo de Especialistas en Mamiferos Acuaticos de America del Sur 12–15 Nov. 1990 Valdivia, Chile Programa y Resumenes, p. 53.

Oporto, J., Brieva, L., Mercado, M. and Turner, A. (1988). Estimacion de la mortalidad accidental (enmallamiento) de pequenos cetaceos causada por la pesqueria artesanal en las IX y X regiones, Chile. *In* Abstracts of the First Chilean Conference on the Management and Conservation of Marine Mammals, 19–20 August 1988, Valdivia, Chile.

Peale, T. R. (1848). United States Exploring Expedition during the years 1838–1842, under the command of Charles Wilkes, U.S.N. Philadelphia, 8, "Mammalia and Ornithology".

Perrin, W. F. (1989). "Dolphins, Porpoises, and Whales. An Action Plan for the Conservation of Biological Diversity: 1988–1992". IUCN/SSC Cetacean Specialist Group and US National Marine Fisheries Service, NOAA, Gland, Switzerland.

Philippi, R. A. (1900). Contribucion a la osteolojia del *Grypotherium domesticum* Roth i un nuevo delfin. Memorias cientificas y literarias. *Anal. Univ. Nac. Chile* 105–114.

Philippi, R. A. (1901). *Tursio? chiloensis* Ph. Eine neue Art chilenscher Delphine. *Arch. Naturg.* **1**, 276–278.

Piana, E., Orquera, L., Goodall R. N. P., Galeazzi, A. R. and Sobral, A. P. (1985). Cetacean remains in Beagle Channel shell middens. *In* Abstracts, Sixth Biennial Conference on the Biology of Marine Mammals.

Poole, A. J. and Schantz, V. S. (1942). Catalogue of the type specimens of mammals in the United States National Museum, including the biological surveys collection. *Bull. U.S. Nat. Mus.* **178**, 1–705.

Schevill, W. E. and Watkins, W. A. (1971). Pulsed sounds of the porpoise *Lagenorhynchus australis*. *Breviora* **366**, 1–10.

Schiavini A. C. M., Goodall R. N. P., Lescrauwaet, A.-K. and Alonso, M. K. (1997). Food habits of the Peale's dolphin, *Lagenorhynchus australis*, review and update. *Rep. Int. Whal. Commn* **47**, 827–834.

Sielfeld, K. W. (1983). "Mamiferos Marinos de Chile". Ediciones de la Universidad de Chile, Santiago, 199 pp.

Sielfeld, K. W. and Venegas, C. (1978). Observaciones de delfinidos en los canales australes de Chile. *Anal. Inst. Patagonia* **9**, 145–151.

Sielfeld, K. W., Venegas, C. and Atalah, A. (1977). Consideraciones acerca del estado de los mamiferos marinos en Chile. *Anal. Inst. Patagonia* **8**, 297–312.

Strange, I. J. (1992). "A Field Guide to the Wildlife of the Falkland Islands and South Georgia". Harper Collins, London, 118 pp.

Van Waerebeek, K., Canto, J., Gonzales, J., Oporto, J. and Brito, J. L. (1991). Southern right whale dolphins, *Lissodelphis peronii*, off the Pacific coast of South America. *Z. Saugetierkunde* **56**, 284–295.

Van Waerebeek, K., van Bree P. J. H. and Best, P. B. (1995). On the identity of *Prodelphinus petersii* Lütken, 1889 and records of dusky dolphins *Lagenorhynchus obscurus* (Gray, 1828) from the southern Indian and Atlantic Oceans. *S. Afr. J. Mar. Sci.* **16**, 25–35.

Webber, M. A. and Leatherwood, S. (1991). Peale's dolphin *Lagenorhynchus australis*. *In* "Whales and Dolphins" (Ed. A.R. Martin) p. 156. Salamander Books, London.

6

Hourglass Dolphin *Lagenorhynchus cruciger* (Quoy and Gaimard, 1824)

Robert L. Brownell, Jr and Meghan A. Donahue

Genus and Species

Taxonomy

The species was first described as *Delphinus cruciger* by Quoy and Gaimard in 1824 during the voyage of the S.M. *l'Uranie* and the *Physicienne*. However, no type specimen exists because the description is based on the distinct colour pattern of some dolphins observed at sea in the South Pacific at latitude 49°S between Cape Horn and Australia.

Quoy and Gaimard (1824) described another new species as *Delphinus albigena* during the same voyage. No type specimen was collected and the species is based on dolphins observed at sea between Australia and Cape Horn. These dolphins were observed some days after sighting the type specimen of *Delphinus cruciger*. Various authors have treated *D. albigena* as a synonym of *L. cruciger* (Mead and Brownell, 1993).

Lesson and Garnot (1826) described a new species, *Delphinus bivittatus*, based on a dolphin observed at sea about 140 leagues west of the Falkland Islands *en route* for Cape Horn during the voyage of *La Coquille*. This species is also considered a synonym of *L. cruciger* (Mead and Brownell, 1993).

Handbook of Marine Mammals, Volume 6
ISBN 0-12-588506-7

Gray (1846, 1849) described a dolphin from the Pacific Ocean as *Lagenorhynchus clanculus*. The holotype (No. 1849.5.25.3-935a) is in the British Museum (Natural History), London. This species has been placed in the genus *Electra* also (Hershkovitz, 1966). However, *Lagenorhynchus clanculus* is a synonym of *L. cruciger* (Fraser and Noble, 1968).

Lillie (1915) described a new dolphin species, *Lagenorhynchus wilsoni*, based on observations he made during the British Antarctic (*Terra Nova*) Expedition of a dolphin seen at sea and on dolphins described and illustrated by Wilson (1907). Based on the drawing made by H. Van Der Lee for Lillie, *L. wilsoni* should be considered a synonym of *L. cruciger*.

The assignment of both generic and specific names to Southern Hemisphere specimens of *Lagenorhynchus* has been very confused since the 1820s (see Hershkovitz, 1966:62–69). Several authors recognized only a single southern species *L. cruciger* (Bierman and Slijper, 1984; Hershkovitz, 1966; Rice and Scheffer, 1968). Three species are now recognized (Fraser, 1966; Mitchell, 1970; Brownell, 1974; Rice, 1977). Confusion over the years because of limited comparative material has led to various specimens being misidentified. A specimen reported as *L. superciliosus* (Schlegel, 1841) was reidentified by Broekema (1983) as *L. cruciger*. This specimen, however, was misidentified by both Schlegel and Broekema and is a *L. obscurus*. This identification can be made from the figures of the skull in Schlegel (1841). One of the present authors (R.L.B.) also examined the specimen (RMNH 24764) in the Rijksmuseum van Natuurlijke Historie, Leiden, the Netherlands. Nicol and Croome (1988) noted that a skull in the Tasmanian Museum (No. A191) had been referred to as *L. obscurus* by Pearson (1935) and *L. cruciger* by Guiler (1978). Van Waerebeek (1993) correctly reidentified this specimen as a southern right whale dolphin, *Lissodelphis peronii*. A delphinid skull collected from Heard Island at 53° 05′S, 73° 30′E was tentatively identified as *L. cruciger* (Puddicombe, 1989). One of the present authors (R.L.B.) examined photographs of this specimen and confirmed the identification. This specimen (SAM 15715) is deposited in the South Australian Museum, Adelaide.

Common names

The specific name *cruciger* is from the Latin *crux*, a cross, and *gero*, to bear, literally, cross-bearing, indicating the hourglass colour pattern of this species.

The most common English name for *Lagenorhynchus cruciger* is the hourglass dolphin. It is also called the Cruciger dolphin, southern whitesided dolphin, dauphin à museau court (French), and delfin cruzado (Spanish) (Hershkovitz, 1966; Cárdenas *et al.*, 1986).

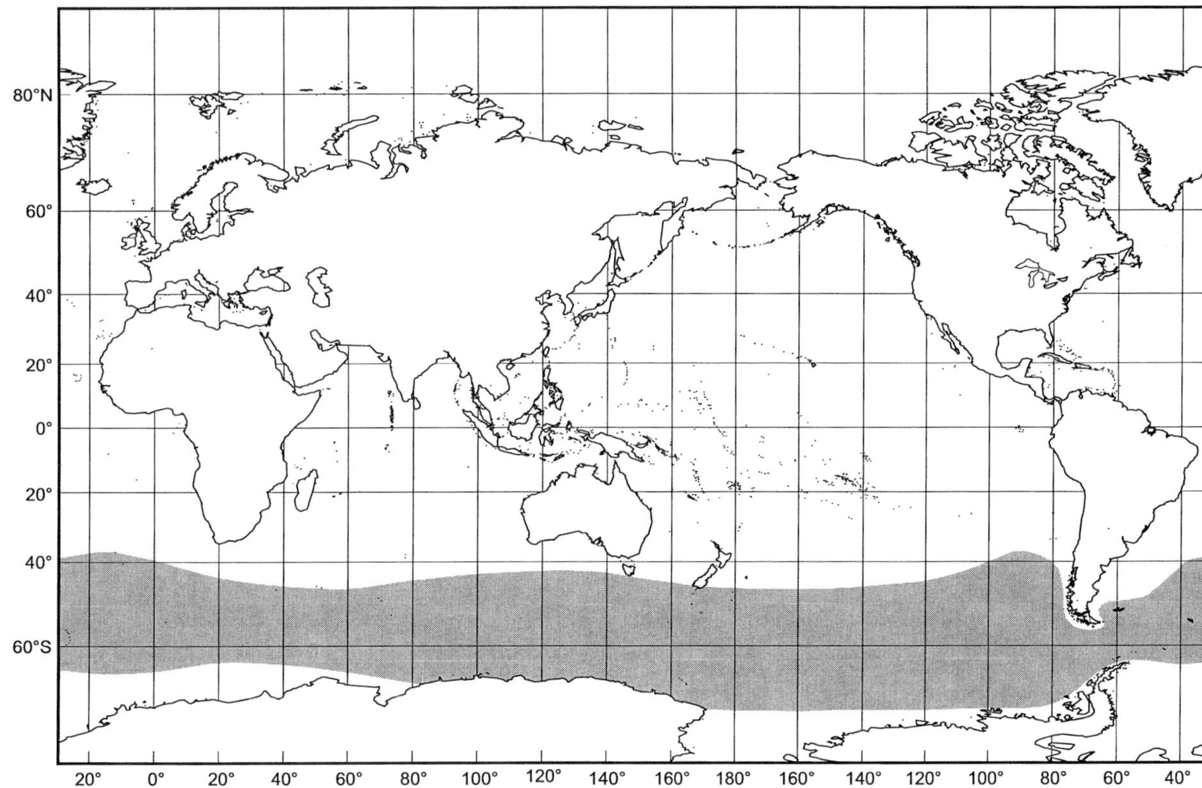

FIG. 1 Distribution of the hourglass dolphin, *Lagenorhynchus cruciger*.

Distribution

Range

The hourglass dolphin is the only small delphinid that is commonly observed south of the Antarctic Convergence (Brownell, 1974). Most sightings of these dolphins are in an area north and south of the Antarctic Convergence between South America and Macquarie Island (Fig. 1). Kasamatsu *et al.* (1988) reported this species (71% of sightings) in surface water temperatures between −0.3 °C and 7.0 °C south of 58°S. In areas just north of the Antarctic Convergence most sightings were made in water temperatures between 5.0 °C and 5.9 °C. One sighting occurred in a water temperature of 13.4 °C. The northernmost sighting is off the coast of Chile at 33° 40′S, 74° 55′W (Clarke, 1962). The southernmost sighting is 67° 38′S, 179° 57′E in the South Pacific (Miyazaki and Kato, 1988).

External Characteristics and Morphology

Colour pattern

At sea *L. cruciger* is distinguished easily from all other species in the genus by the lateral colour pattern from which its common name is derived. Two large lateral white areas are connected with a fine white line separating the black coloration of the dorsum from that of the side below the dorsal fin (Fig. 2). Descriptions of the colour pattern are given in detail by Fraser (1966) and Mitchell (1970). The colour patterns of other individuals photographed at sea are all close to Fraser's description (Fig. 3). Individual differences in the colour patterns can be seen in the dolphins in Fig. 3.

Size and shape

The general body shape of *L. cruciger* is similar to other delphinids (Figs 2 and 3). The body is not robust, has a short ploughshare snout and a very small externally delineated rostrum. The dorsal fin has a prominent falcate shape. The flippers are tapering with a strongly concave posterior border. The tailstock is conspicuously keeled in both sexes.

External measurements were taken from the female British Museum (Natural History) specimen (1960.8.24.1) before being prepared as a skeleton. These measurements (in mm) are: total length (tip of upper jaw to fluke notch) 1829; tip of upper jaw to apex of melon 42, to angle of gape 212, to centre of eye 270, to blowhole 290, to tip of flipper 762; notch of flukes to anus 520; anus to genital opening 950; height of dorsal fin 175; approximate length of dorsal fin along base 310; approximate projection of lower jaw beyond upper 5; length of flipper 295;

FIG. 2 Lateral view of hourglass dolphin, *Lagenorhynchus cruciger*. (Drawing by Pieter A. Folkens)

width of flukes (tip to tip) 335; width of flukes at insertion 170; greatest circumference (anterior to dorsal fin) 1115; length of mammary slit 25.

Lagenorhynchus cruciger may be the smallest species of the genus, but only five specimens have been measured. Nichols (1908) reported a male approximately 163 cm in total length, and Fraser (1966) reported a female 183 cm in length, with a total weight of 88.2 kg. Miyazaki (1986) reported a female specimen 166 cm (NSM M26458) in length (the field measurement was 163.5 cm) col-

FIG. 3 View of hourglass dolphins, *Lagenorhynchus cruciger*, at sea. Note individual differences in colour patterns between dolphins. (Photo by F. Kasamatsu)

lected from the South Atlantic Ocean and a 174 cm adult male (NSM M27421) with a total weight of 94 kg collected from the southeastern South Pacific Ocean (53° 13′S, 106° 20′W) (N. Miyazaki, personal communication). R.S. Whitley collected a stranded male specimen [BM(NH)-ZD 1992.98] in the Falkland Islands measuring 187 cm (Goodall *et al.*, 1997). Malm (1871) describes a specimen captured near Cape Horn; however, only the lengths of the mounted specimen (147 cm) and the skeleton (153 cm) are provided.

Internal Anatomy

Skull

The skull is illustrated in Figs 4 and 5. The largest skull measures 370 mm in condylobasal length and is from a female specimen [BM(NH) 1960.8.24.1] measuring 183 cm in total length. From three specimens described by Fraser and Noble (1968), the cranial measurements ($\bar{x}$, range in mm, and n) are: condylobasal length, 345, 316–370, 3; rostral length, 172, 147–196, 3; basal width of rostrum, 101, 96–106, 3; zygomatic width, 196, 187–207, 3; mandible length, 297, 288–306, 2. Each tooth row has about 30 teeth and the largest teeth have a diameter of about 3 mm.

Postcranial skeleton

The skeleton of *L. cruciger* is poorly known but some details of the postcranial skeleton have been described by Malm (1871). The vertebral count is: C_7, T_{12-13}, L_{19-22}, Ca_{29-33} for a total of 71 vertebrae in each specimen (Malm, 1871 and BM(NH) 1960.8.24.1). The anterior two cervicals are fused. The anterior six ribs have capitular and tubercular attachments. A thirteenth floating rib is present on each side with no thoracic attachment. Eight pairs of sternal ribs are present with the anterior five on the left and four on the right attached directly to the sternum. The phalangeal formula of the specimen is I_3, II_9, III_6, IV_2, V_0 (left).

Organs, tissues and physiology

Cave and Aumonier (1962) noted a morphological peculiarity in the reniculus of this species. Tomlinson and Harrison (1961) reported that *L. cruciger*, unlike most cetaceans, possessed only a single posterior vena cava and lacked a hepatic sinus. No other organ systems have been studied and no studies have been conducted on any aspects of the physiology or biochemistry of *L. cruciger*. Organ weights for the 174 cm male specimen (NSM M27421) collected in the southeastern South Pacific are reported in Goodall *et al.* (1997).

(A)

(B)

(C)

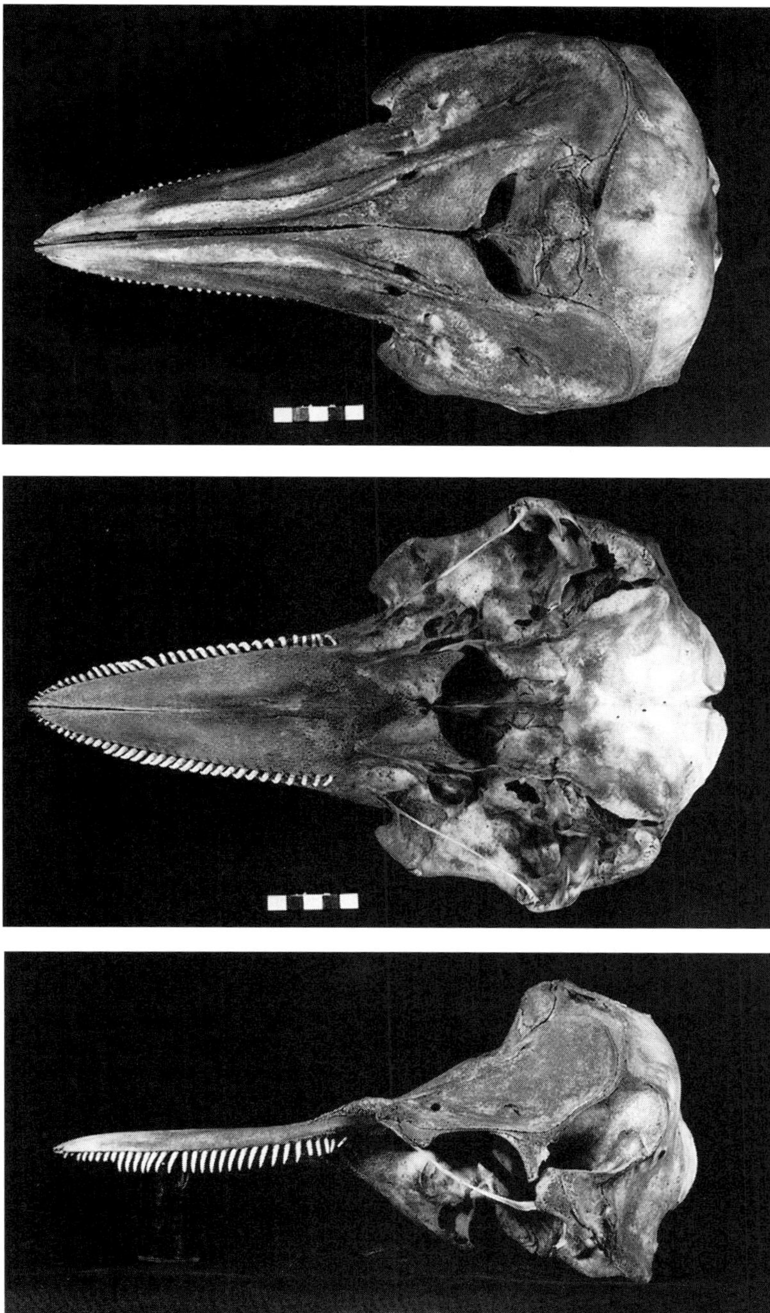

FIG. 4 Dorsal (A), ventral (B) and lateral (C) views of a skull in the National Science Museum, Tokyo (No. M27421) from an adult male hourglass dolphin, *Lagenorhynchus cruciger* (total length 174 cm), from the southeastern South Pacific Ocean. (Photos courtesy of Nobuyuki Miyazaki)

(A)

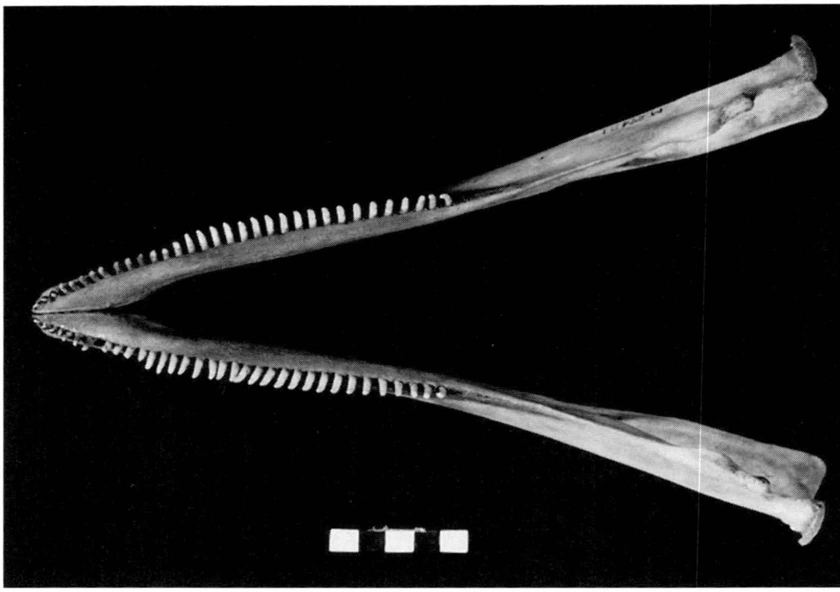

(B)

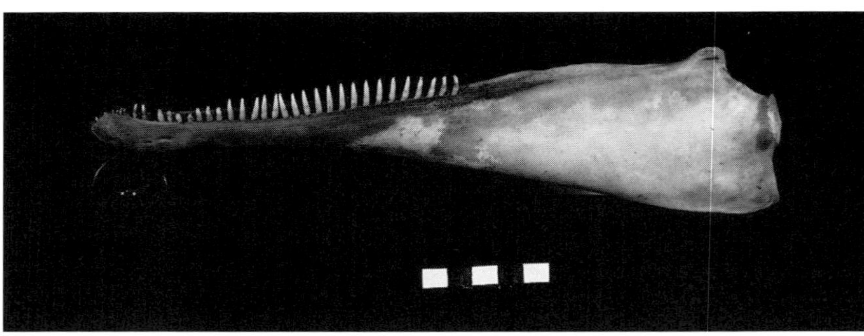

FIG. 5 Dorsal (A) and lateral (B) views of the mandibles of an hourglass dolphin, *Lagenorhynchus cruciger*, in the National Science Museum, Tokyo (No. M27421). (Photos courtesy of Nobuyuki Miyazaki)

Life History and Population Dynamics

Almost no data are available on the life history, mortality, or population dynamics for this species. Few strandings of *L. cruciger* are known, mostly because of its pelagic distribution in the southern oceans. At least six of the 14 confirmed specimens were collected at sea (Table 1). Unfortunately, very few data are available from these specimens. Five additional specimens, four of which stranded in Tierra del Fuego and are thought to be *L. cruciger*, are mentioned in

Goodall *et al.* (1997). Partial crania and/or partial skeletons were collected for three of the Tierra del Fuego specimens. The fourth stranded animal from there was photographed, but no materials were collected. Available data on the fifth specimen [BM(NH)-ZD 1992.98], which stranded in the Falkland Islands and is documented in Goodall *et al.* (1997), are listed in Table 1.

Growth and Reproduction

Richard J. Harrison (personal communication) examined the British Museum (Natural History) specimen (1960.8.24.1) and found a few 5 mm follicles in one ovary, suggesting the dolphin was nearing sexual maturity. Three calves (from three schools) were observed during January and February (Kasamatsu *et al.*, 1990). Combined testis weight for an adult male (174 cm in total length) without the epididymis was approximately 4 kg (N. Miyazaki, personal communication). Nothing else is known about reproduction in this species.

Abundance

During the Southern Hemisphere Minke Whale Assessment Cruises (IWC/IDCR) between 1978/79 and 1987/88, a total of 233 schools of hourglass dolphins (1634 individuals) were observed (Kasamatsu *et al.*, 1990). Kasamatsu and Joyce (1995) combined data from the IWC/IDCR cruises and the Japanese sighting survey programme cruises, which commenced in 1976, to produce an abundance estimate of 144 300 (CV = 0.17) for waters south of the Antarctic Convergence.

Behaviour

No predators are known, but the killer whale, *Orcinus orca*, is a possibility given that *L. cruciger* is the only small delphinid commonly found south of the Antarctic Convergence.

Social organization

Twenty-seven sightings of *L. cruciger*, with school sizes ranging from one to 14 animals, were recorded in the area around southern South America and the Antarctic Peninsula, mostly near islands or banks, in waters of 100 fathoms or less (Goodall and Polkinghorn, 1979). The mean school size from all sightings

TABLE 1 Information on confirmed specimens of *Lagenorhynchus cruciger*, including the date, location, sex, length and material preserved. Specimen collected at sea = *, stranded = +, or unknown origin = ‡

Date	Locality	Lat., Long.	Sex	Total length (cm)	Specimen number[a]	Material and reference
?	Pacific Ocean	?,?	?	?	BM(NH) 1849.5.25.3-935a	skull‡ Gray (1846)
1846	Cape Horn	?,?	?	?	SMNH A558418	skeleton* Malm (1871)
?	New Zealand	?,?	?	?	MNHN-P a3041	skull‡ True (1889)
?	SE of Cape Horn	?,?	?	?	MNHN-P a3045	skull‡ True (1889)
1 Nov 1906	Pacific Ocean, off Chile	49°S, 78°W	M	163	AMNH 35150	skull* Nichols (1908)
May, 1960	S. Indian Ocean	56° 20′S, 40° 09′E	F	183	BM(NH) 1960.8.24.1	skeleton* Fraser and Noble (1968) Fraser (1966)
?	South Shetland Island	?,?	?	?	BM(NH) 1967.7.24.1	skull+ Fraser and Noble (1968)

?	Wycliffe Bay, New Zealand	?,?	?	?	OM A33.11	skull+ A.N. Baker (pers. comm.)
?	"New Zealand"	?,?	?	?	National Museum of New Zealand 1377	skull‡ A.N. Baker (pers. comm.)
25 Dec 1981	SE of Cape Horn	56° 39′S, 60° 41′W	F	166[b]	NSM M26458[c] and ZM40030	skeleton* Miyazaki (1986)
May 1983	Kerguelen Islands	?,?	?	?	MNHN-AC 1985-49	skull+ Robineau (1989)
4 Feb 1987	SE South Pacific	53° 13′S, 106° 20′W	M	174	NSM M27421	skeleton* N. Miyazaki (pers. comm.)
December 1987	Heard Island, Australia	53° 07′S, 73° 44′E	?	?	SAM 1571	skull+ Puddicombe (1989) Anderson (1993)
22 Jan 1981	New Island, Falkland Islands	?,?	M	187	BM(NH) 1992.98	Goodall *et al.* (1997)

[a]Locations of specimens: ZM, South African Science Museum, Cape Town; SAM, South Australian Museum, Adelaide; BM(NH), British Museum (Natural History), London; SMNH, Swedish Museum of Natural History, Stockholm; NSM, National Science Museum, Tokyo; AMNH, American Museum of Natural History, New York; MNHN-AC, Museum National d'Histoire Naturelle—Laboratoire d'Anatomie Comparée, Paris; MNHN-P, Museum National d'Histoire Naturelle, Paris; and OM, Otago Museum, Dunedin, New Zealand.
[b]Total length reported as 163.5 cm in Goodall *et al.* (1997).
[c]Cast only at National Science Museum, Tokyo.

during the 1978/79 to 1987/88 IWC/IDCR cruises was seven, with a range from one to 100 (Kasamatsu *et al.*, 1990).

Feeding

Very little is known about food habits. Nichols (1908) reported that the stomach of his specimen "contained a mass of more or less digested small fish, which may have been mixed with other food matter". Squid beaks from two families (Onychoteuthidae and Enoploteuthidae) were reported from the female BM(NH) specimen (1960.8.24.1) taken southeast of South Africa (Clarke, 1986). Ash (1962) had also reported that this same animal "had been feeding on small squid".

Association with other species

Fraser (1964) reported one school of 40 hourglass dolphins swimming with a pod of 6 fin whales, *Balaenoptera physalus*. Goodall and Polkinghorn (1979) reported two observations of hourglass dolphins with or near long-finned pilot whales, *Globicephala melas*. Webber and Leatherwood (1991) observed hourglass dolphins with Arnoux's beaked whales, *Berardius arnouxi*, and southern bottlenose whales, *Hyperoodon planifrons*. Kasamatsu *et al.* (1988) reported that 14 of their 81 sightings of this species were with other cetaceans, including fin, sei, *Balaenoptera borealis*, killer, minke, *Balaenoptera acutorostrata*, beaked and pilot whales, and southern right whale dolphins, *Lissodelphis peronii*. Most mixed schools contained only two species. Miyazaki and Kato (1988) reported one mixed-species school of long-finned pilot whales, southern right whale dolphins, and hourglass dolphins.

Parasites and Diseases

No endoparasites have been recorded from this species (Dailey and Brownell, 1972). Delyamure (1968) reported *Anisakis simplex* and *A. typica* from *L. cruciger*. However, Delyamure (1968) considered all three southern hemisphere species of *Lagenorhynchus* as one, *L. cruciger*. The parasites he reported from *L. cruciger* were probably collected from *L. obscurus*. Nothing has been recorded about diseases in hourglass dolphin populations.

Live Maintenance

Because of their pelagic distribution in the southern oceans, hourglass dolphins have never been captured for live maintenance.

Human Effects

Directed fisheries and by-catches

No direct fishery for this species has ever existed. One scientific specimen was collected during commercial whaling operations, and several other specimens have been collected during research cruises. At least one hourglass dolphin (NSM M27421) was incidentally caught in an experimental Japanese drift net fishery for squid around 53° 13′S, 106° 20′W (N. Miyazaki, personal communication).

Pollution

No data are available.

Acknowledgments

The authors thank Lawrence G. Barnes, Peter B. Best and William F. Perrin for comments on earlier drafts of the manuscript. Natalie Goodall provided valuable information and suggestions, which greatly added to the manuscript. Alan N. Baker, Peter B. Best, Catherine Kemper, Nobuyuki Miyazaki, Daniel Robineau, Chris Smeenk and Tadasu Yamada provided information on *L. cruciger* specimens in their museums.

References

Anderson, G. R. V. (1993). Australia. Progress report on cetacean research, May 1991 to June 1992. *Rep. Int. Whal. Commn* **43**, 263–269.

Ash, C. (1962). "Whaler's Eye". Macmillan, New York.

Bierman, W. H. and Slijper, E. J. (1948). Remarks upon the species of the genus *Lagenorhynchus*. Pt. II. *Proc. Koninkl. Ned. Akad. Wet.* **51**, 127–133.

Broekema, J. W. (1983). Catalogue of cetacea in the collection of the Rijksmuseum van Natuurlijke Historie, Leiden. *Zool. Mededelingen* **57**, 67–79.

Brownell Jr., R. L. (1974). Small Odontocetes of the Antarctic. *In* "Antarctic Map Folio Series" (Ed. V. C. Bushnell), Folio 18, "Antarctic Mammals". American Geographic Society, New York.

Cárdenas, J. C., Stutzin, M. E., Oporto, J. A., Cabello, C. and Torres, D. (1986). Manual de identificacion de los cetaceos chilenos. Proyecto WH-445 Cetaceos-Chile, World Wildlife Fund and Comite Nacional Pro Defensa de la Fauna y Flora (Chile), 102 pp.

Cave, A. J. E. and Aumonier, F. J. (1962). Morphology of the cetacean reniculus. *Nature* **193**, 799–800.

Clarke, M. R. (1986). Cephalopods in the diet of odontocetes. *In* "Research on Dolphins" (Eds M. M. Bryden and R. Harrison) pp. 281–321. Clarendon Press, Oxford.

Clarke, R. (1962). Whale observations and whale marking off the coast of Chile in 1958 and from Ecuador towards and beyond the Galapagos Islands in 1959. *Nor. Hvalfangst-tid.* **51**, 265–287.

Dailey, M. D. and Brownell Jr., R. L. (1972). A checklist of marine mammal parasites. *In* "Mammals of the Sea: Biology and Medicine" (Ed. S. H. Ridgway) pp. 528–589. Charles C. Thomas, Springfield, IL.

Delyamure, S. L. (1968). "Helminthofauna of Marine Mammals (Ecology and Phylogeny)" (translated from Russian by the Israel Program for Scientific Translations, Jerusalem). Academy of Sciences of the USSR, Laboratory for Helminthology, Moscow.

Fraser, F. C. (1964). Whales and whaling. *In* "Antarctic Research: A Review of British Scientific Achievement in Antarctica" (Eds R. Priestly, R. J. Adie and G. de Q. Robin) pp. 191–205. Butterworths, London.

Fraser, F. C. (1966). Comments on the Delphinoidea. *In* "Whales, Dolphins, and Porpoises" (Ed. K. S. Norris) pp. 7–31. University of California Press, Berkeley, CA.

Fraser, F. C. and Noble, B. A. (1968). Skull of *Lagenorhynchus cruciger* from Livingston Island, South Shetland Islands. *Bull. Brit. Antarct. Surv.* **15**, 29–38.

Goodall, R. N. P. and Polkinghorn, J. T. (1979). Preliminary report on sightings of small cetaceans off southern South America and the Antarctic Peninsula. Paper SC/31/SM2 presented to the International Whaling Commission Scientific Committee.

Goodall, R. N. P., Baker, A. N., Best, P. B., Meyer, M. and Miyazaki, N. (1997). On the biology of the hourglass dolphin, *Lagenorhynchus cruciger* (Quoy and Gaimard, 1824). *Rep. Int. Whal. Commn* **47**, 985–995.

Gray, J. E. (1846). On the cetacean animals. *In* "Zoology of the Voyage of H.M.S. *Erebus* and *Terror* under the Command of Captain Sir James Clarke Ross, R.N., F.R.S., during the Years 1839 to 1843 by the Authority of the Lords Commissioners of the Admiralty (Eds J. Richardson and J. E. Gray), Vol. 1, pp. 13–53. F. W. Janson, London.

Gray, J. E. (1849). Descriptions of three new species of Delphinidae. *Proc. Zool. Soc. Lond.* 1–3.

Guiler, E. R. (1978). Whale strandings in Tasmania since 1945 with notes on some seal reports. *Papers Proc. Roy. Soc. Tasmania* **112**, 189–213.

Hershkovitz, P. (1966). Catalog of living whales. *Bull. U.S. Nat. Mus.* **246**, 1–259.

Kasamatsu, F., Hembree, D., Joyce, G., Tsunoda, L., Rowlett, R. and Nakano, T. (1988). Distribution of cetacean sightings in the Antarctic: results obtained from the IWC/IDCR minke whale assessment cruises, 1978/79 to 1983/84. *Rep. Int. Whal. Commn* **38**, 449–487.

Kasamatsu, F., Joyce, G., Ensor, P. and Mermoz, J. (1990). Current occurrence of cetacea in the Southern Hemisphere minke whale assessment cruises, 1978/79–1987/88. Paper SC/42/015 presented to the International Whaling Commission Scientific Committee.

Kasamatsu, F. and Joyce, G. G. (1995). Current status of odontocetes in the Antarctic. *Antarct. Sci.* **7**, 365–379.

Lesson, R. P. and Garnot, P. (1826). Zoologie. *In* "Voyage autour du Monde, exécute par Ordre du Roi, sur la Corvette de Sa Majesté, *La Coquille*, pendant les Années 1822, 1823, 1824 et 1825", Vol. 1, part 1. Arthus Bertrand, Paris.

Lillie, D. G. (1915). Cetacea. *Brit. Antarct. "Terra Nova" Rept Zool.* **1**, 85–124.

Mead, J. G. and Brownell Jr., R. L. (1993). Order Cetacea. *In* "Mammal Species of the World: A Taxonomic and Geographic Reference" (Eds D. E. Wilson and D. M. Reeder) pp. 349–364. Smithsonian Institution Press, Washington, DC.

Malm, A. W. (1871). Hvaldjur i Sveriges museer, ar 1869. *Kong. Svenska Vetensk-Akad. Handl. Stockholm* **9**, 1–104 + 6 plts.

Mitchell, E. (1970). Pigmentation pattern evolution in delphinid cetaceans: an essay in adaptive coloration. *Can. J. Zool.* **48**, 717–740.
Miyazaki, N. (1986). "Catalogue of Marine Mammal Specimens". National Science Museum, Tokyo, 151 pp.
Miyazaki, N. and Kato, H. (1988). Sighting records of small cetaceans in the Southern Hemisphere. *Bull Natl Sci. Mus.* **14**, 47–65.
Nichols, J. T. (1908). Notes on two porpoises captured on a voyage into the Pacific Ocean. *Bull. Mus. Nat. Hist. NY* **24**, 217–219 + 1 plt.
Nicol, D. J. and Croome, R. L. (1988). Trends in the Tasmanian cetacea stranding record. *In* "Marine Mammals of Australasia, Field Biology and Captive Management" (Ed. M. L. Augree) pp. 59–70. Royal Zoological Society of New South Wales, Sydney.
Pearson, J. (1935). The whales and dolphins of Tasmania. *Papers Proc. Roy. Soc. Tasmania* **1935,** 163–192.
Puddicombe, R. A. (1989). Australia. Progress report on cetacean research, June 1987 to April 1988. *Rep. Int. Whal. Commn* **39**, 171–175.
Quoy, J. R. C. and Gaimard, J. P. (1824). Voyage autour du Monde executé sur les Corvettes de S. M. *L'Uranie* et la *Physicienne*, Paris.
Rice, D. W. (1977). A list of the marine mammals of the world. NOAA Tech. Rep. NMFS SSRF-711, pp. 1–15.
Rice, D. W. and Scheffer, V. B. (1968). A list of the marine mammals of the world. US Fish and Wildlife Service Special Science Report on Fisheries No. 579, pp. 1–16.
Robineau, D. (1989). Les cetaces des Iles Kerguelen. *Mammalia* **53**, 265–278.
Schlegel, H. (1841). "Abhandlungen aus dem Gebiete der Zoologie und Vergleichenden Anatomie", l Heft. A. Arnz und Comp, Leiden.
Tomlinson, J. D. W. and Harrison, R. J. (1961). Venous modifications in two rare marine mammals. *J. Anat.* **95**, 453.
True, F. W. (1889). A review of the Delphinidae. *Bull. U.S. Nat. Mus.* **36**, 191 pp. + 49 plts.
Van Waerebeek, K. (1993). Presumed *Lagenorhynchus* skull at Tasmanian Museum reidentified as *Lissodelphis peronii*. *Austr. Mamm.* **16**, 41–43.
Webber, M. and Leatherwood, S. (1991). Hourglass dolphin *Lagenorhynchus cruciger*. *In* "Whales and Dolphins" (Ed. A. R. Martin) p. 157. Salamander Books, London.
Wilson, E. A. (1907). I. Mammalia: Cetacea. *In* "National Antarctic Expedition 1901–1904: Natural History", Vol. 2, "Zoology (Vertebrata: Mollusca: Crustacea)", pp. 1–9. Trustees of the British Museum, London.

7

Bottlenose Dolphin *Tursiops truncatus* (Montagu, 1821)

Randall S. Wells and Michael D. Scott

Genus and Species

Taxonomy

Bottlenose dolphins were well known to the ancient Greeks and Romans, figuring prominently in their legends, and were described in the writings of Aristotle, Oppian, and Pliny the Elder. At the present, the bottlenose dolphin is the best known of all cetaceans. Several books devoted just to this species have been written for scientific and public audiences (e.g., Caldwell and Caldwell, 1972a; Shane, 1988; Leatherwood and Reeves, 1990; Thompson and Wilson, 1994), and also comprehensive review articles about the species (Tomilin, 1957; Leatherwood and Reeves, 1982; Shane *et al.*, 1986).

Lacépède (1804) first described in the scientific literature a specimen in a Parisian veterinary school as *Delphinus nesarnack*. The specimen disappeared, and the scientific name fell into disuse (nomen oblitum). Montagu (1821) described a dolphin caught in the River Dart in Devonshire, UK, calling the type specimen *Delphinus truncatus*. Interestingly enough, the key characteristic discussed by Montagu was the appearance of the teeth. The teeth were flattened at the tips, which Montagu assumed was characteristic of the species rather than a result of wear (Montagu, 1821; Pilleri and Arvy, 1981). Gray (1843) placed the species in a new genus, *Tursio*, but because *Tursio* was already a synonym for *Physeter*, the genus name was changed to *Tursiops* (Gervais, 1855). The

Handbook of Marine Mammals, Volume 6
ISBN 0-12-588506-7

name *Tursiops* can be translated as dolphin-like, deriving from the Latin "Tursio" (dolphin) and the Greek suffix -ops (appearance); *truncatus* derives from the Latin "trunco-" (truncated), apparently referring to the flattened teeth used by Montagu as an identifying characteristic.

The species is highly polymorphic. The variation in size, coloration, and cranial characteristics related to feeding, and its cosmopolitan distribution have led to the description of at least 20 nominal species of *Tursiops* (Hershkovitz, 1966). The bottlenose dolphin has been placed in several genera in addition to *Tursiops*: *Delphinus*, *Tursio*, *Steno*, *Clymenia*, *Sotalia*, and *Beluga*. Species names have included: *nesarnack*, *compressicauda*, *eurynome*, *metis*, *cymodoce*, *tursio*, *erebennus*, *subridens*, *parvimanus*, *gilli*, *aduncus*, *hamatus*, *nuuanu*, *catalania*, *abusalam*, *perniger*, *gadamu*, *caerulescens*, *gephyreus*, *fergusoni*, *dawsoni*, *maugeanus*, and *obtusus*. Recognition of the polymorphic nature of *Tursiops* and the existence of clinal variation has led to general agreement for many years that *Tursiops* was a single-species genus, but the definitive taxonomic study has yet to be done (Tomilin, 1957; Mitchell, 1975a; Honacki *et al.*, 1982; Ross and Cockcroft, 1990). The taxonomic status of *Tursiops* is made even more confusing by observations of hybridization with several other odontocete species (Sylvestre and Tanaka, 1985). Recent genetic studies suggest that recognition of multiple species within this genus may be warranted (R. LeDuc and B. Curry, personal communications).

Common names

As one would expect from such a widely distributed and polymorphic species, the number of common names is large. The bottlenose dolphin is also known as the porpoise, black porpoise, cowfish (US and UK); delfín negro, tursión, delfín nariz de botella, tonina, delfín de las islas, delfín mular (Spanish-speaking Latin America); boto, golfinho nariz de garrafa (Brazil); souffleur, grand dauphin (France); tursiope (Italy); grosser tümmler (Germany); tumler (Norway); tuimelaar (Netherlands); taiseiyo bando iruka (Japan); afalina, ofalina, bol'shoi del'fin (USSR); roazcorvineiro (Portugal); oresvin (Sweden); afale (Turkey); stokkul (Iceland); and nezarnack (Greenland Eskimo) (Tomilin, 1957; Hershkovitz, 1966; Leatherwood *et al.*, 1988; Sanches, 1989; Reyes, 1990).

Evolution

The fossil record and evolutionary relationships of *Tursiops* have been reviewed by Barnes (1990). Pleistocene fossils of the modern species indicate that *Tursiops* occupied coastal habitats as they do now. It has been suggested that the species originated in the Mediterranean Sea (Pilleri, 1985), but Barnes (1990) cautioned that such an interpretation is premature due to the widespread distributions of *Tursiops* and their kentriodontid ancestors.

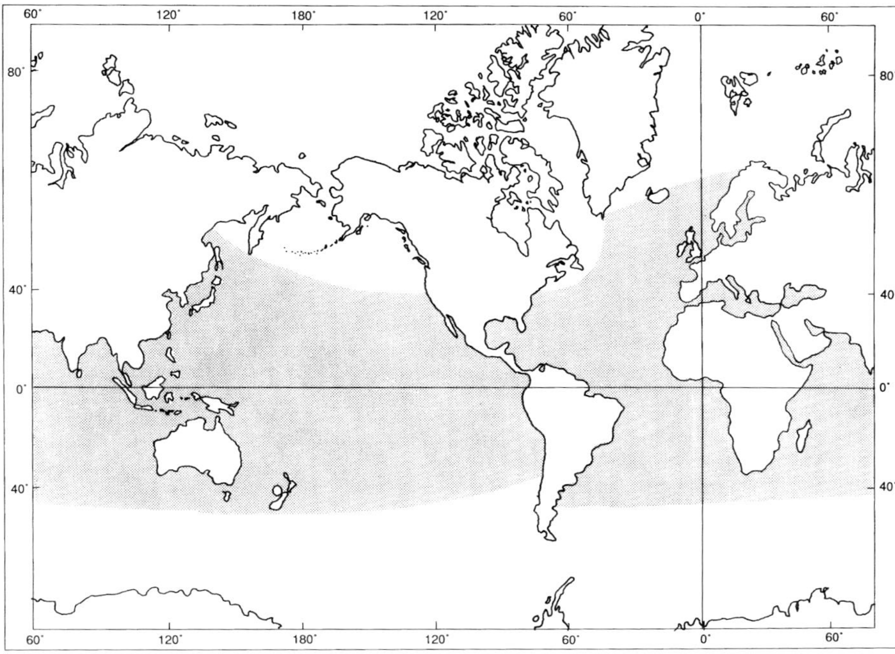

FIG. 1 Species range of the bottlenose dolphin.

Distribution and Abundance

Range

Tursiops is found in temperate and tropical marine waters of the Pacific, Atlantic, and Indian Oceans, and the Mediterranean, Black, and Red Seas (Fig. 1; Mitchell, 1975a). In the western North Pacific they are found as far north as the southern Okhotsk Sea and the Kuril Islands (Arseniev, 1986). In the eastern North Pacific they presently inhabit waters as far north as central California (Wells *et al.*, 1990), although occasional live and stranded specimens have been reported from farther north (Ferrero and Tsunoda, 1989). In the western North Atlantic they are seen inshore during summer months off New England (Kenney, 1990). Farther north, the distribution shifts offshore, where they are known from as far north as Nova Scotia (Leatherwood *et al.*, 1976). In the eastern North Atlantic they have been recorded off Norway (McBrearty *et al.*, 1986), where they have been reported from as far north as the Lofoten Islands (Sars, 1865, as cited in Tomilin, 1957). They are rare in the Baltic Sea, and there is some question as to their occurrence in the Barents Sea (Tomilin, 1957).

In the southern hemisphere, *Tursiops* is common off the coasts of South America (Pinedo *et al.*, 1992), South Africa (Tayler and Saayman, 1972; Ross, 1977), Australia (Ross and Cockcroft, 1990), and New Zealand (Gaskin, 1972). An apparently continuous distribution exists along the west coast of South America, south to at least Concepción, Chile (Aguayo, 1975; Van Waerebeek *et al.*, 1990). In the western South Atlantic they have been observed at least 300 km south of Peninsula Valdez, Argentina (Würsig, 1978), and one dolphin is reported to have stranded in Tierra del Fuego (Goodall and Galeazzi, 1982, as cited in Lichter and Hooper, 1984).

Limits to the species' range appear to be temperature related, either directly, or indirectly through distribution of prey. Off the coasts of North America, they tend to inhabit waters with surface temperatures ranging from about 10 °C to 32 °C. At the northern limit of the species' range in the western North Atlantic they are seasonally migratory, with a more southerly distribution in the winter (Caldwell and Caldwell, 1972a; Mead, 1975a; Kenney, 1990). Along the California coastline, their range expanded several hundred kilometres to the north during an El Niño warm-water event (Wells *et al.*, 1990). Some dolphins remained in the northern waters following return to normal water-temperature patterns, suggesting that they may have been responding more to secondary effects of the warm-water incursion, such as changes in prey distribution, than to the temperature changes themselves.

Habitat

As a result of increased pelagic survey efforts over the last 20 years, researchers have come to recognize *Tursiops* as a truly cosmopolitan species. Although they tend to be primarily coastal, they can also be found in pelagic waters (Kenney, 1990; Scott and Chivers, 1990). They are often associated with oceanic islands (Rice, 1960; Lévêque, 1963; Cárdenas *et al.*, 1986; Scott and Chivers, 1990). They are frequently found over the continental shelf, and especially along the shelf break (Kenney, 1990). *Tursiops* inhabits most warm temperate and tropical shorelines, adapting to a variety of marine and estuarine habitats. They also occasionally range into rivers (Caldwell and Caldwell, 1972a,b).

Abundance and density

Most studies have been of relatively small, local groups of bottlenose dolphins, but a few regional surveys have been conducted. Population size estimates have been reviewed previously by Leatherwood and Reeves (1982), Shane *et al.* (1986), Scott (1990), and in Perrin *et al.* (1994). More recent estimates are also available (Lohoefener *et al.*, 1987; Mullin, 1988; Hansen, 1990; Hansen and Defran, 1990; Wells and Scott, 1990). Density estimates have been published for many parts of the species' range. These estimates range from 0.06 dolphin to

1.22 dolphins per square kilometre (Leatherwood and Reeves, 1982; Shane *et al.*, 1986; Polachek, 1987).

A few abundance estimates are available for *Tursiops* from parts of the species' range. Large-scale research vessel surveys by the US National Marine Fisheries Service (NMFS) have produced an estimate of 243 500 *Tursiops* in the eastern tropical Pacific (Wade and Gerrodette, 1993). Japanese surveys of the northwest Pacific found 316 935 dolphins (Miyashita, 1992). A Russian survey of the Black Sea estimated a population size of 7000 *Tursiops*, although the details of the surveys were not presented (Sokolov *et al.*, 1990, cited in Ivashin and Borodin, 1992).

Two large-scale surveys designed to estimate abundance have been carried out in the USA. The CETAPS aerial surveys in the northwest Atlantic estimated that 10 000–13 000 *Tursiops* lived in the study area; 3–4% of these were coastal animals, the rest were offshore (Kenney, 1990). The NMFS conducted a series of surveys in the northern Gulf of Mexcio, and estimated that 35 000–45 000 *Tursiops* resided inshore of the 100-fathom contour, an area that extends to more than 250 km from shore (Scott *et al.*, 1989).

Several other population size estimates are available from smaller areas. Hansen (1990) estimated the *Tursiops* population along a 155 km coastline in southern California to be about 240. Approximately 900 bottlenose dolphins inhabit the 400 km stretch of coastal waters off Natal, southeast of southern Africa (Ross *et al.*, 1989). Wells and Scott (1990) identified a resident community of 100 individuals inhabiting a 125 km^2 range in Sarasota Bay, FL.

External Characteristics

Tursiops, the "quintessential dolphin" is immediately recognizable (Figs 2 and 3), yet can be surprisingly tricky to identify in areas where other small cetaceans co-occur. They have a generalized appearance — a medium-size, robust body, a moderately falcate dorsal fin, and dark coloration. Field identification, particularly in offshore waters, is often a process of elimination, because key characteristics are few and often not readily visible. Identification is made more difficult by the species' tendency to mix with other cetaceans. Observation of the short rostrum and demarcated melon is typically required to confirm the species identity.

Adult lengths range from under 2 m to about 3.8 m, varying by geographic location (Table 1; see Appendices 1 and 3 in Perrin and Reilly, 1984). Body size appears to vary inversely with water temperature. For example, clinal variation in body length has been reported by Ross and Cockcroft (1990) for dolphins along both coasts of Australia, with larger forms being found in colder waters. A similar pattern has been proposed for dolphins of the Atlantic coast of South America (Barros, 1991). Hersh and Duffield (1990) suggest that in the southeastern United States the larger size of offshore bottlenose dolphins compared with

FIG. 2 Lateral view of an adult male bottlenose dolphin. (Photo by R.S. Wells)

FIG. 3 Ventral view of a bottlenose dolphin. (Photo by R.S. Wells)

TABLE 1 Size data for bottlenose dolphins: comparisons between regions (lengths in cm, weights in kg)

Region	Approximate asymptotic length (m/f)	Maximum length (m/f)	Maximum weight (m/f)	Source
E. coast, South Africa	246	245/240	185/165	Cockcroft and Ross (1990)
Indian Ocean ("*aduncus*") form	239/241	254/252		See Perrin and Reilly (1984)
Japan	280/290	260/270		Kasuya *et al*. (1986)
E. tropical Pacific		266/256		See Perrin and Reilly (1984)
Sarasota, Florida, USA	263/250	283/267	282/263	Read *et al*. (1993)
Florida coasts, USA	270/250	290/276		See Perrin and Reilly (1984)
Central Atlantic coast, USA	250/250	298/306		Mead and Potter (1990)
Black Sea		310/234		See Perrin and Reilly (1984)
E. North Pacific (Offshore)		310/290		See Perrin and Reilly (1984)
W. North Atlantic (not Florida)		310/339		See Perrin and Reilly (1984)
Indian Ocean ("*truncatus*") form		327/280		See Perrin and Reilly (1984)
Mediterranean Sea		330/320		See Perrin and Reilly (1984)
E. North Pacific (Onshore)		333/307		See Perrin and Reilly (1984)
W. North Pacific		335/320–329		See Perrin and Reilly (1984)
E. tropical Atlantic		368/350		See Perrin and Reilly (1984)
Gulf of Mexico (not Florida)		380/367		See Perrin and Reilly (1984)
E. North Atlantic		381/350		See Perrin and Reilly (1984)

coastal dolphins may be related to the deeper, colder waters inhabited by offshore animals.

The flippers are convex with pointed tips. Proportionally larger flippers are found on coastal bottlenose dolphins than on offshore forms (Hersh and Duffield, 1990). The flukes are curved along the rear margin and notched in the centre.

Bottlenose dolphins are coloured light grey to black dorsally and laterally, with a light belly (Fig. 3). This pattern is the most generalized of the delphinids (Mitchell, 1970; Perrin, 1972). A light blaze or brush marking is sometimes observed on their sides. A distinct cape may be visible or may be obscured when the colour pattern is very dark. Offshore forms in the southeastern United States may be darker than coastal forms (Hersh and Duffield, 1990). In the Indian Ocean the dolphins develop ventral spotting at about the time of sexual maturity (Ross and Cockcroft, 1990).

The morphology reflects a generalist strategy. Aleta Hohn (personal communication) has suggested that an examination of the body size of small cetaceans relative to the number of teeth, two characteristics that reflect their ecological niche and food habits, shows *Tursiops* to be midway between the smaller dolphins and the small whales (Fig. 4). This generalized morphology may be a large factor in the wide range of environments that the dolphin inhabits and the wide range of prey species it consumes.

Internal Anatomy

Skull and skeleton

Osteology has been reviewed by Rommel (1990) and Barnes (1990). The skull is broad, with a characteristic short rostrum (Fig. 5). Variation in some skull characteristics appears to be related to ecological differences. The larger offshore form off the southeastern United States tends to have a proportionally shorter rostrum and wider skull than the coastal form (Hersh and Duffield, 1990). Walker (1981) noted that the rostra of offshore forms in the northeastern Pacific taper more acutely and are narrower distally than are those of the coastal forms. Similarly, the more coastal South African form tends to have a narrower cranium and rostrum than the larger offshore form (Ross, 1977; Ross and Cockcroft, 1990). The mandibular symphysis is short and there is no palatal groove (Elliot, 1901). The teeth are stouter and fewer in number (19–26 per quadrant) than in most of the smaller Delphinidae. Unlike *Steno*, which may be closely related to *Tursiops*, the tooth enamel is smooth, and the antorbital air sinus is not enlarged (Barnes, 1990). The number of vertebrae range from 59 to 67, with a vertebral formula of C_7, T_{11-14}, $L_{15-19} + Ca_{25-30}$ (Nishiwaki, 1972; Yablokov *et al.*, 1972; Rommel, 1990; Ross and Cockcroft, 1990). The anatomy of the flippers and flukes has been described by Felts (1966) and Rommel (1990). Barnes (1990) pre-

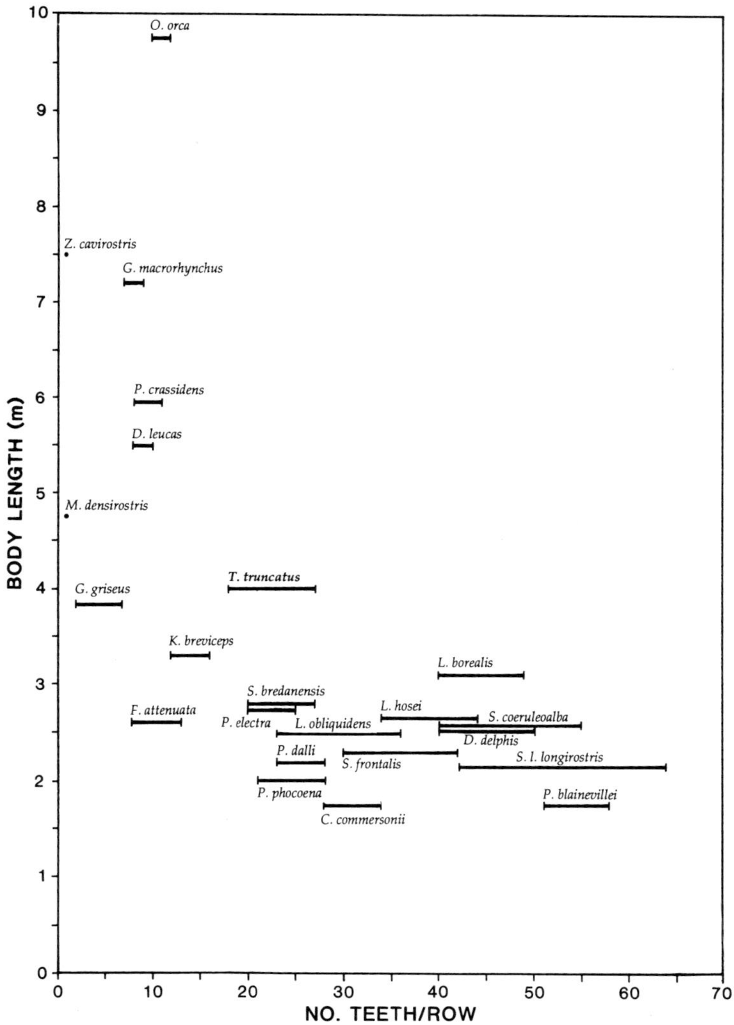

FIG. 4 Number of teeth per tooth row versus the maximum body length for selected odontocetes (data from Perrin and Reilly, 1984).

sents a cladogram that compares the skeletal characteristics with those of other cetaceans.

Organs, tissues and physiology

Organ weights of bottlenose dolphins have been reviewed by Slijper (1958) and Yablokov *et al.* (1972), and physiological, anatomical, and behavioural adaptations to the aquatic environment have been reviewed by Ridgway (1972).

(A)

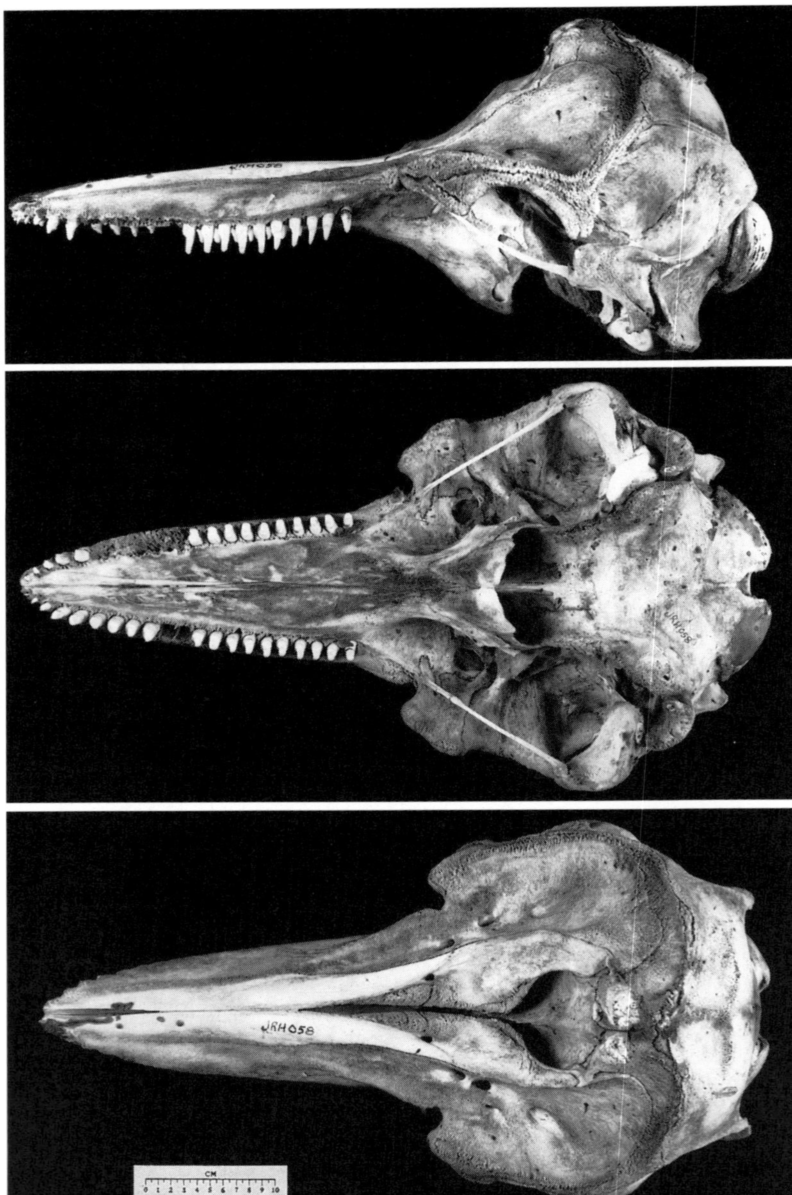

FIG. 5 Skull of a coastal bottlenose dolphin from California (JRH 058 from Southwest Fisheries Center Synoptic Collection): (A) lateral, ventral, and dorsal views of the skull.

(B)

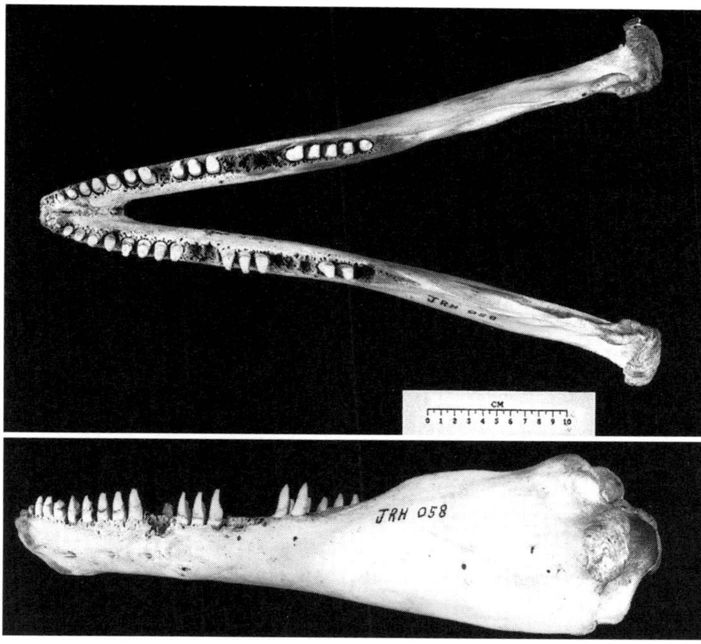

(C)

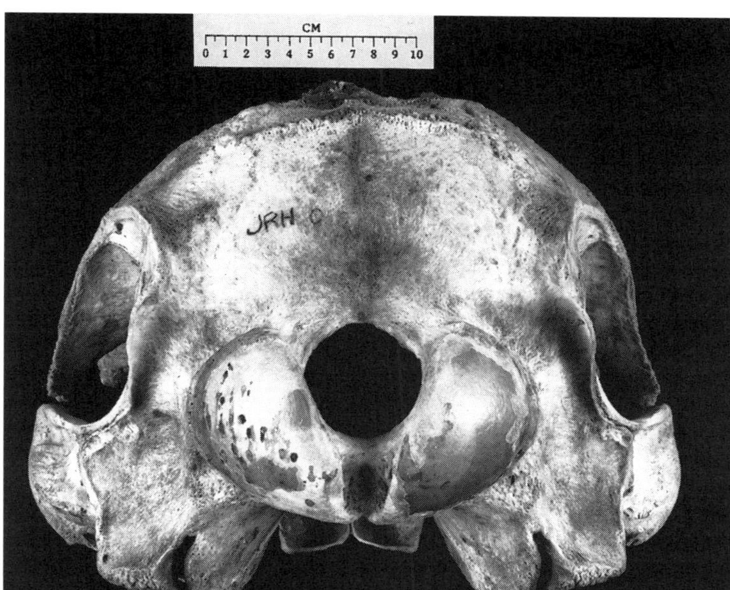

FIG. 5 *Continued* (B) Dorsal and lateral views of the lower jaw, and (C) posterior view of the skull.

Descriptions of the anatomy and function have been made of the central nervous system and brain (Kruger, 1959, 1966; McFarland *et al.*, 1969; Morgane *et al.*, 1980; Ridgway, 1986, 1990), cardiovascular system (Simpson and Gardner, 1972; Cave, 1977), circulatory system (Nagel *et al.*, 1968; Green, 1972; Elsner *et al.*, 1974), nasal passages and facial complex (Mead, 1975b; Cranford *et al.*, 1996), respiratory system (Drabek and Kooyman, 1986; Simpson and Gardner, 1972; Fanning and Harrison, 1974), digestive system (Harrison *et al.*, 1977; Simpson and Gardner, 1972), reproductive system (Harrison *et al.*, 1972), kidney (Cave and Aumonier, 1967; Gaskin, 1986), endocrine organs (Harrison, 1969), thyroid and adrenal hormones (St. Aubin *et al.*, 1996), lacrimal glands (Waller and Harrison, 1978; Tarpley and Ridgway, 1991), tonsillar formation (Cave, 1979), skin (Babenko, 1971; Surkina, 1971; Harrison and Thurley, 1974; Geraci *et al.*, 1986), lipids and esters (Varanasi and Malins, 1970, 1971, 1972), milk composition (Peddemors *et al.*, 1989; Ridgway and Reddy, 1995), eye (Dral, 1977; Dawson, 1980; Dawson *et al.*, 1987; Young *et al.*, 1988), and tongue (Sokolov and Volkova, 1971; Donaldson, 1977).

Recent studies have described the axial musculature and associated connective tissues (Pabst, 1990; Bello *et al.*, 1985), and Pabst (1987) has proposed that the connective-tissue sheaths encompassing the musculature function as a force-transmission system. Another study has proposed a counter-current mechanism for exchanging heat between blood cooled in the superficial venous system of the dorsal fin and flukes and the blood flowing to the testes. This blood could keep the testes relatively cool and maintain sperm viability (Rommel *et al.*, 1992).

The production and reception of sound have been studied extensively, resulting in a number of anatomical studies of the ear, larynx, nasal air sacs, lower jaw tissues, and melon (e.g., see reviews in Busnel, 1967; Bullock and Gurevich, 1979; Busnel and Fish, 1980; Popper, 1980; Johnson, 1986; Morris, 1986; Nachtigall and Moore, 1988). There is still debate about the mechanisms of sound production and reception (Morris, 1986). One school of thought argues that sound is produced in the larynx and received via the external auditory meatus (Fraser and Purves, 1960; Purves, 1966; Purves and Pilleri, 1983). Most other studies have supported a proposed mechanism whereby sound is produced in the nasal air sac system and focused by the melon and received via the lower jaw (e.g., Norris, 1968; Bullock *et al.*, 1968; McCormick *et al.*, 1970; Wever *et al.*, 1971a,b,c; Dormer, 1979; Mackay and Liaw, 1981; Brill *et al.*, 1988).

Life History and Population Dynamics

Age, growth and maturity

Studies in Sarasota, Florida (see Scott *et al.*, 1990) and South Africa have shown that dolphins can live to more than 40 years of age, based on analyses of dentinal and cemental growth layer groups in teeth (Wells *et al.*, 1987; Hohn *et al.*, 1989; Cockcroft and Ross, 1990a). Analyses by Aleta Hohn of growth layer groups in tooth samples from the Sarasota study have shown that a few females live to more than 50 years, and some males have reached 40–45 years of age. Calves achieve most of their growth during the period of suckling, i.e., the first 1.5–2 years of life. Males exhibit a growth spurt in mass at about puberty (Ridgway and Fenner, 1982; Cockcroft and Ross, 1990a; Read *et al.*, 1993). Southern African dolphins reach asymptotic sizes and physical maturity at 12–15 years of age. Males and females from the central US Atlantic coast reached asymptotic length of about 250 cm at about 12 years of age, and physical maturity at about 13 years, based on epiphyseal fusion (Mead and Potter, 1990).

Sexual dimorphism in body size has not been demonstrated for populations from the western North Atlantic or northern Gulf of Mexico (Hersh *et al.*, 1990a; Hohn, 1980; Mead and Potter, 1990; Odell and Schneyer, 1983), although a mild degree of dimorphism is present in the skull and a few body measurements (Hersh *et al.*, 1990a). The age determination techniques used in these analyses, however, have been shown to be biased (Hohn, 1990). Additional biases may arise from the age structure and sources of the samples. Using a more refined age estimation technique (Hohn *et al.*, 1989; Hohn, 1990), Sarasota females were found to grow faster initially and reach asymptotic length at an earlier age than males (Read *et al.*, 1993) (Figs 6–8). Females reached asymptotic sizes at about 12 years of age and males at about 20 years of age, resulting in a significant sexual dimorphism in adult body length, girth, and mass. Further analyses of the Sarasota dolphins found the males to be significantly larger than females in 20 of 29 measurements, and dorsal fin scarring was more pronounced in males than in females (Tolley *et al.*, 1995). Similarly, South African males are slightly longer, considerably heavier, and more robust than females (Cockcroft and Ross, 1990a). It appears that sexual dimorphism may be a result of a diversion of the female's energy from growth to reproduction at an earlier age, whereas males continue to grow for several additional years.

Age at sexual maturity varies by sex and region. In general, females in Florida mature at 5–12 years of age at a length of 220–235 cm and males at 10–13 years at a length of 245–260 cm (Harrison *et al.*, 1972; Sergeant *et al.*, 1973). Off southern Africa, females reach sexual maturity at 9–11 years of age (Cockcroft and Ross, 1990a). Male puberty may begin at 9–12 years of age, with average age of sexual maturity occurring at about 14.5 years. Off Japan, 50% of the females were sexually mature at 7 years and 267 cm, and the youngest mature animal was estimated to be 3.5 years old (Kasuya *et al.*, 1986). Along the central US Atlantic coast, females reach sexual maturity at 7–13 years (Mead and Potter, 1990). In Sarasota, calves have been born to females as young as 6 years old, indicating conception at age 5, but most primiparous females are 8–10 years of age. Sarasota males begin to show elevated testosterone concentrations indicative of puberty at 8–10 years of age (Wells *et al.*, 1987), but sexual maturity may occur

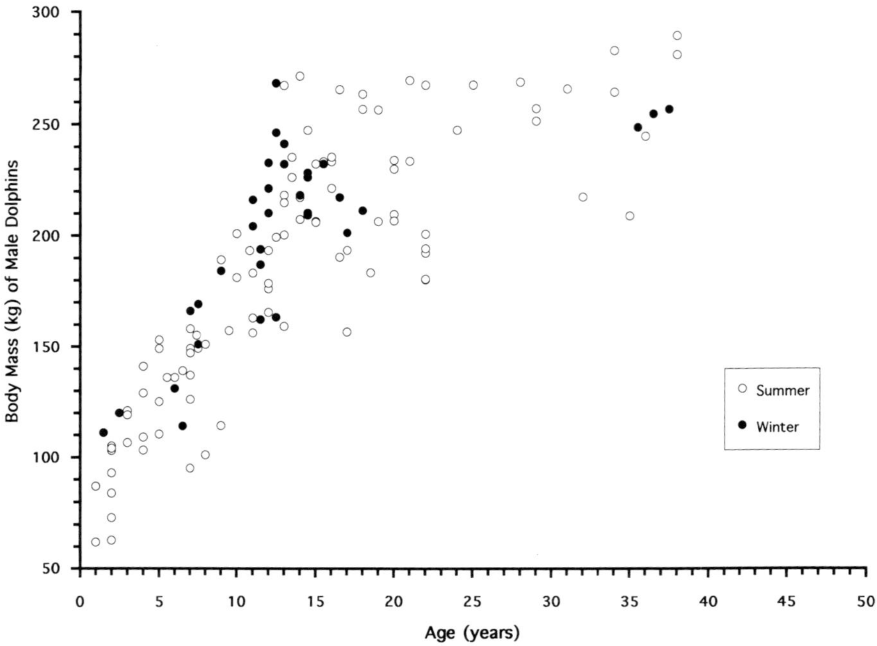

FIG. 6 Age versus mass of male bottlenose dolphins from Sarasota Bay, FL. The different seasons in which data were collected are indicated. Individual dolphins weighed in multiple seasons or years are represented each time they were weighed. Ages were determined from known birthdates or from analysis of tooth sections (Hohn *et al.*, 1989).

many years before males achieve breeding status. Most breeding males in captivity are at least 20 years old (Duffield and Wells, 1991).

Reproduction

Calves may be born at any time of the year, but a tendency for diffusely seasonal calving has been reported from most studies to date, with peaks in calving during warmer months. Hohn (1980) reported one and possibly two calving seasons along the central US Atlantic coast. Mead and Potter (1990) reported a prolonged calving season with a spring peak in the same area. Caldwell and Caldwell (1972a,b) reported a spring to early summer season for northeastern Florida. Wells *et al.* (1987) reported that most births occurred in late spring to early summer, with a secondary peak in early autumn, off Sarasota. A summer peak in births has been observed in Patagonia and South Africa (Würsig, 1978; Cockcroft and Ross, 1990a). Urian *et al.* (1996) found a correlation between the origin of a dolphin population and seasonality of reproduction. Seasonality varied from one geographic location to another, and dolphins transferred from the wild to a zoological park environment retained the seasonality of their wild popu-

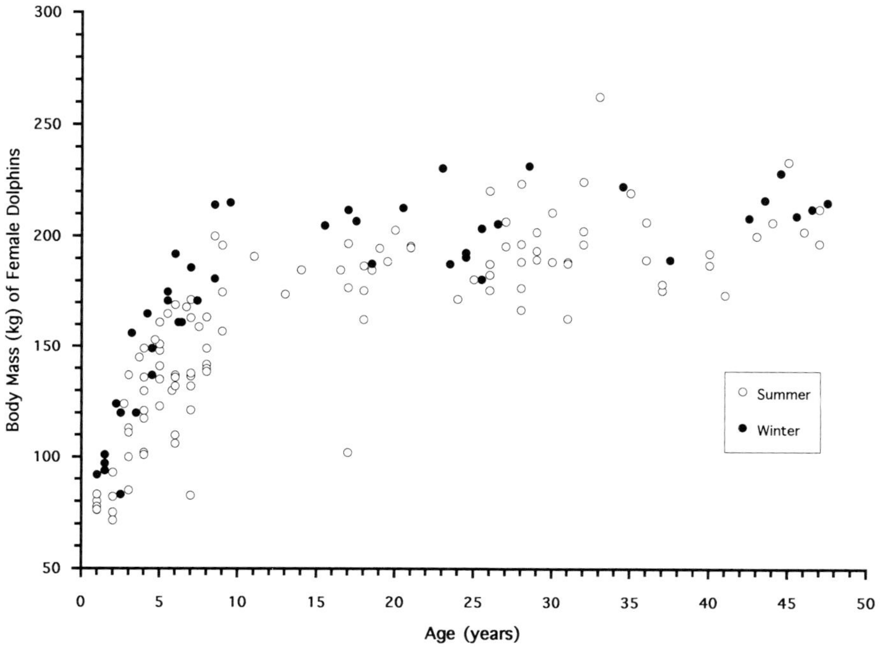

FIG. 7 Age versus mass of female bottlenose dolphins from Sarasota Bay, FL. The different seasons in which data were collected are indicated. Individual dolphins weighed in multiple seasons or years are represented each time they were weighed. Ages were determined from known birthdates or from analysis of tooth sections (Hohn *et al.*, 1989).

lation of origin. Results of hormonal monitoring of captives suggest that females are spontaneous sporadic ovulators, ovulating repeatedly during a given season, while males may be active throughout the year with prolonged elevation of testosterone concentrations over the months that different females may be ovulating (Kirby and Ridgway, 1984; Yoshioka *et al.*, 1986; Schroeder, 1990).

The reproductive lifespan of females is prolonged, with little or no indication of senescence (Cockcroft *et al.*, 1989a). Females up to 45 years of age have successfully given birth and raised young off Sarasota.

Calves are born after a gestation period of about one year (McBride and Kritzler, 1951; Tavolga and Essapian, 1957; Caldwell and Caldwell, 1972a; Saayman and Tayler, 1977; Ross, 1984; Cockcroft and Ross, 1990a). At birth, calves range in length from about 84 cm to 140 cm (Perrin and Reilly, 1984; Kasuya *et al.*, 1986), depending on the geographic region. Newborns average about 103 cm (13.8 kg) off southern Africa (Cockcroft and Ross, 1990a), 128 cm off Japan (Kasuya *et al.*, 1986), and about 117 cm (20.4 g) off the central US Atlantic coast (Hohn, 1980; Mead and Potter, 1990).

Lactation serves as a primary source of nutrition for about one year (Kasuya and Marsh, 1984; Cockcroft and Ross, 1990a; Peddemors *et al.*, 1992), but may

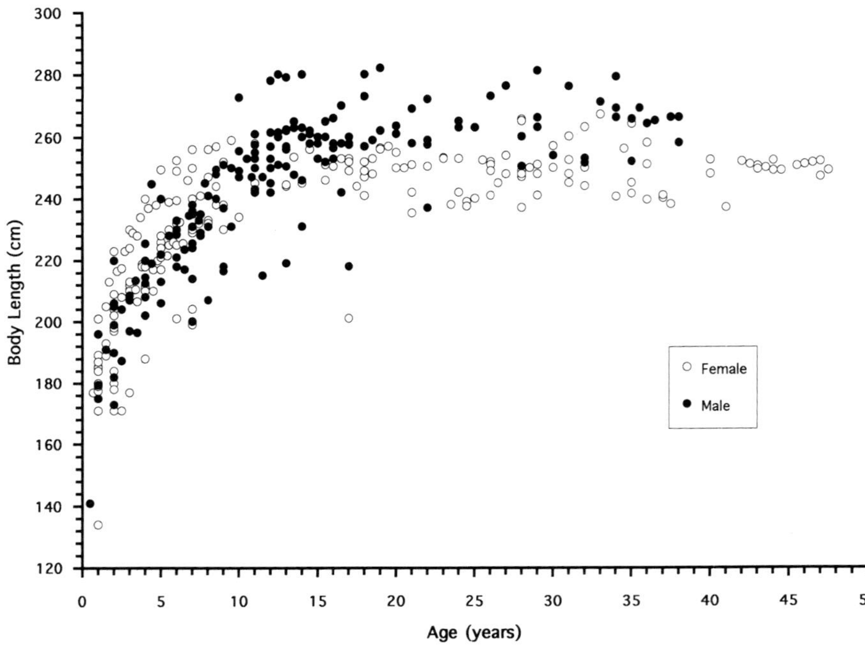

FIG. 8 Age versus standard body length of male and female bottlenose dolphins from Sarasota Bay, FL. Individual dolphins measured in multiple seasons or years are represented each time they were measured. Ages were determined from known birthdates or from analysis of tooth sections (Hohn *et al.*, 1989).

continue for several more years. We have observed calves of up to 7 years of age with lactating mothers off Sarasota, FL, and Cockcroft and Ross (1990b) have noted that lactation continued for up to 3 years. Mammary gland volume for bottlenose dolphins off southern Africa was observed to increase with calf size up to about 50–60 kg at about 18 months of age, then decline as the calves apparently shifted to solid food sources (Cockcroft and Ross, 1990a). Cockcroft and Ross (1990b) calculated a rate of neonatal milk consumption of 4 litres per day. Solid food has been found along with milk in the stomachs of calves 4–11 months old (McBride and Kritzler, 1951; Prescott, 1977; Ross, 1977; Saayman and Tayler, 1977; Leatherwood *et al.*, 1978; Cockcroft and Ross, 1990b). Orphan calves have caused non-related, dry females to lactate and provide months of nursing support (Ridgway *et al.*, 1995).

The period of maternal investment for free-ranging calves extends well beyond the average age of 18–20 months at weaning (Perrin and Reilly, 1984). Data from Sarasota indicate that calves typically remain with their mothers for 3–6 years (mean = 5.4 years, SD = 2.15, range = 2–11 years, $n = 17$; Wells, 1993). Upon separation, calves are about 225 cm in length and weigh about 150 kg. Separation tends to occur at about the time of the birth of the next calf, resulting in a calving interval of at least three years in most cases of successful rearing. Mothers who lose a calf may give birth again in less than three years. Cockcroft and Ross (1990a) estimated a 3-year calving interval for dolphins off southern Africa. Simultaneously pregnant and lactating females have been noted on occasion both in Florida and off southern Africa.

Population structure and vital rates

Little is known about the sex and age structure of free-ranging populations of bottlenose dolphins. Most accounts indicate a sex ratio of about 1:1 (Perrin and Reilly, 1984; Hersh *et al.*, 1990a). Wells *et al.* (1987) found a variable sex ratio of calves that became skewed at a value of approximately 2:1 in favour of adult females due largely to higher juvenile male mortality. The oldest female in the Sarasota dolphin community in 1993 was about 52 years old, 8 years older than the oldest males measured to date. An age distribution for the Sarasota dolphin population members is shown in Fig. 9.

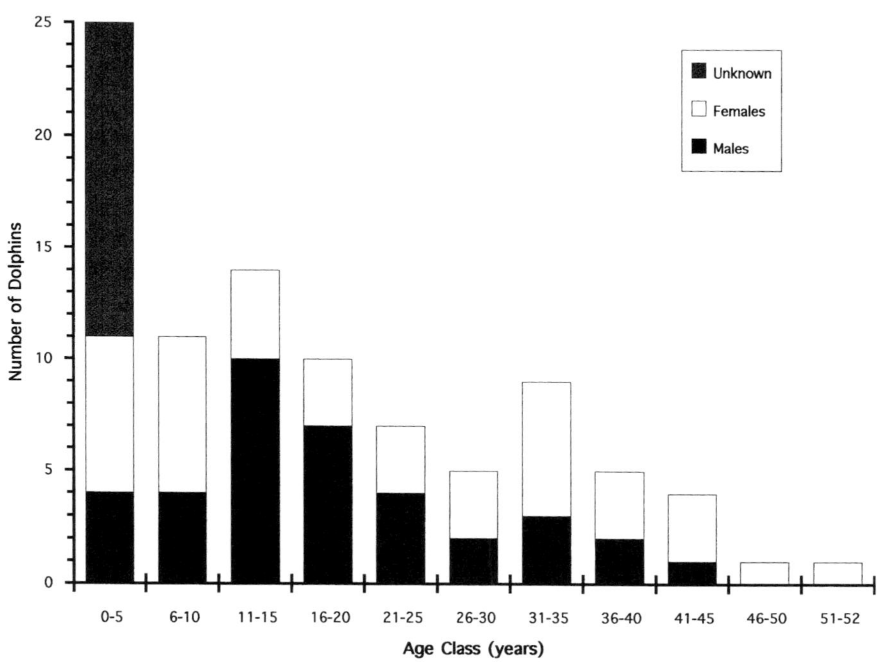

FIG. 9 Age distribution of the Sarasota dolphin community in 1993 (91 of the 99 community members are shown; not included are 3 adult females, 2 adult males, and 2 subadults whose ages were less precisely known). Age data were determined (1) through observations of new calves with recognizable females, or (2) from examination of growth layer groups in a tooth, as provided by Aleta Hohn (Hohn *et al.*, 1989).

Crude birth rates (estimated number of newborn calves divided by the estimated total population) of 0.012–0.156 have been reported for bottlenose dolphin populations (summarized by Leatherwood and Reeves, 1982, and Perrin and Reilly, 1984). In more recent studies, Hansen (1990) reported a crude birth rate of 0.072 for dolphins off southern California. Weigle (1990) reported a crude birth rate of 0.097 for Tampa Bay, Florida. Wells and Scott (1990) found a mean crude birth rate of 0.055 with a recruitment rate to age one of 0.048 for Sarasota dolphins.

Few data are available on rates of exchange between different populations. Wells and Scott (1990) estimated immigration and emigration rates at less than 0.03 per year for the resident Sarasota dolphin community. Transitory movements between populations, especially by males (see below), may occur more frequently than permanent relocations.

Natural mortality rates are difficult to measure accurately. Cockcroft *et al.* (1989b) estimated a mortality rate of 0.022 due to sharks off the Natal coast, but mortality rates from other natural sources were unknown. Wells and Scott (1990) measured a minimum annual mortality rate of 0.010 and a maximum rate of 0.038, including losses of all kinds, for the population of known individuals (>1 year old) in Sarasota. Hersh *et al.* (1990b) estimated annual mortality rates of 0.069–0.092 for dolphins in the Indian and Banana rivers of Florida, depending on the aerial survey population estimate used for the calculations. The apparent difference between these two Florida populations is more probably due to methodological rather than regional differences. Young of the year and subadult males suffer higher mortality rates than other classes. Wells and Scott (1990) reported that young of the year had a minimum annual mortality rate of 0.189 in Sarasota, while Hersh *et al.* (1990b) estimated a rate of 0.112.

Ranging patterns

Little is known about the ranging patterns of pelagic bottlenose dolphins, but coastal dolphins exhibit a full spectrum of movements, including seasonal migrations, year-round home ranges, periodic residency, and a combination of occasional long range movements and repeated local residency. Dolphins living at the high latitude or cold water extremes of the species' range may migrate seasonally, as is the case along the Atlantic coast of the United States (True, 1891; Mead, 1975a). Long term residency may take the form of a relatively permanent home range, or repeated occurrence in a given area over many years. For example, the residents of several dolphin communities along Florida's west coast have maintained relatively stable home ranges during more than 28 years of observations (Wells, 1986; Scott *et al.*, 1990). At Shark Bay, Western Australia, a group of dolphins has returned to the same beach for more than 20 years to interact with humans (Connor and Smolker, 1985). In other areas, residency is long term but more variable. Dolphins seen frequently during 1974–1976 in Golfo San Jose, Argentina, showed a subsequent decline in frequency of occurrence, but were still identified in the area on occasion 8–12 years later (Würsig and Harris, 1990).

The first description of a home range was provided by Caldwell (1955) for a distinctively marked dolphin near Cedar Key, Florida. Subsequently, local residency has been identified for dolphins in many areas, including Georgia and Florida (Caldwell and Golley, 1965; Irvine and Wells, 1972; Irvine *et al.*, 1981; Wells, 1986; Odell and Asper, 1990; Shane, 1990b; Weigle, 1990), Texas (Shane, 1980, 1990b; Gruber, 1981), the Gulf of California (Ballance, 1990, 1992), southern California (Hansen, 1990), and South Africa (Saayman *et al.*, 1973).

Along the central west coast of Florida, communities of resident dolphins appear to inhabit a mosaic of overlapping home ranges (Wells, 1986). The home range of the Sarasota dolphins encompasses an area of about 125 km^2. Most of the activities of the residents are concentrated within the home ranges, but occasional movement between ranges occurs also. Within the home range, habitat use varies with season, with shallow estuarine waters frequented during the summer and coastal waters and passes between barrier islands used during the winter (Irvine *et al.*, 1981). Home range bounds often are demarcated by physiographic features such as passes, or by abrupt changes in water depth. In other areas, underwater topography and water temperature have been suggested as limiting factors for ranging patterns (Würsig and Würsig, 1979; Hansen, 1990).

Caldwell and Caldwell (1972a) suggested that some dolphins may use seasonal home ranges joined by a travelling range. A 4-month cycle of occurrence of dolphins was observed in Golfo San Jose, Argentina (Würsig, 1978; Würsig and Würsig, 1979). Hansen (1990) believed that his southern California study area was part of a seasonal home range for some of the dolphins he identified. Some of the dolphins studied by Shane (1980, 1990a,b) in Florida and Texas were considered to use the study areas on a seasonal rather than year-round basis.

Long distance movements also have been recorded. Dolphins that were residing in southern California in the early 1980s (Hansen, 1983, 1990) subsequently expanded the species' recent range more than 500 km northward in conjunction with an El Niño warm water event (Wells *et al.*, 1990). Following the El Niño, some dolphins remained in northern waters, while others returned to their previous range to the south. Würsig (1978) reported a 600 km roundtrip for several identifiable dolphins in Argentina. Tanaka (1987) reported that a satellite-tracked dolphin off Japan apparently travelled 604 km in 18 days along the Kuroshio Current. Individuals may also move short distances outside their normal home range (Shane, 1980; Wells, 1986; Odell and Asper, 1990).

Population discreteness

In most parts of the species' range *Tursiops* is distributed relatively continuously. This distribution pattern complicates attempts to identify, and assess the discreteness of, individual population units. At a gross level, morphological, ecological, and/or haematological differences have allowed the identification of separate inshore and offshore geographical forms or ecotypes off the northeastern and central seaboard of the United States (Kenney, 1990), southeastern United States (Hersh and Duffield, 1990), southwestern United States and Mexico (Walker, 1981), Peru (Van Waerebeek *et al.*, 1990), South Africa (Ross, 1977), and Australia (Ross and Cockcroft, 1990). Much research and management attention is being focused currently on differentiating between resident, migratory, coastal, and offshore stocks of dolphins in southeastern waters of the United States (Wang *et al.*, 1994).

At a finer level of discrimination, apparent range boundaries for individually identifiable coastal dolphins have been established. This suggests a possible finer-scale geographical basis for population designations; the potential existence of different kinds of ranging patterns within a single region, however, argues against designating population units based on short term home range patterns alone. For example, seasonal changes in abundance of dolphins in some areas (e.g., Shane and Schmidly, 1978; Würsig and Würsig, 1979; Lear and Bryden, 1980; Shane, 1980; Gruber, 1981), or overlap in the ranges of different population units (Wells, 1986) complicate simple geographical models for populations.

Long term studies of social associations and ranging patterns of Sarasota dolphins have been integrated with studies of biochemical genetic information from the same animals to describe the population structure (Duffield and Wells, 1986, 1991; Wells, 1986). Local communities, consisting of dolphins sharing common home ranges and regular associates, tend to have similar genetic profiles; however, there is evidence for genetic exchange with other population units. Thus, although these communities are relatively discrete units in terms of ranging patterns and social interactions, they are not reproductively isolated from other communities.

Behaviour

Social organization

Bottlenose dolphins are social animals. Typically, they are found in groups of 2–15 individuals (Shane *et al.*, 1986), although groups of 1000 have been reported from South Africa (Saayman and Tayler, 1973), and up to 10 000 have been reported from the eastern tropical Pacific Ocean (Scott and Chivers, 1990, but see their caveat about large group-size estimates). Group-size data from a variety of locations have been summarized previously (Wells *et al.*, 1980; Leatherwood and Reeves, 1982; Shane *et al.*, 1986). In general, bottlenose dolphins in bays and estuaries tend to form smaller groups than those in offshore waters, but the trend does not continue with increasing distance from shore (Wells *et al.*, 1980; Scott and Chivers, 1990).

Group composition is dynamic, but subgroupings may be stable or repeated over periods of years (e.g., Würsig, 1978; Wells, 1978; Wells *et al.*, 1987; Shane, 1980; Connor and Smolker, 1985; Smolker *et al.*, 1992; Ballance, 1990; Hansen, 1990; Würsig and Harris, 1990). Sex, age, reproductive condition, familial relationships, and affiliation histories appear to be important factors determining group compositions. Within the Sarasota dolphin community, for example, three kinds of group are common: bands of adult females with their most recent offspring, mixed-sex subadult groups, and adult males that often swim alone or as tightly bonded pairs (Wells *et al.*, 1987). Adult males travel from one female group to the next and associations with females tend to be relatively brief (from minutes to several weeks). Multigenerational female bands tend to be composed of females who are related or who share long histories of association within a common home range. Females at the same stage in their reproductive cycles tend to swim together. Bonds between males of similar age develop early in life, and pair bonds may be maintained for 20 years or more.

Day-to-day behavioural interactions have been studied in greater detail in captivity than in the wild. Dominance hierarchies are established in both sexes. Large adult males dominate all other pool-mates and the largest females dominate smaller males and females (McBride and Hebb, 1948; Brown and Norris, 1956; Caldwell *et al.*, 1965; Tavolga, 1966; Norris, 1967; Caldwell and Caldwell, 1972a,b; Saayman *et al.*, 1973; Östman, 1991; Samuels and Gifford, 1997). These hierarchies are established and maintained through aggressive interactions that include contact and posturing; females display lower levels of aggression than males. Aggressive interactions have also been observed in the wild, especially among males. Violent interactions between known resident males and presumed males from neighbouring home ranges have been observed on occasion along home range borders in Florida (Wells *et al.*, 1987). In Shark Bay, Australia, coalitions of male dolphins fight with other male coalitions and aggressively herd females (Connor, 1990; Connor *et al.*, 1992a,b). Large dolphins have also been observed scouting potentially threatening objects (Bel'kovich *et al.*, 1991a).

Tursiops occurs in both single- and mixed-species groupings. There is a trend for increased frequency of mixed-species groupings progressively farther from shore in the eastern tropical Pacific Ocean (Scott and Chivers, 1990). The most commonly associated species are pilot whales (*Globicephala macrorhynchus*), spotted dolphins (*Stenella attenuata*), Risso's dolphins (*Grampus griseus*), rough-toothed dolphins (*Steno bredanensis*), and humpback whales (*Megaptera novaeangliae*).

Activity patterns

Free-ranging dolphins appear to be active both during the day and at night, interspersing bouts of feeding, travelling, socializing, and idling (Tayler and Saayman, 1972; Shane and Schmidly, 1978; Irvine *et al.*, 1981; Gruber, 1981; Shane *et al.*, 1986; Ballance, 1992). The duration and frequency of activities are influenced by such environmental factors as season, habitat, time of day, and tidal state, and by physiological factors such as reproductive seasonality (Würsig and Würsig, 1979; Shane *et al.*, 1986; Hansen, 1990; Shane, 1990a,b; Ballance, 1992; Waples, 1995).

Food and feeding

The diets of bottlenose dolphins have been described from many regions, including the central US Atlantic coast (Leatherwood *et al.*, 1978; Mead and Potter, 1990), the southeastern USA (Gunter, 1942; Kemp, 1949; Barros and Odell, 1990), South Africa (Ross, 1977, 1984; Cockcroft and Ross, 1990c), the Atlantic coast of Africa (Cadenat, 1959), the Black Sea (Kleinenberg, 1938; Tsalkin, 1940), and Peru (Van Waerebeek *et al.*, 1990). A large variety of fish and/or squid forms most of the diets, although bottlenose dolphins seem to show a consistent preference for sciaenids, scombrids, and mugilids across regions (see summary in Table 4 of Mead and Potter, 1990).

Their diet varies with local prey availability. Along the central US Atlantic coast 31 genera of fish and two species of invertebrates were reported from the stomachs of dolphins (Mead and Potter, 1990). The four most common prey items were fish: *Cynoscion* cf. *C. regalis*, *Micropogonias undulatus*, *Leiostomus xanthurus*, and *Bairdiella chrysoura*. In the southeastern USA the fish found in dolphin stomachs represented 39 genera, with five species of cephalopods. The most important prey, by frequency of occurrence, were fish: *Bairdiella chrysoura*, *Micropogonias undulatus*, *Cynoscion arenarius*, *Mugil* sp., *Leiostomus xanthurus*, and the squid *Lolliguncula brevis* (Barros and Odell, 1990). Most fish prey were bottom dwellers, but some surface dwellers or pelagic fish were also represented. Samples of stomach contents from Texas waters showed mullet (*Mugil cephalus*) to be a primary prey item (Gunter, 1942; Kemp, 1949), although a more recent study indicated that, in general, mullet may not be as important as previously believed (Barros and Odell, 1990). Studies of stomach contents of dolphins caught in gill nets off South Africa found at least 50 genera of fish and at least three genera of cephalopods (Cockcroft and Ross, 1990a). The most important prey included fish: *Trachurus delagoae*, *Pomadasys olivaceum*, *Pagellus bellotti*, and *Scomber japonicus*, and the cephalopods *Sepia officinalis* and *Loligo* sp. This extensive variety of prey inhabits an equally diverse selection of habitats, and includes benthic-reef and sandy-bottom prey and their associated predators, pelagic schooling fish and cephalopods, and deeper-water fish.

Where both inshore and offshore *Tursiops* ecotypes have been identified, differences in diets have been found (Walker, 1981; Ross, 1977, 1984). In the eastern tropical Pacific, coastal *Tursiops* tend to feed primarily on fishes and invertebrates that are year-round inhabitants of the near-shore littoral and sublittoral zones, while offshore dolphins eat epipelagic fish and cephalopods (Walker, 1981). Off Peru, both coastal and offshore dolphins consumed Pacific sardines, anchoveta, and hake, but demersal species such as sciaenids and toadfish were found only in coastal dolphins. By contrast, the offshore animals were found with mesopelagic fishes and squids not found in the stomachs of coastal dolphins (Van Waerebeek *et al.*, 1990). Coastal dolphins from the southeastern USA typically feed on sciaenids and neritic cephalopods, whereas offshore dolphins feed more on pelagic ommastrephid squids (Barros and Odell, 1990). Off Sarasota, Florida, preliminary analyses indicate that dolphins stranded along the Gulf of Mexico beaches had both squid and fish in their stomachs, while known Sarasota Bay residents did not have squid, and fed more extensively on mullet than did the non-bay-residents (Barros and Wells, 1998). Finer-scale division of food resources relative to geography has been reported from South Africa. Cockcroft and Ross (1990c) have shown that bottlenose dolphin groups fed in different areas depending on sex and size. Lactating females and their calves frequented and fed in the nearshore zone, adolescents fed slightly farther offshore, and resting females and adult males fed farther still.

Bottlenose dolphins feed in a large variety of ways (Leatherwood, 1975; Bel'kovich *et al.*, 1978, 1991b; Shane *et al.*, 1986; Shane, 1990b). Individual feeding is perhaps most prevalent, but cooperative herding of schools of prey fish has been reported from a number of regions (Morozov, 1970; Caldwell and Caldwell, 1972a; Tayler and Saayman, 1972; Leatherwood, 1975; Shane, 1977; Bel'kovich *et al.*, 1978; Ballance, 1992). Individual prey capture involves behaviour as diverse as, e.g., high-speed chases with a pin-wheeling capture at the surface (Leatherwood, 1975), "fishwhacking" in which a fleeing fish is struck with the dolphin's flukes and often knocked clear of the water (Wells *et al.*, 1987; Shane, 1990a), pushing fish onto shore and then partially beaching to capture them (Hoese, 1971), and herding and perhaps disorienting fish with percussive leaps and tail lobs (Hamilton and Nishimoto, 1977). A more detailed listing of feeding behaviour has been published by Shane (1990b).

Tursiops takes advantage of human activities in order to facilitate prey capture in a variety of ways. In Mauritania and Brazil, dolphins regularly drive schools of mullet towards fishermen wading with nets in shallow water (Busnel, 1973; Pryor *et al.*, 1990; Simões-Lopes, 1991). Interactions with other fisheries include dolphins feeding behind shrimp trawls and in the vicinity of small purse seines, collecting discarded fish from these operations after the nets are retrieved, and stealing fish from a variety of fishing gear (Gunter, 1942, 1954; Norris and Prescott, 1961; Leatherwood, 1975; Caldwell and Caldwell, 1972a,b; Shane, 1977; Gruber, 1981; Shallenberger, 1981; Leatherwood and Reeves, 1982; Corkeron *et al.*, 1990).

Predation

Sharks are probably the most important dolphin predators. Mutual tolerance during encounters between sharks and dolphins is probably typical, but many dolphins bear shark-bite scars as evidence of occasional encounters with sharks. Wood *et al.* (1970) reported that about half of the Florida dolphins they examined bore shark-bite scars. Definite evidence of shark attack was found on 36.6% of the identifiable dolphins studied by Corkeron *et al.* (1987) in Moreton Bay, Australia. Wells *et al.* (1987) reported that 21.9% of live, non-calf bottlenose dolphins handled in the Sarasota area were scarred by shark bites. Cockcroft *et al.* (1989b) found shark-bite scars or wounds on 10.3% of *Tursiops* recovered from gill nets off Natal. About 4.1% of the stranded dolphins from the central USA Atlantic coast that were examined by Mead and Potter (1990) bore shark-bite wounds or scars. Most wounds and scars from sharks tend to be found on the posterior and ventral regions of the dolphins, suggesting that the dolphins were ambushed from behind and below (Cockcroft *et al.*, 1989b; Mead and Potter, 1990). Some of these attacks may have been something other than a predation attempt (e.g., sharks defending a territory).

Dolphin remains have been recovered in shark stomachs, indicating that at least some interactions are successful predation attempts (Wood *et al.*, 1970). Of more than 6000 sharks caught off Natal only 1.2% contained cetacean remains (Cockcroft *et al.*, 1989b). Thus, in at least some areas *Tursiops* appears to be a relatively minor and occasional part of the diets of sharks. Killer whales also may occasionally prey on bottlenose dolphins (Würsig and Würsig, 1979).

The primary shark species implicated as dolphin predators off Natal are the Zambesi shark (= bull shark, *Carcharhinus leucas*), tiger shark (*Galeocerdo cuvier*), great white shark (*Carcharodon carcharias*), and dusky shark (*Carcharhinus obscurus*). Five other species of sharks in Natal waters were also reported with dolphin remains. Similarly, off the southeastern USA, bull, tiger, and dusky sharks are among the species most often found with dolphin remains in their stomachs (Wood *et al.*, 1970). In Moreton Bay, Australia, tiger and great white sharks were implicated as the source of wounds in those cases where the species could be suggested (Corkeron *et al.*, 1987). Observations of captive and wild dolphins suggest that they may recognize certain species of sharks as potential predators (McBride and Hebb, 1948; Irvine *et al.*, 1973; Connor and Heithaus, 1996).

A number of anecdotal accounts describe dolphins attacking sharks, both in the wild and in captivity, primarily by butting them with their rostra or by striking them with their flukes (summarized by Wood *et al.*, 1970). Such defence may explain the apparently high survival rate indicated by the shark-bite scars on living dolphins. A relatively low incidence of shark-bite wounds is found on calves and juveniles, suggesting either that the young are well-protected by their mothers, or that attacks on young dolphins frequently prove fatal (Wells *et al.*, 1987; Cockcroft *et al.*, 1989b). Cockcroft *et al.* (1989b) found a predominance of young dolphin remains in the stomachs of sharks off Natal. The relatively high incidence of fresh shark bites observed by Corkeron *et al.* (1987) on nursing mothers and their calves in Australian waters, and the avoidance of sharks by mothers with calves, suggest that these adults or their calves may be quite vulnerable to predation attempts.

Although not predators, stingrays have been implicated in the deaths of some dolphins (Walsh *et al.*, 1988). The dolphins were wounded externally, or internally from ingestion of small rays, and deaths resulted from physical trauma, infection, or toxicosis.

Sound production

Tursiops produces three categories of sounds: whistles, echolocation clicks (Au, 1993), and burst-pulse sounds. The form and function of the whistles have been reviewed by Caldwell *et al.* (1990). Much of the recent research on the whistles has centred around the signature-whistle hypothesis proposed by Caldwell and Caldwell (1965). This theory argues that dolphins produce largely stereotypic whistles that are individually specific, and that the whistles could be used to communicate identity, location, and emotional state. Studies of captive and free-ranging dolphins have shown that once the signature whistle develops in neonates it remains stable for at least many years, and probably for life (Caldwell *et al.*, 1990; Sayigh *et al.*, 1990).

The signature whistles of many male calves are similar to the whistles of their mothers, while those of female calves are not (Sayigh *et al.*, 1990, 1995; Sayigh, 1992). This study continues to search for the causes of this difference, but it may reflect the different social patterns displayed by males and females. Both young males and females disperse from the female bands in which they were born, but the females often return to their natal band to raise their own offspring (Wells *et al.*, 1987). Intriguing observations by Tyack (1986, 1991) suggest that dolphins may mimic the signature whistles of other dolphins, apparently as a prelude to social interaction. *Tursiops* has the ability to vocally label objects (Richards, 1986), and it may be that signature whistles function as labels or "names".

Every two weeks for a year, Moore and Ridgway (1996) observed 24-hour patterns of sound production by two females kept together in a pool. Total numbers of calls ranged from just over 5000 to just under 30 000 per day. The animals were generally quiet between midnight and 0400. Springtime blood progesterone spikes indicating ovulation were correlated with an overall lower level of sound production.

Dolphin echolocation has been studied intensively, with most research being done on the bottlenose dolphin. Pulses are predominantly used for echolocation, although some burst pulse sounds apparently have a communicative function. The peak echolocation frequencies are about 40–130 kHz. Recent reviews of dolphin echolocation have been published by Au (1993), Popper (1980), Watkins and Wartzok (1985), Johnson (1986), Moore (1991), and in a series of conference proceedings sponsored by NATO Advanced Science Institute (Busnel, 1967; Busnel and Fish, 1980; Nachtigall and Moore, 1988). Auditory thresholds have been reviewed by Johnson (1986). Ridgway and Carder (1997) found three older male dolphins with high-frequency hearing loss above about 60 kHz; they also found one dolphin that was deaf and mute. Studies on the "language" acquisition capabilities of bottlenose dolphins conducted by Herman and colleagues over the last two decades have demonstrated an apparent ability of bottlenose dolphins to comprehend an artificial "language" (see reviews by Herman, 1980, 1986, 1987, 1991).

Parasites and Diseases

Parasites have been reviewed by Delyamure (1955), Zam *et al.* (1971), Dailey and Brownell (1972), Yablokov *et al.* (1972), Gibson and Harris (1979), Arvy (1982), and Howard *et al.* (1983) (Table 2).

Cetacean diseases have been discussed in numerous publications (e.g., Caldwell and Caldwell, 1968; Ridgway, 1972; Geraci, 1978; Howard, 1983; Dierauf, 1990; Geraci and Ridgway, 1991). Much of the information on dolphin diseases comes from bottlenose dolphins because much veterinary research has been done in oceanaria and research institutions, and because sick or dead dolphins frequently strand. Some of the more common diseases diagnosed are bacterial pneumonia, peripheral arterial infarction, pox virus, osteomyelitis of tooth alveoli, erysipelas, pasteurellosis, hepatitis, systematic mycosis, gastric ulcerations, enteritis, pancreatitis, and hepatic degenerative disease (Sweeney, 1978a,b,c).

During 1987–1988 a die-off affected the "coastal migratory stock" of *Tursiops* off the east coast of the USA. It was estimated that about half of this stock died during the 11-month event (Scott *et al.*, 1988). Identifying the proximate cause has been difficult. A hypothesis proposed by Geraci (1989) suggested poisoning by brevetoxin, a neurotoxin produced by red tide dinoflagellates, *Gymnodinium* (previously *Ptychodiscus*) *brevis*. It was suggested that the toxin triggered other infections such that the dolphins were "eventually infected with a host of bacterial and viral pathogens which produced an array of beguiling clinical signs". High levels of environmental contaminants, such as PCBs and DDEs, also were found (Kuehl *et al.*, 1991). Some indications were discovered that PCBs were being mobilized from the blubber to the liver, perhaps to toxic levels, but Geraci suggested that this mobilization may have been triggered by prior debilitation of the dolphin. Geraci's hypothesis suggested that a red tide bloom originating in Florida drifted north in the Gulf Stream, with brevetoxin becoming concentrated in fish and, eventually, the dolphins.

The presence of morbillivirus in six of 13 specimens (46%), however, raised suspicions about its role in the die-off because this virus was implicated in the high mortality of harbour seals (*Phoca vitulina*) off northern Europe and striped dolphins (*Stenella coeruleoalba*) in the Mediterranean (Mahy *et al.*, 1988; Osterhaus and Vedder, 1988; Domingo *et al.*, 1990). The high levels of environmental contaminants and the uncertainty about their role in the mortality raised concerns as well. Further analyses have concluded that the large scale mortalities in the US Atlantic and Gulf of Mexico were in fact caused by morbillivirus, and that these can be recurrent epizootic events (Hansen, 1992; Duignan *et al.*, 1996; Lipscomb *et al.*, 1996).

TABLE 2 Summary of known parasites of bottlenose dolphins (see also reviews by Delyamure, 1955; Zam *et al.*, 1971; Dailey and Brownell, 1972; Yablokov *et al.*, 1972; Arvy, 1982; and Van Waerebeek *et al.*, 1990

Trematoda (Flukes)	Cestoda (Tapeworms)
Braunina cordiformis	*Diphyllobothrium* sp.
Nasitrema sp.	*Monorygma delphini*
Nasitrema attenuata	*Monorygma grimaldii*
Nasitrema dalli	*Phyllobothrium delphini*
Nasitrema delphini	*Tetrabothrius forsteri*
Synthesium tursionis	Nematoda (Roundworms)
Zalophotrema hepaticum	*Anisakis* sp.
Zalophotrema pacificus	*Anisakis simplex* (=*tursionis*)
Campula palliata	*Anisakis marina*
Campula oblonga (=*campula*)	*Anisakis typica*
Campula delphini	*Crassicauda* sp.
Pholeter gastrophilus	*Crassicauda crassicauda*
	Halocerus sp.
	Halocerus lagenorhynchi
	Stenurus ovatus
	Acanthocephala (Thorny-headed worms)
	Corynosoma cetaceum

Live Maintenance

The most common dolphins held for exhibit or research, the first *Tursiops* were publicly displayed at the Brighton Aquarium in 1883 (Klinowska, 1991) and in the USA at the New York Aquarium in 1914 (Townsend, 1914). The opening of Marine Studios (now Marineland of Florida) in 1938 marked the beginning of the development of the public display industry (McBride, 1940; Lawrence and Schevill, 1954; Caldwell and Caldwell, 1972a). According to a June 1990 NMFS inventory, 34 US facilities held 428 bottlenose dolphins, and another 27 facilities in 16 other countries held an additional 195 bottlenose dolphins. This number is conservative because only those facilities outside the USA that have applied for US permits were included. Within the USA approximately 70% of the dolphins are held primarily for public display while the remainder are used primarily for research or military purposes.

Much has been learned about the requirements for care of dolphins over the last 50 years. This knowledge has resulted in increasing success in the long term maintenance of the animals, to the point where birth and survivorship rates at the better facilities approach and, possibly in a few cases, surpass, those of wild populations (DeMaster and Drevenak, 1988; Wells and Scott, 1990; Small and DeMaster, 1995). Breeding programmes have proved increasingly successful. In spite of the overall indications of improvements in the care of captive dolphins, however, significant variability remains in the quality of the facilities and in survivorship and birth rates (DeMaster and Drevenak, 1988; Small and DeMaster, 1995).

Human Effects

Directed fisheries

Historical catches have been reviewed previously (Mead, 1975a; Mitchell, 1975b; Leatherwood and Reeves, 1982; Northridge, 1984, 1991; Kasuya, 1985; Perrin, 1985; Northridge and Pilleri, 1986). The largest of the fisheries was in the Black Sea where dolphins were caught for oil, meat, and leather (Tomilin, 1957; Danilevskiy and Tyutyunnikov, 1968). The populations of common dolphins (*Delphinus delphis*), harbour porpoises (*Phocoena phocoena*), and bottlenose dolphins declined, and these countries have since outlawed the fishery. The populations are thought to be at risk (Smith, 1982; Brownell *et al.*, 1989; Reyes, 1990; Buckland *et al.*, 1992).

Directed takes still occur in various parts of the world. In Peru, coastal fisheries take *Tursiops* and other cetaceans for human consumption, using gill nets, purse seines, and harpoons (Read *et al.*, 1988; Van Waerebeek *et al.*, 1990). A similar fishery occurs in Sri Lanka (Alling, 1986; Leatherwood and Reeves, 1989; Prematunga *et al.*, 1990). The Japanese drive fishery off Iki Island and the Kii Peninsula takes several hundred *Tursiops* annually to reduce the perceived competition with the commercial fishery for yellowtail, *Seriola* sp. (Miyazaki, 1980; Kasuya, 1985; see also annual reports of the Small Cetacean Sub-Committee of the International Whaling Commission). Occasional dolphin catches for bait, human consumption, or predation control have been reported worldwide (Leatherwood and Reeves, 1982; Northridge, 1984, 1991; Northridge and Pilleri, 1986; Guerra *et al.*, 1987).

Live-capture fisheries for dolphins for public display have existed since near the turn of the century. Up to 20 March 1980, it was estimated that more than 1500 *Tursiops* had been removed from the waters of the USA, Mexico, and the Bahamas for display in many parts of the world or for research; this estimate, however, should be considered very approximate because records prior to 1972 are incomplete (Leatherwood and Reeves, 1982). More than 530 *Tursiops* have been taken from US waters since the passage of the Marine Mammal Protection Act of 1972 (MMPA), particularly from the southeastern USA (Scott, 1990). Present federal regulations limit the annual allowable take to less than 2% of the minimum estimated population in designated management areas, but no bottlenose dolphins have been collected in US waters since 1989. Some small-scale live-capture fisheries continue in other countries.

By-catches

Incidental catches of *Tursiops* have been reported for several fisheries. Purse seine fisheries in the eastern Pacific for tunas (Allen, 1985; Scott and Chivers, 1990) and sardines and anchovetas (Guerra *et al.*, 1987) have been reported to take small numbers of bottlenose dolphins. Dolphins have been killed by fishermen to prevent damage to their fishing gear or stealing of the catch or bait by the dolphins (see reviews by Leatherwood and Reeves, 1982; Kasuya, 1985). A recent conference on gill net fishery interactions with cetaceans summarized the worldwide incidental catches of cetaceans (Perrin *et al.*, 1994). A large incidental take of *Tursiops* has apparently occurred in the Taiwanese gill net fishery off Australia, with an annual mortality perhaps exceeding 2000 animals (Harwood *et al.*, 1984; Northridge, 1984, 1991).

The use of shark nets to protect bathing beaches in South Africa and Australia has caused mortality as well (Bannister, 1977; Ross *et al.*, 1989; Cockcroft, 1990; Cockcroft and Ross, 1991). Dolphins were often found with full stomachs, indicating recent feeding in the vicinity of the nets and that mortalities were correlated with the direction of the prevailing current (Cockcroft and Ross, 1991). Attempts to prevent dolphins from entangling by incorporating active and passive devices in the net were not successful (Peddemors *et al.*, 1991). The relatively high incidental catches of coastal dolphins off South Africa has prompted concerns that the take is not sustainable (Ross *et al.*, 1989; Cockcroft, 1990; Perrin *et al.*, 1994).

Habitat encroachment and pollution

The impacts of habitat alteration and pollution on dolphins have not been studied systematically. Anecdotal accounts suggest that human-caused degradation of bays and coastal waters may have led to declines in some dolphin populations (Gunter, 1942; Odell, 1976; FAO, 1978; Shane *et al.*, 1986).

Extremely high concentrations of chlorinated hydrocarbon residues have been found in the tissues of *Tursiops* (O'Shea *et al.*, 1980; Cockcroft *et al.*, 1989a). Concentrations of polychlorinated biphenyls and dieldrin were measured in the blubber of 105 South African bottlenose dolphins (Cockcroft *et al.*, 1989a). They found that first-born calves received 80% of their mother's body burden of contaminant residues, perhaps leading to increased neonatal mortality. Accumulation of contaminants in tissues of males have reached levels that theoretically could impair testosterone production and thus reduce reproductive ability. Geraci (1989) found high concentrations of a number of pollutants in the tissues of dolphin carcasses recovered during the die-off along the United States Atlantic seaboard. Preliminary results of research by Lahvis *et al.* (1995) indicate that even relatively low levels of PCBs and DDT such as those found in the blood of Sarasota dolphins can result in a decline in immune system function.

The coastal nature of bottlenose dolphins makes them particularly susceptible to human impacts. Recent mass mortalities of bottlenose dolphins have led to increased awareness of the possible cumulative and synergistic effects of habitat alteration, pollution, fisheries, vessel traffic, offshore industrial activity, and other human activities as they might impact bottlenose dolphin populations. The status of the species is listed in the IUCN *Red Data Book* as "insufficiently known", indicating that some populations are suspected, but have not been shown, to be at risk, and that more information is needed (Klinowska, 1991).

Acknowledgments

The late David and Melba Caldwell served as inspiration to both authors as we began our studies of bottlenose dolphins. David and Melba were to have been the original authors of this chapter, and we hope that our efforts in their place have met their high standards. We wish to thank Pamela Parker who assisted us with the literature review. Bill Perrin kindly allowed us to run amok through his reprint collection. Aleta Hohn, Dan Odell and the late Forrest G. Wood provided many suggestions for the improvement of this chapter.

References

Aguayo, A. L. (1975). Progress report of small cetacean research in Chile. *J. Fish. Res. Bd Can.* **32**, 1123–1143.

Allen, R. L. (1985). Dolphins and the purse-seine fishery for yellowfin tuna. *In* "Marine Mammals and Fisheries" (Eds J. R. Beddington, R. J. H. Beverton and D. M. Lavigne) pp. 236–252. Allen and Unwin, Boston.

Alling, A. (1986). Records of odontocetes in the northern Indian Ocean (1981–1982) and off the coast of Sri Lanka (1982–1984). *J. Bombay Nat. Hist. Soc.* **83**, 376–394.

Arseniev, V. A. (1986). "Atlas of Marine Mammals", T. F. H. Publishers, Neptune City, NJ.

Arvy, L. (1982). Phoresies and parasitism in cetaceans: A review. *Invest. Cetacea* **14**, 233–335.

Au, W. L. (1993). "Sonar of Dolphins". Springer-Verlag, New York.

Babenko, V. V. (1971). Some mechanical characteristics of the skin cover of dolphins. *Bionika* **5**, 76–81 (English translation, 1972, JPRS-55982, pp. 115–122, by Joint Publ. Res. Serv., Arlington, VA).

Ballance, L. T. (1990). Residence patterns, group organization, and surfacing associations of bottlenose dolphins in Kino Bay, Gulf of California, Mexico. *In* "The Bottlenose Dolphin" (Eds S. Leatherwood and R. R. Reeves) pp. 267–283. Academic Press, San Diego.

Ballance, L. T. (1992). Habitat use patterns and ranges of the bottlenose dolphin in the Gulf of California, Mexico. *Mar. Mamm. Sci.* **8**, 262–274.

Bannister, J. L. (1977). Incidental catches of small Cetacea off Australia. *Rep. Int. Whal. Commn* **27**, 506.

Barnes, L. G. (1990). The fossil record and evolutionary relationships of the genus *Tursiops*. *In* "The Bottlenose Dolphin" (Eds S. Leatherwood and R. R. Reeves) pp. 3–26. Academic Press, San Diego.

Barros, N. B. (1991). Recent cetacean records for southeastern Brazil. *Mar. Mamm. Sci.* **7**, 296–306.

Barros, N. B. and Odell, D. K. (1990). Food habits of bottlenose dolphins in the southeastern United States. *In* "The Bottlenose Dolphin" (Eds S. Leatherwood and R. R. Reeves) pp. 309–328. Academic Press, San Diego.

Barros, N. B. and Wells, R. S. (1998). Prey and feeding patterns of resident bottlenose dolphins (*Tursiops truncatus*) in Sarasota Bay, Florida. *J. Mamm.* **79**, 1045–1059.

Bel'kovich, V. M., Ivanova, Y., Yefremenkova, O. V., Kozarovitskiy, L. B. and Kharitonov, S. P. (1978). Description of the searching-hunting behavior of dolphins. *In* "Behavior and Bioacoustics of Dolphins" (Ed. V. M. Bel'kovich) pp. 24–56 (English translation, 1979, JPRS-L/8786 by Joint Publ. Res. Serv., Arlington, VA).

Bel'kovich, V. M., Agafonov, A. V., Yefremenkova, O. V., Kozarovitskiy, L. B. and Kharitonov, S. P. (1991a). Dolphin herd structure. *In* "Dolphin Societies: Discoveries and Puzzles" (Eds K. Pryor and K. S. Norris) pp. 17–38, 76–77. University of California Press, Berkeley, CA.

Bel'kovich, V. M., Ivanova, Y., Yefremenkova, O. V., Kozarovitskiy, L. B. and Kharitonov, S. P. (1991b). Searching and hunting behavior in the bottlenose dolphin (*Tursiops truncatus*) in the Black Sea. *In* "Dolphin Societies: Discoveries and Puzzles" (Eds K. Pryor and K. S. Norris) pp. 38–67, 76–77. University of California Press, Berkeley, CA.

Bello, M. A., Roy, R. R., Martin, T. P., Goforth Jr, H. W. and Edgerton, V. R. (1985). Axial musculature in the dolphin (*Tursiops truncatus*): some architectural and histochemical characteristics. *Mar. Mamm. Sci.* **1**, 324–336.

Brill, R. L., Sevenich, M. L., Sullivan, T. J., Sustman, J. D. and Witt, R. E. (1988). Behavioral evidence for hearing through the lower jaw by an echolocating dolphin (*Tursiops truncatus*). *Mar. Mamm. Sci.* **4**, 223–230.

Brown, D. H. and Norris, K. S. (1956). Observations of captive and wild cetaceans. *J. Mamm.* **37**, 311–326.

Brownell Jr, R. L., Ralls, K. and Perrin, W. F. (1989). The plight of the 'forgotten' whales. *Oceanus* **32**, 5–11.

Buckland, S. T., Smith, T. D. and Cattanach, K. L. (1992). Status of small cetacean populations in the Black Sea: Review of current information and suggestions for future research. *Rep. Int. Whal. Commn* **42**, 513–516.

Bullock, T. H., Grinnell, A. D., Ikezono, E., Kameda, K., Katsuki, Y., Nomoto, M., Sato, O., Suga, N. and Yanagisawa, K. (1968). Electrophysiological studies of central auditory mechanisms in cetaceans. *Z. vergl. Physiol.* **59**, 117–156.

Bullock, T. H. and Gurevich, V. S. (1979). Soviet literature on the nervous system and psychobiology of Cetacea. *Int. Rev. Neurobiol.* **21**, 47–127.

Busnel, R.-G. (1967). "Animal Sonar Systems: Biology and Bionics", Vols 1 and 2. Laboratoire de Physiologie Acoustique, Jouy-en-Josas-78, France.

Busnel, R.-G. (1973). Symbiotic relationship between man and dolphins. *Trans. NY Acad. Sci.* **35**, 112–131.

Busnel, R.-G. and Fish, J. F. (1980). "Animal Sonar Systems". Plenum Press, New York.

Cadenat, J. (1959). Notes sur les Delphinides ouest-africains. VI: Le gros dauphin gris (*Tursiops truncatus*) est-il capable de faire des plongées profondes? *Bull. Inst. Francais Afr. Noire* **21**A, 1137–1143.

Caldwell, D. K. (1955). Evidence of home range of an Atlantic bottlenose dolphin. *J. Mamm.* **36**, 304–305.

Caldwell, D. K. and Caldwell, M. C. (1968). "Proceedings of the Second Symposium on Diseases and Husbandry of Aquatic Mammals". Marineland Research Laboratory, St. Augustine, FL.

Caldwell, D. K. and Caldwell, M. C. (1972a). "The World of the Bottlenosed Dolphin". J. B. Lippincott Co., Philadelphia.

Caldwell, D. K. and Caldwell, M. C. (1972b). Senses and communication. *In* "Mammals of the Sea: Biology and Medicine" (Ed. S. H. Ridgway) pp. 466–502. Charles C. Thomas, Springfield, IL.

Caldwell, M. C. and Caldwell, D. K. (1965). Individualized whistle contours in bottlenosed dolphins (*Tursiops truncatus*). *Nature* **207**, 434–435.

Caldwell, D. K. and Golley, F. B. (1965). Marine mammals from the coast of Georgia to Cape Hatteras. *J. Elisha Mitchell Sci. Soc.* **81**, 24–32.

Caldwell, M. C., Caldwell, D. K. and Siebenaler, J. B. (1965). Observations on captive and wild Atlantic bottlenosed dolphins, *Tursiops truncatus*, in the northeastern Gulf of Mexico. *Los Angeles Co. Mus. Contrib. Sci.* **91**, 1–10.

Caldwell, M. C., Caldwell, D. K. and Tyack, P. L. (1990). Review of the signature-whistle hypothesis for the Atlantic bottlenose dolphin. *In* "The Bottlenose Dolphin" (Eds S. Leatherwood and R. R. Reeves) pp. 199–234. Academic Press, San Diego.

Cárdenas, J. C., Stutzin, M. E., Oporto, J. A., Cabello, C. and Torres, D. (1986). "Manual de Identificación de Los Cetáceos Chilenos". Project WH-445, World Wildlife Fund, Santiago, Chile.

Cave, A. J. E. (1977). The coronary vasculature of the bottlenosed dolphin (*Tursiops truncatus*). *In* "Functional Anatomy of Marine Mammals", Vol. 3 (Ed. R. J. Harrison) pp. 199–215. Academic Press, New York.

Cave, A. J. E. (1979). Tonsillar formation in the bottle-nosed dolphin (*Tursiops truncatus*). *Invest. Cetacea* **10**, 229–243.

Cave, A. J. E. and Aumonier, F. J. (1967). The reniculus of *Tursiops truncatus*, *Stenella longirostris*, and other cetaceans. *J. Roy. Microscop. Soc.* **86**, 323–342.

Cockcroft, V. G. (1990). Dolphin catches in the Natal shark nets, 1980 to 1988. *S. Afr. Tydskr. Natuurnav.* **20**, 44–51.

Cockcroft, V. G. and Ross, G. J. B. (1990a). Age, growth, and reproduction of bottlenose dolphins *Tursiops truncatus* from the east coast of southern Africa. *Fish. Bull.* **88**, 289–302.

Cockcroft, V. G. and Ross, G. J. B. (1990b). Observations on the early development of a captive bottlenose dolphin calf. *In* "The Bottlenose Dolphin" (Eds S. Leatherwood and R. R. Reeves) pp. 461–478. Academic Press, San Diego.

Cockcroft, V. G. and Ross, G. J. B. (1990c). Food and feeding of the Indian Ocean bottlenose dolphin off southern Natal, South Africa. *In* "The Bottlenose Dolphin" (Eds S. Leatherwood and R. R. Reeves) pp. 295–308. Academic Press, San Diego.

Cockcroft, V. G. and Ross, G. J. B. (1991). Bottlenose dolphins in Natal shark nets, 1980 through 1987: catch rates and associated contributing factors. *In* "Cetaceans and Cetacean Research in the Indian Ocean Sanctuary" (Eds S. Leatherwood and G. P. Donovan) pp. 115–127. United Nations Environmental Programme Marine Mammal Technical Report No. 3.

Cockcroft, V. G., De Kock, A. C., Lord, D. A. and Ross, G. J. B. (1989a). Organochlorines in bottlenose dolphins *Tursiops truncatus* from the east coast of South Africa. *S. Afr. J. Mar. Sci.* **8**, 207–217.

Cockcroft, V. G., Cliff, G. and Ross, G. J. B. (1989b). Shark predation on Indian Ocean bottlenose dolphins *Tursiops truncatus* off Natal, South Africa. *S. Afr. J. Zool.* **24**, 305–309.

Connor, R. C. (1990). Alliances among male bottlenose dolphins and a comparative analysis of mutualism. Doctoral dissertation, University of Michigan.

Connor, R. C. and Smolker, R. S. (1985). Habituated dolphins (*Tursiops* sp.) in Western Australia. *J. Mamm.* **66**, 398–400.

Connor, R. C., Smolker, R. and Richards, A. F. (1992a). Two levels of alliance formation among male bottlenose dolphins (*Tursiops* sp.). *Proc. Natl Acad. Sci. U.S.A.* **89**, 987–990.

Connor, R. C., Smolker, R. and Richards, A. F. (1992b). Dolphin alliances and coalitions. *In* "Coalitions and Alliances in Humans and Other Animals" (Eds A. H. Harcourt and F. B. M. de Waal) pp. 415–442. Oxford University Press, New York.

Connor, R. C. and Heithaus, M. R. (1996). Approach by great white shark elicits flight response in bottlenose dolphins. *Mar. Mamm. Sci.* **12**, 602–606.

Corkeron, P. J., Morris, R. J. and Bryden, M. M. (1987). Interactions between bottlenose dolphins and sharks in Moreton Bay. *Aquat. Mamm.* **13**, 109–113.

Corkeron, P. J., Bryden, M. M. and Hedstrom, K. E. (1990). Feeding by bottlenose dolphins in association with trawling operations in Moreton Bay, Australia. *In* "The Bottlenose Dolphin" (Eds S. Leatherwood and R. R. Reeves) pp. 329–336. Academic Press, San Diego.

Cranford, T. W., Amundin, M. and Norris, K. S. (1996). Functional morphology and homology in the odontocete nasal complex: implications for sound generation. *J. Morph.* **228**, 223–285.

Dailey, M. D. and Brownell Jr, R. L. (1972). A checklist of marine mammal parasites. *In* "Mammals of the Sea: Biology and Medicine" (Ed. S. H. Ridgway) pp. 528–589. Charles C. Thomas, Springfield, IL.

Danilevskiy, N. N. and Tyutyunnikov, V. P. (1968). Present state of Black Sea dolphin described. *Rybn. Khoz. (Moscow)* **11**, 25–27 (English translation, 1969, JPRS-47438, pp. 1–6, by Joint Publ. Res. Serv., Arlington, VA).

Dawson, W. W. (1980). The cetacean eye. *In* "Cetacean Behavior: Mechanisms and Functions" (Ed. L. M. Herman) pp. 53–100. Wiley, New York.

Dawson, W. W., Schroeder, J. P. and Dawson, J. F. (1987). The ocular fundus of two cetaceans. *Mar. Mamm. Sci.* **3**, 1–13.

Delyamure, S. L. (1955). "Helminthfauna of Marine Mammals (Ecology and Phylogeny)". Academy of Science USSR, Moscow (English translation, 1968, Israel Program for Scientific Translations, Jerusalem).

DeMaster, D. P. and Drevenak, J. K. (1988). Survivorship patterns in three species of captive cetaceans. *Mar. Mamm. Sci.* **4**, 297–311.

Dierauf, L. A. (1990). "Handbook of Marine Mammal Medicine: Health, Disease, and Rehabilitation". CRC Press, Boca Raton, FL.

Domingo, M., Ferrer, L., Pumarola, M., Marco, A., Plana, J., Kennedy, S., McAliskey, M. and Rima, B. K. (1990). Morbillivirus in dolphins. *Nature* **348**, 21.

Donaldson, B. J. (1977). The tongue of the bottlenosed dolphin (*Tursiops truncatus*). *In* "Functional Anatomy of Marine Mammals", Vol. 3 (Ed. R. J. Harrison) pp. 175–197. Academic Press, New York.

Dormer, K. J. (1979). Mechanism of sound production and air recycling in delphinids: Cineradiographic evidence. *J. Acoust. Soc. Am.* **65**, 229–239.

Drabek, C. M. and Kooyman, G. L. (1986). Bronchial morphometry of the upper conductive zones of four odontocete cetaceans. *In* "Research on Dolphins" (Eds M. M. Bryden and R. Harrison) pp. 109–148. Oxford University Press, Oxford.

Dral, A. D. G. (1977). On the retinal anatomy of Cetacea (mainly *Tursiops truncatus*). *In* "Functional Anatomy of Marine Mammals", Vol. 3 (Ed. R. J. Harrison) pp. 81–134. Academic Press, New York.

Duffield, D. A. and Wells, R. S. (1986). Population structure of bottlenose dolphins: (2) Genetic studies of dolphins along the central west coast of Florida. NMFS Contract Rep. 45-WCNF-5-00366.

Duffield, D. A. and Wells, R. S. (1991). The combined application of chromosome, protein and molecular data for investigation of social unit structure and dynamics in *Tursiops truncatus*. *In* "Genetic Ecology of Whales and Dolphins" (Ed. A. R. Hoelzel) pp. 155–169 (*Rep. Int. Whal. Commn*, Spec. Issue No. 13, Cambridge).
Duignan, P. J., House, C., Odell, D. K., Wells, R. S., Hansen, L. J., Walsh, M. T., St. Aubin, D. J., Rima, B. K. and Geraci, J. R. (1996). Morbillivirus infection in bottlenose dolphins: Evidence for recurrent epizootics in the western Atlantic and Gulf of Mexico. *Mar. Mamm. Sci.* **12**, 499–515.
Elliot, D. G. (1901). A synopsis of the mammals of North America and the adjacent seas. *Publ. Field Columbian Mus. (Zool. Ser.)* **2**, 1–471.
Elsner, R., Pirie, J., Kenney, D. D. and Schemmer, S. (1974). Functional circulatory anatomy of cetacean appendages. *In* "Functional Anatomy of Marine Mammals", Vol. 2 (Ed. R. J. Harrison) pp. 143–159. Academic Press, New York.
Fanning, J. C. and Harrison, R. J. (1974). The structure of the trachea and lungs of the South Australian bottlenosed dolphin. *In* "Functional Anatomy of Marine Mammals", Vol. 2 (Ed. R. J. Harrison) pp. 231–252. Academic Press, New York.
Felts, W. J. L. (1966). Some functional and structural characteristics of cetacean flippers and flukes. *In* "Whales, Dolphins, and Porpoises" (Ed. K. S. Norris) pp. 252–276. University of California Press, Berkeley, CA.
Ferrero, R. C. and Tsunoda, L. M. (1989). First record of a bottlenose dolphin (*Tursiops truncatus*) in Washington State. *Mar. Mamm. Sci.* **5**, 302–305.
Food and Agriculture Organization of the UN (FAO) (1978). "Mammals in the Seas", Vol. 1, FAO Fish. Ser. No. 5.
Fraser, F. C. and Purves, P. E. (1960). Hearing in cetaceans: evolution of the accessory air sacs and the structure and function of the outer and middle ear in recent cetaceans. *Bull. Brit. Mus. (Nat. Hist.)* **7**, 1–140.
Gaskin, D. E. (1972). "Whales, Dolphins, and Seals". Heinemann, Auckland, NZ.
Gaskin, D. E. (1986). Kidney and water metabolism. *In* "Research on Dolphins" (Eds M. M. Bryden and R. Harrison) pp. 129–148. Oxford Science Publishers, Oxford.
Geraci, J. R. (1978). Marine mammals (Cetacea, Pinnipedia, and Sirenia). *In* "Zoo and Wild Animal Medicine" (Ed. M. E. Fowler) pp. 554–610. Saunders, Philadelphia.
Geraci, J. R. (1989). Clinical investigations of the 1987–88 mass mortality of bottlenose dolphins along the US central and south Atlantic coast. Final report to the Nat. Mar. Fish. Serv., US Navy, Office of Naval Research, and Marine Mammal Commission.
Geraci, J. R., St. Aubin, D. J. and Hicks, B. D. (1986). The epidermis of odontocetes: a review from within. *In* "Research on Dolphins" (Eds M. M. Bryden and R. Harrison) pp. 3–21. Clarendon Press, Oxford.
Geraci, J. R. and Ridgway, S. H. (1991). On disease transmission between cetaceans and humans. *Mar. Mamm. Sci.* **7**, 191–194.
Gervais, P. (1855). "Histoire Naturelle des Mammiferes". Paris.
Gibson, D. I. and Harris, E. A. (1979). The helminth-parasites of cetaceans in the collection of the British Museum (Natural History). *Invest. Cetacea* **10**, 309–324.
Green, R. F. (1972). Observations on the anatomy of some cetaceans and pinnipeds. *In* "Mammals of the Sea: Biology and Medicine" (Ed. S. H. Ridgway) pp. 247–297. Charles C. Thomas, Springfield, IL.
Gray, J. E. (1843). "List of the Mammals of the British Museum". London.
Gruber, J. A. (1981). Ecology of the Atlantic bottlenosed dolphin (*Tursiops truncatus*) in the Pass Cavallo area of Matagorda Bay, Texas. Masters thesis, Texas A. & M. University.

Guerra, C. G., Van Waerebeek, K., Kandora, G. P. and Jorquera, G. L. (1987). Presencia de cetaceos frente a la segunda region de Chile (The presence of cetaceans off the northern Chilean coast). *Estud. Oceanol.* **6**, 87–96 (English abstract).

Gunter, G. (1942). Contributions to the natural history of the bottlenose dolphins, *Tursiops truncatus* (Montague), on the Texas coast, with particular reference to food habits. *J. Mamm.* **23**, 267–276.

Gunter, G. (1954). Mammals of the Gulf of Mexico. *In* "Gulf of Mexico: Its Origin, Waters, and Marine Life" (Ed. P. S. Galtsoff) pp. 543–551. US Fish and Wildlife Service Fisheries Bulletin No. 55.

Hamilton, P. V. and Nishimoto, R. T. (1977). Dolphin predation on mullet. *Fla. Sci.* **40**, 251–252.

Hansen, L. J. (1983). Population biology of the coastal bottlenose dolphin (*Tursiops truncatus*) of southern California. Masters thesis, California State University, Sacramento.

Hansen, L. J. (1990). California coastal bottlenose dolphins. *In* "The Bottlenose Dolphin" (Eds S. Leatherwood and R. R. Reeves) pp. 403–420. Academic Press, San Diego.

Hansen, L. J. (1992). Report on investigation of 1990 Gulf of Mexico bottlenose dolphin strandings. Nat. Mar. Fish. Serv. Southeast Fisheries Center Contribution No. MIA-92/93-21.

Hansen, L. J. and Defran, R. H. (1990). A comparison of photo-identification studies of California coastal bottlenose dolphins. *In* "Individual Recognition of Cetaceans: Use of Photo-Identification and Other Techniques to Estimate Population Parameters" (Eds P. S. Hammond, S. A. Mizroch and G. P. Donovan) pp. 101–104. *Rep. Int. Whal. Commn*, Spec. Issue 12, Cambridge.

Harrison, R. J. (1969). Endocrine organs: hypophysis, thyroid, and adrenal. *In* "The Biology of Marine Mammals" (Ed. H. T. Andersen) pp. 349–390. Academic Press, New York.

Harrison, R. J. and Thurley, K. W. (1974). Structure of the epidermis in *Tursiops*, *Delphinus*, *Orcinus* and *Phocoena*. *In* "Functional Anatomy of Marine Mammals", Vol. 2 (Ed. R. J. Harrison) pp. 45–71. Academic Press, New York.

Harrison, R. J., Brownell Jr, R. L. and Boice, R. C. (1972). Reproduction and gonadal appearances in some odontocetes. *In* "Functional Anatomy of Marine Mammals", Vol. 2 (Ed. R. J. Harrison) pp. 361–429. Academic Press, New York.

Harrison, R. J., Johnson, F. R. and Young, B. A. (1977). The small intestine of dolphins (*Tursiops*, *Delphinus*, *Stenella*). *In* "Functional Anatomy of Marine Mammals", Vol. 3 (Ed. R. J. Harrison) pp. 297–331. Academic Press, New York.

Harwood, M. B., McNamara, K. J. and Anderson, G. R. V. (1984). Incidental catch of small cetaceans in a gillnet fishery in northern Australian waters. *Rep. Int. Whal. Commn* **34**, 555–559.

Herman, L. M. (1980). Cognitive characteristics of dolphins. *In* "Cetacean Behavior: Mechanisms and Functions" (Ed. L. M. Herman) pp. 363–429. Wiley, New York.

Herman, L. M. (1986). Cognition and language competencies of bottlenosed dolphins. *In* "Dolphin Cognition and Behavior: A Comparative Approach" (Eds R. J. Schusterman, J. A. Thomas and F. G. Wood) pp. 221–252. Lawrence Erlbaum, Hillsdale, NJ.

Herman, L. M. (1987). Receptive competencies of language-trained animals. *In* "Advances in Animal Behavior", Vol. 17 (Eds J. S. Rosenblatt, C. Beer, M.-C. Busnel, and P. J. B. Slater) pp. 1–60. Academic Press, Orlando.

Herman, L. M. (1991). What the dolphin knows, or might know, in its natural world. *In* "Dolphin Societies: Discoveries and Puzzles" (Eds K. Pryor and K. S. Norris) pp. 349–363. University of California Press, Berkeley, CA.

Hersh, S. L. and Duffield, D. A. (1990). Distinction between northwest Atlantic offshore and coastal bottlenose dolphins based on hemoglobin profile and morphometry. *In*

"The Bottlenose Dolphin" (Eds S. Leatherwood and R. R. Reeves) pp. 129–139. Academic Press, San Diego.

Hersh, S. H., Odell, D. K. and Asper, E. D. (1990a). Sexual dimorphism in bottlenose dolphins from the east coast of Florida. *Mar. Mamm. Sci.* **6**, 305–315.

Hersh, S. L., Odell, D. K. and Asper, E. D. (1990b). Bottlenose dolphin mortality patterns in the Indian/Banana River system of Florida. *In* "The Bottlenose Dolphin" (Eds S. Leatherwood and R. R. Reeves) pp. 155–164. Academic Press, San Diego.

Hershkovitz, P. (1966). Catalog of Living Whales. U.S. Nat. Mus. Bull. No. 246. Smithsonian Institution, Washington, DC.

Hoese, H. D. (1971). Dolphins feeding out of water in a salt marsh. *J. Mamm.* **52**, 222–223.

Hohn, A. A. (1980). Age determination and age related factors in the teeth of western North Atlantic bottlenose dolphins. *Sci. Rep. Whales Res. Inst.* **32**, 39–66.

Hohn, A. A. (1990). Reading between the lines: analysis of age estimation in dolphins. *In* "The Bottlenose Dolphin" (Eds S. Leatherwood and R. R. Reeves) pp. 575–585. Academic Press, San Diego.

Hohn, A. A., Scott, M. D., Wells, R. S., Sweeney, J. C. and Irvine, A. B. (1989). Growth layers in teeth from known-age, free-ranging bottlenose dolphins. *Mar. Mamm. Sci.* **5**, 315–342.

Honacki, J. H., Kinman, K. E. and Koeppl, J. W. (1982). "Mammal Species of the World". Allen Press and The Association of Systematics Collections, Lawrence, KS.

Howard, E. B. (1983). "Pathobiology of Marine Mammal Diseases", Vols. 1 and 2. CRC Press, Boca Raton, FL.

Howard, E. B., Britt Jr, J. O. and Matsumoto, G. K. (1983). Parasitic diseases. *In* "Pathobiology of Marine Mammal Diseases", Vol. 1 (Ed. E. B. Howard) pp. 119–232. CRC Press, Boca Raton, FL.

Irvine, B. and Wells, R. S. (1972). Results of attempts to tag Atlantic bottlenosed dolphins (*Tursiops truncatus*). *Cetology* **13**, 1–5.

Irvine, B., Wells, R. S. and Gilbert, P. W. (1973). Conditioning an Atlantic bottle-nosed dolphin, *Tursiops truncatus*, to repel various species of sharks. *J. Mamm.* **54**, 503–505.

Irvine, B., Scott, M. D., Wells, R. S. and Kaufmann, J. H. (1981). Movements and activities of the Atlantic bottlenose dolphin, *Tursiops truncatus*, near Sarasota, Florida. *Fish. Bull.* **79**, 671–688.

Ivashin, M. V. and Borodin, R. G. (1992). U.S.S.R. Progress Report on Cetacean Research, March 1990 to April 1991. *Rept. Int. Whal. Commn* **42**, 374–376.

Johnson, C. S. (1986). Dolphin audition and echolocation capacities. *In* "Dolphin Cognition and Behavior: A Comparative Approach" (Eds R. J. Schusterman, J. A. Thomas and F. G. Wood) pp. 115–139. Lawrence Erlbaum, Hillsdale, NJ.

Kasuya, T. (1985). Fishery–dolphin conflict in the Iki Island area of Japan. *In* "Marine Mammals and Fisheries" (Eds J. R. Beddington, R. J. H. Beverton and D. M. Lavigne) pp. 253–272. George Allen & Unwin, London.

Kasuya, T. and Marsh, H. (1984). Life history and reproductive biology of the short-finned pilot whale, *Globicephala macrorhynchus*, off the Pacific coast of Japan. *In* "Reproduction in Whales, Dolphins and Porpoises" (Eds W. F. Perrin, R. L. Brownell and D. P. DeMaster) pp. 259–310. *Rep. Int. Whal. Commn*, Spec. Issue No. 6, Cambridge.

Kasuya, T., Tobayama, T., Saiga, T. and Kataoka, T. (1986). Perinatal growth of delphinoids: information from aquarium reared bottlenose dolphins and finless porpoises. *Sci. Rep. Whales Res. Inst.* **37**, 85–97.

Kemp, R. J. (1949). Report on stomach analysis — Delphininae. Annual Report of the Marine Laboratory of the Texas Game, Fish and Oyster Community for the fiscal year 1948–1949, pp. 111–112, 126–127.

Kenney, R. D. (1990). Bottlenose dolphins off the northeastern United States. *In* "The Bottlenose Dolphin" (Eds S. Leatherwood and R. R. Reeves) pp. 369–386. Academic Press, San Diego.

Kirby, V. L. and Ridgway, S. H. (1984). Hormonal evidence of spontaneous ovulation in captive dolphins, *Tursiops truncatus* and *Delphinus delphis*. *In* "Reproduction in Whales, Dolphins and Porpoises" (Eds W. P. Perrin, R. L. Brownell and D. P. DeMaster) pp. 459–464. *Rep. Int. Whal. Commn*, Spec. Issue No. 6, Cambridge.

Kleinenberg, S. E. (1938). Nekotorye dannye o pitanii *Tursiops tursio* Fabr. v Chernom more (Some data on feeding of *Tursiops tursio* Fabr. in the Black Sea). *Bull. Soc. Nat. Moscow* **47**, 406–413.

Klinowska, M. (1991). "Dolphins, Porpoises and Whales of the World": "The IUCN Red Data Book". IUCN, Cambridge.

Kruger, L. (1959). The thalamus of the dolphin (*Tursiops truncatus*) and comparison with other mammals. *J. Comp. Neurol.* **111**, 133–194.

Kruger, L. (1966). Specialized features of the cetacean brain. *In* "Whales, Dolphins and Porpoises" (Ed. K. S. Norris) pp. 232–254. University of California Press, Berkeley, CA.

Kuehl, D. W., Haebler, R. and Potter, C. (1991). Chemical residues in dolphins from the U.S. Atlantic coast including Atlantic bottlenose obtained during the 1987/1988 mass mortality. *Chemosphere* **22**, 1071–1084.

Lacépède, B. G. (1804). "Histoire Naturelle des Cetacés". Paris.

Lahvis, G. P., Wells, R. S., Kuehl, D. W., Stewart, J. L., Rhinehart, H. L. and Via, C. S. (1995). Decreased lymphocyte responses in free-ranging bottlenose dolphins (*Tursiops truncatus*) are associated with increased concentrations of PCBs and DDT in peripheral blood. *Envir. Hlth Perspect.* **103**, 67–72.

Lawrence, B. and Schevill, W. E. (1954). *Tursiops* as an experimental subject. *J. Mamm.* **35**, 225–232.

Lear, R. J. and Bryden, M. M. (1980). A study of the bottlenose dolphin *Tursiops truncatus* in eastern Australian waters. Australian National Parks and Wildlife Service Occasional Paper No. 4, pp. 1–25.

Leatherwood, S. (1975). Some observations of feeding behavior of bottlenosed dolphins (*Tursiops truncatus*) in the northern Gulf of Mexico and (*Tursiops* cf. *gilli*) off southern California, Baja California, and Nayarit, Mexico. *Mar. Fish. Rev.* **37**, 10–16.

Leatherwood, S. and Reeves, R. R. (1982). Bottlenose dolphin (*Tursiops truncatus*) and other toothed cetaceans. *In* "Wild Mammals of North America: Biology, Management, Economics" (Eds J. A. Chapman and G. A. Feldhamer) pp. 369–414. Johns Hopkins University Press, Baltimore, MD.

Leatherwood, S. and Reeves, R. R. (1989). Marine mammal research and conservation in Sri Lanka, 1985–1986. United Nations Environmental Program Marine Mammal Technical Report No. 1, vi + 138 pp.

Leatherwood, S. and Reeves, R. R. (1990). "The Bottlenose Dolphin". Academic Press, San Diego.

Leatherwood, S., Caldwell, D. K. and Winn, H. E. (1976). "Whales, Dolphins, and Porpoises of the Western North Atlantic: A Guide to Their Identification". NOAA Technical Report NMFS CIRC-396.

Leatherwood, S., Deerman, M. W. and Potter, C. W. (1978). Food and reproductive status of nine *Tursiops truncatus* from the northeastern United States coast. *Cetology* **28**, 1–6.

Leatherwood, S., Reeves, R. R., Perrin, W. F. and Evans, W. E. (1988). "Whales, Dolphins, and Porpoises of the Eastern North Pacific and Adjacent Arctic Waters: A Guide to Their Identification". Dover, New York.

Lévêque, R. (1963). Notes sur quatre cétacés de l'Océan Pacifique (Ecuador et Galapagos). *Mammalia* **27**, 608–609.

Lichter, A. and Hooper, A. (1984). "Guia para el Reconocimiento de Cetáceos del Mar Argentino". Fundación Vida Silvestre Argentina, Buenos Aires.

Lipscomb, T. P., Kennedy, S., Moffett, D., Krafft, A., Klaunberg, B. A., Lichy, J. H., Regan, G. T., Worthy, G. A. J. and Taubenberger, J. K. (1996). Morbilliviral epizootic in bottlenose dolphins of the Gulf of Mexico. *J. Vet. Diagn. Invest.* **8**, 283–290.

Lohoefener, R., Hoggard, W., Ford, R. and Benigno, J. (1987). Studies of Mississippi Sound bottlenose dolphins. Final report to the US Marine Mammal Commission.

Mackay, R. S. and Liaw, H. M. (1981). Dolphin vocalization mechanisms. *Science* **212**, 676–678.

Mahy, B. W. J., Barrett, T., Evans, S., Anderson, E. C. and Bostock, C. J. (1988). Characterization of a seal morbillivirus. *Nature* **336**, 115.

McBrearty, D. A., Message, M. A. and King, G. A. (1986). Observations on small cetaceans in the northwest Atlantic Ocean and the Mediterranean Sea: 1978–1982. *In* "Research on Dolphins" (Eds M. M. Bryden and R. J. Harrison) pp. 225–249. Oxford University Press.

McBride, A. F. (1940). Meet Mr. Porpoise. *Nat. Hist.* **45**, 16–29.

McBride, A. F. and Hebb, D. O. (1948). Behavior of the captive bottle-nose dolphin, *Tursiops truncatus*. *J. Comp. Physiol. Psych.* **41**, 111–123.

McBride, A. F. and Kritzler, H. (1951). Observations on pregnancy, parturition, and post-natal behavior in the bottlenose dolphin. *J. Mamm.* **32**, 251–266.

McCormick, J. G., Wever, E. G., Palin, J. and Ridgway, S. H. (1970). Sound conduction in the dolphin ear. *J. Acoust. Soc. Am.* **48**, 1418–1428.

McFarland, W. L., Morgane, P. J. and Jacobs, M. S. (1969). Ventricular system of the brain of the dolphin, *Tursiops truncatus*, with comparative anatomical observations and relations to brain specializations. *J. Comp. Neurol.* **135**, 275–368.

Mead, J. G. (1975a). Preliminary report on the former net fisheries for *Tursiops truncatus* in the western North Atlantic. *J. Fish. Res. Bd Can.* **32**, 1155–1162.

Mead, J. G. (1975b). Anatomy of the external nasal passages and facial complex in the Delphinidae (Mammalia: Cetacea). *Smithsonian Contrib. Zool.* **207**, 1–72.

Mead, J. G. and Potter, C. W. (1990). Natural history of bottlenose dolphins along the central Atlantic coast of the United States. *In* "The Bottlenose Dolphin" (Eds S. Leatherwood and R. R. Reeves) pp. 165–195. Academic Press, San Diego.

Mitchell, E. (1970). Pigmentation pattern evolution in delphinid cetaceans: an essay in adaptive coloration. *Can. J. Zool.* **48**, 717–740.

Mitchell, E. (1975a). Report of the meeting on smaller cetaceans, Montreal, April 1–11, 1974. *J. Fish. Res. Bd Can.* **32**, 889–983.

Mitchell, E. (1975b). Porpoise, dolphin and small whale fisheries of the world: status and problems. International Union for Conservation of Nature, Monograph No. 3, pp. 1–129.

Miyashita, T. (1992). Abundance of dolphin stocks in the western North Pacific taken by the Japanese drive fishery. Document SC/44/SM15 presented to the IWC Scientific Committee.

Miyazaki, N. (1980). Catch records of cetaceans off the coast of the Kii Peninsula. *Mem. Nat. Sci. Mus. Tokyo* **13**, 69–82.

Montagu, G. (1821). Description of a species of *Delphinus* which appears to be new. *Mem. Wernerian Nat. Hist. Soc.* **3**, 75–82.

Moore, P. W. B. (1991). Dolphin psychophysics: Concepts for the study of dolphin echolocation. *In* "Dolphin Societies: Discoveries and Puzzles" (Eds K. Pryor and K. S. Norris) pp. 365–382. University of California Press, Berkeley, CA.

Moore, S. E. and Ridgway, S. H. (1996). Patterns of sound production and ovulation. *Aquat. Mamm.* **22**, 175–184.

Morgane, P. J., Jacobs, M. S. and McFarland, W. L. (1980). The anatomy of the brain of the bottlenose dolphin (*Tursiops truncatus*): surface configurations of the telencephalon of the bottlenose dolphin with comparative anatomical observations in four other cetacean species. *Brain Res. Bull.* **5**, 1–108.

Morozov, D. A. (1970). Dolphin hunting. *Rybnoe Khoziaistvo* **46**, 16–17.

Morris, R. J. (1986). The acoustic facility of dolphins. *In* "Research on Dolphins" (Eds M. M. Bryden and R. Harrison) pp. 369–399. Oxford University Press.

Mullin, K. D. (1988). Comparative seasonal abundance and ecology of bottlenose dolphins (*Tursiops truncatus*) in three habitats of the north-central Gulf of Mexico. Doctoral dissertation, Mississippi State University.

Nachtigall, P. E. and Moore, P. W. B. (1988). "Animal Sonar: Processes and Performance". Plenum Press, New York.

Nagel, E. L., Morgane, P. J., McFarland, W. L. and Galliano, R. E. (1968). Rete mirabile of dolphin: its pressure-damping effect on cerebral circulation. *Science* **161**, 898–900.

Nishiwaki, M. (1972). General biology. *In* "Mammals of the Sea" (Ed. S. H. Ridgway) pp. 3–204. Charles C. Thomas, Springfield, IL.

Norris, K. S. (1967). Aggressive behavior in Cetacea. *In* "Aggression and Defense" (Eds C. D. Clemente and D. B. Lindsley) pp. 225–241. University of California Press, Berkeley, CA.

Norris, K. S. (1968). The evolution of acoustic mechanisms in odontocetes. *In* "Evolution and Environment" (Ed. E. T. Drake) pp. 297–324. Yale University Press, New Haven, CT.

Norris, K. S. and Prescott, J. H. (1961). Observations on Pacific cetaceans of Californian and Mexican waters. University of California Publications in Zoology No. 63, pp. 291–402.

Northridge, S. P. (1984). World review of interactions between marine mammals and fisheries. *FAO Tech. Paper* **251**, 1–190.

Northridge. S. P. (1991). An updated world review of interactions between marine mammals and fisheries. *FAO Tech. Paper* **251** (Suppl. 1), 1–58.

Northridge, S. P. and Pilleri, G. (1986). A review of human impact on small cetaceans. *Invest. Cetacea* **18**, 221–260.

Odell, D. K. (1976). Distribution and abundance of marine mammals in south Florida: Preliminary results. University of Miami Sea Grant Special Report No. 5, pp. 203–212.

Odell, D. K. and Schneyer, A. L. (1983). Age estimation and hormone analyses for bottlenose dolphins, *Tursiops truncatus*, from Mississippi. *In* "Atlantic Bottlenose Dolphin, *Tursiops truncatus*, Herd Studies in the Mississippi Sound, USA: Capture, Freeze Marking and Biological Sampling" (Eds M. A. Solangi and G. E. Dukes) pp. 131–152. National Marine Fisheries Service Contract Report No. NA 82-GA-C-00023.

Odell, D. K. and Asper, E. D. (1990). Distribution and movements of freeze-branded bottlenose dolphins in the Indian and Banana Rivers, Florida. *In* "The Bottlenose Dolphin" (Eds S. Leatherwood and R. R. Reeves) pp. 515–540. Academic Press, San Diego.

Osterhaus, A. D. M. E and Vedder, E. J. (1988). Identification of virus causing recent seal deaths. *Nature* **335**, 20.

O'Shea, T. J., Brownell, R. L., Clark, D. R., Walker, W. A., Gay, M. L. and Lamont, T. G. (1980). Organochlorine pollutants in small cetaceans from the Pacific and South Atlantic Oceans, November 1968–June 1976. *Pesticides Monitoring J.* **14**, 35–46.

Östman, J. (1991). Changes in aggressive and sexual behavior among two male bottlenose dolphins (*Tursiops truncatus*) in a captive colony. *In* "Dolphin Societies: Discoveries and Puzzles" (Eds K. Pryor and K. S. Norris) pp. 305–317. University of California Press, Berkeley, CA.

Pabst, D. A. (1987). A membranous force transmission system in dolphins. *Am. Zool.* **27**, 86A.

Pabst, D. A. (1990). Axial muscles and connective tissues of the bottlenose dolphin. *In* "The Bottlenose Dolphin" (Eds S. Leatherwood and R. R. Reeves) pp. 51–67. Academic Press, San Diego.

Peddemors, V. M., de Muelenaere, H. J. H. and Devchand, K. (1989). Comparative milk composition of the bottlenosed dolphin (*Tursiops truncatus*), humpback dolphin (*Sousa plumbea*), and common dolphin (*Delphinus delphis*) from southern African waters. *J. Comp. Biochem. Physiol.* **94**A, 639–641.

Peddemors, V. M., Cockcroft, V. G. and Wilson, R. B. (1991). Incidental dolphin mortality in the Natal shark nets: A preliminary report on prevention measures. *In* "Cetaceans and Cetacean Research in the Indian Ocean Sanctuary" (Eds S. Leatherwood and G. P. Donovan) pp. 129–137. United Nations Environmental Program on Marine Mammal Technology Report No. 3.

Peddemors, V. M., Fothergill, M. and Cockcroft, V. G. (1992). Feeding and growth in a captive-born bottlenose dolphin *Tursiops truncatus*. *S.-Afr. Tydskr. Dierk.* **27**, 74–80.

Perrin, W. F. (1972). Colour patterns of spinner porpoises (*Stenella* cf. *S. longirostris*) of the eastern Pacific and Hawaii, with comments on delphinid pigmentation. *Fish. Bull.* **70**, 983–1003.

Perrin, W. F. (1985). The former dolphin fishery at St. Helena. *Rep. Int. Whal. Commn* **35**, 423–428.

Perrin, W. F. and Reilly, S. B. (1984). Reproductive parameters of dolphins and small whales of the family Delphinidae. *In* "Reproduction in Whales, Dolphins and Porpoises" (Eds W. F. Perrin, R. L. Brownell and D. P. DeMaster) pp. 97–133. *Rep. Int. Whal. Commn*, Spec. Issue No. 6, Cambridge.

Perrin, W. F., Donovan, G. P. and Barlow, J. (eds) (1994). "Gillnets and Cetaceans". *Rep. Int. Whal. Commn*, Spec. Issue No. 15, Cambridge.

Perrine, D. (1991). JoJo: Rogue dolphin? *Sea Front* **36**, 32–41.

Pilleri, G. (1985). Second record of *Tursiops osennae* (Cetacea, Delphinidae) in a pliocene horizon of the Romagna Apennines, central Italy, and the phylogeny of *Tursiops*. *Invest. Cetacea* **17**, 11–30.

Pilleri, G. and Arvy, L. (1981). The precursors in cetology: from Guillaume Rondelet to John Anderson. *Invest. Cetacea* **12**, 1–258.

Pinedo, M. C., Rosas, F. C. W. and Marmontel, M. (1992). "Cetáceos e Pinípedes do Brasil". UNEP/FUA, Manaus, Brazil.

Polachek, T. (1987). Relative abundance, distribution and inter-specific relationship of cetacean schools in the eastern tropical Pacific. *Mar. Mamm. Sci.* **3**, 54–77.

Popper, A. N. (1980). Sound emission and detection by delphinids. *In* "Cetacean Behavior: Mechanisms and Functions" (Ed. L. M. Herman) pp. 1–52. Wiley, New York.

Prematunga, W. P., Leatherwood, S. and Alling, A. (1990). Bycatches of small cetaceans in gillnets off Trincomalee, Sri Lanka, 1984–1986. *In* "Cetaceans and Cetacean Research in the Indian Ocean Sanctuary" (Eds S. Leatherwood and G. P. Donovan). United Nations Environmental Program on Marine Mammal Technical Paper No. 3.

Prescott, J. (1977). Comments on captive births of *Tursiops truncatus* at Marineland of the Pacific (1957–1972). *In* "Breeding Dolphins: Present Status, Suggestions for the Future" (Eds S. H. Ridgway and K. W. Benirschke) pp. 71–76. PB-273673, NTIS, Washington, DC.

Pryor, K., Lindbergh, J., Lindbergh, S. and Milano, R. (1990). A dolphin-human fishing cooperative in Brazil. *Mar. Mamm. Sci.* **6**, 77–82.

Purves, P. E. (1966). Anatomy and physiology of the outer and middle ear in cetaceans. *In* "Whales, Dolphins and Porpoises" (Ed. K. S. Norris) pp. 320–380. University of California Press, Berkeley, CA.

Purves, P. E. and Pilleri, G. E. (1983). "Echolocation in Whales and Dolphins". Academic Press, New York.

Read, A. J., Van Waerebeek, K., Reyes, J. C., McKinnon, J. S. and Lehman, L. C. (1988). The exploitation of small cetaceans in coastal Peru. *Biol. Cons.* **46**, 53–70.

Read, A. J., Wells, R. S., Hohn, A. H. and Scott, M. D. (1993). Patterns of growth in wild bottlenose dolphins, *Tursiops truncatus*. *J. Zool. Lond.* **231**, 107–123.

Reyes, J. C. (1990). The Bonn Convention and the conservation of small cetaceans: A review. Report to the Secretariat of the Convention on the Conservation of Migratory Species of Wild Animals.

Rice, D. W. (1960). Distribution of the bottle-nosed dolphin in the Leeward Hawaiian Islands. *J. Mamm.* **41**, 407–408.

Richards, D. G. (1986). Dolphin vocal mimicry and vocal object labeling. *In* "Dolphin Cognition and Behavior: A Comparative Approach" (Eds R. J. Schusterman, J. A. Thomas and F. G. Wood) pp. 273–288. Lawrence Erlbaum, Hillsdale, NJ.

Ridgway, S. H. (1972). Homeostasis in the aquatic environment. *In* "Mammals of the Sea: Biology and Medicine" (Ed. S. H. Ridgway) pp. 590–747. Charles C. Thomas, Springfield, IL.

Ridgway, S. H. (1986). Physiological observations on dolphin brains. *In* "Dolphin Cognition and Behavior: A Comparative Approach" (Eds R. J. Schusterman, J. A. Thomas and F. G. Wood) pp. 31–59. Lawrence Erlbaum, Hillsdale, NJ.

Ridgway, S. H. (1990). The central nervous system of the bottlenose dolphin. *In* "The Bottlenose Dolphin" (Eds S. Leatherwood and R. R. Reeves) pp. 69–97. Academic Press, San Diego.

Ridgway, S. H. and Reddy, M. (1995). Residue levels of several organochlorines in *Tursiops truncatus* milk collected at varied stages of lactation. *Mar. Poll. Bull.* **30**, 609–614.

Ridgway, S. H. and Carder, D. A. (1997). Hearing deficits measured in some *Tursiops truncatus*, and discovery of a deaf/mute dolphin. *J. Acoust. Soc. Am.* **101**, 590–593.

Ridgway, S. H. and Fenner, C. A. (1982). Weight–length relationships of wild-caught and captive Atlantic bottlenose dolphins. *J. Am. Vet. Med. Assoc.* **181**, 1310–1315.

Ridgway, S., Kamolnick, T., Reddy, M., Curry, C. and Tarpley, R. J. (1995). Orphan-induced lactation in *Tursiops* and analysis of collected milk. *Mar. Mamm. Sci.* **11**, 172–182.

Rommel, S. (1990). Osteology of the bottlenose dolphin. *In* "The Bottlenose Dolphin" (Eds S. Leatherwood and R. R. Reeves) pp. 29–49. Academic Press, San Diego.

Rommel, S.A, Pabst, D.A, McLellan, W. A., Mead, J. G. and Potter, C. W. (1992). Anatomical evidence for a countercurrent heat exchanger associated with dolphin testes. *Anat. Rec.* **232**, 150–156.

Ross, G. J. B. (1977). The taxonomy of bottlenosed dolphins *Tursiops* species in South African waters, with notes on their biology. *Ann. Cape Prov. Mus. (Nat. Hist.)* **11**, 135–194.

Ross, G. J. B. (1984). The smaller cetaceans of the southeast coast of southern Africa. *Ann. Cape Prov. Mus. (Nat. Hist.)* **15**, 173–410.

Ross, G. J. B. and Cockcroft, V.G. (1990). Comments on Australian bottlenose dolphins and the taxonomic status of *Tursiops aduncus* (Ehrenberg, 1832). *In* "The Bottlenose Dolphin" (Eds S. Leatherwood and R. R. Reeves) pp. 101–128. Academic Press, San Diego.

Ross, G. J. B., Cockcroft, V. G., Melton, D. A. and Butterworth, D. S. (1989). Population estimates for bottlenose dolphins *Tursiops truncatus* in Natal and Transkei waters. *S. Afr. J. Mar. Sci.* **8**, 119–129.

Saayman, G. S. and Tayler, C. K. (1977). Observations on the sexual behavior of Indian Ocean bottlenosed dolphins (*Tursiops aduncus*). *In* "Breeding Dolphins: Present Status, Suggestions for the Future" (Eds S. H. Ridgway and K. W. Benirschke) pp. 113–129. PB-273673, NTIS, Washington, DC.

Saayman, G. S., Tayler, C. K. and Bower, D. (1973). Diurnal activity cycles in captive and free-ranging Indian Ocean bottlenose dolphins (*Tursiops aduncus* Ehrenberg). *Behaviour* **44**, 212–233.

St. Aubin, D. J., Ridgway, S. H., Wells, R. S. and Rhinehart, H. (1996). Dolphin thyroid and adrenal hormones: circulating levels in wild and semidomesticated *Tursiops truncatus*, and influence of sex, age, and season. *Mar. Mamm. Sci.* **12**, 1–13.

Samuels, A. and Gifford, T. (1997). A quantitative assessment of dominance relations among bottlenose dolphins. *Mar. Mamm. Sci.* **13**, 70–99.

Sanches, J. G. (1989). Nomenclatura Portuguesa de Organismos Aquáticos". Publicações Avulsas do Instituto Nacional de Investigação das Pescas No. 14, pp. 1–322.

Sayigh, L. S. (1992). Development and functions of signature whistles of free-ranging bottlenose dolphins, *Tursiops truncatus*. Ph.D. dissertation, Woods Hole Oceanographic Institution and Massachusetts Institute of Technology, 344 pp.

Sayigh, L. S., Tyack, P. L., Wells, R. S. and Scott, M. D. (1990). Signature whistles of free-ranging bottlenose dolphins *Tursiops truncatus*: stability and mother-offspring comparisons. *Behav. Ecol. Sociobiol.* **26**, 247–260.

Sayigh, L. S., Tyack, P. L., Wells, R. S., Scott, M. D. and Irvine, A. B. (1995). Sex difference in signature whistle production of free-ranging bottlenose dolphins, *Tursiops truncatus*. *Behav. Ecol. Sociobiol.* **36**, 171–177.

Schroeder, J. P. (1990). Breeding bottlenose dolphins in captivity. *In* "The Bottlenose Dolphin" (Eds S. Leatherwood and R. R. Reeves) pp. 435–446. Academic Press, San Diego.

Scott, G. P. (1990). Management-oriented research on bottlenose dolphins by the Southeast Fisheries Center. *In* "The Bottlenose Dolphin" (Eds S. Leatherwood and R. R. Reeves) pp. 623–639. Academic Press, San Diego.

Scott, G. P., Burn, D. M. and Hansen, L. J. (1988). The dolphin dieoff: long-term effects and recovery of the population. *In* Proceedings of the Oceans '88 Conference, pp. 819–823.

Scott, G. P., Burn, D. M., Hansen, L. J. and Owen, R. E. (1989). Estimates of bottlenose dolphin abundance in the Gulf of Mexico from regional aerial surveys. National Marine Fisheries Service Southeast Fisheries Center Contribution No. CRD-88/89-07.

Scott, M. D. and Chivers, S. J. (1990). Distribution and herd structure of bottlenose dolphins in the eastern tropical Pacific Ocean. *In* "The Bottlenose Dolphin" (Eds S. Leatherwood and R. R. Reeves) pp. 387–402. Academic Press, San Diego.

Scott, M. D., Wells, R. S. and Irvine, A. B. (1990). A long-term study of bottlenose dolphins on the west coast of Florida. *In* "The Bottlenose Dolphin" (Eds S. Leatherwood and R. R. Reeves) pp. 235–244. Academic Press, San Diego.

Sergeant, D. E., Caldwell, D. K. and Caldwell, M. C. (1973). Age, growth, and maturity of bottlenosed dolphin (*Tursiops truncatus*) from northeast Florida. *J. Fish. Res. Bd Can.* **30**, 1009–1011.

Shallenberger, E. W. (1981). The status of Hawaiian cetaceans. Rep. No. PB82-109398, NTIS, Washington, DC.

Shane, S. H. (1977). The population biology of the Atlantic bottlenose dolphin, *Tursiops truncatus*, in the Aransas Pass area of Texas. Masters thesis, Texas A. & M. University.

Shane, S. H. (1980). Occurrence, movements and distribution of bottlenose dolphin, *Tursiops truncatus*, in southern Texas. *Fish. Bull.* **78**, 593–601.

Shane, S. H. (1988). "The Bottlenose Dolphin in the Wild". S. H. Shane, Felton, CA.

Shane, S. H. (1990a). Behavior and ecology of the bottlenose dolphin at Sanibel Island, Florida. *In* "The Bottlenose Dolphin" (Eds S. Leatherwood and R. R. Reeves) pp. 245–265. Academic Press, San Diego.

Shane, S. H. (1990b). Comparison of bottlenose dolphin behavior in Texas and Florida, with a critique of methods for studying dolphin behavior. *In* "The Bottlenose Dolphin" (Eds S. Leatherwood and R. R. Reeves) pp. 541–558. Academic Press, San Diego.

Shane, S. H. and Schmidly, D. J. (1978). The population biology of the Atlantic bottlenose dolphin, *Tursiops truncatus*, in the Aransas Pass area of Texas. National Technical Information Services, PB-283-393, US Department of Commerce, Springfield, VA, 130 pp.

Shane, S. H., Wells, R. S. and Würsig, B. (1986). Ecology, behavior and social organization of the bottlenose dolphin: a review. *Mar. Mamm. Sci.* **2**, 34–63.

Simões-Lopes, P. C. (1991). Interaction of coastal populations of *Tursiops truncatus* (Cetacea, Delphinidae) with the mullet artisanal fisheries in southern Brazil. *Biotemas* **4**, 83–94.

Simpson, J. G. and Gardner, M. B. (1972). Comparative microscopic anatomy of selected marine mammals. *In* "Mammals of the Sea: Biology and Medicine" (Ed. S. H. Ridgway) pp. 298–418. Charles C. Thomas, Springfield, IL.

Slijper, E. J. (1958). Organ weights and symmetry problems in porpoises and seals. *Arch. Nér. Zool.* **13** (Suppl. 1), 97–113.

Small, R. J. and DeMaster, D. P. (1995). Survival of five species of captive marine mammals. *Mar. Mamm. Sci.* **11**, 209–226.

Smith, T. D. (1982). Current understanding of the status of small cetacean populations in the Black Sea. *In* "Mammals in the Seas", Vol. 4, pp. 121–130. UN Food and Agriculture Organization, Rome.

Smolker, R. A., Richards, A. F., Connor, R. C. and Pepper, J. W. (1992). Sex differences in patterns of association among Indian Ocean bottlenose dolphins. *Behaviour* **123**, 38–69.

Sokolov, V. E. and Volkova, O. V. (1971). Structure of the dolphin's tongue. *In* "Morphology and Ecology of Marine Mammals" (Eds K. K. Chapskii and V. E. Sokolov) pp. 119–127. Wiley, New York (English translation, 1973, of "Morfologiya i Ekologiya Morskikh Mlekopitayushchikh" (1971) and of "Prisposoblenie Tyulenei K Obitaniyu V Arktike" (1971) by Israel Program for Scientific Translations, Jerusalem).

Surkina, R. M. (1971). Structure and function of the skin muscles of dolphins. *Bionika* **5**, 81–87 (English translation, 1972, by Joint Publ. Res. Serv., JPRS-55982, pp. 123–133, Arlington, VA).

Sweeney, J. (1978a). Infectious diseases. *In* "Zoo and Wild Animal Medicine" (Ed. M. E. Fowler) pp. 587–588. Saunders, Philadelphia.

Sweeney, J. (1978b). Infectious diseases of body systems. *In* "Zoo and Wild Animal Medicine" (Ed. M. E. Fowler) pp. 589–592. Saunders, Philadelphia.

Sweeney, J. (1978c). Noninfectious diseases. *In* "Zoo and Wild Animal Medicine" (Ed. M. E. Fowler) pp. 599–601. Saunders, Philadelphia.

Sylvestre, J.-P. and Tanaka, S. (1985). On the inter-generic hybrids in cetaceans. *Aquat. Mamm.* **11**, 101–108.

Tanaka, S. (1987). Satellite radio tracking of bottlenose dolphins *Tursiops truncatus*. *Nippon Suisan Gakkaishi* **53**, 1327–1338.

Tarpley, R. J. and Ridgway, S. H. (1991). Orbital gland structure and secretions in the Atlantic bottlenose dolphin (*Tursiops truncatus*). *J. Morphol.* **207**, 173–184.
Tavolga, M. C. (1966). Behavior of the bottlenose dolphin (*Tursiops truncatus*): social interactions in a captive colony. *In* "Whales, Dolphins, and Porpoises" (Ed. K. S. Norris) pp. 718–730. University of California Press, Berkeley, CA.
Tavolga, M. C. and Essapian, F. S. (1957). The behavior of the bottlenosed dolphin (*Tursiops truncatus*): mating, pregnancy, parturition and mother-infant behavior. *Zoologica* **42**, 11–31.
Tayler, C. K. and Saayman, G. S. (1972). The social organisation and behavior of dolphins (*Tursiops aduncus*) and baboons (*Papio ursinus*): Some comparisons and assessments. *Ann. Cape Prov. Mus. (Nat. Hist.)* **9**, 11–49.
Thompson, P. and Wilson, B. (1994). "Bottlenose Dolphins". Voyageur Press, Stillwater, MN.
Tolley, K. A., Read, A. J., Wells, R. S., Urian, K. W., Scott, M. D., Irvine, A. B. and Hohn, A. A. (1995). Sexual dimorphism in wild bottlenose dolphins (*Tursiops truncatus*) from Sarasota, Florida. *J. Mamm.* **76**, 1190–1198.
Tomilin, A. G. (1957). "Mammals of the U.S.S.R. and Adjacent Countries", Vol. IX, "Cetacea" (Ed. V. G. Heptner). Nauk, USSR, Moscow (English translation, 1967, Israel Program for Scientific Translations, Jerusalem).
Townsend, C. H. (1914). The porpoise in captivity. *Zoologica* **1**, 289–299.
True, F. W. (1891). Observations on the life history of the bottlenose porpoise. *U.S. Nat. Mus. Proc.* **13**, 197–203.
Tsalkin, V. I. (1940). Nekotorye nablyudeniya nad biologiei delfinov Azobskovo i Chernovo morei (Some observations of the biology of the Azov and Black Sea dolphins). *Bull. Soc. Nat. Moscow* **49**, 61–70.
Tyack, P. (1986). Whistle repertoires of two bottlenosed dolphins, *Tursiops truncatus*: mimicry of signature whistles? *Behav. Ecol. Sociobiol.* **18**, 251–257.
Tyack, P. (1991). Use of a telemetry device to identify which dolphin produces a sound. *In* "Dolphin Societies: Discoveries and Puzzles" (Eds K. Pryor and K. S. Norris) pp. 319–344. University of California Press, Berkeley, CA.
Urian, K. W., Duffield, D. A., Read, A. J., Wells, R. S. and Shell, E. D. (1996). Seasonality of reproduction in bottlenose dolphins, *Tursiops truncatus*. *J. Mamm.* **77**, 394–403.
Van Waerebeek, K., Reyes, J. C., Read, A. J. and McKinnon, J. S. (1990). Preliminary observations of bottlenose dolphins from the Pacific coast of South America. *In* "The Bottlenose Dolphins" (Eds S. Leatherwood and R. R. Reeves) pp. 143–154. Academic Press, San Diego.
Varanasi, U. and Malins, D. C. (1970). Unusual wax esters from the mandibular canal of the porpoise (*Tursiops gilli*). *Biochem.* **9**, 3629–3631.
Varanasi, U. and Malins, D. C. (1971). Unique lipids of the porpoise (*Tursiops gilli*): Differences in triacyl glycerols and wax esters of acoustic (mandibular canal and melon) and blubber tissues. *Biochim. Biophys. Acta* **231**, 415–418.
Varanasi, U. and Malins, D. C. (1972). Triacylglycerols characteristic of porpoise acoustic tissues: molecular structures of diisovaleroylglycerides. *Science* **176**, 926–928.
Wade, P. R. and Gerrodette, T. (1993). Estimates of cetacean abundance and distributions in the eastern tropical Pacific. *Rep. Int. Whal. Commn* **43**, 477–493.
Waller, G. H. and Harrison, R. J. (1978). The significance of eyelid glands in delphinids. *Aquat. Mamm.* **6**, 1–9.
Walker, W. A. (1981). Geographical variation in morphology and biology of bottlenose dolphins (*Tursiops*) in the eastern North Pacific. National Marine Fisheries Service Southwest Fisheries Center Administrative Report No. LJ-81-03C.

Walsh, M. T., Beusse, D., Bossart, G. D., Young, W. G., Odell, D. K. and Patton, G. W. (1988). Ray encounters as a mortality factor in Atlantic bottlenose dolphins (*Tursiops truncatus*). *Mar. Mamm. Sci.* **4**, 154–162.

Wang, K. R., Payne, P. M. and Thayer, V. G. (compilers) (1994). Coastal stock(s) of Atlantic Bottlenose Dolphin: status review and management. Proceedings and Recommendations from a Workshop held in Beaufort, NC, 13–14 September 1993, NOAA Technical Memorandum NMFS-OPR-4, 120 pp.

Waples, D. M. (1995). Activity budgets of free-ranging bottlenose dolphins (*Tursiops truncatus*) in Sarasota Bay, Florida. M.Sc. thesis, University of California at Santa Cruz.

Watkins, W. A. and Wartzok, D. (1985). Sensory biophysics of marine mammals. *Mar. Mamm. Sci.* **1**, 219–260.

Weigle, B. (1990). Abundance, distribution and movements of bottlenose dolphins (*Tursiops truncatus*) in Lower Tampa Bay, Florida. *In* "Individual Recognition of Cetaceans: Use of Photo-identification and Other Techniques to Estimate Population Parameters" (Eds P. S. Hammond, S. A. Mizroch and G. P. Donovan) pp. 195–201. *Rep. Int. Whal. Commn*, Spec. Issue No. 12, Cambridge.

Wells, R. S. (1978). Home range characteristics and group composition of Atlantic bottlenosed dolphins, *Tursiops truncatus*, on the west coast of Florida. Masters thesis, University of Florida.

Wells, R. S. (1986). Structural aspects of dolphin societies. Doctoral dissertation, University of California at Santa Cruz.

Wells, R. S. (1993). Parental investment patterns of wild bottlenose dolphins. *In* "Proceedings of the 18th International Marine Animal Trainers Association Conference" (Ed. N. F. Hecker) pp. 58–64.

Wells, R. S. and Scott, M. D. (1990). Estimating bottlenose dolphin population parameters from individual identification and capture-release techniques. *In* "Individual Recognition of Cetaceans: Use of Photo-identification and Other Techniques to Estimate Population Parameters" (Eds P. S. Hammond, S. A. Mizroch and G. P. Donovan) pp. 407–415. *Rep. Int. Whal. Commn*, Spec. Issue No. 12, Cambridge.

Wells, R. S., Irvine, A. B. and Scott, M. D. (1980). The social ecology of inshore odontocetes. *In* "Cetacean Behavior: Mechanisms and Functions" (Ed. L. M. Herman) pp. 263–317. Wiley, New York.

Wells, R. S., Scott, M. D. and Irvine, A. B. (1987). The social structure of free-ranging bottlenose dolphins. *In* "Current Mammalogy", Vol. 1 (Ed. H. H. Genoways) pp. 247–305. Plenum Press, New York.

Wells, R. S., Hansen, L. J., Baldridge, A., Dohl, T. P., Kelly, D. L. and Defran, R. H. (1990). Northward extension of the range of bottlenose dolphins along the California coast. *In* "The Bottlenose Dolphins" (Eds S. Leatherwood and R. R. Reeves) pp. 421–431. Academic Press, San Diego.

Wever, E. G., McCormick, J. G., Palin, J. and Ridgway, S. H. (1971a). The cochlea of the dolphin *Tursiops truncatus*: general morphology. *Proc. Natl Acad. Sci. U.S.A.* **68**, 2381–2385.

Wever, E. G., McCormick, J. G., Palin, J. and Ridgway, S. H. (1971b). The cochlea of the dolphin *Tursiops truncatus*: the basilar membrane. *Proc. Natl Acad. Sci. U.S.A.* **68**, 2708–2711.

Wever, E. G., McCormick, J. G., Palin, J. and Ridgway, S. H. (1971c). The cochlea of the dolphin *Tursiops truncatus*: hair cells and ganglion cells. *Proc. Natl Acad. Sci. U.S.A.* **68**, 2908–2912.

Wood, F. G., Jr., Caldwell, D. K. and Caldwell, M. C. (1970). Behavioral interactions between porpoises and sharks. *Invest. Cetacea* **2**, 264–277.

Würsig, B. (1978). Occurrence and group organization of Atlantic bottlenose porpoises (*Tursiops truncatus*) in an Argentine bay. *Biol. Bull.* **154**, 348–359.

Würsig, B. and Würsig, M. (1979). Behavior and ecology of bottlenose porpoises, *Tursiops truncatus*, in the South Atlantic. *Fish. Bull.* **77**, 399–412.

Würsig, B. and Harris, G. (1990). Site and association fidelity in bottlenose dolphins off Argentina. *In* "The Bottlenose Dolphins" (Eds S. Leatherwood and R. R. Reeves) pp. 361–365. Academic Press, San Diego.

Yablokov, A. V., Bel'kovich, V.M. and Borisov, V. I. (1972). "Kity i Del'finy" ("Whales and Dolphins"), Parts I and II. Nauka, Moscow (English translation, 1974, Joint Publ. Res. Serv. JPRS-62150-1 & 2, Arlington, VA).

Yoshioka, M., Mohri, E., Tobayama, T., Aida, K. and Hanyu, I. (1986). Annual changes in serum reproductive hormone levels in the captive female bottle-nosed dolphins. *Bull. Jap. Soc. Fish.* **52**, 1939–1946.

Young, N. M., Hope, G. M., Dawson, W. W. and Jenkins, R. L. (1988). The tapetum fibrosum in the eyes of two whales. *Mar. Mamm. Sci.* **4**, 281–290.

Zam, S. G., Caldwell, D. K. and Caldwell, M. C. (1971). Some endoparasites from small odontocete cetaceans collected in Florida and Georgia. *Cetology* **2**, 1–11.

8

Risso's Dolphin
Grampus griseus (G. Cuvier, 1812)

S. Kruse, D.K. Caldwell*, and M.C. Caldwell*

Genus and Species

Taxonomy

The type specimen of *Delphinus griseus* was collected in Brest, France, and described to G. Cuvier by M. Risso in 1812. The species was assigned to the subgenus *Grampus* by Gray (1828), which was raised to genus level by Hamilton (1837). For the 61 years following Gray's (1828) work, new specimens of Risso's dolphin often were described as separate species if they were collected from different areas or varied in size and/or coloration from the type specimen (e.g. *D. rissoanus* Desmarest, *G. cuvieri* Gray, *G. souverbianus* Fisher, *G. sakamata* Gray, *G. sternsii* Dall, *G. richardsoni* Gray, *Globicephalus rissii* Hamilton, and *G. chinensis* Gray). Based on the morphometric data available at that time, True (1889) stated that " . . . there can be no reasonable doubt that the grampuses of American and European waters are identical", thus eliminating all prior nomenclature in favour of the monotypic *Grampus griseus*. Iredale and Troughton (1936) reassigned the name *Grampus* to killer whales and erected the genus *Grampidelphis* for the Risso's dolphin. They justified their classification on the basis that *Grampus* was a "complex synonym which has left the respective names (of killer whales and Risso's dolphin) in doubt for over a century". Although some authors dutifully exchanged *Grampidelphis* for *Grampus* (e.g. Miller and Kellogg, 1955), the

*Posthumous contribution

Handbook of Marine Mammals, Volume 6
ISBN 0-12-588506-7

proposed change was later refuted by Schevill (1954) and Hershkovitz (1961), the latter referring to *Grampidelphis* as "a gratuitous synonym".

Nishiwaki (1972) proposed splitting the family Delphinidae into Delphinidae, Globicephalidae, and Grampidae, basing the split on number of fused vertebrae, dentition, and several other morphological distinctions. However, this classification is not popular, and most authors still recognize the single family Delphinidae (e.g. Barnes, 1978; Heyning, 1989; Mead, 1975).

Basing their conclusions on the similarities of air sinus development between *Tursiops* and *Grampus*, Fraser and Purves (1960) classified *Grampus* within the subfamily Delphininae. Although Mead (1975) remarked that he "would be tempted to put (*Grampus*) onto a subfamily by itself", he supported Fraser and Purves' conclusions, electing to preserve *Grampus* within the Delphininae, citing similarities in osteology, structure of pterygoid air sinuses and existence of hybrids between *Tursiops* and *Grampus* (e.g. Fraser, 1940; Sezaki *et al.*, 1984; Sylvestre and Tasaka, 1985). Basing systematic distinctions on tympano-periotic bone morphology, Kasuya (1973) placed Risso's dolphins within the subfamily Globicephalinae. Electrophoretic studies by Gallien *et al.* (1970) seem to support Kasuya's interpretation by showing a greater similarity in blood proteins between the long-finned pilot whale, *Globicephala melaena* (= *G. melas*) and *Grampus* compared with the common dolphin, *Delphinus delphis*, and the striped dolphin, *Stenella styx* (= *S. coeruleoalba*). However, Gallien *et al.*'s (1970) data are limited and more work is needed to better interpret affinities between these taxa.

Common names

The term "grampus" was derived from "grand poisson" and "gras poisson" (Beddard, 1900) (originally "crassus piscis" — Harrison and King, 1965) and historically was applied by whalers to all medium-sized toothed whales (e.g. Scammon, 1874; True, 1889; Iredale and Troughton, 1936; Gilmore, 1951). Today, "grampus" refers exclusively to Risso's dolphin, *Grampus griseus*. Other vernacular names include: the gray, whiteheaded or mottled grampus; dauphin de Risso (French); calderón de Risso, delfin gris, delfin de Risso (Spanish); Gramper, Rundkopfdelfin (German); halvgrindehval (Danish); grijze dolfijn (Dutch); grampo (Portugese); seryy delfin (Russian); and hana-gondo, matsuba-iruka, kamabire-sakamata, and shiro-shachi (Japanese).

Distribution

Range

Risso's dolphins occur pan-globally in tropical and temperate seas (Leatherwood *et al.*, 1980). Sighting records indicate this species occurs roughly between 60°N and 60°S latitudes (Fig. 1), where surface water temperature is over roughly

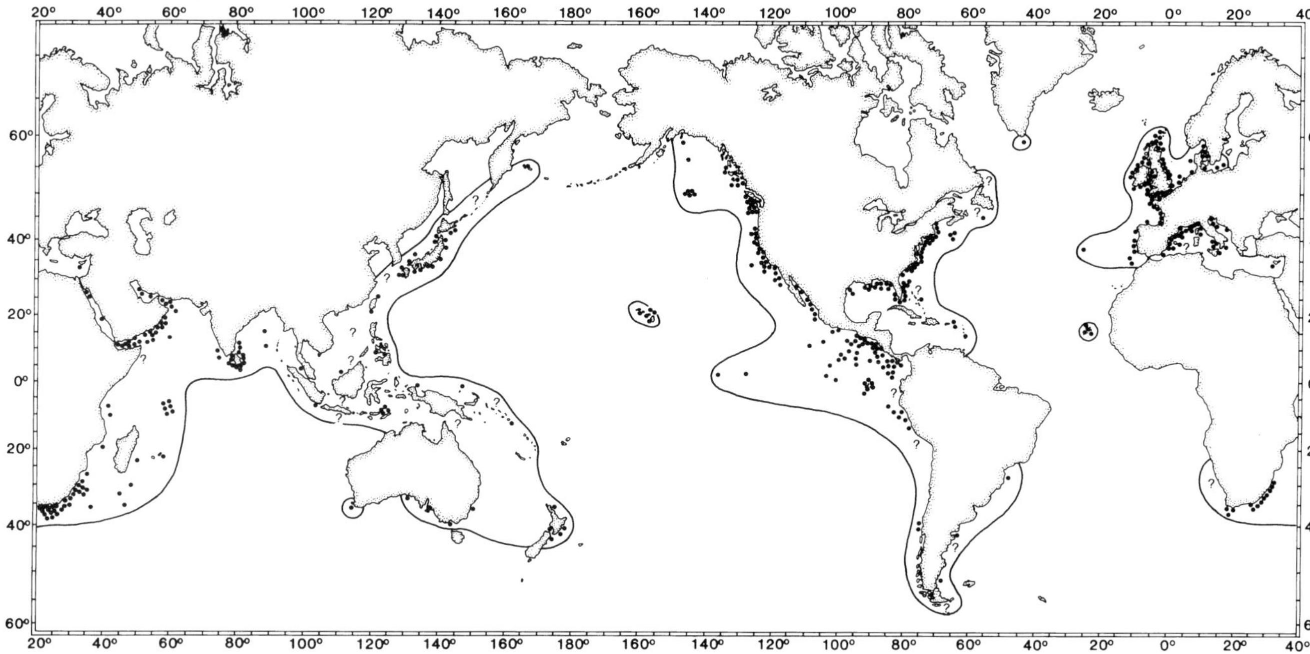

FIG. 1 The known distribution of Risso's dolphins. Dots represent actual sighting and stranding records. One dot may represent multiple records. Question marks indicate records where accurate geographical location was not given. Heavy outlines portray the author's prediction of distribution based solely on available records and known habitat preferences. Probable distribution is thought to be throughout tropical and temperate seas.

10 °C. In the western Atlantic, Risso's dolphins have been observed from Greenland (about 60°N, Vibe, 1950), south to Cape Horn at approximately 53°S (Leatherwood *et al.*, 1980). Sightings and specimens have been recorded along the eastern coasts of the Americas from Newfoundland, south to the Argentine coast (Schevill, 1954; Erdman, 1970; Mitchell, 1975; Caldwell and Caldwell, 1977; Schmidly, 1981; Katona *et al.*, 1983; Geise and Borobia, 1987; Lichter and Goodall, 1988).

In the region of the eastern Atlantic, this species has been recorded from the Shetland Islands, Scotland (about 60°N, Fraser, 1974) and the North Sea (Lepiksaar, 1966; Schultz, 1970; Wolff, 1982; Oen, 1987), down to South Africa (to about 34°S) (Casinos and Vericad, 1976; Teixeira, 1979; Clarke, 1981; Ross, 1984). Risso's dolphins have been recorded throughout the Mediterranean Sea complex, but not in the Black Sea (Aguayo, 1978; Viale, 1985; Raga *et al.*, 1986; Bayed and Beaubrun, 1987). They have been sighted throughout the studied portions of the Red Sea, Indian Ocean and Indo-Australian archipelago (Ellerman and Morrison-Scott, 1951; Medway, 1977; Alling, 1986; Kruse *et al*, 1991; Weitkowitz, 1992).

In the southwestern Pacific, Risso's dolphins have been recorded from New Zealand (about 45°S) (Baker, 1974) and Australia (Troughton, 1931; Mörzer-Bruyns, 1971). Interestingly, no records have been reported from Tasmanian waters (Nicol and Croome, 1988). In the western tropical Pacific, Risso's dolphins have been recorded from New Guinea up through the South China Sea (Maxwell, 1952; Dawbin, 1966; Nishiwaki, 1972; Hammond and Leatherwood, 1984). They are known to occur throughout the northwest Pacific from Japan to the Commander Islands (approximately 55°N; Leatherwood *et al.*, 1980). In the eastern Pacific, these widely distributed animals range from the Gulf of Alaska to waters off Chile (Aguayo, 1975; Leatherwood *et al.*, 1980; Reimchen, 1980; Braham, 1983; Au and Perryman, 1985; Polacheck, 1987; Vida *et al.*, 1993; Forney *et al.*, 1995). Occasional sightings have been made off Hawaii (Shallenberger, 1981).

Anecdotal evidence indicates that changes in local distribution and abundance occur occasionally. Since the late 1800s, the recorded abundance of Risso's dolphins in Monterey Bay, California, has fluctuated from common to non-existent to common again (Kruse, 1989). Scammon (1847) reported observations of "whiteheaded grampus" and collected the type specimens of *Grampus stearnsii* Dall from Monterey Bay (Dall, 1873). No sightings of these animals were logged during repeated surveys conducted during 1950–1955 (Barham, 1982). Leatherwood *et al.* (1980) reported that during 1958–1975, records of Risso's dolphins were rare from this area. Risso's dolphins gradually have become more abundant in Monterey Bay since the 1960s (A. Baldridge, personal communication, 1987). Sighting records from 1986–1987 indicate that these animals again are common in Monterey Bay (Kruse, 1989). Also, see Forney *et al.* (1995) for more recent information.

Another example of long term changes in abundance and distribution has occurred in the Southern California Bight. In the late 1950s, Risso's dolphins were rarely encountered in this area (Norris and Prescott, 1961). Between 1975 and 1978, these animals were considered to be a "minor" constituent of the cetacean fauna of the Bight, representing 3% of the cetaceans observed (Dohl *et al.*, 1980). Since the El Niño of 1982/83, numbers of Risso's dolphins have increased in the Southern California Bight, especially around Santa Catalina Island where they now are considered to be common (Shane, 1990, 1995). To date, these changes in distribution and abundance are unexplained.

Habitat

Risso's dolphins are pelagic, encountered most commonly seaward of the continental slope (Baker, 1974; Leatherwood *et al.*, 1980; Hain *et al.*, 1981; Ross, 1984; Alling, 1986; Kruse, 1989). They are known to frequent subsurface seamounts and escarpments where they are thought to feed on vertically migrant and mesopelagic cephalopods (Madsen and Herman, 1980; Norris and Dohl, 1980; Clarke and Pascoe, 1985). In Monterey Bay, California, abundance of Risso's dolphins was most concentrated over areas with steep bottom topography (Kruse, 1989). Bathymetrically induced circulation causing local increases in marine productivity may enhance feeding opportunities, resulting in the patchy distribution and local abundance of these delphinids world-wide (Au *et al.*, 1979; Breaker and Broenkow, 1989).

External Characteristics

Colour pattern

Risso's dolphins are coloured strikingly (Flower, 1872, details the coloration of a mature female and a calf). The youngest calves range in colour from iridescent gunmetal grey to fawn-brown dorsally and are creamy-white ventrally. Pale ochre-yellow highlights accentuate the muzzle. A white anchor-shape patch between the flippers resembles the chest chevron seen on pilot whales but is typically brighter and more extensive (Fig. 2). Calves become silver-grey, then darken to nearly black, retaining the ventral patches of white. As animals age further, their heads, abdomens, and flanks lighten. In older animals, lip colour frequently contrasts with the surrounding background (i.e., a narrow, white lip line bordered ventrally by a band of black or dark grey).

One of the most striking characteristics of Risso's dolphins is the prominent scar pattern which criss-crosses the animals' bodies. These distinctive linear scars accumulate primarily on the animals' dorsal and lateral surfaces (Fig. 3) and have been suggested, but not proven, to result from the combined effects of

FIG. 2 Ventral view of a Risso's dolphin illustrating the individually distinctive white anchor patch. This individual, shown jumping up to receive a squid from the trainer, was a rehabilitated stranded dolphin kept at Marineland of Florida. The animal below in this photograph is a bottlenose dolphin, *Tursiops truncatus*, the species with which Risso's dolphins have produced hybrid offspring in Japan.

lack of repigmentation of damaged tissue and a slower healing process than that observed in animals such as bottlenose dolphins (Nemoto, 1955; White and Norris, 1978). Measurements made of scar widths and the distances between scars on several captive animals suggest that these marks are made primarily by the teeth of conspecifics (White and Norris, 1978). Less common are circular white scars located anywhere on the body, which have been attributed to bites of lampreys, cephalopods and cookie-cutter sharks (*Isistius* spp.), infestations of parasites, and bacterial infections (Pike, 1951; Nemoto, 1955; McCann, 1974; White and Norris, 1978) (Fig. 3). Ontogenetic lightening, in conjunction with the characteristic body scars accumulated throughout life, make some older animals appear almost pure white (Nishiwaki, 1972; Fraser, 1974). Sexually dimorphic coloration has been suggested by Flower (1872) and Mizue and Yoshida (1962), but again, lack of data limits confidence in these suggestions.

Size and shape

Risso's dolphins are moderately large, blunt-headed delphinids. The longest specimen reported was a 408.9 cm female (Fraser, 1974); but no voucher specimen was collected. Reports of other specimens at least 400 cm long (e.g. Mörzer-Bruyns, 1971; Leatherwood *et al.*, 1979) appear to be field estimates rather than accurate measurements. We have been unable to locate reliable published

FIG. 3 Two adult size Risso's dolphins with a calf. Note the difference in body markings between adults and the calf. (Photo from the Ray Gilmore Archives at the San Diego Museum of Natural History)

accounts of animals exceeding 400 cm total length. Perrin and Reilly (1984) report a maximum length for males at 383 cm and for females at 366 cm. A Chi-square test of a sample of 216 Risso's dolphins of all ages and from various geographic locations revealed no significant difference in frequency distributions of total lengths of females and males ($t = -0.0009$, $p > 0.5$; Fig. 4). Additional data presented by Pilleri and Gihr (1969) and Ross (1984) indicated no significant sexual dimorphism in body dimensions.

Risso's dolphins are robust, particularly anterior to the dorsal fin (Fig. 5). The peduncle narrows rapidly, terminating in broad, lunate flukes. Flippers are long (about 19% TL) and sickle-shaped. The dorsal fin is centred on the body, is tall (about 12% TL) and moderately falcate in adults. Reported total body weights range from 38.1 kg to 400 kg (Table 1).

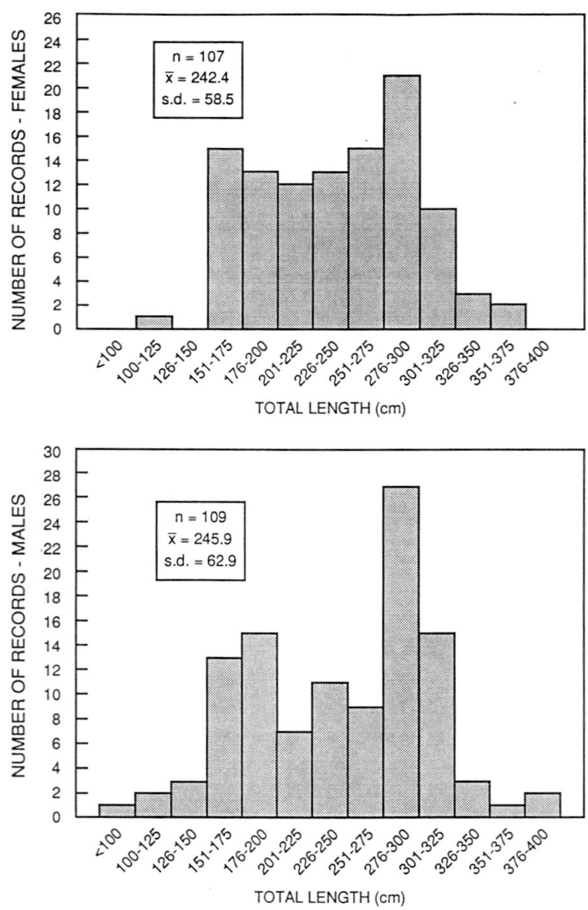

FIG. 4 Length distribution of Risso's dolphins. Statistics comparing males with females were not significant ($t = 0.0009$, $p > 0.5$), indicating no obvious sexual dimorphism in body length.

FIG. 5 Lateral view of a Risso's dolphin; note blunt head with cleft in melon (facial cleft), long flippers and erect, falcate dorsal fin.

A shallow, vertical cleft is present on the anterior surface of the melon (Fig. 6). White and Norris (1978) reported that the cleft becomes more prominent with age and that the skin within it pulses during vocalization. The relatively large blowhole is crescentic (concavity forward) and is offset slightly left of the midline

TABLE 1 Total lengths and weights of Risso's dolphins

Total length (cm)	Weight (kg)	Sex	Source
163.0	38.1	M	Ross (1984)
212.0	84.8	F	Ross (1984)
242.5	127.0	M	Ross (1984)
254.0	161.5	F	Ross (1984)
256.0	186.0	F	Ross (1984)
275.0	234.0	F	Pilleri and Busnel (1969)
275.0	260.0	M	Sezaki *et al.* (1984)
286.0	221.0	M	Sezaki *et al.* (1984)
304.8	318.0	M	Fahlgren-Torres (1988)
307.0	386.0	M	Harrison *et al.* (1972)
308.0	291.0	F	Teixeira (1979)
320.0	400.0	?	Ridgway and Brownson (1984)
335.5	385.5	?	Anon (1967)

(A)

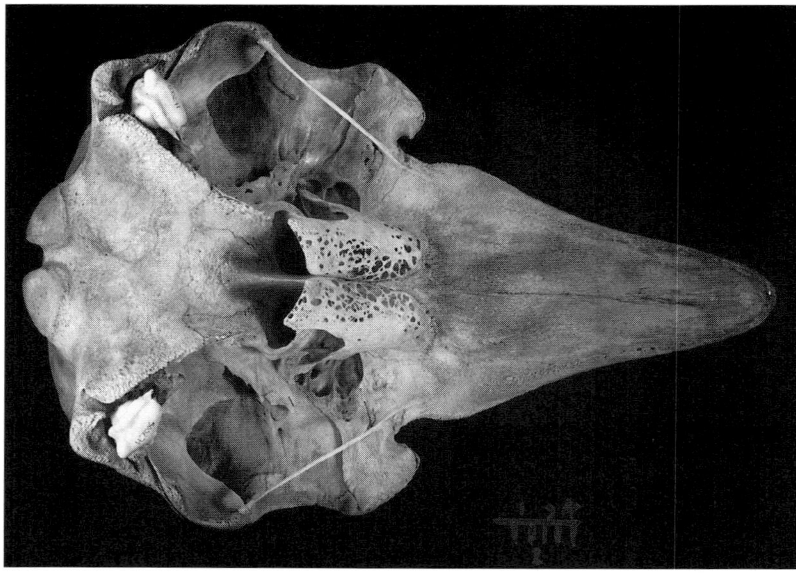

(B)

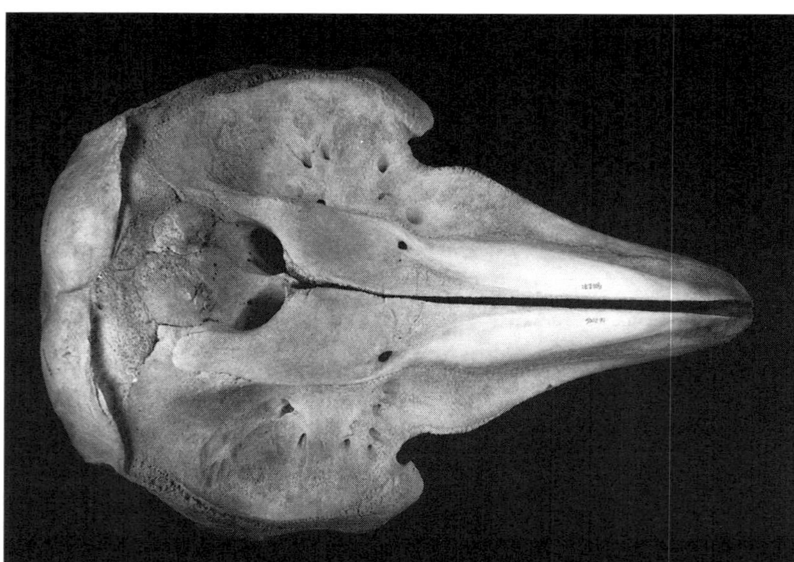

FIG. 6 Female Risso's dolphin (257 cm) skull (USNM 500271) from Smithsonian Institution, courtesy of J. Mead and C. Potter: (A) Ventral view with centimetre marks and (B) dorsal view with right side toward bottom of page.

(C)

(D)

(E)

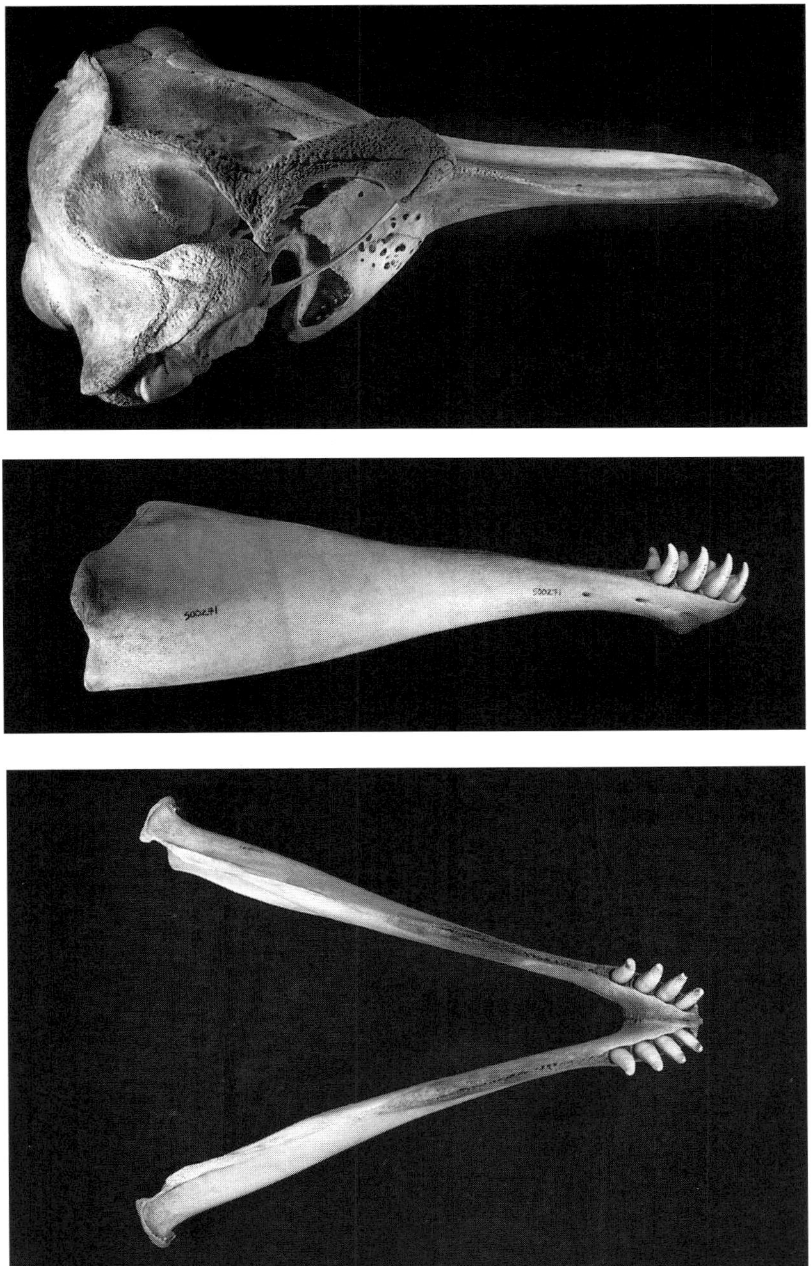

FIG. 6 *Continued* (C) Lateral view of skull, (D) lateral view of mandible and (E) dorsal view of lower jaw and mandibular symphysis. This specimen was from the North Atlantic Ocean, Saint Johns County, Florida, USA, collected by D.K. Caldwell and maintained alive at Marineland of Florida after stranding with a group of six animals. (Photo by Chip Clark)

(Murie, 1871; White and Norris, 1978). The gape sweeps down in profile and then curves up, ending below and in line with the anterior edge of the eye. The eye is somewhat small and elliptical and in many specimens is surrounded by a dark grey or black patch.

Internal Anatomy

Skull

Portions of the skull have been described by Fraser and Purves (1960) and Ross (1984) (Fig. 6). In general, the broad rostrum expands just anterior to the antorbital notches and tapers to an obtuse termination. The area of the rostrum anterior to the external nares is raised, triangular and convex. The rostral portions of the premaxillae are broad and slightly rounded. Ventrally, the pterygoids are joined. The cranial vertex is low. The skull is asymmetrical and the degree of "skew" (the ratio of asymmetry ×100 to the skull length) is similar to that observed in *Globicephala* spp., and is greater than that reported for other delphinids (Ness, 1967). The mandibular symphysis is short. Dentition is reduced (0/2–7), with three or four anteriad mandibular teeth being typical in each jaw (Fig. 6D, E). The teeth are conical and relatively large (3.5–4.0 cm long and about 1.5 cm in diameter). Occasionally, small teeth confined to the terminal half of the rostrum are found in the upper jaws of some specimens (Fraser, 1953; Nishiwaki, 1972). Tooth reduction has been suggested as an adaptation for taking cephalopod prey (Clarke, 1986).

The osteology of a hybrid (see Live Maintenance) between *Tursiops* and *Grampus* was studied by Miyazaki *et al.* (1992).

Postcranial skeleton

The skeleton was well described by Flower (1872), who observed that general external similarities between *Grampus* and *Globicephala* are mirrored in the proportions of the different regions of the skeleton, including the pectoral limbs, the tympano-periotic bones, and aspects of the skull. The vertebrae of *Grampus* are more numerous than those of *Globicephala*, especially in the lumbar region, where they are also shorter. The spinous and transverse processes are long and slender, more similar to those of *Delphinus* and *Lagenorhynchus* than to *Globicephala*, *Orcinus*, or other bulbous-headed delphinids. He notes that "the most noticeable peculiarity of the vertebral column . . . is the very feeble development of the metapophyses" (Flower, 1872). The vertebral count is 68–69, divided into C_7, T_{12-13}, L_{18-19}, Ca_{30-31} (Flower, 1872; Lee, 1877; Flower and Lydekker, 1891; Nishiwaki, 1972).

Organs, tissues and physiology

The soft facial anatomy, including the nasal passages, their diverticula and the melon are described in detail by Mead (1975) and Heyning (1989). The general plan of facial soft anatomy of Risso's dolphins is similar to that of other delphinids. An interesting exception is the left anterior portion of the nasofrontal sac, reported to be missing in about half of the examined specimens (Mead, 1975; Heyning, 1989; Curry, 1991). Unfortunately, few specimens have been examined, and the significance of this variability is unknown.

The melon is hypertrophied, especially in the posterolateral portions, resulting in a slightly bilobate, bulbous forehead (Mead, 1975). Little corresponding increase in heavy dermal connective tissue has occurred, and the melon is soft in comparison with those in other bulbous-headed delphinidae genera (Mead, 1975). The bilobed melon is manifested in the facial cleft that is seen easily in Fig. 5. This distinctly bifurcate melon is the basis for a suggestion by Nachtigall *et al.* (1995) that the special anatomy might create two sources of sound that interact to propagate highly directional echolocation signals.

According to Murie (1871), the laryngeal anatomy is similar to that described for the pilot whale. However, a glandular sac, resembling that observed in baleen whales, opens into the interior of the larynx, and nearly fills the angle of junction between the large epiglottis and the thyroid cartilage (Murie, 1871). The walls of the sac are thick and composed of glandular tissue. A thick muscle borders the sac and the inner wall of the cartilagenous thyroid plate. Although this structure is compared with those observed in some mysticetes, Murie (1871) does not elaborate on its function. He also describes parallel folds stretching longitudinally down the larynx as analogues to vocal cords and suggests that the animals can alter air currents within the channels of the larynx. Anatomy of the hyoid apparatus has been discussed by Reidenberg and Laitman (1994).

General descriptions of musculature and viscera may be found in Murie (1871) (young female) and Turner (1892) (male). Risso's dolphins possess "marked adaptations" in their vascular systems for deep diving and undergoing long apnoea — traits in accordance with their continental shelf-edge habitat and slow-swimming, deeper-water prey (Clarke, 1986). The spleen is very small and has been described as "primitive" (Nakamine *et al.*, 1992).

An ophthalmological examination described in Dawson *et al.* (1987, 1992) provides additional support for the contention that Risso's dolphins are deep divers. The individual examined displayed a remarkable accommodative stability to ambient light levels and also an increased transmittance or sensitivity to blue (short wavelength) light when compared with *Tursiops*. Colour fundus photographs showed a yellow fundus for *Tursiops* and a blue-red fundus for the *Grampus*. These colour differences imply that Risso's dolphins are adapted for enhanced visual acuity in low light levels, an important consideration for a deep diving predator.

Pilleri and Gihr (1969) reported organ descriptions, weights, and extensive brain and pituitary measurements from a single specimen. The spleen and kidney were studied by Arvy and Pilleri (1970) and Gihr and Kraus (1970). The brain-to-body weight ratio appears to be similar to that of bottlenose dolphins, based on the observation that a single female specimen weighing 234 kg had a brain weight of 2270 g (Pilleri and Busnel, 1969). An individual of unknown sex, 320 cm long and weighing 400 kg had a brain weight of 2551 g and an encephalization quotient (EQ) of 3.92 (Ridgway and Brownson, 1984).

The haematology of Risso's dolphin is detailed by De Monte and Pilleri (1968). Blood values from a captive specimen are presented by Nachtigall *et al.* (1990). Characteristics of melon, jaw and blubber fats are described by Litchfield and Greenberg (1974) (comparison of fats with those of other odontocetes), Litchfield *et al.* (1975) (composition of melon, jaw, and blubber fats), and Wedmid and Litchfield (1976) (locations of isovaleroyl triglycerides).

Life History and Population Dynamics

Growth and reproduction

Little is known about the life history of Risso's dolphins. Reported lengths at birth range from 110 cm (Mizue and Yoshida, 1962) to 150 cm (Tomilin, 1957; see Perrin and Reilly, 1984, for summary). The length of the smallest mature male recorded was 262 cm, and a range of 260–264 cm was listed for females (Perrin and Reilly, 1984). Limited life history data suggest that morphological differences occur between populations (True, 1889; Ross, 1984). Based on a data set which included unverified (i.e. not represented by osteological voucher specimens) animals reported by Fraser (1974), Ross (1984) reported regional differences in body lengths in samples from the northeast Atlantic, the Mediterranean, and South Africa. The occurrence of different "races" of Risso's dolphins off Japan was suggested by Mizue and Yoshida (1962). The data are sparse, and comparison among animals collected from different areas should be interpreted cautiously until the question of population differences has been resolved.

The average length of sexually mature females collected from the northwest Pacific was 277 cm ($n = 13$, SD $= 6.3$; Kasuya and Izumizawa, 1981) and 332 cm ($n = 5$) from the northeast Atlantic (Perrin and Reilly, 1984). Working from data collected from South African specimens, Ross (1984) suggested that females may attain sexual maturity at 277 cm: similar to values suggested by data from the North Pacific.

The average length of mature males collected in the western North Pacific was 286 cm ($n = 9$) (Kasuya, 1985). Ross (1984) suggested that males begin maturation between 260 cm and 300 cm. A single specimen from Florida was mature at 283 cm (Paul, 1968).

There is little documentation on longevity of Risso's dolphins. The famous New Zealand dolphin "Pelorus Jack" was observed escorting ships for 24 years (Alpers, 1960). The standard delphinid technique of estimating age by counting dental growth layer groups (GLGs) has not been tried on a large sample of Risso's dolphins, but several studies have used this technique with varying degrees of success (Stroud, 1968; Anonymous, 1980; Pierce and Kajimura, 1980; Ross, 1984; Kruse, 1989). Teeth from a 361 cm male had 30+ GLGs and a 309 cm female had 14+ GLGs (Kruse, 1989). Ross (1984) estimated that a 277 cm long female had over 17 GLGs. Calibration of GLGs in *Grampus* has not been done, but each layering group has been determined to represent one year in other delphinid species (e.g. *Tursiops truncatus*, Sergeant, 1959; Hohn, 1980a,b; *Globicephala macrorhynchus* and *G. melas* (= *G. melaena*), Kasuya and Matsui, 1984; Sergeant, 1962; and *Stenella longirostris*, Myrick *et al.*, 1984). We assume that GLG deposition in Risso's dolphins follows this established, layer-per-year, delphinid pattern. Available age-at-length data are summarized in Fig. 7.

Behaviour

Social organization

Risso's dolphins are gregarious animals. Schools can comprise from one to over several hundred individuals, and they average about 30 animals (Dohl *et al.*, 1980, 1983; Leatherwood *et al.*, 1980; Kruse, 1989; Kruse *et al.*, 1991). A single aggregation of some 4000 animals was reported off the central California coast by Dohl *et al.* (1983). The largest aggregations may form in response to abundant food resources (Tomilin, 1957; Tsutsumi *et al.*, 1961; Mizue and Yoshida, 1962). Mullin *et al.* (1994) reported a mean herd size of 12.8 animals (range 1–48) in their Gulf of Mexico observations from July 1989 to June 1990.

School composition appears to be extremely labile. Fluid associations were implied strongly by fluctuations of school sizes in which identifiable individuals and subgroups were observed (Kruse, 1989). Small subgroups of individuals within schools display the highest degree of group fidelity. In Monterey Bay, Kruse (1989) identified several cohesive social groups, including a subgroup of eight identifiable individuals which were observed repeatedly together over a 15 month period. Sex and reproductive condition of subgroup members were not determined. Limited data on subgroup composition obtained from mass strandings and observations of captive animals suggest that cohesive subgroups may be composed of same-sex and similar-age individuals (Caldwell and Caldwell, 1972; Leatherwood *et al.*, 1979; Baker, 1983).

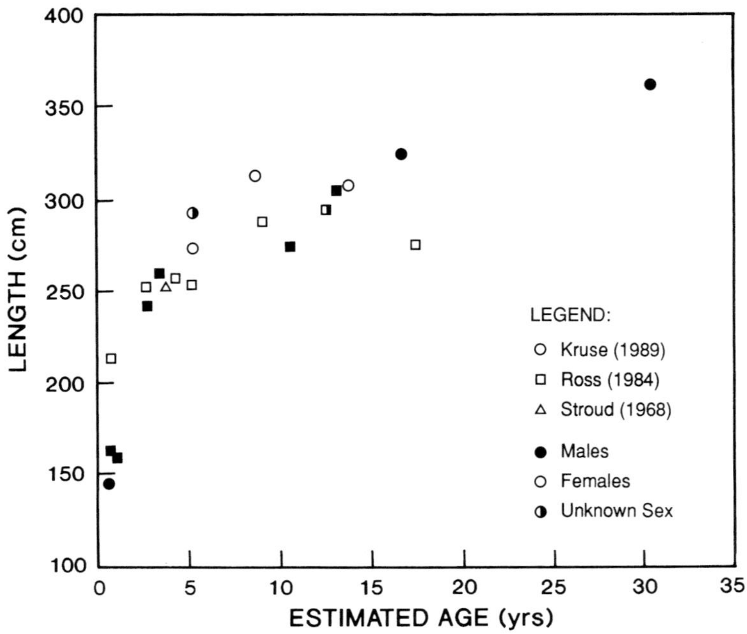

FIG. 7 Age-at-length curve for Risso's dolphins based on data from the authors listed in the key. Ages were estimated by counting growth layer groups (GLGs) in teeth. One GLG is thought to represent one year's growth.

Movements

Although no supporting data are offered, some authors report that Risso's dolphins are migratory in some areas (Australia, Troughton, 1931; eastern Atlantic, Paul, 1968). Mörzer-Bruyns (1971) suggested that they were more abundant around northern Scotland in the summer and in the Mediterranean in the winter. Similar seasonal shifts in abundance have been reported from the northwest Atlantic (Hain *et al.*, 1981; Katona *et al.*, 1983), British coastal waters (Evans, 1980), and the southeast coast of South Africa (Ross, 1984). Summer "reproductive migrations" (characterized by schools of 20–30 animals with empty stomachs and females carrying large foetuses), and winter "feeding migrations" (characterized by schools of nearly 200 animals with full stomachs and females carrying smaller foetuses) have been observed off Japan (Tsutsumi *et al.*, 1961; Mizue and Yoshida, 1962). Details concerning data samples were unavailable, so it is difficult to determine the significance of the use of the term "migration" by the authors. Others suggest that Risso's dolphins are equally common throughout the year, showing no evidence of seasonal migration (UK, Fraser, 1953, 1974; Cowan and Guiget, 1956; western Atlantic, Caldwell and Caldwell,

1977). Systematic studies of the distribution and abundance of Risso's dolphins in localized areas are required to resolve this conflict.

Water temperature appears to be a factor that affects the distribution of Risso's dolphins. These animals have been observed in waters only as cold as 7.5 °C, and captives have perished when the water temperature fell below 5 °C (Nishiwaki, 1967). No upper temperature limits have been published, but Risso's dolphins have been maintained successfully in Hawaii and Florida, where sea surface temperatures reach 25–35 °C (White and Norris, 1978; Nachtigall *et al.*, 1990), and have been reported from the Gulf of the Suez and the Gulf of Aquaba where water temperatures exceed 30 °C (Beadon, 1991). In their studies of the cetacean fauna of California, Dohl *et al.* (1980, 1983) reported increasing numbers of Risso's dolphin and a shoreward shift in their distribution occurring during periods of warm water. They suggested that seasonal patterns of distribution and abundance were associated with changing sea surface temperatures. In the eastern North Pacific, Leatherwood *et al.* (1980) reported that animals were sighted in waters ranging from 10 °C to 28 °C. Kruse (1989) noted that in Monterey Bay, Risso's dolphins were observed in waters ranging between 10.0 °C and 17.9 °C but were most abundant when sea surface temperatures rose to 15.0–16.9 °C. All accounts support the contention that the species primarily occurs in tropical and warm temperate seas and occasionally extends its range into higher latitudes, probably in response to protracted or unseasonal warm water events (Leatherwood *et al.*, 1980). Clearly, more extensive and detailed studies of the movement patterns and ecology of Risso's dolphin must be completed in order to elucidate the relationship between temperature effects and seasonal "migrations" recorded in this species.

Feeding

Risso's dolphins are known to prey on a mix of neritic, oceanic, and occasionally bottom dwelling cephalopods (see Lee, 1897; Clarke, 1966, 1986; Orr, 1966; Stroud, 1968; White and Norris, 1978; Turner, 1982; Ross, 1984; Clarke and Pascoe, 1985; Cockcroft *et al.*, 1993). Shane (1995) observed daily activity patterns off Santa Catalina Island, California and concluded that Risso's dolphins are mainly nocturnal feeders in the study area.

Association with other species

Risso's dolphins are commonly observed in mixed-species aggregations. Associates include gray whales (Sheldon *et al.*, 1995), Pacific white-sided dolphins (*Lagenorhynchus obliquidens*), northern right whale dolphins (*Lissodelphis borealis*), Dall's porpoises (*Phocoenoides dalli*), sperm whales (*Physeter macrocephalus*), short-finned pilot whales (*Globicephala macrorhynchus*), bottlenose dolphins (*Tursiops truncatus*), common dolphins (*Delphinus delphis*), striped dolphins

(*Stenella coeruleoalba*), spotted dolphins (*Stenella attenuata*), false killer whales (*Pseudorca crassidens*), and pygmy killer whales (*Feresa attenuata*) (Hubbs, 1960; Fiscus and Niggol, 1965; Leatherwood *et al.*, 1980, 1982, 1984; Braham, 1983; Dohl *et al.*, 1983; Viale, 1985; Alling, 1986; Kruse, 1989).

Swimming and diving

In the wild, Risso's dolphins engage in types of behaviour that are typical of other delphinids. They have been observed swimming at speeds of up to 28–32 km/hr, though typically they swim 6–7 km/hr (Mörzer-Bruyns, 1971; White and Norris, 1978). Commonly they surface to breathe about every 7 sec during typical surfacing sequences, but may remain submerged on longer dives for 30 min (Clarke, 1986; Mörzer-Bruyns, 1971).

Aerial and other behaviour

Risso's dolphins frequently engage in percussive behaviour, such as lob-tailing and breaching. Kruse (1989) documented aggressive physical contact, such as flipper slapping between individuals, striking with flukes and dorsal fins, and full body blows.

Epimeletic or succorant behaviour has not been reported for this species (Caldwell and Caldwell, 1966). Pilleri and Gihr (1969) noted that after a female was struck with a harpoon, the rest of her school swam around the wounded animal "for some time" before disappearing rapidly. Present authors D.K.C. and M.C.C. observed a large male pull a stillborn *Tursiops* foetus from its mother's vagina at Marineland of Florida. Whether the male's behaviour was an act of aggression or succorance towards the female is unclear. Two male Risso's dolphins did not display any succorant behaviour at the time of their two female tankmates' deaths.

Sound production

Risso's dolphins produce a wide range of sounds, including echolocation clicks, squeals, cries, and rasps (Poulter, 1967). Caldwell *et al.* (1969), suggested the presence of "signature whistles" in the call repertoires of this species.

During a 10-day cruise of the Scripps Institute research vessel Alpha Helix, Cummings and Fish (1972) recorded sounds from a group of about 25 animals. These Risso's dolphins emitted a variety of sounds with energy distributed from 200 Hz to over 30 kHz. The principal energy of clicks was centred at about 24 kHz. Most whistle-type sounds were confined to a band of 6–13 kHz.

Hearing

Hearing sensitivity was investigated in an older female by Nachtigall *et al.* (1995). In Kaneohe Bay, Hawaii, where the tests were conducted, the animal had thresholds near the background bay noise in the range of 4–64 kHz. Reported thresholds were 124 dB at 1.6 kHz, 71.7 dB at 4 kHz, 63.7 dB at 8 kHz, 63.8 dB at 16 kHz, 66.5 dB at 32 kHz, 67.3 dB at 64 kHz, 74.3 dB at 80 kHz, 124.2 dB at 100 kHz, and 122.9 dB at 110 kHz.

Parasites and Disease

Baker (1992) examined 18 dolphins from British waters including a female *Grampus* and a "pup". He suggested that when dead marine mammals are thoroughly examined, a high prevalence of disease can often be found.

The following internal parasites have been reported in stranded and captive animals: cestodes, *Phyllobothrium delphini* (Meggitt, 1924) and *Tetrabothrius* (Shults *et al.*, 1982); and nematodes, *Crassicauda grampicola* (Daily and Brownell, 1972; Raga *et al.*, 1982), *Stenurus minor* (Daily and Brownell, 1972) and "*Anisakis* (?) *simplex*" (identification and punctuation as in Ross, 1984). Morimitsu *et al.* (1992) reported on mass-stranded Risso's dolphins with *Nasitrema* and *Crassicauda*.

External parasites, including the cyamid *Isocyamus delphini* and *Xenobalanus globicipitis*, have been recorded (Leung, 1967; Ross, 1984). Due to limited experience with these animals, knowledge of disease conditions is relatively sparse. Dental abscesses have been observed in stranded individuals (Fraser, 1953). Gastroenteritis and diabetes were diagnosed in a stranded specimen which died shortly after capture (Anonymous, 1967).

Bacterial species including *Vibrio alginolyticus* and *Vibrio parahaemolyticus* (both species are pathogens in humans), *Pseudomonas putrefaciens*, *Escherichia coli*, *Edwardsiella tarda*, and *Staphylococcus aureus* were isolated from the blowhole of a newly stranded specimen (Nachtigall *et al.*, 1990). The same authors report *Vibrio* spp. occurring on the skin and in skin lesions. In Japan, captives have been afflicted with pneumonia, bacterial infections caused by *Staphylococcus albus* and *S. aureus*, and various skin lesions (White and Norris, 1978). A scoliosis condition resulting in paralysis and death developed in several animals following capture (White and Norris, 1978). Eversion of the first or second stomach (agonal gastro-oesophageal prolapse) has caused the deaths of a number of newly captured Risso's dolphins, probably resulting from capture trauma (White and Norris, 1978). The incidence of prolapse and scoliosis in wild populations is unknown.

Live Maintenance

Perhaps around 50 Risso's dolphins taken after drive fisheries or incidental to other fishery operations have been displayed in Japan (Klinowska, 1991). In the United States, a few have been maintained for display or research after stranding. Maintenance of these animals has met with varying levels of success (Defran and Pryor, 1980). One individual, rehabilitated at the New England Aquarium after stranding near Boston, Massachusetts, was later flown to Hawaii. This female, called Lou, was kept for a total of at least 16 years (Nachtigall *et al.*, 1990). Thirteen *Grampus* × *Tursiops* hybrids have been born in Japan but no calves have survived longer than 6 years and 7 months (see Sezaki *et al.*, 1984; Sylvestre and Tasaka, 1985; Klinowska, 1991). Although apparently there have been no non-hybrid births, it appears that Risso's dolphins may adapt well under proper conditions. Klinowska (1991) suggests that captive breeding is an option and that existing and future captive animals be studied to provide information relevant to conservation and management of the species.

In captivity, Risso's dolphins have been characterized as "slow, deliberate and nonaggressive" (White and Norris, 1978). Although they take an average of 20% more training time than bottlenose dolphins to learn a new behaviour pattern, they are responsive to training and can learn the same behavioural repertoires (White and Norris, 1978). Present authors D.K.C. and M.C.C. observed "mild" food competition between two male captives during which one of the animals "teased" the other by trailing food items from the side of its mouth and preventing the other from grasping it. Later, this behaviour evolved into a deliberate game of "keep away" between an adult female *Tursiops* and one of the male Risso's dolphins. The male repeatedly tempted the bottlenose dolphin to obtain a large carangid fish (blue runner) he withheld from his feedings. The game lasted as long as 15 min each session, with the bottlenose dolphin eventually making off with the fish. D.K.C. and M.C.C. observed aggressive interactions between the male Risso's dolphin and large bottlenose dolphins (especially males) in the tank. They also noted that initially the male *Grampus* responded aggressively to newly introduced pilot whales.

Human Effects

Directed fisheries and by-catches

Although they have never been the basis of a large-scale fishery, Risso's dolphins have been taken periodically in small-whale fisheries and as by-catches in other fisheries throughout the world (cf. Perrin *et al.*, 1994; Read, 1996). They have occasionally been taken in the Northeast Atlantic (Duguy and Hussenot, 1982), the Northwest Atlantic (Mitchell, 1975), the southern Caribbean off St. Vincent (Caldwell and Caldwell, 1975), off the Azores (Clarke, 1981), Peru (Van Waerebeek and Reyes, 1990), and in the Solomon Islands (Dawbin, 1966). They are a rare by-catch in the US tuna purse seine industry (Perrin *et al.*, 1979), and are taken occasionally in coastal gill net and squid seining industries off the US coast (Heyning *et al.*, 1990), or shot as a nuisance to fishermen (Guiget and Pike, 1965; Stroud, 1968). One survey (Hill and Barlow, 1992) was apparently motivated by the need to estimate the size of populations impacted by gill nets off California.

In Japan, Risso's dolphins are taken periodically for food and fertilizer in set nets and as a limited catch in the small-type whaling industry (Ohsumi, 1972, 1975). They are killed in the drive fishery (oikomi) in response to competition with commercial fisheries, and some are collected for live exhibition (Kasuya *et al.*, 1984; Kasuya, 1985). Risso's dolphins are considered to be the most difficult to capture of the seven small cetacean species most commonly caught in Japanese drive fisheries (White and Norris, 1978). Indeed, attempts to drive schools in Newfoundland often failed because "the animals became more excited than pilot whales and were difficult to contain" (Mitchell, 1975, p. 925).

Between 1983 and 1986, 241 individuals (representing one quarter of the small cetaceans reported) were landed in the Sri Lankan gill net fishery (Leatherwood and Reeves, 1989; Kruse *et al.*, 1991). An estimated 1300 Risso's dolphins may be landed annually as a result of this fishery (IWC, 1990). Rough population estimates of Risso's dolphins inhabiting Sri Lankan waters range from 5500 to 13 000 animals (Kruse *et al.*, 1991). These data indicate that this population may be at risk of depletion if fishery-related mortalities remain at current levels (IWC, 1990; Kruse *et al.*, 1991).

Pollution

Accumulation of butyltin compounds has been reported in Risso's dolphins from the Pacific Coast of Japan (Kim *et al.*, 1996). Organochloride pesticide levels have been analysed in tissue samples from a single specimen (Aucamp *et al.*, 1971). Levels of DDT (0.256 ppm in blubber, 0.140 ppm in liver) and dieldrin (0.068 ppm in blubber, 0.048 ppm in liver) were reported from this animal, which was collected "far from any obvious source of contamination" off the South African coast. Also off South Africa, de Kock *et al.* (1994) obtained specimens from one male 3.14 m in length which had higher levels of DDT.

Necropsies of specimens at Enoshima Marineland (Japan) revealed that they had eaten foreign materials such as plastic bags, soda cans, and pieces of rope, which may have been fatal (White and Norris, 1978; J.E. Heyning, personal communication, 1990). Bits of reeds, rags, roots, and fragments of chestnut wood were reported from stranded specimens in Europe (Pilleri and Knuckey, 1969). Captive and wild delphinids from several genera are known to ingest objects accidentally or as part of their "play" behaviour (cf. Caldwell and

Caldwell, 1972; Ridgway, 1972). Increasing levels of plastics and other "throw-aways" at sea may pose a threat to wild populations.

Acknowledgments

We thank the following people for their helpful criticisms of early drafts of this paper: J. Barlow, D. DeMaster, E. Edwards, and W. Perrin, NMFS/SWFSC, La Jolla, CA; J.E. Heyning, Natural History Museum of Los Angeles County; S. Leatherwood, Oceans Unlimited, San Diego, CA; and T. Jefferson, Texas A&M University. J.E. Heyning kindly provided skull photographs. The San Diego Natural History Museum generously shared their archival photographs of stranded Risso's dolphins. R. Allen created the figures.

References

Aguayo, L. A. (1975). Progress report on small cetacean research in Chile. *J. Fish. Res. Bd Can.* **32**, 1123–1143.

Aguayo, L. A. (1978). Smaller cetaceans in the Baltic Sea. *Rep. Int. Whal. Commn* **28**, 131–146.

Alling, A. (1986). Records of odontocetes in the northern Indian Ocean (1981–82) and off the coast of Sri Lanka (1982–84). *J. Bombay Nat. Hist. Soc.* **83**, 376–394.

Alpers, A. (1960). "Dolphins: the Myth and the Mammal". Houghton Mifflin, Boston, MA.

Anonymous (1967). Rare dolphin recuperates at Sea World. *Seahorse News of Oceanography and Limnology, San Diego* **2**, 1.

Anonymous (1980). Report of the workshop. *In* "Age Determination of Toothed Whales and Sirenians" (Eds W. F. Perrin and A. C. Myrick Jr) pp. 1–50. *Rep. Int. Whal. Commn*, Spec. Issue No. 3, Cambridge.

Arvy, L. and Pilleri, G. (1970). Some characteristics of the cetacean spleen. *Invest. Cetacea* **2**, 165–167.

Au, D. W. K., Perryman, W. L. and Perrin, W. F. (1979). Dolphin distribution and the relationship to environmental features in the eastern tropical Pacific. NMFS/SWFC Administrative Report LJ-79-43.

Au, W. K. and Perryman, W. L. (1985). Dolphin habitats in the eastern tropical Pacific. *Fish. Bull.* **83**, 623–643.

Aucamp, P. J., Henry, J. L. and Stander, G. H. (1971). Pesticide residues in South African marine mammals. *Mar. Pollut. Bull.* **2**, 190–191.

Baker, A. N. (1983). "Whales and Dolphins of New Zealand and Australia, an Identification Guide". Victoria University Press, Wellington, NZ.

Baker, A. N. (1974). Risso's dolphin in New Zealand waters and the identity of "Pelorus Jack". *Records of the Dominican Museum* **8**, 267–276.

Baker, J. R. (1992). Causes of mortality and parasites and incidental lesions in dolphins and whales from British waters. *Vet. Rec.* **130**, 569–572.

Barham, E. G. (1982). Marine mammals in Monterey Bay, California, during the years 1950–1955. *Calif. Fish Game* **68**, 213–223.

Barnes, L. G. (1978). A review of *Lophocetus* and *Liolithax* and their relationship to the delphinoid family Kentriodontidae (Cetacea: Odontoceti). *Nat. Hist. Mus. Los Angeles Co. Mus. Sci. Bull.* **38**, 1–35.

Bayed, A. and Beaubrun, P. C. (1987). The marine mammals of Morocco — Preliminary Inventory. *Mammalia* **51**, 437–446.

Beadon, J. (1991). Cetaceans seen and caught in the Gulf of Aquaba and the Gulf of Suez 13 September 1980 through 1 September 1981. *In* "Cetaceans and Cetacean Research in the Indian Ocean Sanctuary" (Eds S. Leatherwood and G. Donovan) pp. 111–114. United Nations Environmental Programme, Nairobi, Kenya.

Beddard, F. E. (1900). "A Book of Whales". G. P. Putnam's Sons, New York.

Braham, H. W. (1983). Northern records of Risso's dolphin, *Grampus griseus*, in the northeast Pacific. *Can. Field-Naturalist* **97**, 89–90.

Breaker, L. C. and Broenkow, W. W. (1989). The circulation of Monterey Bay and related processes. Moss Landing Marine Laboratories, Technical Publication No. 89-1, 91 pp.

Caldwell, D. K. and Caldwell, M. C. (1966). Epimeletic (care-giving) behavior in cetacea. *In* "Whales, Dolphins and Porpoises" (Ed. K. S. Norris) pp. 755–789. University of California Press, Berkeley, CA.

Caldwell, M. C. and Caldwell, D. K. (1972). Behavior of marine mammals. *In* "Mammals of the Sea, Biology and Medicine" (Ed. S. H. Ridgway) pp. 419–465. Charles C. Thomas, Springfield, IL.

Caldwell, D. K. and Caldwell, M. C. (1975). Dolphin and small whale fisheries off the Caribbean and West Indies: occurrence, history, and catch statistics — with special reference to the Lesser Antillean island of St. Vincent. *J. Fish. Res. Bd Can.* **32**, 1105–1110.

Caldwell, D. K. and Caldwell, M. C. (1977). The distribution, migrations, and relative abundance of Risso's dolphins, *Grampus griseus*, in the western north Atlantic. Report to the Naval Oceans Systems Center, San Diego, CA.

Caldwell, D. K., Caldwell, M. C. and Miller, J. F. (1969). Three brief narrow-band sound emissions by a captive male Risso's dolphin, *Grampus griseus*. *Bull. So. Calif. Acad. Sci.* **68**, 252–256.

Casinos, A. and Vericad, J. R. (1976). The cetaceans of the Spanish coasts: a survey. *Mammalia* **40**, 267–289.

Clarke, M. R. (1966). A review of the systematics and ecology of oceanic squids. *Adv. Mar. Biol.* **4**, 91–300.

Clarke, M. R. (1986). Cephalopods in the diet of odontocetes. *In* "Research on Dolphins" (Eds M. M. Bryden and R. Harrison) pp. 282–321. Clarendon Press, Oxford.

Clarke, M. R. and Pascoe, P. L. (1985). The stomach contents of a Risso's dolphin (*Grampus griseus*) stranded at Thurlestone, South Devon. *J. Mar. Biol. Assoc. U.K.* **65**, 663–665.

Clarke, R. (1981). Whales and dolphins of the Azores and their exploitation. *Rep. Int. Whal. Commn* **31**, 607–615.

Cockcroft, V. G., Haschick, S. L. and Klages, N. T. W. (1993). The diet of Risso's dolphin, *Grampus griseus* (Cuvier, 1812), from the east coast of South Africa. *Z. Saugetierkunde* **58**, 286–293.

Cowan, I. Mc. and Guiget, C. J. (1956). "The Mammals of British Columbia". British Columbia Provincial Museum, Victoria, B.C., Canada.

Cummings, W. C. and Fish, J. F. (1972). Bioacoustics of Cetaceans. *In* "Alpha Helix Whale Cruise, Phase 1 (13–23 October, 1971)" p. 29. Scripps Institution of Oceanography, La Jolla, CA.

Curry, B. E. (1991). Facial anatomy of the harbor porpoise (*Phocoena phocoena*) and Dall's porpoise (*Phocoenoides dalli*). M.Sc. thesis, Moss Landing Marine Laboratory. California State University at San Jose, CA.

Dailey, M. D. and Brownell Jr, R. L. (1972). A checklist of marine mammal parasites. *In* "Mammals of the Sea: Biology and Medicine" (Ed. S. H. Ridgway) pp. 528–589. Charles C. Thomas, Springfield, IL.

Dall, W. H. (1873). Descriptions of three new species of cetacea, from the coast of California. *Proc. Calif. Acad. Sci.* **5**, 13–14.

Dawbin, W. H. (1966). Porpoises and porpoise hunting in Malaita. *Aust. Nat. Hist.* **15**, 207–211.

Dawson, W. W., Schroeder, J. P. and Dawson, J. F. (1987). The ocular fundus of two cetaceans. *Mar. Mamm. Sci.* **3**, 1–13.

Dawson, W. W., Schroeder, J. P., Dawson, J. C. and Nachtigall, P. E. (1992). Cyclic ocular hypertension in cetaceans. *Mar. Mamm. Sci.* **8**, 135–142.

Defran, R. H. and Pryor, K. (1980). The behavior and training of cetaceans in captivity. *In* "Cetacean Behavior: Mechanisms and Functions" (Ed. L. M. Herman) pp. 319–362. Wiley, New York.

De Monte, T. and Pilleri, G. (1968). Plasmaproteins and some hematochemical data related to *Grampus griseus* Cuvier (Cetacea, Delphinidae) of the western Mediterranean. *Mar. Biol.* **1**, 182–184.

Dohl, T. P., Norris, K. S., Guess, R. C., Bryant, J. D. and Honig, M. W. (1980). Cetacea of the Southern California Bight. *In* "Marine Mammal and Sea Bird Surveys of the Southern California Bight Area, 1975–1978", Vol. III, "Investigator's Reports", Part II. PB81-248-71, NTIS.

Dohl, T. P., Guess, R. C., Duman, M. L. and Helm, R. C. (1983). Cetaceans of central and northern California, 1980–1983: status, abundance, and distribution. Contract #14-12-0001-29090, Minerals Management Service, US Department of the Interior.

Duguy, R. and Hussenot, E. (1982). Occasional captures of delphinids in the northeast Atlantic. *Rep. Int. Whal. Commn* **32**, 461–462.

Ellerman, J. R. and Morrison-Scott, T. C. S. (1951). "Checklist of Palearctic and Indian Mammals 1758 to 1946". British Museum (Natural History), London.

Erdman, D. S. (1970). Marine mammals from Puerto Rico to Antigua. *J. Mamm.* **51**, 636–638.

Evans, P. G. H. (1980). Cetaceans in British waters. *Mamm. Rev.* **10**, 1–52.

Fahlgern-Torres, S. (1988). Remains of a rare dolphin wash ashore at island area. *Corpus Christi Caller Times* 14 February.

Fiscus, C. H. and Niggol, K. (1965). Observations of cetaceans off California, Oregon, and Washington. US Fisheries and Wildlife Service Special Science Report on Fisheries No. 498, pp. 1–27.

Flower, W. H. (1872). On Risso's dolphin, *Grampus griseus* (Cuv.). *Trans. Zool. Soc. Lond.* **8**, 1–21 + 2 plts.

Flower, W. H. and Lydekker, R. (1891). "An Introduction to the Study of Mammals Living and Extinct". Adam and Charles Black, London.

Forney, K. A., Barlow, J. and Carretta, J. V. (1995). The abundance of cetaceans in California waters. Part II: Aerial surveys in winter and spring of 1991 and 1992. *Fish. Bull.* **93**, 15–26.

Fraser, F. C. (1953). "Report on Cetacea Stranded on the British Coasts from 1938–1947". British Museum (Natural History), London.

Fraser, F. C. (1940). Three anomalous dolphins from Blacksod Bay, Ireland. *Proc. Roy. Irish Acad.* **45**B, 413–455.

Fraser, F. C. (1974). "Report on Cetacea Stranded on British Coasts from 1948–1966". British Museum (Natural History), London.

Fraser, F. C. and Purves, P. E. (1960). Hearing in the cetaceans. *Bull. Brit. Mus. Zool.* **7**, 1–140.

Gallien, C. L., Chalumeau-Le-Foulgoc, M.Th. and Fine, J. M. (1970). Comparative study of serum proteins in four dolphin species (Cetaces-Odontocetes). *Comp. Biochem. Physiol.* **37**, 375–385.

Geise, L. and Borobia, M. (1987). New Brazilian records for *Kogia, Pontoporia, Grampus,* and *Sotalia* (Cetacea, Physeteridae, Platanistidae, and Delphinidae). *J. Mamm.* **68**, 873–875.

Gihr, M. and Kraus, C. (1970). Quantitative investigations on the cetacean kidney. *Invest. Cetacea* **2**, 168–179.

Gilmore, R. M. (1951). The whaling industry, whales, dolphins, and porpoises. *In* "Marine Products of Commerce" (Eds D. K. Tressler and J. McW. Lemon) pp. 680–715. Reinhold, New York.

Gray, J. E. (1828). Original figures and short systematic descriptions of new and unfigured animals. *In* "Spicilegia Zoologica", Vol. 1, pp. 1–8. Treuttel, Wurz and Co., London.

Guiguet, C. J. and Pike, G. C. (1965). First specimen of the grey grampus or Risso's dolphin, *Grampus griseus* (Cuvier) from British Columbia. *Murrelet* **46**, 16.

Hain, J. H. W., Edel, R. K., Hays, H. E., Katona, S. K. and Roanowicz, J. D. (1981). General distribution of cetaceans in the continental shelf waters of the northeastern US. *In* "Cetacean and Turtle Assessment Program: a Characterization of Marine Mammals and Sea Turtles in the Mid and North Atlantic Areas of the US Outer Continental Shelf. Annual Report for 1979". US Bureau of Land Management, Washington, DC, 345 pp.

Hamilton, R. (1837). Mammalia, whales, etc. *In* "The Naturalist's Library", Vol. 26 (Ed. Sir W. Jardine). W. H. Lizars, Edinburgh, and H. G. Bohn, London.

Hammond, D. D. and Leatherwood, S. (1984). Cetaceans live captured for Ocean Park, Hong Kong: April 1974–February 1983. *Rep. Int. Whal. Commn* **34**, 491–495.

Harrison, R. J., Brownell Jr, R. L. and Boice, R. C. (1972). Reproduction and gonadal appearances in some odontocetes. *In* "Functional Anatomy of Marine Mammals", Vol. I (Ed. R. J. Harrison), pp. 362–429. Academic Press, New York.

Harrison, R. J. and King, J. E. (1965). "Marine Mammals". Hutchinson, London.

Hershkovitz, P. (1961). On the nomenclature of certain whales. *Fieldiana Zool.* **39**, 547–550.

Heyning, J. E. (1989). Comparative facial anatomy of beaked whales (Ziphiidae) and a systematic revision among the families of extant Odontoceti. *Contrib. Sci. Los Angeles City Mus.* **405**, 1–64.

Heyning, J. E., Lewis, T. D. and Woodhouse, C. D. (1990). Odontocete mortality from fishing gear entanglements off southern California. Report presented at the International Whaling Commission Workshop on Mortality of Cetaceans in Passive Fishing Nets and Traps, La Jolla, CA, 22–25 October.

Hill, P. S. and Barlow, J. (1992). "Report of a Marine Mammal Survey of the California Coast Aboard the Research Vessel McArthur July 28–November 5, 1991". NOAA-NMFS, San Diego, CA.

Hohn, A. A. (1980a). Age determination and age related factors in the teeth of western north Atlantic bottlenose dolphins. *Sci. Rep. Whales Res. Inst.* **32**, 29–66.

Hohn, A. A. (1980b). Analysis of growth layers in the teeth of *Tursiops truncatus* using light microscopy, microradiography, and SEM. *In* "Age Determination of Toothed Whales and Sirenians" (Eds W. F. Perrin and A. C. Myrick) pp. 155–160. *Rep. Int. Whal. Commn*, Spec. Issue No. 3, Cambridge.

Hubbs, C. L. (1960). The marine vertebrates of the outer coast. *Systematic Zool.* **9**, 134–147.

International Whaling Commission (1990). "Workshop Report on Mortality of Cetaceans in Passive Fishing Nets and Traps, La Jolla, CA 22–25 October", IWC.

Iredale, T. and Troughton, E.LeG. (1936). The correct generic names for the grampus or killer whale, and the so-called grampus or Risso's dolphin. *Rec. Aust. Mus.* **19**, 28–36.

Kasuya, T. (1985). Fishery–dolphin conflict in the Iki Island area of Japan. *In* "Marine Mammals and Fisheries" (Eds J. R. Beddington, R. J. H. Beverton and D. M. Lavigne) pp. 253–272. George Allen and Unwin, London.

Kasuya, T. (1973). Systematic consideration of recent toothed whales based on the morphology of tympano-periotic bone. *Sci. Rep. Whales Res. Inst.* **25**, 1–103.

Kasuya, T. and Izumisawa, Y. (1981). The fishery–dolphin conflict in the Iki Island, Japan area. Report to the US Marine Mammal Commission NM 1533791-7, NTIS, 31 pp.

Kasuya, T. and Matsui, S. (1984). Age determination and growth of the short-finned pilot whale off the Pacific coast of Japan. *Sci. Rep. Whales Res. Inst.* **35**, 57–91.

Kasuya, T., Tobayama, T. and Matsui, S. (1984). Review of the live capture of small cetaceans in Japan. *Rep. Int. Whal. Commn* **34**, 587–602.

Katona, S. K., Rough, V. and Richardson, D. T. (1983). "A Field Guide to the Whales, Porpoises, and Seals of the Gulf of Maine and Eastern Canada — Cape Cod to Newfoundland", 3rd Edn. Charles Scribner's Sons, New York.

Kim, G. B., Tanabe, S., Iwakiri, R., Tatsukawa, R., Amano, M., Miyazaki, N. and Tanaka, H. (1996). Accumulation of butyltin compounds in Risso's dolphin (*Grampus griseus*) from the Pacific Coast of Japan: comparison with organochlorine residue pattern. *Environ. Sci. Technol.* **30**, 2620–2625.

Klinowska, M. (1991). "Dolphins, Porpoises, and Whales of the World". IUCN, Gland, Switzerland.

de Kock, A. C., Best, P. B., Cockcroft, V. and Bosma, C. (1994). Persistent organochlorine residues in small cetaceans from the east and west coasts of southern Africa. *Sci. Total Environ.* **154**, 153–162.

Kruse, S. L. (1989). Aspects of the biology, ecology, and behavior of Risso's dolphins (*Grampus griseus*) off the California coast. M.Sc. thesis, University of California at Santa Cruz, CA.

Kruse, S., Leatherwood, S., Prematunga, W. P., Mendes, C. and Gamage, A. (1991). Records of Risso's dolphins, *Grampus griseus*, in the Indian Ocean, 1891–1986. *In* "Cetaceans and Cetacean Research in the Indian Ocean Sanctuary" (Eds S. Leatherwood and G. Donovan) pp. 66–77. United Nations Environmental Programme, Nairobi, Kenya.

Leatherwood, S. (1977). Risso's dolphin, *Grampus griseus*, a summary of knowledge about the species. Report to the Naval Oceans Systems Center, San Diego, CA.

Leatherwood, S., Hubbs, C. L. and Fisher, M. (1979). First records of Risso's dolphin (*Grampus griseus*) from the Gulf of California with detailed notes on a mass stranding. *Trans. San Diego Soc. Nat. Hist.* **19**, 45–52.

Leatherwood, S., Perrin, W. F., Kirby, V. L., Hubbs, C. L. and Dahlheim, M. (1980). Distribution and movements of Risso's dolphin, *Grampus griseus*, in the eastern North Pacific. *Fish. Bull.* **77**, 951–963.

Leatherwood, S., Peters, B. B., Santerre, R., Santerre, M. and Clarke, J. T. (1984). Observations of cetaceans in the northern Indian Ocean Sanctuary, November 1980–May 1983. *Rep. Int. Whal. Commn* **34**, 509–520.

Leatherwood, S., Reeves, R. R., Perrin, W. F. and Evans, W. E. (1982). Whales, dolphins, and porpoises of the eastern North Pacific and adjacent Arctic waters: a guide to their identification. NOAA Technical Report NMFS Circular 444.

Leatherwood, S. and Reeves, R. R. (1989). "Marine Mammal Research and Conservation in Sri Lanka 1985–1986". United Nations Environmental Programme, Nairobi, Kenya.

Lee, H. (1877). On the capture of a specimen of Risso's grampus at Sidlesham, near Chichester. *Proc. Zool. Soc. Lond.* 808–809.

Lepiksaar, J. (1966). Zahnwalfunde in Schweden. *Bijdragen tot de dierkunde* **36**, 3–16 (Translation, 1974, Encounters with toothed whales in Sweden (Ed. K. Hollingshead), Department of the Navy Oceanography Office #NOO T-13).

Leung, Y. M. (1967). An illustrated key to the species of whale lice (Amphipoda, Cyamidae) ectoparasites on Cetacea, with a guide to the literature. *Crustaceana* **12**, 279–291.

Lichter, A. A. and Goodall, R. N. P. (1988). Argentina. Progress report on cetacean research, June 1984 to April 1987. *Rep. Int. Whal. Commn* **38**, 173–176.

Litchfield, C. and Greenberg, A. J. (1974). Comparative lipid patterns of the melon fats of dolphins, porpoises, and toothed whales. *Comp. Biochem. Physiol.* **47**B, 401–407.

Litchfield, C., Greenberg, A. J., Caldwell, D. K., Caldwell, M. C., Sipos, J. C. and Ackman, R. G. (1975). Comparative lipid patterns in acoustical and nonacoustical fatty tissues of dolphins, porpoises, and toothed whales. *Comp. Biochem. Physiol.* **50**B, 591–597.

Madsen, C. J. and Herman, L. M. (1980). Social and ecological correlates of cetacean vision and visual appearance. *In* "Cetacean Behavior: Mechanisms and Functions" (Ed. L. M. Herman) pp. 101–147. Wiley, New York.

Maxwell, G. (1952). "Harpoon Venture". Viking Press, New York.

McCann, C. (1974). Body scarring on Cetacea–Odontocetes. *Sci. Rep. Whales Res. Inst.* **26**, 145–155.

Mead, J. G. (1975). Anatomy of the external nasal passages and facial complex in the Delphinidae (Mammalia: Cetacea). *Smithsonian Contrib. Zool.* **207**.

Medway, Lord (1977). "Mammals of Borneo: Field Keys and Annotated Checklist". Monographs of the Malaysian Branch of the Royal Asiatic Society, No. 7, Kuala Lumpur, Malaysia, 172 pp.

Meggitt, F. J. (1924). "The Cestodes of Mammals". H. Pohle, London.

Miller, G. S. and Kellogg, R. (1955). List of North American recent mammals. *U.S. Natl Mus. Bull.* No. 205.

Mitchell, E. (1975). Review of biology and fisheries for smaller cetaceans. Report of the Meeting on Smaller Cetaceans, Montreal, April 1–11, 1974. *J. Fish. Res. Bd Can.* **32**, 889–983.

Miyazaki, N., Hirosaki, Y., Kinuta, T. and Omura, H. (1992). Osteological study of a hybrid between *Tursiops truncatus* and *Grampus griseus*. *Bull. Natl Sci. Mus.* **18**, 79–94.

Mizue, K. and Yoshida, K. (1962). Studies on the little toothed whales in the west sea area of Kyushu. *Bull. Faculty Fish. Nagasaki Univ.* **12**, 45–52.

Mörzer-Bruyns, W. F. J. (1971). "Field Guide to Whales and Dolphins". Mees, Amsterdam.

Morimitsu, T., Kawano, H., Torihara, K., Kato, E. and Kogno, M. (1992). Histopathology of eighth cranial nerve of mass stranded dolphins at Goto Islands, Japan. *J. Wildlife Diseases* **28**, 656–658.

Mullin, K. D., Hoggard, W., Roden, C. L., Lohoefener, R. R., Rogers, C. M. and Taggart, B. (1994). Cetaceans on the upper continental slope in the north-central Gulf of Mexico. *Fish. Bull.* **92**, 773–786.

Murie, J. (1871). On Risso's grampus, *G. rissoanus* (Desm.). *J. Anat. Physiol.* **5**, 118–138.

Myrick Jr, A. C., Shallenberger, E. W., Kang, I. and Mackay, D. B. (1984). Calibration of dental layers in seven captive Hawaiian spinner dolphins (*Stenella longirostris*), based on tetracycline labels. *Fish. Bull.* **82**, 207–225.

Nachtigall, P. E., Pawloski, J. L., Schroeder, J. P. and Sinclair, S. (1990). Successful maintenance and research with a formerly stranded Risso's dolphin (*Grampus griseus*). *Aquat. Mamm.* **161**, 8–13.

Nachtigall, P. E., Au, W. L., Pawloski, J. L. and Moore, P. W. (1995). Risso's dolphin (*Grampus griseus*) hearing thresholds in Kaneohe Bay, Hawaii. *In* "Sensory Systems of Aquatic Mammals" (Eds R. A. Kastelein, J. A. Thomas and P. E. Nachtigall) pp. 49–53. De Spil, Woerden, The Netherlands.

Nemoto, T. (1955). White scars on whales (1) lamprey marks. *Sci. Rep. Whales Res. Inst.* **10**, 69–77.

Ness, A. R. (1967). A measure of the asymmetry of the skulls of odontocete whales. *J. Zool. Lond.* **153**, 209–221.

Nakamine, H., Nagata, S., Yonezawa, M. and Tanaka, Y. (1992). The whale (Odontoceti) spleen: a type of primitive mammalian spleen. *Acta Anat. Nippon* **67**, 69–82.

Nicol, D. J. and Croome, R. L. (1988). Trends in the Tasmanian stranding record. *In* "Marine Mammals of Australasia, Field Biology and Captive Management" (Ed. M. L. Augee) pp. 59–70. Royal Zoological Society of New South Wales, Sydney.

Nishiwaki, M. (1967). Distribution and migration of marine mammals in North Pacific area. *Bull. Univ. Tokyo Ocean Res. Inst.* **1**, 1–64.

Nishiwaki, M. (1972). General biology. *In* "Mammals of the Sea: Biology and Medicine" (Ed. S. H. Ridgway) pp. 3–204. Charles C. Thomas, Springfield, IL.

Norris, K. S. and Dohl, T. P. (1980). The structure and function of cetacean schools. *In* "Cetacean Behavior: Mechanisms and Functions" (Ed. L. M. Herman) pp. 211–261. Wiley, New York.

Norris, K. S. and Prescott, J. H. (1961). Observations on Pacific cetaceans of Californian and Mexican waters. *Univ. Calif. Publ. Zool.* **63**, 291–402.

Oen, E. O. (1987). An observation of Risso dolphins in Norwegian waters. *Fauna (Oslo)* **39**, 173.

Ohsumi, S. (1972). Catch of marine mammals, mainly of small cetaceans, by local fisheries along the coast of Japan. *Far Seas Fish. Res. Lab. Bull.* **7**, 137–166.

Ohsumi, S. (1975). Review of Japanese small-type whaling. *J. Fish. Res. Bd Can.* **32**, 1111–1121.

Orr, R. T. (1966). Risso's dolphins on the Pacific coast of North America. *J. Mamm.* **47**, 341–343.

Paul, J. R. (1968). Risso's dolphin, *Grampus griseus*, in the Gulf of Mexico. *Mamm.* **49**, 746–748.

Perrin, W. F., Lo, N. C. and Wahlen, B. E. (1979). Progress in reducing incidental kill of dolphins in the U.S. tuna purse-seine fishery in the eastern tropical Pacific. NMFS/SWFC Administrative Report LJ-79-19.

Perrin, W. F. and Reilly, S. B. (1984). Reproductive parameters of dolphins and small whales of the family Delphinidae. *In* "Reproduction in Whales, Dolphins and Porpoises" (Eds W. F. Perrin, R. L. Brownell and D. P. DeMaster) pp. 97–134. *Rep. Int. Whal. Commn*, Spec. Issue No. 6, Cambridge,

Perrin, W. F., Donovan, G. P. and Barlow, J. (1994). "Gillnets and Cetaceans". International Whaling Commission, Cambridge, 629 pp.

Pierce, K. V. and Kajimura, H. (1980). Acid etching and highlighting for defining growth layers in cetacean teeth. *In* "Age Determination of Toothed Whales and Sirenians" (Eds W. F. Perrin and A. C. Myrick Jr) pp. 99–103. *Rep. Int. Whal. Commn*, Spec. Issue No. 3, Cambridge.

Pike, G. C. (1951). Lamprey marks on whales. *J. Fish. Res. Bd Can.* **8**, 275–280.

Pilleri, G. and Busnel, R. G. (1969). Brain/body weight ratios in Delphinidae. *Acta Anat.* **73**, 92–97.

Pilleri, G. and Gihr, M. (1969). On the anatomy and behavior of Risso's dolphin (*Grampus griseus* G. Cuvier). *Invest. Cetacea* **1**, 74–93.

Pilleri, G. and Knuckey, J. (1969). Behaviour patterns of some Delphinidae observed in the western Mediterranean. *Z. Tierpsychol.* **26**, 48–72.

Polacheck, T. (1987). Relative abundance, distribution, and inter-specific relationship of cetacean schools in the eastern tropical Pacific. *Mar. Mamm. Sci.* **3**, 54–77.

Poulter, T. C. (1967). Communication in selected groups. *In* "Animal Communication, Techniques of Study and Results of Research" (Ed. T. A. Sebeok) pp. 405–465. Indiana University, Bloomington, IN.

Raga, J. A., Casinos, A., Filella, S. and Raduan, A. (1982). Notes on the cetaceans of the Iberian coasts. V. *Crassicauda grampicola* Johnston and Mawson, 1941 (Nematoda) cause of injuries in the pterygoids of some specimens of *Grampus griseus*. Part 4, pp. 315–318. BLV Verlagsgesellschaft mbH, Munich.

Raga, J. A., Rudvan, M. A. and Blanco, C. (1986). Contributions to the study of the distribution of cetaceans in Mediterranean and Iberian waters. *Misc. Zool.* **9**, 361–366.

Read, A. J. (1996). Incidental catches of small cetaceans. *In* "The Conservation of Whales and Dolphins: Science and Practice" (Eds M. P. Simmonds and J. D. Huchinson) pp. 109–128. Wiley, New York.

Reidenberg, J. S. and Laitman, J. T. (1994). Anatomy of the hyoid apparatus in odontoceti (Toothed whales): specializations of their skeleton and musculature compared with those of terrestrial mammals. *Anat. Rec.* **240**, 598–624.

Reimchen, T. (1980). Sightings of Risso's dolphins (*Grampus griseus*) off Queen Charlotte Islands, British Columbia. *Murrelet* **61**, 44–45.

Ridgway, S. H. (1972). Homeostasis in the aquatic environment. *In* "Mammals of the Sea: Biology and Medicine" (Ed. S. H. Ridgway) pp. 590–747. Charles C. Thomas, Springfield, IL.

Ridgway, S. H. and Brownson, R. H. (1984). Relative brain sizes and cortical surface areas in odontocetes. *Acta Zool. Fennica* **172**, 149–152.

Ross, G. J. B. (1984). The smaller cetaceans of the south east coast of southern Africa. *Ann. Cape Prov. Mus. (Nat. Hist.)* **15**, 147–400.

Scammon, C. M. (1874). "The Marine Mammals of the Northwestern Coast of North America, Together with an Account of the American Whale-Fishery". J. H. Carmany and Co., San Francisco; G. P. Putnam's Sons, New York.

Schevill, W. E. (1954). Sight records of the gray grampus, *Grampus griseus* (Cuvier). *J. Mamm.* **35**, 123–124.

Schmidly, D. J. (1981). Marine mammals of the southeastern United States coast and the Gulf of Mexico. US Bureau of Land Management/US Fish and Wildlife Service Biological Services Program Report FWS/OBS-80/41.

Schultz, W. (1970). Üeber das vorkommen von walen in der Nord-und Ostsee (ordn Cetacea). *Zool. Anz.* **185**, 172–264.

Sergeant, D. E. (1959). Age determination of odontocete whales from dentinal growth layers. *Norsk. Hvalfangst-Tid.* **48**, 273–288.

Sergeant, D. E. (1962). The biology of the pilot or pothead whale (*Globicephala melaena*) (Traill) in Newfoundland waters. *Bull. Fish. Res. Bd Can.* **132**, 1–84.

Sezaki, K., Hirosaki, Y., Watabe, S. and Hashimoto, K. (1984). Electrophoretic characters of the hybrids between two dolphins *Tursiops truncatus* and *Grampus griseus*. *Bull. Jap. Soc. Sci. Fish.* **50**, 1771–1776.

Shallenberger, E. W. (1981). The status of Hawaiian cetaceans. NTIS Report MMC-77/23, US Department of Commerce.

Shane, S. H. (1990). Case study: the effects of natural phenomena. *In* "Proceedings of the American Cetacean Society Fourth Biennial Conference. Monterey, California,

November 9–11" (Ed. S. Taylor) pp. 34–39. American Cetacean Society, San Pedro, CA.

Shane, S. H. (1995). Behavior patterns of pilot whales and Risso's dolphins off Santa Catalina Island, California. *Aquat. Mamm.* **21**, 195–197.

Sheldon, K. E. W., Baldridge, A. and Withrow, D. E. (1995). Observations of Risso's dolphins *Grampus griseus* with gray whales, *Eschrichtius robustus. Mar. Mamm. Sci.* **11**, 231–240.

Shults, L. M., Fay, F. H. and Hall, J. D. (1982). Helminths from Stejneger's beaked whale, *Mesoplodon stejnegeri*, and Risso's dolphin, *Grampus griseus*, in Alaska. *Proc. Helminthol. Soc. Wash.* **49**, 146–147.

Stroud, R. K.(1968). Risso's dolphin in Washington state. *J. Mamm.* **49**, 347–348.

Sylvestre, J. and Tasaka. S. (1985). On the intergeneric hybrids in cetaceans. *Aquat. Mamm.* **11**, 101–108.

Teixeira, A. M. A. P. (1979). Marine mammals of the Portuguese coast. *Z. Säugetierkunde* **44**, 221–238.

Tomilin, A. G. (1957). "Mammals of USSR and Adjacent Countries", Vol. IX, "Cetacea" (Ed. V. G. Heptner). Nauk USSR, Moscow (English translation, 1967, Israel Program for Scientific Translations, Jerusalem).

Troughton, E. G. (1931). The occurrence of a male and female *Grampus griseus* (Delphinidae) at Sydney, NSW. *Proc. Zool. Soc. Lond.* Pt. 2, 265–269.

True, F. W. (1889). A review of the family Delphinidae. *U.S. Nat. Mus. Bull.* No. 39.

Tsutsumi, T., Kamimura, Z. and Mizue, K. (1961). Studies of the little toothed whales in the west sea areas of Kyusyu-V. *Bull. Nagasaki. Univ. Fac. Fish.* **11**, 19–28.

Turner, W. (1892). Notes on some of the viscera of Risso's dolphin (*Grampus griseus*). *J. Anat. Physiol.* **26**, 258–270.

Van Waerebeek, K. and Reyes, J. C. (1990). Catch of small cetaceans at Pucusana port, central Peru, during 1987. *Biol. Conserv.* **51**, 15–22.

Viale, D. (1985). Cetaceans in the northwestern Mediterranean: their place in the ecosystem. *Oceanogr. Mar. Biol. Ann. Rev.* **23**, 491–571.

Vibe, C. (1950). The marine mammals and the marine fauna of the Thule district (NW Greenland), with observations on ice conditions in 1939–41. *Medd. om Grønland* No. 150.

Vidal, O., Findley, L. T. and Leatherwood, S. (1993). Annotated checklist of the marine mammals of the Gulf of California. *Proc. San Diego Soc. Nat. Hist.* **28**, 1–16.

Wedmid, Y. and Litchfield, C. (1976). Positional analysis of isovaleroyl triglycerides using proton magnetic resonance with $Eu(fod)_3$ and $Pr(fod)_3$ Shiff reagents: II. Cetacean triglycerides. *Lipids* **11**, 189–193.

White Jr, M. J. and Norris, J. (1978). A report on the *Grampus griseus* of Japan. Hubbs/Sea World Technical Report 78-113.

Wolff, W. J. (1982). The status of marine mammals in the Wadden Sea area. *In* "Marine Mammals of the Wadden Sea" (Eds P. J. H. Reijnders and W. J. Wolff) pp. 7–14. A. A. Balkema, Rotterdam.

Weitkowitz, W. (1992). Sightings of whales and dolphins in the Middle East (Cetacea). *Zool. Mid. East* **6**, 5–12.

9

False Killer Whale
Pseudorca crassidens (Owen, 1846)

Daniel K. Odell and Karen Miller McClune

Genus and Species

Taxonomy

Owen's (1846) original description of the false killer whale, *Phocoena crassidens* Owen, 1846, was based on subfossil material from the upper Pliocene (Marcuzzi and Pilleri, 1971). A living specimen was not documented until 1861 (Mobius, 1862). Gray (1846) considered Owen's specimen to belong to the genus *Orca* (*Orca crassidens* Gray, 1846). Reinhardt (1862, 1866) examined the original specimen in addition to several others that became available in the interim and determined that the genus was neither *Phocoena* nor *Orca*. He named the new genus *Pseudorca* (Reinhardt, 1862). Purves and Pilleri (1978) and Stacey *et al.* (1994) present a more detailed review of the taxonomic history of the false killer whale.

Common names

Common English names which have been used for the species include false killer whale and blackfish (Leatherwood *et al.*, 1988). Other names include faux-orque (French; Busnel and Dziedzic, 1968), pseudorque (French; Hoyt, 1984), oki gondō kujira (Japanese; Leatherwood *et al.*, 1988), malaya or chornaya kosatka (Russian; Tomilin, 1957) and false Orca (Portuguese/Spanish: Sielfeld *et al.*, 1977; Pinedo *et al.*, 1992; Hetzel and Lodi, 1993).

Handbook of Marine Mammals, Volume 6
ISBN 0-12-588506-7

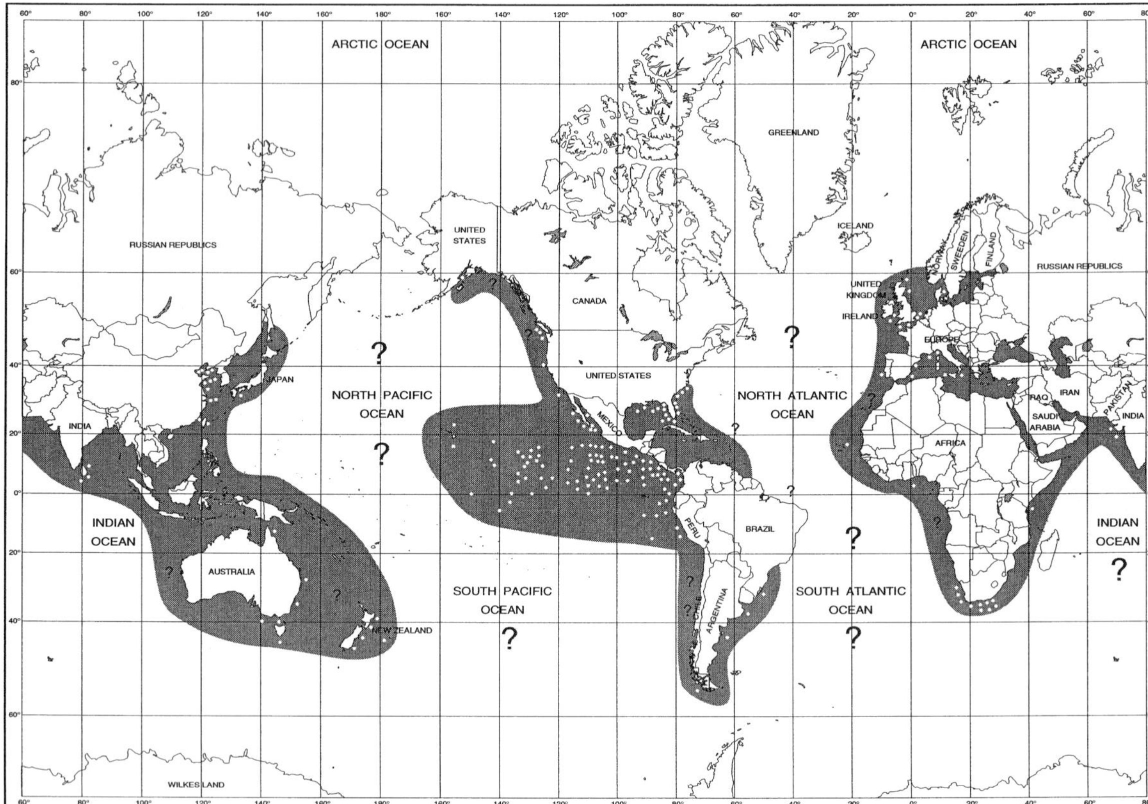

FIG. 1 The approximate world distribution of the false killer whale. Since most records come from world-wide stranding records and sightings in the central part of the range (e.g. eastern tropical Pacific, courtesy W.F. Perrin), the northern and southern limits are approximate. Density within the range should not be assumed to be uniform.

Distribution

Range

Pseudorca occurs in all tropical and temperate oceans (see Fig. 1) generally between 50°S and 50°N latitude with a few records north of 50°N in the Pacific (Leatherwood *et al.*, 1988) and in the Atlantic (Tomilin, 1957; Nishiwaki, 1967; Sielfeld *et al.*, 1977; Purves and Pilleri, 1978; Ross, 1984; Evans, 1987; Reiner, 1990; Stacey and Baird, 1991; Pinedo *et al.*, 1992; Simões-Lopes and Ximenez, 1992, 1993; Hetzel and Lodi, 1993; Stacey *et al.*, 1994). Goodall and Schiavini (1989a,b, 1990) and Oporto *et al.* (1994) report a mass stranding between 52° 27′S and 52° 55′S in the Straits of Magellan, Chile. Most of the distributional records and many of the data available for the species are the result of strandings.

Habitat

Pseudorca is found primarily in deep water and offshore areas, although there are sightings from the coasts of oceanic islands, such as the Hawaiian Islands, and some other coastal areas (Mitchell, 1965; Leatherwood and Reeves, 1983). Shallenberger (1981) reported both shallow (less than 200 m) and deep water (greater than 2000 m) habitats for the species in Hawaiian waters. *Pseudorca* appears to be relatively common in Japanese coastal waters (Nishiwaki, 1967). With the exception of sightings from the eastern tropical Pacific (Fig. 1), data on distribution are lacking for most oceanic areas. In addition, relative abundance is not known for different parts of the known range.

Migration is not well documented, although Kasuya (1971) suggested that globicephalid whales including *Globicephala*, *Pseudorca* and *Grampus* species in the western North Pacific move from warmer, southern waters in winter to cooler, northern waters in summer. Apparent seasonal movements in the western North Pacific may be related to prey distribution (Tomilin, 1957; Kasuya, 1971).

External Characteristics

Colour pattern

The body is generally uniform black or slate grey (Hinton, 1928; Ross, 1984) with a light grey thoracic blaze (Hinton, 1928) (Fig. 2) and sometimes light patches on the sides of the head (Mitchell, 1970; Leatherwood and Reeves, 1983).

Size and shape

Pseudorca is a slim delphinid (Fig. 2) reaching a maximum recorded length of 610 cm in males (Leatherwood and Reeves, 1983) and 506 cm in females

(A)

(B)

FIG. 2 External appearance of the false killer whale: (A) general body profile showing relative size and position of the dorsal fin and (B) partial body profile showing shape of head and distinctive flipper shape.

(C)

(D)

FIG. 2 *Continued* (C) Relative size of the teeth and (D) ventral view showing the thoracic blaze. (Photos courtesy Sea World of Florida)

(Perrin and Reilly, 1984). Ranges of male and female body lengths from several mass strandings are given in Table 1. The head is relatively small, a beak is absent and, in males, the forehead extends further forward than in females (Mead, 1975). Some proportional body measurements summarized from Purves and Pilleri (1978) are given in Table 2 for animals from the eastern North Atlantic. Mizue and Yoshida (1961) give real and proportional body measurements for animals from the western North Pacific. The dorsal fin is tall and falcate (Fig. 2) and is located approximately at the midpoint of the back. The characteristic flippers (Fig. 2) have a broad hump on the anterior edge near the middle of the flipper (Leatherwood *et al.*, 1976).

Weight

Data on body weight are rare (see Table 3). Scheffer and Slipp (1948) reported a 523 cm individual (sex undetermined) weighing 1990 lb (902 kg). Haynes-Lovell (1993) reported that individuals from Australian waters with body lengths of 280–305 cm weighed about 250 kg and were estimated to be about 6 years old. Others estimate that a large individual may weigh about 1300–1400 kg (Leatherwood and Reeves, 1983; Leatherwood *et al.*, 1988).

Internal Anatomy and Physiology

Skull

Purves and Pilleri (1978) present an extensive discussion of the *Pseudorca* skull. The skull (Fig. 3) is massive with a short, broad rostrum (Tomilin, 1957; Ross, 1984). The maximum width of the skull between the squamosals is about 60% of the condylobasal length (CBL) (Tables 4 and 5). Cowley (1944) compared the skull with that of other large delphinids. Pinedo and Rosas (1989) and Kitchener *et al.* (1990) compared skull morphology of *Pseudorca* stranded in several different geographic locations. The mandibles are robust (Ross, 1984) and generally have 7–10 conical teeth in each upper arcade and 8–10 in the lower (mandibular) arcades (Table 5). Purves and Pilleri (1978) list the dental formula as 7-7/8-8 to 11-11/12-12. Growth layer groups (GLGs) in the teeth (Fig. 4) can be used to estimate age (Pierce and Kajimura, 1980; Kasuya, 1986) but GLG (dentine and cementum) deposition patterns have not been studied in this species (Perrin and Myrick, 1980).

Postcranial skeleton

The total vertebral counts are in the range 47–52 (Purves and Pilleri, 1978) with the "average" being 49–50. The "average" vertebral formula is C_7, T_{10-11}, L_{9-11}, Ca_{18-26} (Purves and Pilleri, 1978). Some anatomical studies provide

TABLE 1 Body length data from four false killer whale mass strandings

Males				Females				M+F			
N	Mean	SD	Range	*N*	Mean	SD	Range	*N*	Mean	SD	Range
34	413.8	78.0	236–523	20	354.7	53.7	234–409	62	384.2	79.4	137–523
Source: Scott and Green (1975)								(includes animals of unknown sex)			
12	457.9	48.0	377–534	17	418.8	42.4	328–493	29	435.0	48.2	328–534
Source: Odell *et al*. (1979)											
12	407.8	57.9	272–479	15	349.3	80.6	122–415	27	375.3	76.2	122–479
Source: Odell *et al*. (unpublished data; 1986 mass stranding in Florida)											
17	—	—	~ 185–~ 547	20	—	—	~ 205–~ 484	—	—	—	—
Source: Goodall and Schiavini (1989b, 1990)											

TABLE 2 Summary of selected proportional (% total length) body measurements for male and female false killer whales (after Hinton, 1928; Purves and Pilleri, 1978)

Measurement		Males	Females
Tip of rostrum to centre of dorsal fin	(Range)	41.6–53.0%	43.0–52.0%
	(Mean)	45.86	46.41
	(*n*)	47	60
Fluke notch to centre of anus		30.6–36.6	30.2–39.0
		33.52	32.64
		46	60
Fluke notch to centre genital slit		36.8–46.8	32.2–39.0
		42.92	37.00
		46	60
Width of flukes		18.4–24.5	19.0–23.7
		21.2	21.58
		47	59
Height of dorsal fin		6.25–8.8	6.7–10.3
		10.46	7.89
		31	56
Length of flipper		9.4–13.6	9.7–14.5
		12.17	11.86
		49	59
Tip of rostrum to blowhole		9.2–14.1	9.5–14.5
		11.13	11.33
		47	58
Tip of rostrum to eye		8.0–11.5	8.1–13.0
		9.83	9.94
		48	60
Tip of rostrum to angle of jaw		7.8–11.8	7.6–12.5
		9.61	9.37
		46	60

detailed descriptions of the flipper and tympano-periotic bones (Kasuya, 1973; Gihr *et al.*, 1982).

Organs, tissues and physiology

Little information is available on organ weights and other aspects of the soft anatomy. Some organ weights are listed in Table 3, but care should be exercised in interpreting these data since they came from mass-stranded animals and may not be "normal". Purves and Pilleri (1978) present the most detailed description of the musculature and soft tissues of the head they considered to be involved in sound production. Mead (1975) described details of nasal and facial musculature, dermal tissue and structure, noting differences between the sexes in the amount of connective tissue anterior to the melon. McFarland (1971) found that the retina of the eye is composed primarily of rods, and he found only one retinal pigment, rhodopsin.

In terms of reproductive systems, corpora lutea and the female system were characterized from a 1935 stranding in Scotland (Comrie and Adam, 1938), in subsequent works by Slijper (1949, 1961, 1966) and reviewed by Harrison *et al.* (1972). Male reproductive organ morphology also was discussed by Slijper (1966).

Physiological studies, too, have been limited for the species. Slijper (1961) noted some details of cetacean locomotion characteristic of *Pseudorca*. The internal deep body temperature is in the range 36–37.2 °C, as measured with an ingested radio transmitter (Whittow *et al.*, 1974). Body temperature dropped during rest and feeding, and increased during active swimming.

Blood chemistry data from strandings and individuals in oceanaria are available. Brown *et al.* (1966) reported the first blood values based on one female in captivity. Whole blood and serum parameters from a mass stranding were summarized by Odell *et al.* (1980), and additional data from a 1986 stranding have been gathered (Walsh *et al.*, 1990). Duffield *et al.* (1979) found haemoglobin polymorphisms similar to *Orcinus orca*.

Pseudorca has the typical cetacean diploid number of chromosomes ($2n = 44$) (Kulu, 1972; Duffield *et al.*, 1991).

Life History and Population Dynamics

Growth and reproduction

Almost everything that is known about *Pseudorca* life history comes from the examination of strandings, individuals taken in fishery-related activities (e.g. Kasuya, 1985, 1986), and individuals in marine zoological parks. Migrations to specific breeding and calving areas have not been documented (Leatherwood and Reeves, 1983). Purves and Pilleri (1978) speculated that calving may occur between January and March in tropical or subtropical waters, while others report that calving probably occurs year-round (Leatherwood and Reeves, 1983; Kasuya, 1986).

Length at birth has been noted as about 160 cm (Smithers, 1938; Cowley, 1944, near-term foetus; Tomilin, 1957; Scott and Green, 1975, largest foetus; Ross, 1984), about 175 cm (Kasuya, 1985, 1986, range 170–179 cm) and about

TABLE 3 Organ weights (g) of false killer whales stranded in Florida in 1976 (see Odell *et al.*, 1979)

	Field No.					
	C-76-18	C-76-19	SWF-PC-7627-B	SWF-PC-7628-B	SWF-PC-7629-B	SWF-PC-7630-B
Sex	F	M	F	F	F	F
Total length (cm)	440	520	338	297	358	475
Body weight (kg)	—	—	327	250	359	773
Heart	4400	6000	2700	1990	2540	5740
			(0.83%)[a]	(0.80)	(0.71)	(0.74)
Liver	—	22 500	13 930	9230	11 870	16 120
			(4.26)	(3.69)	(3.31)	(2.09)
Lungs						
Left	—	8750[b]	5720	3960	4530	8550
			(1.76)	(1.58)	(1.26)	(1.11)
Right	—	6750[b]	6810	4720	4240	7760
			(2.08)	(1.89)	(1.18)	(1.00)
Spleen	—	—	270	200	140	190
			(0.08)	(0.08)	(0.04)	(0.02)

Kidneys						
Left	2300	3700	1680	1119	1670	3080
			(0.51)	(0.45)	(0.47)	(0.40)
Right	2500	3500	1700	1300	1580	3020
			(0.52)	(0.52)	(0.44)	(0.39)
Adrenals						
Left	—	—	40		40	—
			(0.01)		(0.01)	
Right	—	—	40		40	—
			(0.01)		(0.01)	
Pancreas	—	—	370	400	530	—
			(0.11)	(0.16)	(0.15)	
Brain	—	5000	—	—	3610	—
					(1.01)	
Ovaries						
Left	28	—	7.0	6.8	8.7	23.4
Right	37	—	7.0	6.5	6.4	23.4
Testes						
Left	—	3700	—	—	—	—
Right	—	4500	—	—	—	—

[a]Organ weight as percentage of body weight.
[b]Left and right lungs not differentiated in this specimen.

TABLE 4 Summary of *Pseudorca* skull proportions (%CBL) based on data from Purves and Pilleri (1978)

Measurement	Males			Females			Males + Females	
	N	Mean	SD	*N*	Mean	SD	*N*	Mean
1. Condylobasal length (cm)	42	59.21	4.60	34	58.77	5.12	76	59.01
2. Width of cranium	42	44.13	2.86	34	44.43	2.91	76	44.26
3. Condylar notch to mesethmoid	42	42.20	0.96	34	41.87	1.43	76	42.05
4. Tip of rostrum to antorbital notch	42	48.42	1.81	34	49.59	1.32	76	48.94
5. Greatest width between squamosals	42	57.11	2.45	33	60.17	1.85	75	58.46
6. Greatest width between frontals	42	56.33	1.74	34	57.90	2.49	76	57.03
7. Width of rostrum at antorbital notch	42	30.70	1.14	34	32.83	2.09	76	31.65
8. Width of rostrum at mid-point	42	27.22	1.52	34	29.42	1.59	76	28.20
9. Length of right maxilla	42	81.64	2.32	34	83.09	2.58	76	82.29
10. Length of left maxilla	42	80.48	2.72	34	81.73	1.46	76	81.04
11. Width of right maxilla	42	17.02	0.76	34	16.93	1.36	76	16.98
12. Width of left maxilla	42	16.71	0.91	34	16.77	1.30	76	16.74
13. Width of right premaxilla	42	12.72	0.98	34	13.22	0.61	76	12.94
14. Width of left premaxilla	42	8.62	0.37	34	8.99	0.70	76	8.79
15. Anterioposterior width of interparietal	42	6.88	1.25	34	7.18	2.55	76	7.01
16. Lateral width of interparietal	42	13.04	6.61	34	13.81	8.39	76	13.38
17. Length of mandible	35	79.45	1.80	31	82.61	3.22	66	80.93

190 cm (Peacock *et al.*, 1936; Slijper, 1949; Purves and Pilleri, 1978). Calves nurse for about 18 months to 2 years (Purves and Pilleri, 1978; Kasuya and Izumisawa, 1981; Kasuya, 1985).

Annual growth rate from one large stranding was estimated at 0.34 ft (10.4 cm) per year after the first few years, based on reported lengths and estimated age from GLGs in sectioned teeth, with males reaching lengths of about 1 ft (30 cm) longer than females (Purves and Pilleri, 1978). One female resident in a marine zoological park grew 35 cm (from 343 cm to 378 cm) in length in less than two years (Brown *et al.*, 1966). Kasuya (1986) reported growth patterns for *Pseudorca* from Japan. Both males and females grow 40–50% in body length during their first year. After year 1, males grow at a faster rate.

Length at sexual maturity in females has been reported as follows: North Atlantic, 366–427 cm (Purves and Pilleri, 1978); and North Pacific, 380–399 cm and 340–380 cm (Kasuya and Izumisawa, 1981 as reported in Perrin and Reilly, 1984 and Kasuya, 1986, respectively). Length at sexual maturity in males has been reported as follows: North Atlantic, 396–457 cm (Purves and

TABLE 5 Summary of *Pseudorca crassidens* skull measurements from mass stranding in Florida (USA) in 1976 (measurements after Perrin, 1975)

Measurement	N[a]	Mean (cm)	SD	Range (cm)	Proportion of CBL (%)			
					N	Mean	SD	Range
1. Condylobasal length	23	59.66	3.09	54.8–63.4				
2. Width cranium	21	26.91	1.00	25.5–29.2	20	44.79	1.96	41.5–47.5
3. Condylar notch to mesethmoid	23	20.38	1.38	18.1–22.4	23	34.16	1.37	31.6–36.2
4. Tip rostrum to antorbital notch	24	30.36	1.67	26.9–33.3	23	50.73	1.29	49.2–52.4
5. Max. width between squamosals	21	36.34	1.95	32.2–39.5	20	60.50	2.47	56.2–65.1
6. Max. width between frontals	24	35.31	1.79	32.1–38.3	23	59.15	2.28	54.4–64.1
7. Greatest width of rostrum	24	20.31	1.52	16.7–23.6	23	34.00	1.82	28.9–37.6
8. Width rostrum at mid-point	24	16.65	1.23	14.7–19.1	23	27.86	1.40	25.1–30.4
9. Length right maxilla	24	50.79	2.61	46.1–55.0	23	84.90	2.14	80.6–87.7
10. Length left maxilla	24	49.62	2.55	44.0–54.2	23	82.93	2.36	78.6–86.2
11. Width right maxilla	24	9.22	0.56	8.0–10.2	23	15.47	0.86	14.4–16.9
12. Width left maxilla	24	9.22	0.81	7.7–10.2	23	15.47	1.30	13.3–17.9
13. Width right premaxilla	24	6.60	0.42	6.0–7.9	23	11.07	0.60	10.2–12.6
14. Width left premaxilla	24	5.70	0.40	5.2–7.0	23	9.57	0.60	8.8–11.2
15. Length right mandible	20	49.95	3.20	44.2–55.5	20	83.06	3.23	75.5–86.9
16. Length left mandible	20	49.92	3.04	45.0–55.1	20	83.01	2.85	76.5–85.6
17. Width right mandible	20	13.92	0.85	12.3–15.2	20	23.16	0.98	21.8–25.1
18. Width left mandible	20	13.79	0.80	12.0–15.2	20	22.94	0.86	21.5–23.8
19. Tooth alveoli	Mean	Mode	Range					
Upper R	8.2	8	7–10					
Upper L	7.8	8	7–10					
Lower R	9.3	9	8–10					
Lower L	9.4	9	9–10					

[a]Carcass condition prevented determination of sex.

(A)

(B)

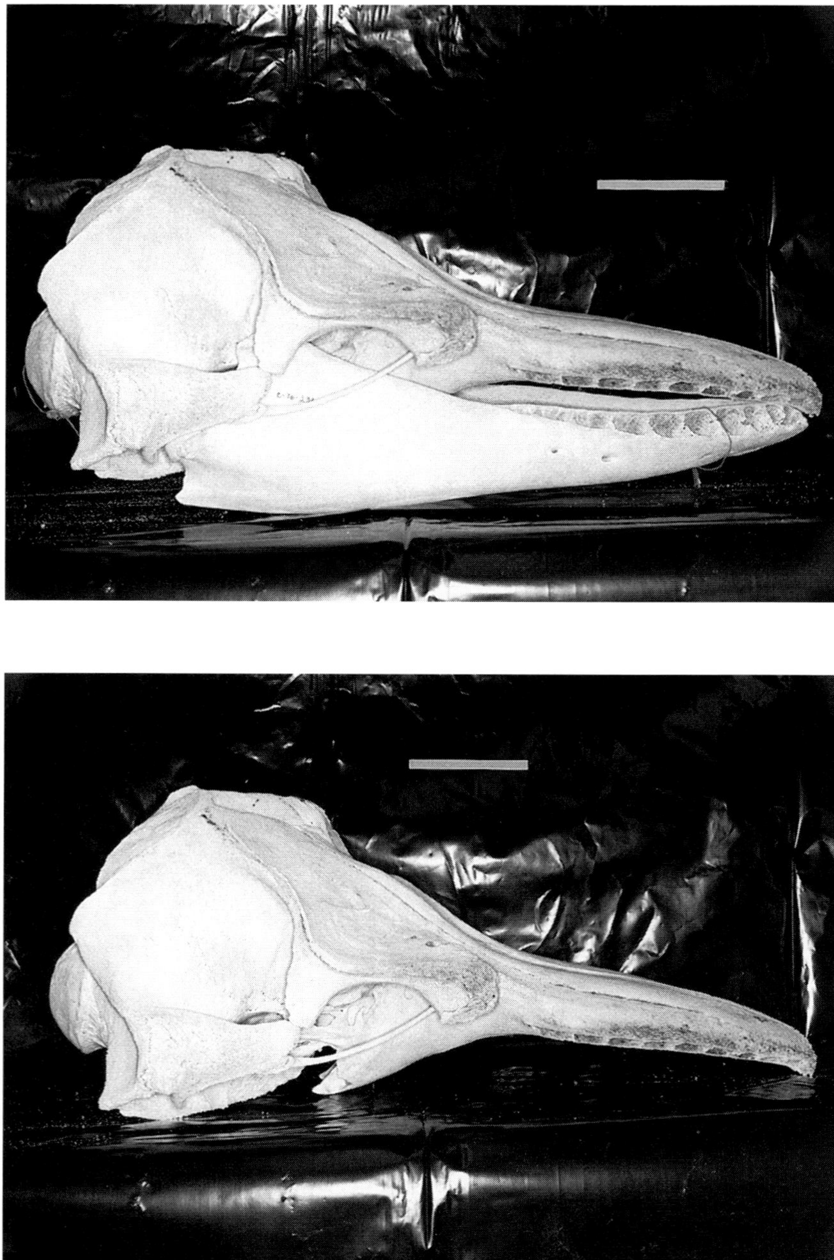

FIG. 3 The skull of the false killer whale. This specimen has a condylobasal length of 61.9 cm. White scale bars = 10 cm. It was part of a mass stranding in southern Florida in 1976 (see Odell *et al.*, 1979). Field ID: C-76-23A. (Photos by D.K. Odell)

(C)

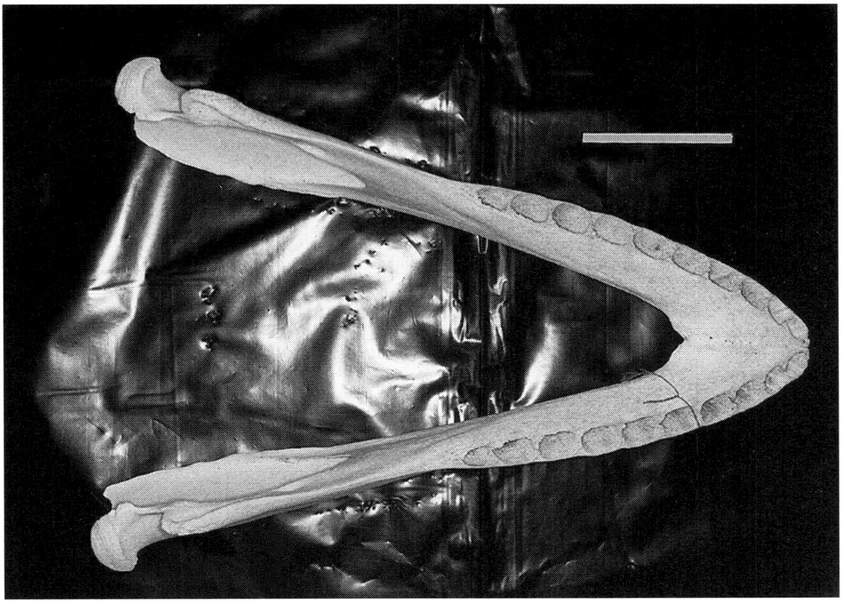

(D)

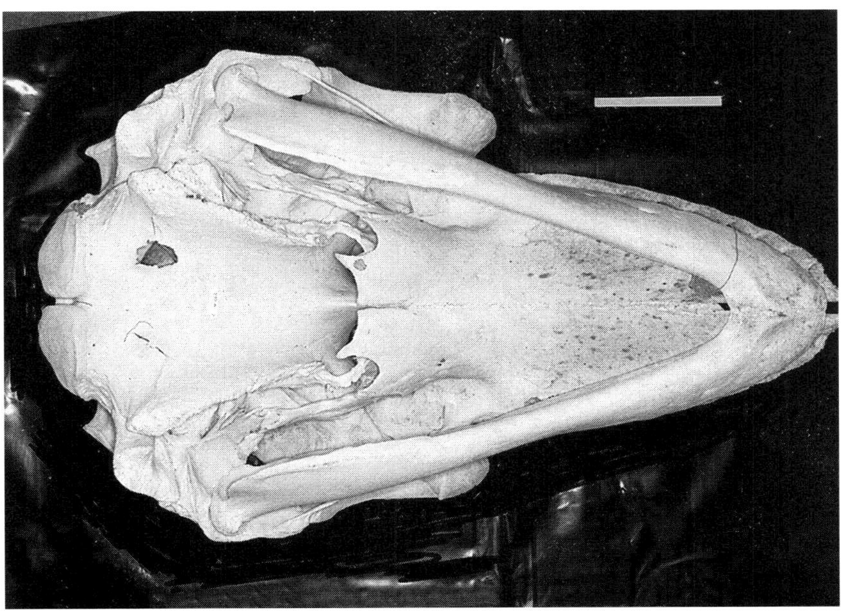

FIG. 3 *Continued*

(E)

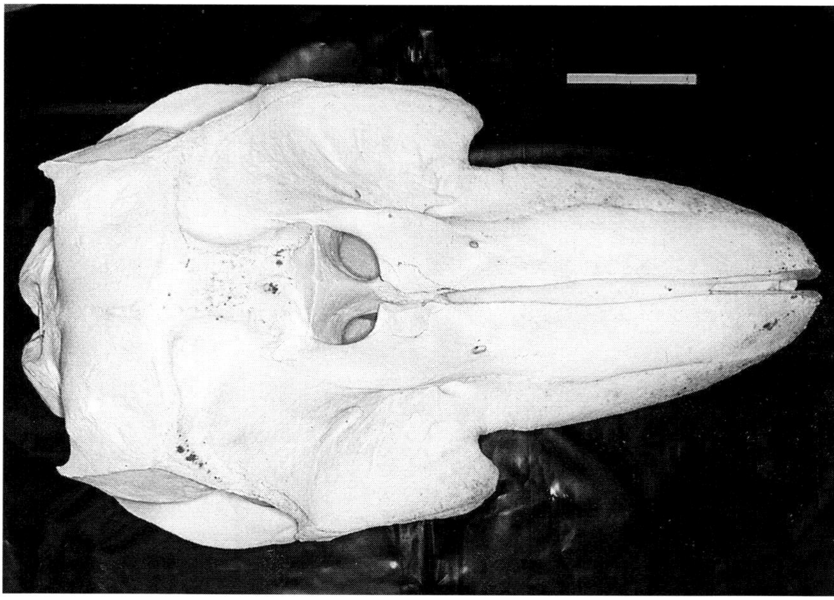

(F)

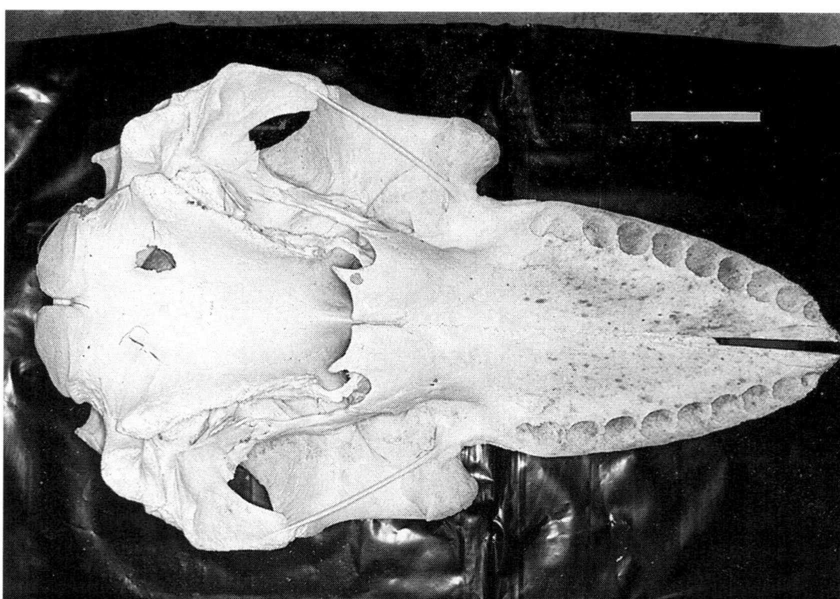

FIG. 3 *Continued*

Pilleri, 1978); and North Pacific, about 430 cm (Kasuya, 1986). Scott and Green (1975) examined post-partum females measuring 349 cm and 361 cm in length, and one 360 cm female with a foetus. This report suggests the minimum size might be lower than previously reported. Physical maturity based on fusion of epiphyseal plates occurs in females at 427–457 cm (approximately 15–19 GLGs) and in males at 518–549 cm (approximately 16–20 GLGs) (Purves and Pilleri, 1978). Kasuya (1986) estimated that females cease growth in length at about 20 years (GLGs) and males at 25–30 years (GLGs).

According to Purves and Pilleri (1978), both males and females attain sexual maturity at about 8–14 years. Kasuya (1986) estimated that females mature at 8–11 years and males at about 18 years.

Maximum reported age for false killer whales based on tooth sectioning studies is given by Kasuya (1986). The oldest female in Kasuya's (1986) sample was estimated to be 63 years (GLGs) and the oldest male 58 years (GLGs). Kasuya (1986) found that females over 45 years (GLGs) ($n = 3$) were post-reproductive. Kasuya's estimates update earlier estimated maximum ages of 22 years for females and 20 years for males given by Purves and Pilleri (1978). Baird *et al.* (1988) reported approximately 26 GLGs in the teeth of one male stranded in British Columbia. This animal was considered physically mature based on osteological evidence.

Age determination studies for false killer whales based on tooth sectioning (Purves and Pilleri, 1978; Pierce and Kajimura, 1980; IWC, 1984; Ross, 1984; Kasuya, 1986; Baird *et al.*, 1988; Pinedo and Rosas, 1989) may assist in correlating age and length at sexual maturity, although such estimates of age should be viewed with care until GLGs are calibrated with actual age in years (Perrin and Reilly, 1984; Myrick *et al.*, 1988). In addition, the techniques vary among investigators. For example, Kasuya (1986) used decalcified, haematoxylin stained 40–50 μm sections and counted GLGs in both dentine and cementum. Other investigators used etched (i.e. partially decalcified), longitudinally bisected teeth and counted GLGs in the dentine (see Fig. 4).

There is little information on reproductive rates for false killer whales, but we can report some related parameters. Ovulation has been reported as spontaneous (Comrie and Adam, 1938; Slijper, 1966; Harrison *et al.*, 1972; Perrin and Reilly, 1984) and as occurring several times a year (Comrie and Adam, 1938) or once a year (Purves and Pilleri, 1978). Several investigators have estimated gestation: 11–12 months (Harrison, 1969), 15.5 months (based on foetal growth rate and comparisons with *Globicephala*, Purves and Pilleri, 1978), and 15.7 months (based on mean neonatal length, Kasuya, 1985, 1986). Robeck *et al.* (1994) used urine and blood serum hormone analyses to determine that the gestation period was 14 months for the calf born at Sea World of Florida in December 1993. The percentage of pregnant females in a population reported from strandings and fishery-related takes has been 14.0% in the North Atlantic (Perrin and Reilly, 1984 citing Purves and Pilleri, 1978) and 21.1% in the North Pacific (Perrin and Reilly, 1984 citing Kasuya and Izumisawa, 1981). Annual reproduc-

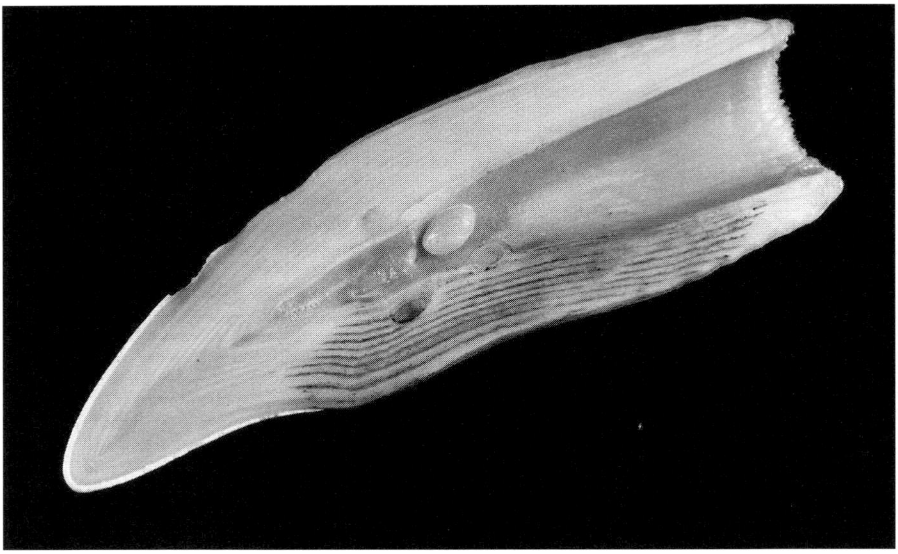

FIG. 4 Acid-etched, longitudinally bisected tooth of a 379 cm female false killer whale from Japan showing distinct growth layer groups highlighted by rubbing with pencil lead. Note the osteodentine "pearl". (Photo courtesy Sea World of Florida)

FIG. 5 Kekaimalu, the female *Pseudorca* × *Tursiops* born at Sea Life Park, Hawaii, on 15 May 1985. This photo was taken in June 1994. (Photo courtesy Sea Life Park, Hawaii)

tive rate, based on the gestation period and the percentage of pregnant females in the Iki Island population was estimated at 5–6% by Kasuya (1985, 1986).

Mortality

Kasuya (1986) estimated annual survival rates (±95% CI) were 0.9412 ± 0.050 for females and 0.9574 ± 0.057 for males. However, we would expect annual survival rates for males to be lower than for females. Kasuya (1986) estimated the gross annual reproductive rate to be 0.0668 and the net annual reproductive rate to be 0.0138 ± 0.0381 (95% CI). Note that Kasuya's estimates were based on a sample in which all age classes were represented.

The best known aspect of *Pseudorca* behaviour is its propensity to strand in large numbers at many places around the world (see review in Sergeant, 1982). Herd size in some recent mass strandings has ranged from 28 to over 1000 animals (Caillet-Bois, 1948; Barnard, 1954; Tomilin, 1957; Mitchell, 1965; Cummings and Fish, 1971; Odell *et al.*, 1979; Ross, 1984). One hundred and fifty-seven herds sighted in the eastern tropical Pacific during 1971–1986 ranged from one to 89 individuals (mean = 18, SD = 28) (W.F. Perrin, unpublished data). Kasuya (1971) reported a mean herd size of 55 in Japanese waters. Mass stranded herds have about equal numbers of males and females of various sizes (Table 1). Little is known about behaviour during mass strandings since the animals usually are found already on the beach. Two mass strandings on the southwest coast of Florida in 1976 and 1986 (Odell *et al.*, 1979; Walsh *et al.*, 1990) suggest that the animals are being carried by the southward coastal currents. The cause(s) of mass strandings remain speculative. It is clear that the majority of animals from Florida (Walsh *et al.*, 1990) were quite ill at the time of stranding. The geographic stranding pattern and health condition also applies to short-finned pilot whale (*Globicephala macrorhynchus*) mass strandings in southwestern Florida (Bossart *et al.*, 1991; Walsh *et al.*, 1991).

Abundance

There is no estimate of worldwide abundance (Northridge, 1984; NMFS, 1994). Population estimates of 16 000 have been reported for the coastal waters of China and Japan (IWC, 1994).

Behaviour

Social organization

From what we know about the false killer whales, the species is very "social" in the sense that they occur in large herds, can develop strong social bonds and are inquisitive and somewhat malleable in their behaviour (Brown *et al.*, 1966). Sightings of groups of 10–20 individuals are common (e.g. Miyazaki and Wada, 1978) and group sizes as high as 300 have been reported (Brown *et al.*, 1966). Some authors speculate that very large herds may be aggregations of many smaller groups converging when food is abundant. Observations of female *Pseudorca* in captivity showed that the species may even bond (and reproduce) with other species in the absence of conspecifics (Brown *et al.*, 1966; Ryan, 1985).

Movements

False killer whales have been seen travelling in line formation (Alling *et al.*, 1982) and one large herd of about 300 individuals was distributed over an area several miles long and half a mile wide (Brown *et al.*, 1966). Reported travelling speeds are 3–6 knots (5.6–11.1 km hr^{-1}) (Alling *et al.*, 1982, 1983) and as high as 10 knots (18.5 km hr^{-1}) (Brown *et al.*, 1966; Busnel and Dziedzic, 1968).

Feeding

False killer whales primarily feed on deep sea cephalopods and fishes (Madsen and Herman, 1980) and are considered teuthoichthyophagic (Tomilin, 1954) or ichthyobenthophagic (Defran and Pryor, 1980) (also see Stacey and Baird, 1991; Stacey *et al.*, 1994). Scheffer and Slipp (1948) reported salmon vertebrae in the stomach of a single animal shot in Puget Sound on 15 May 1937. Barry *et al.* (1989) reported a single false killer whale feeding on salmon (*Oncorhynchus* sp.) in Barkley Sound, Vancouver Island in July and August 1987. A 462 cm male stranded on Denman Island, British Columbia in May 1987 had remnants of salmon (*Oncorhynchus* sp.) and squid (*Berryteuthis magister* or *Gonatopsis borealis*) in its stomach (Baird *et al.*, 1989). Stomachs of mass stranded animals are almost always empty. Single strandings in south Brazil had remains of sciaenid and carangid fishes in their stomachs (Pinedo and Rosas, 1989). False killer whales have been observed feeding on bonito (*Sarda* sp.), mahi mahi or dolphin-fish (*Coryphaena* sp.) (Brown *et al.*, 1966; Shallenberger, 1981) and yellowfin tuna (*Thunnus albacares*) (Shallenberger, 1981). Perryman and Foster (1980) reported that *Pseudorca* fed on and chased other dolphins in the eastern tropical Pacific during chase and backdown operations of tuna purse seine fishing, a habit that has also been attributed to the pygmy killer whale, *Feresa attenuata*. Stomach contents of non-stranded individuals have revealed digested squid, yellowtail (*Pseudosciana* spp.) and perch (*Lateolabrax japonicus*) in western North Pacific waters (JFA, 1968; Mizue and Yoshida, 1961; Kasuya, 1985), *Coryphaena* in Hawaiian waters (Brown *et al.*, 1966); amberjacks (*Seriola alandi*) off Florida (Bullis and Moore, 1956); and barracuda, carangid fishes and catfish (*Tachysurus* spp.) in the Indian Ocean (Silas and James, 1984; Silas *et al.*, 1984). Individuals in oceanaria have been fed mackerel, herring, smelt and squid (Brown *et al.*, 1966; Odell *et al.*, 1980), in quantities of 15–25 kg per day (Odell *et al.*, 1980; Thomas *et al.*, 1988). Sergeant (1969) calculated a mean daily feeding rate of 4.5% body weight based on a single animal.

Association with other species

False killer whales scored high in "affiliative/social/contact behaviour" in a survey of dolphin trainers (Defran and Pryor, 1980) and are known to adapt easily to captivity (Brown *et al.*, 1966). In the wild, herds may associate with other species (Mizue and Yoshida, 1961; Tsutsumi *et al.*, 1961; Palacios and Mate, 1994), perhaps the result of shared or overlapping feeding grounds (Defran and Pryor, 1980). Brown *et al.* (1966) reported interactions of one female at Marineland of the Pacific with a Pacific white-sided dolphin (*Lagenorhynchus obliquidens*), bottlenose dolphin (*Tursiops truncatus*), Atlantic spotted dolphin (*Stenella frontalis*), common dolphin (*Delphinus delphis*), and pilot whale (*Globicephala macrorhynchus*). These interactions included bonding, close associations during parturition, removal of afterbirth, response to vocalization and subsequent "protection" of the dolphin from humans by the *Pseudorca*. In captivity, copulations and other sexual activity occur with *Globicephala* and *Tursiops*, though only the latter has produced hybrid offspring (Brown *et al.*, 1966; Nishiwaki and Tobayama, 1984; Ryan, 1985). Defran and Pryor's (1980) survey of dolphin trainers indicates that *Pseudorca* is as "trainable" as *Tursiops*; however, since it is also the most aggressive species surveyed, it thus may not be used as often. Norris (1967) and Defran and Pryor (1980) cite false killer whale aggression towards humans and other marine mammals in captivity. Palacios and Mate (1994, 1996) described an attack by false killer whales on sperm whales (*Physeter*) in the Galapagos Islands.

Swimming and diving

Details on the swimming and diving behaviour of false killer whales are sparse. Cummings and Fish (1971) estimate maximum dive depth at about 500 m.

Sound production

Acoustic work with the species has revealed that, like many other delphinids, false killer whales can vocalize and use sound to detect objects in their environment. Schevill and Watkins (1962) made what is probably the first recording and sonogram for the species. Subsequently, Mizue *et al.* (1969) reported other recordings and described some components of the vocalizations. More recent works by Thomas *et al.* (1986a,b; 1990), Thomas and Turl (1990) and Turl and Thomas (1992) have shown that echolocation clicks during tests over an open water range in Kaneohe Bay, Hawaii had a peak frequency of 100–105 kHz. In a concrete pool the peak frequency range was 20–65 kHz. The mean distance (in a sea pen) for maximum detection of a 7.62 cm diameter metal sphere was 117 m from the animal. Best hearing sensitivity was around 16–64 kHz (defined as within 10 dB of maximum sensitivity), corresponding to the range of echolocation pulses produced by the species in the concrete pool (Thomas *et al.*, 1988, 1990). Au *et al.* (1995) documented the echolocation signals and transmission beam patterns for a female false killer whale in the open water range. Species-specific sound signatures have been used to identify false killer whales passing by a vessel (Thomas *et al.*, 1986a).

Parasites and Disease

Known parasites of *Pseudorca* include: trematodes, *Nasitrema attenuata*, *N. globicephalae* and *N. gondo* from the air sinuses (Neiland *et al.*, 1970; Dailey and Brownell, 1972; Morimitsu *et al.*, 1987); a nematode, *Stenurus globicephalus*, from the pterygoid sinus complex and lungs (Hall and Schimpf, 1979; Odell *et al.*, 1980); two whale lice, *Syncyamus pseudorcae* and *Isocyamus delphini* (Bowman, 1955; Leung, 1967); two nematodes from the stomach, *Anisakis simplex* and *A. typica* (Zam *et al.*, 1971; Dailey and Brownell, 1972; Baird *et al.*, 1988, 1989); and one acanthocephalan, *Bolbosoma capitatum*, from the intestine (von Linstow, 1880; Dailey and Brownell, 1972). Morimitsu *et al.* (1987) proposed that trematode damage to the eighth cranial nerve caused the mass stranding that they studied. The acanthocephalan has been observed by the hundreds in the intestines of false killer whales stranded in Florida in 1976 and 1986 (Odell *et al.*, 1979; D.K. Odell, unpublished data) and in Chile (Goodall and Schiavini, 1989b). An unidentified crassicaudid nematode also was found in the pterygoid sinus complex of one individual (Odell *et al.*, 1980). The pseudostalked barnacle *Xenobalanus globicipitus* was found on *Pseudorca* by Caldwell *et al.* (1970).

Less is known about non-parasitic disease in false killer whales. Staeger and Reeder (1951) found one specimen with a severe osteoma in which the first eight vertebrae were fused to the cranium. Baird *et al.* (1988, 1989) found another type of bone disease, advanced osteoarthritis, in an old male stranded in British Columbia in 1987. Douglas (1993) reported dental erosion and the problems associated with it in a single, captive false killer whale. A similar dental erosion problem has been observed in several whales from a mass stranding in Florida (Odell *et al.*, 1980; D.K. Odell, unpublished data). It is not surprising that bone (i.e. hard tissue) rather than other types of disease are represented in the literature for the species known best to us through its stranding record. Pneumonia also is known as a final cause of death for some stranded individuals taken into captivity for recovery (Odell *et al.*, 1980). Haematological data (suggestive of multi-organ dysfunction) from mass-stranded animals are given by Walsh *et al.* (1990). Duignan *et al.* (1995) demonstrated positive morbillivirus neutralizing titres in serum from false killer whales that had mass stranded in Florida.

Live Maintenance

Asper *et al.* (1990) reported on the relatively small numbers (10 in 1990) of false killer whales maintained in North America. Live captures have taken place in United States (Hawaii), Japanese and Australian waters (IWC, 1984; Haynes-Lovell, 1993). They are also kept in Japanese aquaria (Nishiwaki and Tobayama, 1982), the Netherlands, Hong Kong and Australia. Female false killer whales have crossbred with *Tursiops* in Japanese (Nishiwaki and Tobayama, 1982, 1984) and American oceanaria (Ryan, 1985), occasionally producing live young. The female hybrid calf (*Pseudorca* female × *Tursiops* male) born at Sea Life Park, Hawaii (see Fig. 5) in May 1985 (Ryan, 1985) was still alive in 1998. This female actually gave birth to two live calves (19 July 1990 and 8 November 1991; both sired by a *Tursiops*). The first calf died at four days of age (Breese, 1990). The mother did not nurse the second calf and it was hand-reared and weaned in April 1992 (Breese, 1992). The first successful *Pseudorca* × *Pseudorca* calf bred and born in captivity was born on 1 May 1986 at Sea Life Park, Hawaii (Sea Life Park, 1987). A healthy female calf (*Pseudorca* × *Pseudorca*) was born to a 410 cm female at Sea World of Florida on 23–24 December 1993 and successfully nursed by the mother (Robeck *et al.*, 1994). A second *Pseudorca* × *Pseudorca* female calf was born at Sea World of Florida on 3 September 1994 to a 390 cm female. This calf was successfully nursed by the mother (Sea World of Florida, unpublished data). The first birth was not directly observed but occurred "in the middle of the night" (sometime between 22:00 hr and 06:00 hr, EST). The second birth occurred at about 01:00 hr, EDT and presentation was tail-first. The December 1993 calf died in May 1998. The second remained healthy as of June 1998.

Human Effects

Direct fisheries

Northridge (1984) and Mitchell (1975) summarized fisheries interactions. *Pseudorca* are occasionally taken in Japan for food (Miyazaki, 1983) and in St. Vincent Island, Caribbean for meat and cooking oil according to observations from 1969 to 1970 (Caldwell *et al.*, 1970). The species is known to cause heavy damage to tuna longline fisheries in Japan (e.g. Mizue *et al.*, 1969) and, to a lesser degree, in Hawaiian waters (Shallenberger, 1981) and other areas (Brown *et al.*, 1966). The largest documented fisheries interaction is in the waters around Iki Island, Japan (Ohsumi, 1972), where over 900 false killer whales were killed in drive fisheries from 1965 to 1980 in an attempt to reduce interactions with the yellowtail (*Pseudosciaena* spp.) fishery (Kasuya, 1985).

By-catches

Incidental take of small numbers of false killer whales in gill nets has occurred off northern Australia (Harwood *et al.*, 1984), off the Andaman Islands in the Andaman Sea (James, 1984), in the south of Brazil (Simões-Lopes and Ximenez, 1992) and in tuna purse seines in the eastern tropical Pacific (Perrin and Oliver, 1981). Dolphin entrapment in tuna purse seine nets may be providing artificial feeding opportunities for *Pseudorca* on other marine mammals (Perryman and Foster, 1980).

Pollution

Pollution of the environment by humans is affecting the habitat of false killer whales in at least three documented cases. A physically mature male *Pseudorca* that stranded on Denman Island, British Columbia in 1987 showed high levels of pesticides (blubber DDE = 1400 ppm wet weight) and heavy metals (mercury = 728 ppm wet weight) (Baird *et al.*, 1988, 1989). Another individual that stranded on Vancouver Island in September 1989 had 1614 ppm (wet weight) of mercury in the liver (Langelier *et al.*, 1990). Barros *et al.* (1990) reported finding the remains of a plastic jug in the stomach of a false killer whale found stranded in Florida in 1988.

Acknowledgments

N.B. Barros, J. Thomas and the anonymous reviewers provided many helpful comments on the manuscript. M. Breese kindly provided the photograph in Figure 5.

References

Alling, A., Gordon, J., Rotton, L. and Whitehead, H. (1982). WWF-Netherlands Indian Ocean sperm whale study, 1981–1982, Interim Report. Document SC/34/SP 9, 43 pp., International Whaling Commission.

Alling, A., Gordon, J., Rice, M. and Whitehead, H. (1983). WWF/IUCN Indian Ocean Sperm whale project 1983. Interim Report WWF/IUCN.

Asper, E. D., Duffield, D. A., Dimeo-Ediger, N. and Shell, D. (1990). Marine mammals in zoos, aquaria and marine zoological parks in the United States and Canada: 1990 census report. *Int. Zoo Yearbook* **29**, 179–187.

Au, W. W. L., Pawloski, J. L., Nachtigall, P.E., Blonz, M. and Gisner, R. C. (1995). Echolocation signals and transmission beam pattern of a false killer whale (*Pseudorca crassidens*). *J. Acoust. Soc. Am.* **98**, 51–59.

Baird, R. W., Langelier, K. M. and Stacey, P. J. (1988). Stranded whale and dolphin program of B.C. *Brit. Columbia Vet. Med. Assoc. Wildlife Vet. Rep.* **1**, 9–12.

Baird, R. W., Langelier, K. M. and Stacey, P. J. (1989). First records of false killer whales (*Pseudorca crassidens*) in Canada. *Can. Field-Nat.* **103**, 368–371.

Barnard, K. H. (1954). "A Guide Book to South African Whales and Dolphins". South African Museum, Capetown.

Barros, N. B., Odell, D. K. and Patton, G. W. (1990). Ingestion of plastic debris by stranded marine mammals from Florida. *In* "Proceedings of the Second International Conference on Marine Debris" (Eds R. S. Shomura and M. L. Godfrey) 2–7 April 1989, Honolulu, Hawaii, p. 746. NOAA Technical Memo NMFS, NOAA-TM-NMFS-SWFSSC-154, US Department of Commerce.

Barry, L. M., Baird, R. W., Hall, J., Gonzales, R. and Stacey, P. J. (1989). Observations on the behavior of a lone false killer whale (*Pseudorca crassidens*) in British Columbia. *In* Abstracts of the 8th Biennial Conference on the Biology of Marine Mammals, 7–11 December 1989, Pacific Grove, CA, p. 5.

Bossart, G. D., Walsh, M. T., Odell, D. K., Lynch, J. D., Beusse, D. O., Friday, R. and Young, G. (1991). Histopathologic findings of a mass stranding of pilot whales (*Globicephala macrorhynchus*). *In* "Marine Mammal Strandings in the United States", Proceedings of the Second Marine Mammal Stranding Workshop, Miami, Florida, 3–5 December 1987 (Eds J. E. Reynolds III and D. K. Odell) pp. 85–90. NOAA Technical Report NMFS 98.

Bowman, T. E. (1955). A new genus and species of whale-louse (Amphipoda-Cyamidae) from the false killer whale. *Bull. Mar. Sci. Gulf Carib.* **5**, 315–320.

Breese, M. (1990). History of a *Pseudorca/Tursiops* hybrid. *In* Abstracts of the Annual Meeting of the Marine Animal Trainers Association, Chicago, IL.

Breese, M. (1992). Intervention in a non-nursing mother/calf pair. *In* Abstracts of the 20th Annual Conference of the International Marine Animal Trainers Association, 1–6 November 1992.

Brown, D. H., Caldwell, D. K. and Caldwell, M. C. (1966). Observations on the behavior of wild and captive false killer whales, with notes on associated behavior of other genera of captive delphinids. *Los Angeles Co. Mus. Contrib. Sci.* **95**, 1–32.

Bullis Jr, H. R. and Moore, J. C. (1956). Two occurrences of false killer whales and a summary of American records. *Am. Mus. Novitates*, 1–5.

Busnel, R. -G. and Dziedzic, A. (1968). Caracteristiques physiques des signaux acoustiques de *Pseudorca crassidens* Owen (Cetace, Odontocete). *Mammalia* **32**, 1–5.

Caillet-Bois, T. (1948). Las Pseudorcas de Mar del Plata. *Revista Geog. Am.* **29**, 5–10.

Caldwell, D. K., Caldwell, M. C. and Walker Jr, C. M. (1970). Mass and individual strandings of false killer whale, *Pseudorca crassidens*, in Florida. *J. Mammal.* **51**, 634–636.

Comrie, L. C. and Adam, A. B. (1938). The female reproductive system and corpora lutea of the false killer whale *Pseudorca crassidens* Owen. *Trans. Roy. Soc. Edinb.* **59**, 521–531.

Cowley, L. F. (1944). The skull of the false killer dolphin, *Pseudorca crassidens* (Owen). *Proc. Zool. Soc. Lond.* **114**, 382–387.

Cummings, W. C. and Fish, J. F. (1971). A synopsis of marine animal underwater sounds in eight geographic areas. Naval Undersea Center Special Report, San Diego, CA, 94 pp.

Dailey, M. D. and Brownell Jr, R. L. (1972). A checklist of marine mammal parasites. *In* "Mammals of the Sea: Biology and Medicine" (Ed. S. H. Ridgway) pp. 528–589. Charles C. Thomas, Springfield, IL.

Defran, R. H. and Pryor, K. (1980). The behavior and training of cetaceans in captivity. *In* "Cetacean Behavior: Mechanisms and Functions" (Ed. L. M. Herman) pp. 319–362. Wiley-Interscience, New York.

Douglas, A. (1993). Dental erosion in *Pseudorca crassidens* and subsequent health related concerns. *In* Proceedings of the 19th Annual Conference of the International Marine Animal Trainers Association, 4–8 November 1991, pp. 76–79.

Duffield, D. A., Sparkes, M., Sparkes, R., Asper, E. and Cornell, L. (1979). Physiological differences of polymorphic hemoglobins in the Cetacea. *In* Abstracts of the Third Biennial Conference on the Biology of Marine Mammals, Seattle, WA, p. 15.

Duffield, D. A., Chamberlin-Lea, J., Sweeney, J. C., Odell, D. K., Asper, E. D. and Gilmartin, W. G. (1991). Use of corneal cell culture for R-band chromosome studies on stranded cetaceans. *In* "Marine Mammal Strandings in the United States", Proceedings of the Second Marine Mammal Stranding Workshop, Miami, Florida, 3–5 December 1987 (Eds J. E. Reynolds III and D. K. Odell) pp. 91–100. NOAA Technical Report NMFS 98.

Duignan, P. J., House, C., Geraci, J. R., Duffy, N., Rima, B. K., Walsh, M. T., Early, G., St. Aubin, D. J., Sadove, S., Koopman, H. and Rhinehart, H. (1995). Morbillivirus infection in cetaceans of the Western Atlantic. *Vet. Microbiol.* **44**, 241–249.

Evans, P. G. H. (1987). Cetaceans in U. K. waters. *In* "The European Cetacean Society Report of the 1987 Meeting" (Eds J. W. Broekema and C. Smeenk) pp. 45–46. European Cetacean Society.

Gihr, M., Kraus, C. and Pilleri, G. (1982). The manus of *Pseudorca crassidens* (Owen): a study of variability. *Invest. Cetacea* **14**, 101–124.

Gray, J. E. (1846). On the cetaceous animals. *In* "The Zoology of the Voyage of the H.M.S. *Erebus* and *Terror* under the command of Cpt. SM James Clark-Ross, R.N., F.R.S., during the Years 1839–1843", Vol. 1 (Eds J. Richardson and J. E. Gray) pp. 13–53. London.

Goodall, R. N. P. and Schiavini, A. C. M. (1989a). Preliminary report on a mass stranding of false killer whales, *Pseudorca crassidens*, along the Strait of Magellan. *In* Abstracts of the Eighth Biennial Conference on the Biology of Marine Mammals, Pacific Grove, CA, 7–11 December 1989, p. 23.

Goodall, R. N. P. and Schiavini, A. C. M. (1989b). Preliminary report on a mass stranding of false killer whales, *Pseudorca crassidens*, along the Strait of Magellan, available from the authors, 18 pp.

Goodall, R. N. P. and Schiavini, A. C. M. (1990). Informe preliminar sobre un varamiento masivo de falsas orcas, *Pseudorca crassidens*, en el Estrecho de Magallanes, Chile. *In* Abstracts of the 4ª Reunion de Trabajo de Especialistas en Mammiferos Acuaticos de America del Sur, Instituto de Biologia Marina, Universita Austral de Chile, p. 28.

Hall, N. R. and Schimpf, R. D. (1979). Mass stranded whales. *In* "Biology of Marine Mammals: Insights Through Strandings" (Eds J. Geraci and D. J. St. Aubin) pp. 236–242. US Department of Commerce, PB 293380, NTIS, Springfield, VA.

Harrison, R. J. (1969). Reproduction and Reproductive Organs. *In* "The Biology of Marine Mammals" (Ed. H. T. Andersen) pp. 253–348. Academic Press, New York.

Harrison, R. J., Brownell Jr, R. L. and Boice, R. C. (1972). Reproduction and gonadal appearances in some cetaceans. *In* "Functional Anatomy of Marine Mammals", Vol. 1 (Ed. R. J. Harrison) pp. 361–429. Academic Press, London.

Harwood, M. B., McNamara, K. J., Anderson, G. R. V. and Walter, D. G. (1984). Incidental catch of small cetaceans in a gillnet fishery in northern Australian waters. *Rep. Int. Whal. Commn* **34**, 555–559.

Haynes-Lovell, K. E. (1993). *Pseudorca* downunder: from dreamtime to prime time. *In* Proceedings of the 19th Annual Conference of the International Marine Animal Trainers Association, 4–8 November 1991, pp. 58–75.

Hetzel, B. and Lodi, L. (1993). "Baleias, Botos e Golfinhos: guia de Identificação para o Brasil". Editora Nova Fronteira, Rio de Janeiro, Brazil, 279 pp.

Hinton, M. A. C. (1928). Stranded whales at Dornoch Firth. *Nat. Hist. Mag.* **1**, 131.

Hoyt, E. (1984). "The Whale Watcher's Handbook". Doubleday, Garden City, NY.

IWC (International Whaling Commission) (1984). Report of the Sub-Committee on Small Cetaceans, Annex H, Report of the Scientific Committee. *Rep. Int. Whal. Commn* **34**, 144–159.

IWC (International Whaling Commission) (1994). Report of the Workshop on Mortality of Cetaceans in Passive Fishing Nets and Traps. *Rep. Int. Whal. Commn*, Spec. Issue No. 15, pp. 6–57, Cambridge.

JFA (Japanese Fisheries Agency) (1968). Report of the basic survey in search of countermeasures of the damage by dolphins on the fisheries in western Japan (in Japanese).

James, D. B. (1984). Capture of a false killer whale, *Pseudorca crassidens*, at Port Blair, Andamans. *Mar. Fish. Infor. Ser. (India) T&E Ser.* **55**, 17.

Kasuya, T. (1971). Consideration of distribution and migration of toothed whale off the Pacific coast of Japan based on aerial sighting records. *Sci. Rep. Whales Res. Inst.* **23**, 37–60.

Kasuya, T. (1973). Systematic consideration of recent toothed whales based on the morphology of tympano-periotic bone. *Sci. Rep. Whales Res. Inst.* **25**, 1–103.

Kasuya, T. (1985). Fishery–dolphin conflict in the Iki Island area of Japan. *In* "Marine Mammals and Fisheries" (Eds J. R. Beddington, R. J. H. Beverton and D. M. Lavigne) pp. 253–272. Allen and Unwin, London.

Kasuya, T. (1986). False killer whales. *In* "Report of Investigation in Search of Solution for Dolphin–Fishery Conflict in the Iki Island Area" (Eds T. Tamura, S. Ohsumi and S. Arai) pp. 178–187. Japan Fisheries Agency, 285 pp.

Kasuya, T. and Izumisawa, Y. (1981). The fishery–dolphin conflict in the Iki Island, Japan area. US Marine Mammal Commission Report Contract NMF 1533791-7(80/02), p. 31.

Kitchener, D. J., Ross, G. J. B. and Caputi, N. (1990). Variation in skull and external morphology in the false killer whale, *Pseudorca crassidens*, from Australia, Scotland and South Africa. *Mammalia* **54**, 119–135.

Kulu, D. D. (1972). Evolution and Cytogenetics. *In* "Mammals of the Sea: Biology and Medicine" (Ed. S. H. Ridgway) pp. 503–527. Charles C. Thomas, Springfield, IL.

Langelier, K. M., Stacey, P. J. and Baird, R. W. (1990). Stranded whale and dolphin program of B.C. — 1989 report. *Wildl. Vet. Rep.* **3**, 10–11.

Leatherwood, S., Caldwell, D. K. and Winn, H. E. (1976). Whales, dolphins, and porpoises of the western North Atlantic. NOAA Technical Report NFMS Cir. 396.

Leatherwood, S. and Reeves, R. R. (1983). "The Sierra Club Handbook of Whales and Dolphins". Sierra Club Books, San Francisco.

Leatherwood, S., Reeves, R. R., Perrin, W. F. and Evans, W. E. (1988). "Whales, Dolphins, and Porpoises of the Eastern North Pacific and Adjacent Arctic Waters". Dover, New York.

Leung, Y. -M. (1967). An illustrated key to the species of whale lice (Amphipoda, Cyamidae), ectoparasites of Cetaceans, with a guide to the literature. *Crustaceana* **12**, 279–291.

Linstow, O. (1880). Helminthologische Untersuchunge. *Arch. Naturg. (Berlin)* **46**, 41–54.

Madsen, C. J. and Herman, L. M. (1980). Social and ecological correlates of cetacean vision and visual appearance. *In* "Cetacean Behavior: Mechanisms and Function" (Ed. L. M. Herman) pp. 101–147. Wiley-Interscience, New York.

Marcuzzi, G. and Pilleri, G. (1971). On the zoogeography of Cetacea. *Invest. Cetacea* **3**, 101–170.

McFarland, W. N. (1971). Cetacean visual pigments. *Vision Res.* **11**, 1065–1076.

Mead, J. G. (1975). Anatomy of the external nasal passages and facial complex in the Delphinidae (Mammalia: Cetacea). *Smithsonian Contrib. Zool.* **207.**

Mitchell, E. (1965). Evidence for mass strandings of the false killer whale (*Pseudorca crassidens*) in the eastern North Pacific Ocean. *Norsk Hvalfangst-Tid.* **54**, 172–177.

Mitchell, E. (1970). Pigmentation pattern evolution in delphinid cetaceans: an essay in adaptive coloration. *Can. J. Zool.* **48**, 717–740.

Mitchell, E. D. (1975). Review of biology and fisheries for smaller cetaceans. *J. Fish. Res. Bd Can.* **32**, 875–983.

Miyazaki, N. (1983). Catch statistics of small cetaceans taken in Japanese waters. *Rep. Int. Whal. Commn* **33**, 621–631.

Miyazaki, N. and Wada, S. (1978). Observation of Cetacea during whale marking cruise in the western tropical Pacific. *Sci. Rep. Whales Res. Inst.* **30**, 179–195.

Mizue, K. and Yoshida, K. (1961). Studies on the little toothed whales in the west sea area of Kyushu — VII. About *Pseudorca crassidens* caught at Arikawa in Goto Is., Nagasaki Pref. *Bull. Faculty Fish. Nagasaki Univ.* **11**, 39–48.

Mizue, K., Takemura, A. and Nakasai, K. (1969). Studies on the little toothed whales in the west sea area of Kyushu — XVI. Underwater sound of the false killer whale. *Bull. Faculty Fish. Nagasaki Univ.* **28**, 19–29.

Mobius (1862). Brief uber die Strandung von *Pseudorca. Der Zool. Garten* **3**, 39.

Morimitsu, T., Nagai, T., Ide, M., Kawano, H., Naichuu, A., Koono, M. and Ishii, A. (1987). Mass stranding of odontoceti caused by parasitogenic eighth cranial neuropathy. *J. Wildl. Dis.* **23**, 586–590.

Myrick Jr, A. C., Yochem, P. K. and Cornell, L. H. (1988). Toward calibrating dentinal layers in captive killer whales by use of tetracycline. *Rit Fiskideildar* **11**, 285–296.

Neiland, K. A., Rice, D. W. and Holden, B. (1970). Helminths of marine mammals, I. The genus *Nasitrema*, air sinus flukes of delphinid Cetacea. *J. Parasitol.* **56**, 305–316.

Nishiwaki, M. (1967). Distribution and migration of marine mammals in the North Pacific area. *Bull. Ocean. Res. Inst. Univ. Tokyo* **1**, 1–64.

Nishiwaki, M. and Tobayama, T. (1982). Morphological study of the hybrid between *Tursiops* and *Pseudorca. Sci. Rep. Whales Res. Inst.* **34**, 109–121.

Nishiwaki, M. and Tobayama, T. (1984). Hybrids between *Pseudorca crassidens* and *Tursiops truncatus gilli. In* "Reproduction in Whales, Dolphins and Porpoises" (Eds W. F. Perrin, R. L. Brownell Jr and D. P. DeMaster) p. 481. *Rep. Int. Whal. Commn*, Spec. Issue No. 6, Cambridge.

NMFS (1994). "Marine Mammal Protection Act of 1972: Annual Report, January 1, 1992 to December 31, 1993". US Department of Commerce, National Oceanic and Atmospheric Administration, National Marine Fisheries Service, Office of Protected Species, Silver Spring, MD, 166 pp.

Norris, K. S. (1967). Aggressive behavior in Cetacea. *In* "Aggression and Defense: Natural Mechanisms and Social Patterns" (Eds C. D. Clemente and D. B. Lindsley) pp. 225–241. University of California Press, Berkeley, CA.

Northridge, S. P. (1984). World review of interactions between marine mammals and fisheries. FAO Fisheries Paper 251.

Odell, D. K., Asper, E. D., Baucom, J. and Cornell, L. H. (1979). A summary of information derived from the recurrent mass stranding of a herd of false killer whales, *Pseudorca crassidens* (Cetacea: Delphinidae). *In* "Biology of Marine Mammals: Insights Through Strandings" (Eds J. Geraci and D. J. St Aubin) pp. 207–222. US Department of Commerce PB 293380, NTIS, Springfield, VA.

Odell, D. K., Asper, E. D., Baucom, J. and Cornell, L. H. (1980). A recurrent mass stranding of the false killer whale, *Pseudorca crassidens*, in Florida. *Fish. Bull. U.S.* **78**, 171–177.

Ohsumi, S. (1972). Catch of marine mammals, mainly of small cetaceans, by local fisheries along the coast of Japan. *Far Seas Fish. Res. Lab. Bull.* **7**, 137–166.

Oporto, J. A., Lescrauwaet, A. C., Maslow, N., Matus, R., Canto, J. and Turina, O. (1994). Primeros antecedentes ambientales de un varamiento masivo de falsa orca (*Pseudorca crassidens*), ocurrido en el estrecho de Magallanes, Chile. *In* Anais da 6a Reunião de Trabalho de Especialistas em Mamíferos Aquáticos da América do Sul, 24–28 October 1994, Florianópolis, Brazil (Eds A. Ximenez and P. C. Simões-Lopes) pp. 277–285.

Owen. R. (1846). "A History of British Fossil Mammals and Birds". London, 516 pp.

Palacios, D. M. and Mate, B. R. (1994). Ataque de orcas falsas a cachalotes en las Islas Galápagos, Ecuador. *In* Anais da 6a Reunião de Trabalho de Especialistas em Mamíferos Aquáticos da América do Sul, 24–28 October 1994, Florianópolis, Brazil (Eds A. Ximenez and P. C. Simões-Lopes) pp. 82–83.

Palacios, D. M. and Mate, B. R. (1996). Attack by false killer whales (*Pseudorca crassidens*) on sperm whales (*Physeter macrocephalus*) in the Galapagos Islands. *Mar. Mamm. Sci.* **12**, 582–587.

Peacock, A. D., Comrie, L. and Greenshields, F. (1936). The false killer whale stranded in the Tay Estuary. *Scottish Nat.* **220**, 93–104.

Perrin, W. F. (1975). Variation of spotted and spinner porpoise (genus *Stenella*) in the eastern Pacific and Hawaii. *Bull. Scripps Inst. Oceanog.* **21**, 1–206.

Perrin, W. F. and Myrick Jr, A. C. (1980). Report of the workshop. *In* "Age Determination of Toothed Whales and Sirenians" (Eds W. F. Perrin and A. C Myrick Jr) pp. 1–50. *Rep. Int. Whal. Commn*, Spec. Issue No. 3, Cambridge.

Perrin, W. F. and Oliver, C. W. (1981). Time/area distribution and composition of the incidental kill of dolphins and small whales in the U.S. purse seine fishery for tuna in the eastern tropical Pacific, 1979–1980. *Rep. Int. Whal. Commn* **32**, 429–444.

Perrin, W. F. and Reilly, S. B. (1984). Reproductive parameters of dolphins and small whales of the family Delphinidae. *In* "Reproduction in Whales, Dolphins and Porpoises" (Eds W. F. Perrin, R. L. Brownell Jr and D. P. DeMaster) pp. 97–133. *Rep. Int. Whal. Commn*, Spec. Issue No. 6, Cambridge.

Perryman, W. L. and Foster, T. C. (1980). Preliminary report on predation by small whales, mainly the false killer whale, *Pseudorca crassidens*, on dolphins (*Stenella* spp. and *Delphinus delphis*) in the eastern tropical Pacific. NOAA/NMFS SWFC Administrative Report LJ-80-05.

Pierce, K. V. and Kajimura, H. (1980). Acid etching and highlighting for defining growth layers in cetacean teeth. *In* "Age Determination of Toothed Whales and Sirenians" (Eds W. F. Perrin and A. C. Myrick Jr) pp. 99–103. *Rep. Int. Whal. Commn*, Spec. Issue No. 3, Cambridge.

Pinedo, M. C. and Rosas, F. C. W. (1989). Novas ocorrências de *Pseudorca crassidens* (Cetacea, Delphinidae) para o Atlântico Sul ocidental, com observações sobre medidas cranianas e alimentação. *Atlântica* **11**, 77–83.

Pinedo, M. C., Rosas, F. C. W. and Marmontel, M. (1992). "Cetáceos e Pinípedes do Brasil: uma revisão dos registros e guia para identificação das espécies". United

Nations Environment Programme/Fundação Universidade do Amazonas, Manaus, Brazil, 213 pp.

Purves, P. E. and Pilleri, G. (1978). The functional anatomy and general biology of *Pseudorca crassidens* (Owen) with a review of the hydrodynamics and acoustics in Cetacea. *Invest. Cetacea* **9**, 67–227.

Reiner, F. (1990). Records of marine mammals of the Azorean islands. *Garcia de Orta Sér. Zool.* **15**, 21–36.

Reinhardt, J. (1862). Om en for den Danske Fauna ny Delphinart (*Pseudorca crassidens* Owen). *Oversigt over d. Kongl. Dansk. Vetenskab. Selsk. Forhande*, 103–152.

Reinhardt, J. (1866). *Pseudorca crassidens*, a cetacean hitherto unknown in the Danish fauna. *In* "Recent Memoirs on the Cetacea" (Ed. W. H. Flower) pp. 189–218. The Ray Society, London, Publ. No. 40, 213 pp. + 6 plts (translation of Reinhardt, 1862).

Robeck, T. R., Gross, T., Walsh, M., Campbell, T. and McBain, J. (1994). Preliminary results of radioimmunoassay determination of post enzyme hydrolysis urinary progestin concentrations in the false killer whale (*Pseudorca crassidens*). *In* Abstracts of the 25th Annual Conference of the International Association for Aquatic Animal Medicine, 10–14 May 1994, Napa, CA.

Ross, G. J. B. (1984). The smaller cetaceans of the south east coast of southern Africa. *Ann. Cape Prov. Mus. (Nat. Hist.)* **15**, 173–410.

Ryan, T. (1985). Crossed cetaceans. *Oceans* **6**, 50.

Scheffer, V. B. and Slipp, J. W. (1948). The whales and dolphins of Washington State with a key to the cetaceans of the west coast of North America. *Am. Midl. Nat.* **39**, 257–337.

Schevill, W. E. and Watkins, W. A. (1962). Whale and porpoise voices: a phonograph record. *Woods Hole Contrib.* 1320.

Scott, E. O. G. and Green, R. H. (1975). Recent whale strandings in northern Tasmania. *Papers Proc. Roy. Soc. Tasmania* **109**, 91–96.

Sea Life Park (1987). "Sea Life Park's Two Most Famous Babies and a History of their Interesting Family". Sea Life Park, Hawaii, 2 pp.

Sergeant, D. E. (1969). Feeding rates of Cetacea. *Fisk. Dir. Skr. Ser. Havunders* **15**, 246–258.

Sergeant, D. E. (1982). Mass strandings of toothed whales (Odontoceti) as a population phenomenon. *Sci. Rep. Whales Res. Inst.* **34**, 1–47.

Shallenberger, E. W. (1981). "The Status of Hawaiian Cetaceans". U.S. Marine Mammal Commission Final Report, Contract MM7AC028, PB82-109398, NTIS, Springfield, VA.

Sielfeld, W. M., Venegas, C. U. and Atalah, A. (1977). Concideraciones acerca del estado de los mamíferos marinos en Chile. *Ann. Inst. Patagonia* **8**, 297–315.

Silas, E. G. and James, D. B. (1984). Capture of false killer whale *Pseudorca crassidens* at Port Blair, Andamans. *Mar. Fish. Info. Ser. T.S.E.* **55**, 17.

Silas, E. G., Lal Mohan, R. S., Somasekharan, K. U., Nair, K. V. S. and Ramadoss, P. (1984). Stranding of *Pseudorca crassidens* at Calicut, Kerala. *Mar. Fish. Info. Ser. T.S.E.* **55**, 15.

Simões-Lopes, P. C. and Ximenez, A. (1992). *Pseudorca crassidens* (Owen, 1846) — Morte acidental em rede de pesca artesanal no Sul do Brasil. *In* Annales, III Reunión de Trabajo de Especialistas en Mamíferos Acuáticos de América del Sul, pp. 67–69.

Simões-Lopes, P. C. and Ximenez, A. (1993). Annotated list of cetaceans of Santa Catarina coastal waters, southern Brazil. *Biotemas* **6**, 67–92.

Slijper, E. J. (1949). On some phenomena concerning pregnancy and parturition of the Cetacea. *Bijdragen Tot Dierkunde* **28**, 416–448.

Slijper, E. J. (1961). Foramen ovale and ductus arteriosus botalli in aquatic mammals. *Mammalia* **25**, 528–570.

Slijper, E. J. (1966). Functional morphology of the reproductive system in Cetacea. *In* "Whales, Dolphins and Porpoises" (Ed. K. S. Norris) pp. 277–318. University of California Press, Berkeley, CA.

Smithers, R. H. N. (1938). Notes on the stranding of a school of *Pseudorca crassidens* at the Berg River mouth. *Trans. Roy. Soc. S. Afr.* **25**, 403.

Stacey, P. J. and Baird, R. W. (1991). Status of the false killer, *Pseudorca crassidens*, in Canada. *Can. Field-Naturalist* **105**, 189–197.

Stacey, P. J., Leatherwood, S. and Baird, R. W. (1994). "*Pseudorca crassidens*". Mammalian Species No. 456, American Society of Mammalogists.

Staeger, K. E. and Reeder, W. G. (1951). Occurrence of the false killer whale *Pseudorca* on the California coast. *So. Calif. Acad. Sci.* **50**, 14–20.

Thomas, J., Fischer, S. R., Fern, L. M. and Holt, R. S. (1986a). Acoustic detection of cetaceans using a towed array of hydrophones. *In* "Behavior of Whales in Relation to Management" (Ed. G. P. Donovan) pp. 139–148. *Rep. Int. Whal. Commn*, Spec. Issue No. 8, Cambridge.

Thomas, J., Stoermer, M., Bowers, C., Anderson, L. and Garver, A. (1986b). Detection abilities and signal characteristics of echolocating false killer whales (*Pseudorca crassidens*). *In* "Animal Sonar: Processes and Performance" (Eds P. E. Nachtigall and P. W. B. Moore) pp. 323–328. Plenum Press, New York.

Thomas, J., Chun, N., Au, W. and Pugh, K. (1988). Underwater audiogram of a false killer whale (*Pseudorca crassidens*). *J. Acoust. Soc. Am.* **84**, 936–940.

Thomas, J. A., Pawloski, J. L. and Au, W. W. L. (1990). Masked hearing abilities in a false killer whale (*Pseudorca crassidens*). *In* "Sensory Abilities of Cetaceans" (Eds J. Thomas and R. Kastelein) pp. 395–404. Plenum Press, New York.

Thomas, J. A. and Turl, C. W. (1990). Echolocation characteristics and range detection threshold in a false killer whale (*Pseudorca crassidens*). *In* "Sensory Abilities of Cetaceans" (Eds J. Thomas and R. Kastelein) pp. 321–334. Plenum Press, New York.

Tomilin, A. G. (1954). Adaptive types in the order Cetacea (the problem of an ecological classification of Cetacea). *Moscow Inst. Furs Peltries* **33**, 677–692.

Tomilin, A. G. (1957). *In* "Mammals of the U.S.S.R. and Adjacent Countries", Vol. IX, "Cetacea" (Ed. V. G. Heptner). Nauk USSR, Moscow (English translation, 1967, Israel Program for Scientific Translations, Jerusalem).

Tsutsumi, T., Kamimura, Z. and Mizue, K. (1961). Studies on the little toothed whales in the West Sea areas of Kyushu — V. *Bull. Fac. Fish. Nagasaki Univ.* **11**, 19–28.

Turl, C. W. and Thomas, J. A. (1992). Possible relationship between oceanographic conditions and long-range target detection by a false killer whale. *In* "Marine Mammal Sensory Systems" (Eds J. A. Thomas *et al.*) pp. 421–432. Plenum Press, New York.

Walsh, M. T., Beusse, D. O., Young, W. G., Lynch, J. D., Asper, E. D. and Odell, D. K. (1991). Medical findings in a mass stranding of pilot whales (*Globicephala macrorhynchus*) in Florida. *In* "Marine Mammal Strandings in the United States", Proceedings of the Second Marine Mammal Stranding Workshop, Miami, Florida, 3–5 December 1987 (Eds J. E. Reynolds III and D. K. Odell) pp. 75–83. NOAA Technical Report NMFS 98.

Walsh, M. T., Odell, D. K., Young, G., Asper, E. D. and Bossart, G. (1990). Mass strandings of cetaceans. *In* "CRC Handbook of Marine Mammal Medicine" (Ed. L. Dierauf) pp. 673–683. CRC Press, Boca Raton, FL.

Whittow, G. C., Hampton, I. F. G., Matsura, D. T., Ohta, C. A., Smith, R. M. and Allen, J. F. (1974). Body temperature of three species of whales. *J. Mammal.* **55**, 653–656.

Zam, S. G., Caldwell, D. K. and Caldwell, M. C. (1971). Some endoparasites from small odontocete cetaceans collected in Florida and Georgia. *Cetology* **2**, 1–11.

10

Pilot Whales
Globicephala Lesson, 1828

Hannah J. Bernard and Stephen B. Reilly

Genus and Species

Taxonomy

The genus *Globicephala* was so named (a marriage of Latin "globus" = round ball or globe, and Greek "kephale" = head) because of the characteristic globe-shaped melon. While a number of specific names have been proposed for the genus, van Bree (1971) demonstrated that only two species are recognizable osteologically: *G. macrorhynchus* Gray, 1846, and *G. melaena* (Traill, 1809). Rice and Scheffer (1968), Rice (1977) and van Bree (1971) published lists of recent assignment of other proposed species names. Most recently, Jones *et al.* (1986) proposed a reversion of the specific name for the long-finned pilot whale to its original form, *melas*, and since this usage has gained world-wide acceptance, we shall use it throughout the text.

Subspecific status was proposed for the southern form of *G. melas* in the Atlantic (*G. melaena edwardii*) by Davies (1960), but the validity of this distinction is questionable (Rice, 1977). The nominal species, *G. scammonii* Cope, 1869, from the eastern Pacific is now believed to be a geographical form of *G. macrorhynchus* (Polisini, 1980). Geographic forms of *G. macrorhynchus* also exist in the northwest Pacific Ocean off the coast of Japan (Kasuya *et al.*, 1986a). A northern form is found between latitudes 35°N and 42°N, and a southern form south of this range. The two forms from Japanese waters are segregated both geographically

Handbook of Marine Mammals, Volume 6
ISBN 0-12-588506-7

and thermally (12–24 °C in the north and over 20 °C in the south) and are genetically isolated stocks (Wada, 1986). Both the northeast and northwest Pacific forms appear to be generally larger than the temperate-tropical or southern forms (Polisini, 1980; Kasuya *et al.*, 1986a). Kasuya (1986a) speculated that the larger body size of the northwestern form is an adaptation to the colder waters it inhabits.

Evolution

Barnes (1984) and Barnes *et al.* (1986) proposed the most recent classification of pilot whales within the Delphinidae in the subfamily Globicephalinae. Although the killer whale is also placed in Globicephalinae, Barnes believes that the pilot whale is more closely related to the false killer whale than to *Orcinus*.

The oldest fossil (*Delphinus (phocoenae) uncidens* Lankester, 1864) was found in Pliocene or Pleistocene crag deposits in England. Several fossils have been discovered in North America. A globicephalid fossil was found in a Pleistocene deposit in Florida (*Globicephala baereckeii* Sellards, 1960), and a fossil believed to be a globicephalid was found in late Pliocene strata of California (Barnes, 1976).

Common names

Numerous common names have been assigned to both species of *Globicephala* (Table 1). Both species are referred to as pilot whales (derived from the theory

TABLE 1 A list of common names for the pilot whale

	G. macrorhynchus	*G. melas*
English	Short-finned pilot whale	Long-finned pilot whale
Japanese	Gondo kujira, Siho-goto, Kobire gondo, Tappa naga (northern form), Magondo (southern form), Sibo-goto, Naiso-goto	
Mexican	Bufeo	
Indians of Washington, USA	Pat-so-ol-qual	
Shetland Islands		Caa'ing whale
Faroe Islands		Grind, Grindhval
Danish		Grindehval
Norwegian		Grinde, Grinzehval
German		Grindwal, Schwartzal
French		Chaudron
Uruguayan Spanish, S. America		Calderon
Russian		Grinda, Chernyi del'fin

that a leader guides or "pilots" a pod), blackfish, and pothead whales. The common name for *Globicephala melas* is the long-finned pilot whale, while *G. macrorhynchus* is known as the short-finned pilot whale.

External Characteristics

Colour pattern

Pilot whales are predominantly dark grey, brown or black dorsally, except for a light grey saddle behind the dorsal fin. Other light areas more or less apparent are an anchor-shaped patch under the chin, a faint post-orbital blaze, a mid-ventral blaze and genital patch (Figs 1 and 2). Yonekura *et al.* (1980) found the genital patch of long-finned pilot whales to be noticeably lighter than in the sample of the short-finned pilot whales they examined, but variation in colour pattern has not proved a reliable taxonomic character (Evans *et al.*, 1986). However, it may be that variation in the post-dorsal saddle is useful for distinguishing stocks

FIG. 1 A group of long-finned pilot whales near Drake Passage (62° 15′S latitude, 62° 20′W longitude), showing dorsal saddles and post-orbital blaze. (Photo by A. Bowles)

FIG. 2 A group of short-finned pilot whales off Hawaii. (Photo by J. Kilbride, courtesy of Earthviews)

(Evans *et al.*, 1986; Kasuya *et al.*, 1986a, 1988). Davies (1960), Sergeant (1962b), and Norris and Prescott (1961) provided detailed discussions of colour pattern.

Sightings of anomalous white pilot whales on two occasions in the northwest Atlantic indicate the possibility of albinism (Hain and Leatherwood, 1982). The rarity of this condition is emphasized by Bloch (1992) in her review of the Faroe Island drive fishery: only 3 albino whales were recorded in the estimated cumulative catch of over 100 000 pilot whales since 1985.

Size and shape

Pilot whales are among the largest of the delphinids, males reaching 7.2 m in the western North Pacific (*G. macrorhynchus*, Kasuya, 1986a), and 6.17 m in the western North Atlantic (de Kock, 1956). In the northeast Atlantic, in the Faroe Islands area, males have attained a maximum length of 6.25 m (Bloch *et al.*, 1993a). The Indian Ocean stock attains a size intermediate (5.15 m) between the northern and southern Japanese stocks (Kasuya and Matsui, 1984).

In general, the long-finned pilot whale is larger than the short-finned pilot whale (Kasuya and Marsh, 1984; Yonekura *et al.*, 1980). The body shape is considerably more robust than most delphinids, with a large, bulbous melon and long-based, falcate dorsal fin, set far forward on the body. The flippers are long and sickle shaped, and the tail stock laterally compressed and quite deep dorsoventrally (Fig. 3).

There is a marked sexual dimorphism in size, males being slightly larger at birth and reaching overall greater size by accelerated growth at puberty (Sergeant, 1962b). In the northern Pacific, the largest adult female short-finned pilot whales are shorter than the largest adult males by 2.1 m (Kasuya, 1986a). In the North Atlantic, this difference is 1.06 m (Sergeant, 1962b). Long-finned male pilot whales develop more bulbous melons than do females; however, in a sample of short-finned pilot whales off the Pacific coast of Japan, the size of the melon did not necessarily differ when comparing males and females of the same body length (Yonekura *et al.*, 1980). The large melon of the male may result from a generalized increase in size and not be a separate secondary sexual characteristic (Yonekura *et al.*, 1980). Apparently, there is sexual dimorphism in the flippers and flukes of pilot whales taken in the Faroe Islands area. Bloch *et al.* (1993b) report that male pilot whales there have slightly longer flippers than do females, and longer and wider flukes.

Table 2 lists selected measurements and proportions for comparison of short-finned and long-finned pilot whales. The external characters which have been used most often to distinguish between the two species are length of the pectoral flippers, relative to total body length, and number of teeth (Fraser, 1950; van Bree, 1971). The traditional "rules of thumb" regarding these characters are that the long-finned pilot whale has 9–12 teeth in each row and flippers 1/5 total body length, while the short-finned pilot whale has 7–9 teeth in each row and flippers to 1/6 body length. Recent information from a large series of Atlantic specimens of both species shows considerable overlap in these features, and there appear to be clines rather than distinct groupings as suggested in past literature (personal communication from J.G. Mead of the Marine Mammal Program, Smithsonian Institution). For this reason, both relative flipper length and tooth count are probably unreliable taxonomic characters. The most reliable characteristics for distinguishing between the two species are differences in the skulls (see Table 3 and later in Figs 5 and 6).

Weight

Detailed morphometrics of adult and juvenile pilot whales can be found in Norris and Prescott (1961), Sergeant (1962b), and Yonekura *et al.* (1980). Weights were not reported for these specimens, but Sergeant (1962a) derived a weight–length relationship for long-finned pilot whales in the North Atlantic: $W = 0.000025 \times L^{2.895}$ ($n = 30$), where weight (W) is in kg and length (L) is in centimetres. Lockyer (1993b) presented weights of long-finned pilot whales taken in the Faroes, and a slightly different weight–length curve: $W = 0.00023 \times L^{2.501}$ ($n = 620$), where weight (W) (whales were all weighed whole) is in kg and length (L) is in centimetres. This curve is for both sexes,

TABLE 2 Comparison of selected proportions of *G. melas* and *G. macrorhynchus*[a] (southern form)

Measurement		*G. macrorhynchus* *n*	*G. macrorhynchus* Range	*G. macrorhynchus* Mean	*G. melas* *n*	*G. melas* Range	*G. melas* Mean
Standard length (cm)	M	13	424–491	452.6	10	500–610	551.8
	F	73	334–392	357.5	10	405–472	440.7
Tip of snout to eye (%)	M	13	8.8–10.7	9.5	9	7.2–9.3	8.3
	F	63	8.4–11.0	9.7	10	9.0–10.6	9.8
Tip of snout to ant. ins. of flipper (%)	M	12	15.1–17.5	15.8	9	12.6–14.9	13.7
	F	53	14.3–19.0	17.0	10	14.4–16.3	15.5
Notch of flukes to apex of dorsal fin (%)	M	11	54.2–61.5	57.4	8	52.1–57.8	54.6
	F	47	51.3–60.1	56.1	9	48.2–53.1	51.2
Notch of flukes to umbilicus (%)	M	1	60.2	60.2	8	58.2–68.0	61.2
	F	16	54.8–59.9	56.3	9	57.7–63.2	60.3
Notch of flukes to anus (%)	M	13	32.7–36.4	34.8	10	33.9–41.4	38.2
	F	65	28.5–35.8	33.0	10	33.8–37.0	36.1
Flipper, ant. insertion to tip (%)	M	0			9	24.1–27.4	26.2
	F	18	15.8–18.9	16.3	10	21.9–26.2	23.7

[a] After Yonekura *et al.* (1980).

TABLE 3 Distinguishing skull characteristics for *G. melas* and *G. macrorhynchus*[a]

	G. melas	*G. macrorhynchus*
Rostrum	More elongate with rather narrow premaxillae leaving uncovered 1 cm lateral margin of the maxillae	Rather short and broad with broad premaxillae completely covering the maxillae, or leaving a small margin on both sides
Teeth[b]	Usually 9–12 in each row	Usually 7–9 in each row

[a] According to van Bree (1971).
[b] See discussion of teeth in Size and shape section.

which do not differ significantly in weight–length relationship. Kasuya and Matsui (1984) found a nearly identical weight–length relationship to that of Sergeant's (1962b) in their examination of short-finned pilot whales taken by driving fisheries off the coast of central Japan. They presented weights and the weight–length curve: $\log W = 2.8873 \times \log L + \log(2.377 \times 10^{-5})$ $(n = 21)$, where weight (W) is in kg and length (L) is in centimetres.

FIG. 3 Comparison of the short-finned pilot whale, *G. macrorhynchus* (upper), and long-finned pilot whale, *G. melas* (lower). (Drawings courtesy of P. Folkens)

Distribution

In general, pilot whales are distributed throughout the world's temperate and tropical oceans, both near shore and in the mid ocean (Fig. 4). The short-finned pilot whale is distributed in the tropical and warm temperate waters of the Atlantic, Pacific and Indian Oceans, while the long-finned pilot whale occurs "antitropically" (Hubbs, 1952; Davies, 1963) in the temperate North Atlantic, and all southern temperate oceans (Rice, 1977). Although the long-finned pilot whale is not known to occur at present in the North Pacific, subfossil material from Japan (Kasuya, 1975) indicates that this species lived in the North Pacific as recently as the tenth century A.D. Polisini (1980) hypothesized that the short-finned pilot whale is in the process of expanding to fill the former range of long-finned pilot whales in the North Pacific.

The distributions of the two recognized species appear to overlap in the temperate waters off the mid-Atlantic coast of the USA (Rice and Scheffer, 1968; Leatherwood *et al.*, 1976; Payne and Heinemann, 1993), off southern Europe and the Azores (Clarke, 1981), South Africa (van Bree *et al.*, 1978), off the coast of Peru and Ecuador (Reilly, 1977; Van Waerebeek and Reyes, 1986), and possibly some parts of the Indian Ocean, and off southeast South America (Ross, 1984; Castello and Pinedo, 1985; J.S. Leatherwood, personal communication).

Marked seasonality in the distribution of pilot whales has been observed in at least five areas: the coastal waters of Newfoundland (Sergeant, 1962a; Mercer, 1975); near New Zealand (Gaskin, 1968, 1972); off southern California (Brown and Norris, 1956; Norris and Prescott, 1961); in the eastern tropical Pacific (Reilly, 1977); and off the coast of Japan (Kasuya, 1975). Along the northeast coast of the USA, there is an apparent seasonal shift of pilot whales from offshore onto the edge of the continental shelf and then northward to Georges Bank (Payne and Heinemann, 1993). In at least two of the above areas (Newfoundland and southern California) the seasonal abundance of pilot whales appears to be correlated with the seasonal abundance of spawning squid (Sergeant, 1962a; Leatherwood *et al.*, 1973; Mercer, 1975). During years of low squid abundance (1983–85), fewer pilot whales were sighted near Catalina Island off the coast of California (C.W. Oliver, personal communication).

In both the coastal and pelagic waters of the eastern tropical Pacific, the density of population centres appears to change seasonally in response to major changes in the current structure of the area (Reilly, 1977). For the rest of the world's oceans, little is known of the details of the distribution of this genus. Leatherwood and Dahlheim (1978) summarized the gross aspects of worldwide pilot whale distribution. Their distribution map is adapted and updated in Fig. 4. Hui (1985) found that the occurrence of short-finned pilot whales was associated with high relief topography in the southern California Bight. Dohl *et al.* (1980) found a seasonal distribution with depth where pilot whales were found in significantly shallower water during winter ($\bar{x} = 375.83$ m) than summer

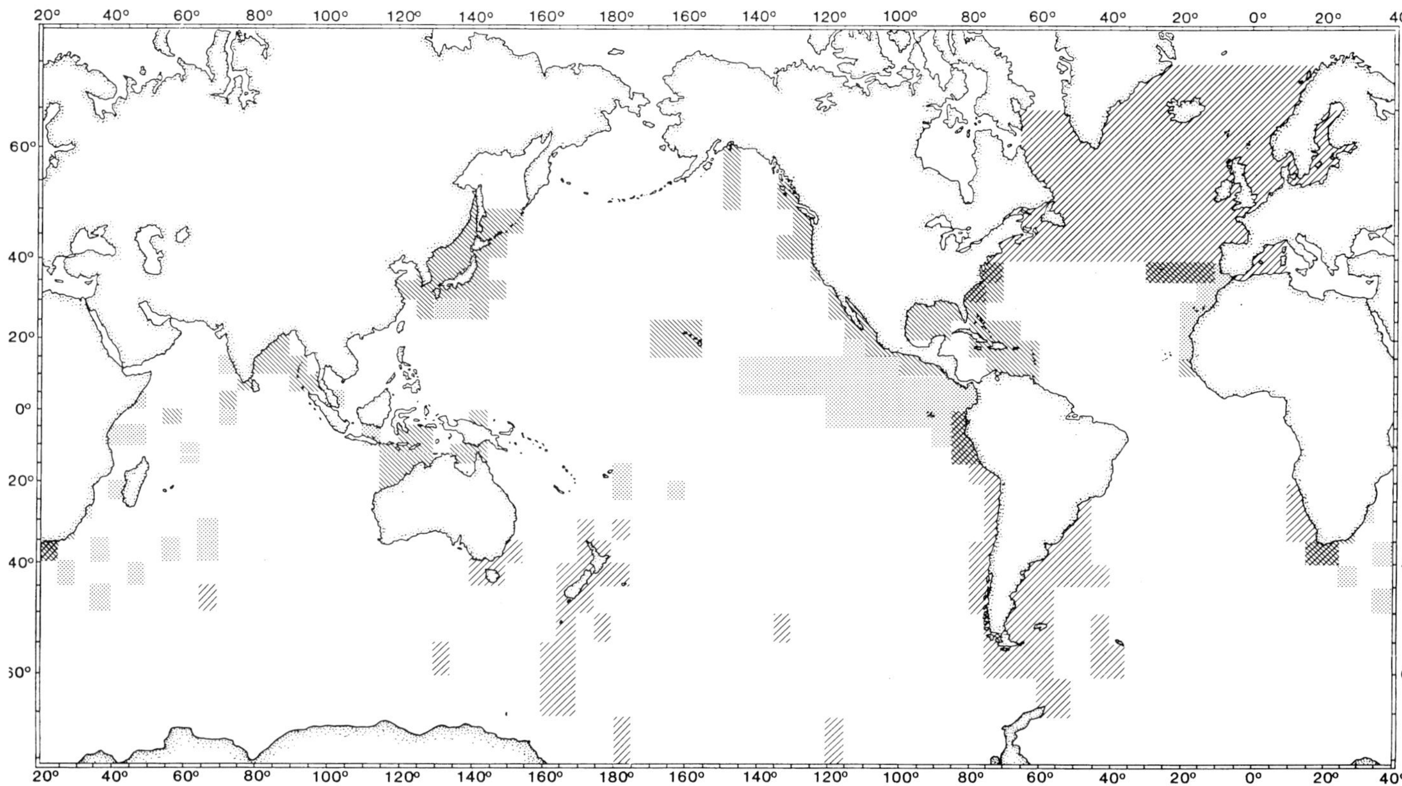

FIG. 4 Distribution of pilot whales (*Globicephala* spp.). *Globicephala melas* approximate distribution represented by leftward slanting parallel lines. *G. macrorhynchus* approximate distribution represented by rightward slanting parallel lines. Areas where the two species overlap are represented by cross-hatching. Dots represent the presence of pilot whales reported but not identified to species. Lined area denotes regions where pilot whales have been observed. Straight borders should not be taken to suggest abrupt limitations of range.

($\bar{x} = 798.98$ m). However, north of Pt. Conception, Dohl *et al.* (1983) found the occurrences of pilot whales correlated with deeper water (> 1092 m).

Internal Anatomy

Skull

The skull is wide and slightly depressed dorso-ventrally, with a wide and relatively short rostrum (slightly shorter than the braincase). The rostrum is characterized by a notable expansion in width of the premaxillaries. These bones are concave in front of the nares, and, in the mesorostral region, their width is at least 3/4 the total rostrum width (the amount varies between the two species). The relative length of the rostrum decreases with age, with a growth spurt in male short-finned pilot whales at a standard length of 285–375 cm and in females at 195–285 cm (Yonekura *et al.*, 1980). There are also changes with age reflected in relative increases of the orbital width and mesorostral width, and the basal width of the rostrum.

The type skulls of the two species are illustrated in Figs 5 and 6, and a comparison of skull characteristics is presented in Table 3.

Postcranial skeleton

The vertebral formula for the genus is C_7, T_{11}, L_{12-24}, Ca_{28-29} (Tomilin, 1967). Nishiwaki (1972) mentioned specific differences, but their consistent occurrence has not been confirmed. The first 5–6 cervical vertebrae are fused. The phalangeal formula is I_{3-4}, II_{9-14}, III_{9-11}, IV_{2-3}, V_{1-2}. The ribs are in 11 pairs, 6 pairs are bicipital, and 5 pairs articulate with the sternum on each side. The sternum is in 3–4 parts (Tomilin, 1967).

Organs and tissues

Murie (1874) presented the most complete treatment of pilot whale soft anatomy, but current sources are listed in Table 4.

Life History and Population Dynamics

Growth

Sergeant (1962a) plotted a growth curve for the long-finned pilot whale, but did not linearize the axes or report the parameters of the regression. In another publication, Sergeant (1962b) investigated the morphometrics of growth in more

TABLE 4 Some references on the internal anatomy of pilot whales (*Globicephala* spp.)

Organ	References
Skeleton	Spurrell (1965, 1966), Tomilin (1967), Mead (1975), Nishiwaki (1972)
Musculature	Murie (1874), Lawrence and Schevill (1965), Mead (1975)
Brain	Murie (1874), Elias *et al.* (1968), Ridgway and Brownson (1984), Haug (1987), Glezer and Morgane (1990), Glezer *et al.* (1993, 1995), Tarpley and Ridgway (1994), Hof *et al.* (1995)
Thyroid	Ridgway and Patton (1971)
Organ weights	Cowan (1966a,b)
Cardiovascular system	Elsner *et al.* (1966), Medway and Moldovan (1966), Olson *et al.* (1967, 1969), de Kock (1969), Ridgway *et al.* (1970), Ridgway (1972)
Anatomy, histology, molecular biology and biochemistry	Walen and Modin (1965), Hashimoto *et al.* (1967), Elias *et al.* (1968), Harrison and Young (1969), Pilson (1970), Kulu (1972), Simpson and Gardner (1972), Wedmid *et al.* (1973), Wada (1986), Andersen and Friedrich (1988)
Reproduction	Kasuya and Marsh (1984), Marsh and Kasuya (1984), Perrin and Reilly (1984), Donovan *et al.* (1993)
Testicle	Desportes (1994)
Teeth	Lockyer (1993a)
Skull	Oelschlager (1990)
Ear	Ketten and Wartzok (1990)

detail, again for long-finned pilot whales from Newfoundland waters. In their re-examination of life history parameters for long-finned pilot whales, Kasuya *et al.* (1986b) found that after the age of 2.5 years, males grow at a slower but constant rate until about 20 years of age. After the age of 6, the female growth rate declines rapidly and growth stops at 21–25 years.

Growth rates for long-finned pilot whales estimated from some mass strandings in the northeast Atlantic (Martin *et al.*, 1986), and from the Faroe Islands drive fishery (Bloch *et al.*, 1993a) appear to be comparable with those of long-finned pilot whales from the northwest Atlantic (Sergeant, 1962a).

Yonekura *et al.* (1980) plotted growth rates for segments of the body and appendages of the short-finned pilot whale from 211 whales taken in the Taiji drive fishery. They also compared body sizes of whales from their study with those of Sergeant (1962b) (see Table 2). The short-finned pilot whale has a slower growth rate than the long-finned pilot whale and reaches a smaller asymptotic body length. Once again utilizing whales from the drive fisheries of Taiji and Futo, Kasuya and Matsui (1984) derived a mean growth curve which they divided into four phases. The first phase comprises rapid neonatal growth,

(A)

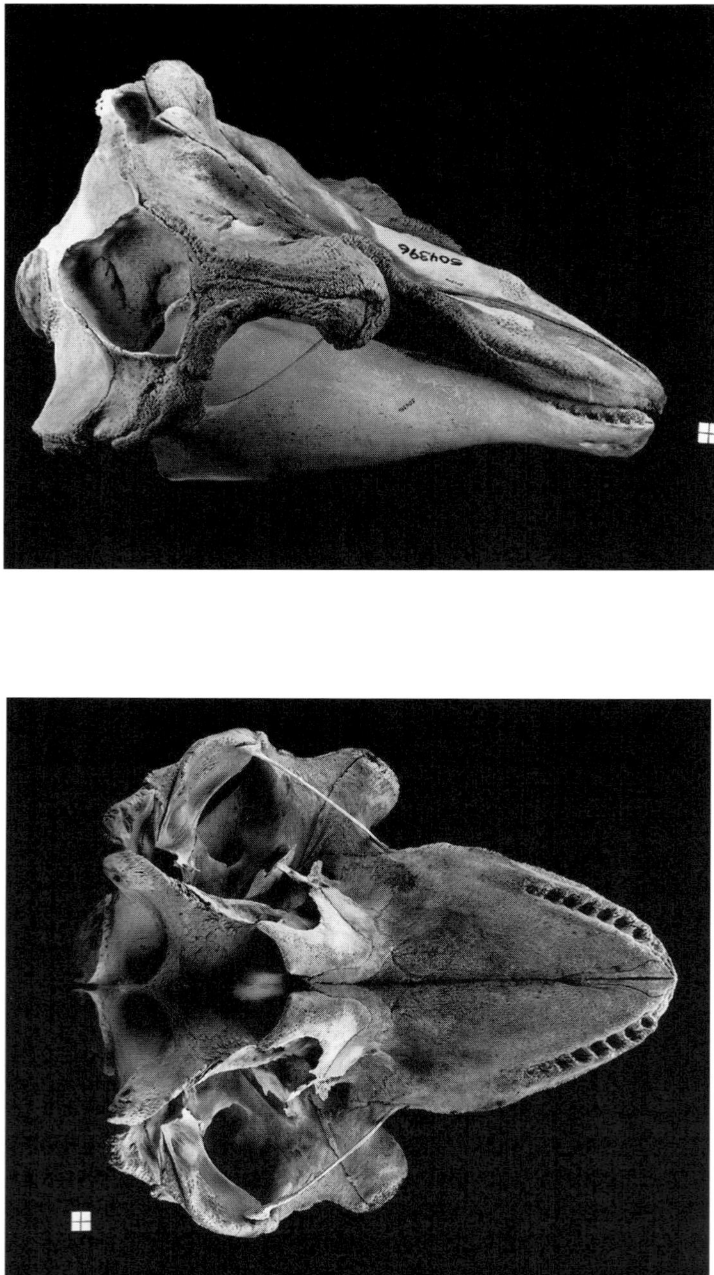

(B)

FIG. 5 Type skull of *Globicephala macrorhynchus* (USNM 504396): (A) lateral view and (B) ventral view.

(C)

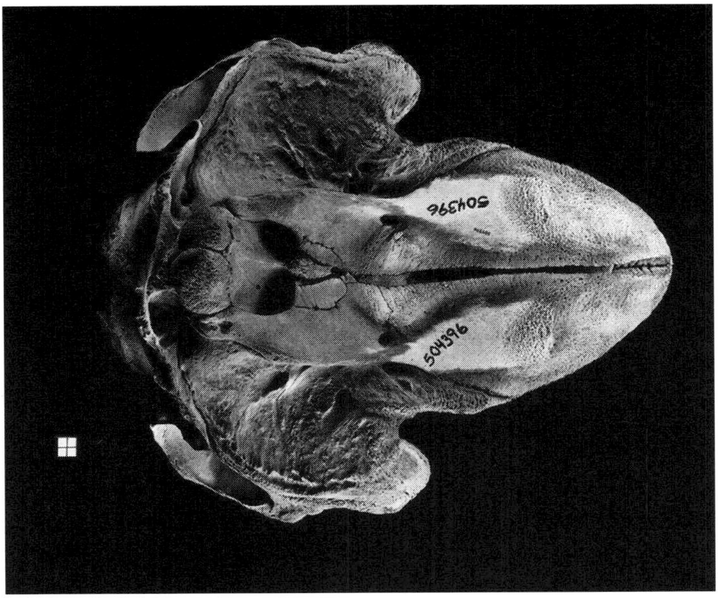

FIG. 5 *Continued* (C) Dorsal view. (Courtesy of C. Potter, USNM; photos by Chip Clark)

which is similar for both males and females and ends at about 1.25 years. The second or juvenile phase is characterized by less rapid but nearly linear growth ending at about the age of first ovulation for females (9 years) and at about 10 years for males. The females tend to be smaller than the males by several centimetres. The third phase encompasses that between puberty and mean asymptotic length. Mean asymptotic length is about 22 years for females. Males undergo a slightly accelerated prepubertal growth followed by a gradual decrease until about the age of 27. The fourth phase is when growth stops.

Reproduction

Reproduction and birth are distinctly seasonal in long-finned pilot whales off Newfoundland, with matings peaking during the period April–May, and calvings peaking during July–September (Sergeant, 1962a). Gestation was calculated to average 15–16 months, and calves are nursed for up to 22 months. Rarely does lactation overlap with pregnancy. The overall reproductive cycle lasts an average of 40 months, with a female producing a maximum of 9 calves during her lifetime.

The long-finned pilot whales off the Faroes have a similar breeding and calving season to those off Newfoundland, with matings extending into July and births through October (Martin and Desportes, 1987). Martin and Rothery (1993)

(A)

(B)

(C)

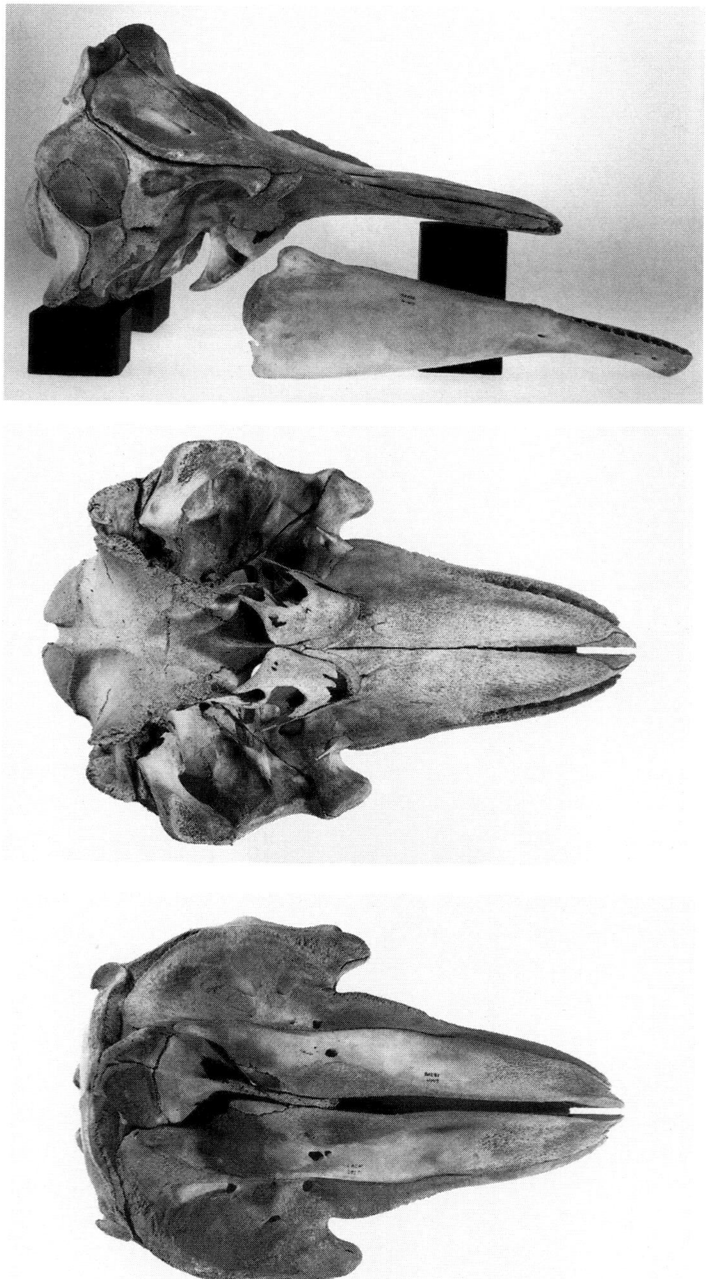

FIG. 6 Type skull of *Globicephala melas* (LACM 28256) from Tasmania: (A) lateral view, (B) ventral view and (C) dorsal view. (Photo courtesy of J. Heyning, Los Angeles County Museum of Natural History)

examined nearly 2000 pilot whales killed in the Faroes drive fishery between 1986–88. Although there may be some biases in this dataset, it is probably the most representative sample available from a pilot whale population. From these data Martin and Rothery (1993) estimated a gestation period of 12 months, significantly lower than estimates from other studies (15–16 months). They believe that previous calculations of gestation period for both long-finned and short-finned pilot whales (Sergeant, 1962a; Kasuya and Marsh, 1984) overestimated gestation length because those studies assumed a linear rather than nonlinear foetal growth rate.

In short-finned pilot whales of the southern form off Japan, reproduction and birth are seasonal, but not distinctly, with a single peak of conception in April–May, parturition during July–August, and a small number of births occurring year-round. The gestation period was calculated to be the same as the longer estimate for the long-finned pilot whale, approximately 15–16 months, with weaning occurring after 3.5–5.5 years. An older female, in her 30s, may nurse her last calf (presumably male calves, because of their older age at maturation) for a much longer time, possibly up to 15 years (Kasuya and Marsh, 1984). This phenomenon, which may demonstrate the tendency of a female to invest more in her current offspring as she becomes less likely to bear additional offspring, is discussed by Marsh and Kasuya (1991).

Kasuya (1986a), and Kasuya and Tai (1993) examined morphology and life history parameters of the northern form of the short-finned pilot whales off Japan and determined that it differs from the southern form in body size (see Section on Taxonomy), dorsal saddle mark (Kasuya *et al.*, 1988) and breeding season. The northern form has a distinctly seasonal breeding pattern, with a peak of conception in September and presumed parturition in December. Kasuya (1986a) speculated that the northern yearlings would be better able to switch from nursing to solid food in the more productive summer/autumn season.

In both species, the females can remain alive long past last ovulation (at least 14 years) (Sergeant, 1962a; Kasuya and Marsh, 1984; and Marsh and Kasuya, 1984). The proportion of postreproductive females is significant in short-finned pilot whales (Kasuya and Marsh, 1984), whereas long-finned pilot whales may remain reproductively active throughout their lifespan (Martin and Rothery, 1993). Long-finned pilot whale females of up to 59 years are reported off the Faroe Islands (Bloch *et al.*, 1993), and one 55 year old female examined by Martin and Rothery (1993) was pregnant.

Table 5 shows a comparison of reproductive parameters for short-finned (both northern and southern stocks) and long-finned pilot whales.

Although the long-finned pilot whale of the western North Atlantic generally is larger than the short-finned, the long-finned pilot whale reaches sexual maturity at a lesser age (northwest Atlantic females, 6 years; northeast Atlantic females, 8 years; males, 12 years) than the short-finned (females, 9 years; males, 16 years). Kasuya and Marsh (1984) speculated that differences in exploitation

TABLE 5 Reproductive and growth parameters of pilot whales (values for males in parentheses)

	G. melas[a]	*G. macrorhynchus*	
		Northern form[b]	Southern form[a,c]
Average length at sexual mat. (cm)	365 (490)	390–400 (550)	316 (422)
Average age at sexual mat. (yrs)	6 (11) 8.3[d]	9 (17.0–17.5)	9 (17)
Age of oldest individual (yrs)	59[e] (35.5)	61.5 (44.5)	62.5 (45.5)
Mean gestation period (months)	15 (12)[d]	15–16	14.9
Annual preg. rate	30%	18–24%	10%
Mean calving interval (yrs)	3.3	4.58–5.70	4.8
Average age at weaning (yrs)	2	2.08–2.78	2.03
Mean length at birth (cm)	177	185	140

[a]Sergeant (1962a).
[b]Kasuya *et al.* (1986b).
[c]Kasuya and Marsh (1984).
[d]Martin and Rothery (1993).
[e]Bloch *et al.* (1993a).

pressure or food availability may have influenced the maturation rate of the long-finned pilot whale in the western North Atlantic.

Sergeant (1962a) found that, in the Newfoundland area, approximately 40% of the females were sexually mature during any one year; nearly 40% of these were pregnant.

Mortality

As with other cetaceans, it has proved very difficult to obtain unbiased estimates of natural mortality for pilot whales. The specimen series available (long-finned from the northeast Atlantic, Sergeant, 1962a; Kasuya *et al.*, 1986b; Bloch *et al.*, 1993a; and short-finned from the northwest Pacific, Kasuya and Marsh, 1984) are from fisheries, and are from populations which cannot be assumed to be either stationary in size or stable in age distribution. Consequently, the available estimates represent apparent mortality. Data do not exist to remove the influence of exploitation or recruitment rate changes from any series. However, data from the Faroese drive fishery may provide unbiased representations of sex and age structure within individual pods because usually entire pods were captured.

Given the above limitations, some patterns do emerge. For both species, males appear to have higher mortality rates than females at all ages. Long-finned pilot whales (from the northeast Atlantic) have higher mortality rates (also shown by shorter lifespans) than short-finned pilot whales (from the northwest Pacific). The southern and northern forms of short-finned pilot whales in the northwest Pacific do not appear to have different mortality schedules.

Kasuya *et al.* (1986b) estimated the apparent annual mortality rate of male long-finned pilot whales as about 0.07 below 25 years of age, and 0.15 for the 21–36 year age class. The female mortality rate from this sample was slightly over 0.02 for the below 25 years class, and 0.1 for those 21 years and older. Kasuya and Marsh (1984) reported apparent mortality rates for female short-finned pilot whales of 0.025 for 18–47 year olds, and 0.157 for those aged 45–63 years. For males, they reported apparent rates of 0.044 for 9–30 year olds, and 0.102 for whales aged 27–46 years. Bloch *et al.* (1993a) estimated mortality for northeast Atlantic pilot whales. For males, they show a mean range of 0.0745–0.0817, and for females of 0.0622–0.0737. As in the other forms, males have a consistently higher mortality rate than females.

Abundance

Few quantitatively derived population estimates exist for pilot whales. For long-finned pilot whales fished off Newfoundland, Mercer (1975) used the correlation between the inshore abundance of squid (*Illex illecebrosus*) and pilot whales as a basis for a modified Leslie–DeLury population estimate of approximately 62 000 for 1947, before the start of the intensive fishery which occurred during 1952–1968. Hay (1982) conducted an aerial line-transect survey of long-finned pilot whales in the Newfoundland–Labrador region. His population estimate of long-finned pilot whales is 6731–19 603 for 1980 for over 150 km of the coastal shelf. Sergeant (1986) gave an estimate of 40 000–80 000 long-finned pilot whales for all North Atlantic waters. In 1987 and 1989, joint surveys of the northeast Atlantic were conducted under the auspices of the North Atlantic Sightings Survey (NASS-87 and NASS-89). Buckland *et al.* (1993) analysed the data and estimated 778 000 (CV = 0.295) long-finned pilot whales for most of the north-eastern North Atlantic Ocean.

Clarke *et al.* (1978) compared the numbers of long-finned pilot whales sighted per track mile between ship surveys off Chile in 1958 and 1964. They recorded 13.2 per 100 miles in 1958 and 11.0 per 100 miles in 1964.

Using the line-transect methodology developed for the minke whale population in the Antarctic by the IWC/IDCR (for the years 1976–77 and 1988), Kasamatsu and Joyce (1995) estimated 200 000 (CV = 0.35) long-finned pilot whales for this region.

Several aerial line-transect surveys of short-finned pilot whales conducted off the coast of California during the late 70s–early 80s (Dohl *et al.*, 1980) placed the minimum population at between 200 and 4000, but it is not clear if this represents one or more than one stock.

There have been no systematic studies of home range or migration of individuals of this genus. Opportunistic observations in the southern California Bight have indicated that a pod of 20–30 individuals, identified by scars, unusual marks, etc., lived in the area year-round (Walker, 1975). Following a strong El Niño event in 1982–83, subsequent surveys throughout the 1980s turned up few sightings (Shane, 1986; Forney, 1994), and documented the absence of all but one pod of pilot whales near Catalina Island. Shipboard surveys along the entire California coast using line-transect methodology were conducted in 1991 and 1993 within 550 km of shore (Barlow and Gerrodette, 1996). They documented an apparent return of this stock and calculated an abundance estimate of 1004 (CV = 0.37).

Wade and Gerrodette (1993) estimated abundance of short-finned pilot whales in the eastern tropical Pacific to be 160 200 (CV = 0.138). Their data came from an extensive series of line-transect sightings surveys directed at dolphin populations affected by the tuna purse seine fishery. The cruises searched an area of approximately 19×10^6 km^2, using two ships for 4 months each year, 1986–1990. They noted that a small but undetermined portion of this estimate may apply to long-finned pilot whales in the southeast portion of their study area off Ecuador and northern Peru, where probably the two species are sympatric.

Miyashita (1986) estimated the abundance of short-finned pilot whales off the coast of Japan, using line-transect analysis of sightings data from 1984–1985. He estimated the northern form at 5300 (CV = 0.43) and the southern form at 53 000 (CV = 0.33).

Behaviour

Association with other species

Pilot whales are often seen in association with other cetaceans. In the eastern tropical Pacific, approximately 15% of pilot whale sightings include other cetaceans. They are sighted with bottlenose dolphins (*Tursiops truncatus*) and with tuna–dolphin aggregations (*Thunnus albacares* and *Stenella* spp., Reilly, 1977) and *S. coeruleoalba* (Donovan, 1984). The most common associate in coastal waters is the bottlenose dolphin; pilot whales have been sighted also with common dolphins (*Delphinus delphis*), Pacific white-sided dolphins (*Lagenorhynchus obliquidens*), gray whales (*Eschrichtius robustus*) (Norris and Prescott, 1961; Leatherwood, 1974), fin and sperm whales (*Balaenoptera physalus* and *Physeter catodon*) (Clark *et al.*, 1978; Weller *et al.*, 1996) and with killer whales (*Orcinus orca*) in Faroese waters (Bloch and Lockyer, 1988). Recently, there has been a report of a pilot whale biting and towing a human diver (Shane and Costello, 1993).

Group size and social organization

Pod size appears to vary greatly, with the median size (in the eastern tropical Pacific) being near 10 individuals, while the mean is about 25, with a CV of 2.6. Off southern California, the median pod size is 15, with a mean of 43 and a CV of 2.8 (Reilly, 1977). Shipboard observations in the northwestern Pacific Ocean have found schools of 5–50, with a mean of 20.6 and a CV of 0.62 (Kasuya and Marsh, 1984). Near Newfoundland, pods driven ashore in the years 1951–1959 ranged from 10 to 250 animals, with a mean of 85 (Sergeant, 1962a). Off the Faroe Islands, schools driven ashore between July 1986 and July 1988 ranged from five to 220 animals with a mean of 84 (43 schools represented, $n = 3618$) (Bloch *et al.*, 1990).

Several schools may form loose aggregations to exchange individuals or to mate. These aggregations may also serve a protective function as they become more cohesive in certain situations (Sergeant, 1962a; Kasuya and Marsh, 1984). The lower CV from the estimate by Kasuya and Marsh (1984) may have resulted from their explicit exclusion of aggregations from their estimate of mean school size.

Pilot whales are among the most socially structured of the odontocetes. The short-finned pilot whale is loosely polygynous (Kasuya and Marsh, 1984) like the long-finned (Sergeant, 1962a), with more reproductive females than males per school. Like killer whales, the offspring usually reach sexual maturity in their mother's herd and probably remain there for life; yet there appears to be no inbreeding within a pod. In fact, a high level of outbreeding of male pilot whales has been ascertained through DNA fingerprinting in a study of the Faroes drive fishery (Amos *et al.*, 1993). There is scant evidence to support geographical or social segregation of "bachelor" herds (Kasuya and Marsh, 1984; Amos *et al.*, 1993), and the genetic evidence is supported by field observations of large aggregations of several pods. Therefore, it is likely that the males move between schools when several come together in these temporary aggregations (Amos *et al.*, 1993). Kasuya (1986a) considered older, post-reproductive females to be significant members of the school, contributing to the survival of younger individuals. The social structure revealed by Amos *et al.* (1993) suggests that male pod members may contribute to the survival of their related pod members as well.

Three types of social organization for pilot whale pods off southern California were described by Norris and Prescott (1961) and Brown and Norris (1956); travelling/hunting groups, feeding groups, and loafing groups. The travelling/hunting groups have also been appropriately described as "chorus lines" (Leatherwood *et al.*, 1973) as the animals in these are oriented in a broad rank of up to 2 miles in width, but only a few animals deep. Sexual and age–class segregation also have been observed in chorus lines (Norris and Prescott, 1961). In the second type of group described, the feeding group, there was sometimes a " . . . general movement of whales in a given direction, but individual animals tended to remain fairly independent of one another" (p. 345). The third type of pod, the "loafing group", was described as an almost stationary aggregation of 12–30 or more individuals, floating at the surface, nearly or actually touching one another. A wide variety of types of behaviour, including mating, was reported to occur in loafing groups.

Feeding

Squids are the predominant food item (Sergeant, 1962a; Mercer, 1975; Hacker, 1986; Gales and Pemberton, 1992; Desportes and Mouritsen, 1993; Nelson and Lien, 1996). Widely observed co-occurrence of pilot whales and squid also supports this (see section on Distribution). Fish are also occasionally consumed. In the western North Atlantic, fish prey species reported are Atlantic cod (*Gadus morhua*), Greenland turbot (*Rheinhardtius hippoglossoides*) (Mercer, 1975). Atlantic mackerel (*Scomber scombris*) may be an important component of the winter diet (Waring *et al.*, 1990). Although squids are the predominant prey around the Faroe Islands, some fish, such as *Argentina silus* and *Micromesistius poutassou*, are taken too. The whales in this region do not appear to select cod, herring or mackerel, although they are periodically abundant (Desportes and Mouritsen, 1993). In captivity, pilot whales prefer squid but sometimes will also accept fish (Kritzler, 1952; Brown, 1960).

The daily feeding rate of pilot whales has been reported as 4.0–5.9% of the body weight, both from examination of wild animals (Sergeant, 1962a) and from observation of captive animals (Brown, 1960, 1962; Gilmore, 1962). Captive animals reportedly consume slightly more than those which Sergeant studied near Newfoundland. An annual consumption rate was extrapolated by Mercer (1975) to be between 15 and 18 times body weight. For an "average" size animal of 3.96 m and 830 kg, this is equivalent to 12 410–14 965 kg of prey per year. Lockyer (1993b) calculated a 1000 kg female of reproductive age would consume 11 300 kg of squid per year.

Diving depth and duration

Both in captivity and in open-ocean training, pilot whales have been observed in order to determine how long and how deeply they can submerge themselves. A captive male was observed underwater for 10 min (Caldwell *et al.*, 1963); a radio-tagged whale remained down for up to 12 min and reached a depth of 45.5 m (Leatherwood *et al.*, 1973). At the onset of darkness, the pattern of respirations and frequency of protracted drives increased. During open-ocean training by the US Navy, one short-finned pilot whale made dozens of dives greater than 300 m, with a maximum measured depth of 504 m. Although the depth-of-dive transmitter was malfunctioning when this whale was sent to a target at 610 m, it is assumed that he made it to this depth because of his past performance and dive time (13 min). This whale's maximum time submerged was 14 min 49 sec (Bowers and Henderson, 1972).

Sound production

Pilot whales are known to produce non-harmonic sounds (including echolocating clicks) and harmonic sounds (Evans, 1967). Studies conducted to determine differences in vocalizations between pods of different regions show that only mean whistle duration is useful for distinguishing between long-finned and short-finned pilot whales (Taruski, 1976) and between regional groups of short-finned pilot whales in the Pacific (Evans *et al.*, 1986). Pilot whales can emit variations of their whistle resulting in "shrill or plaintive cries" when in distressing situations such as strandings, captures, and injuries (Brown, 1960; Evans, 1967; and Alagarswami *et al.*, 1973).

Mass strandings

Both species of pilot whale strand singly and *en masse* worldwide (see Dudok von Heel, 1962; Alagarswami *et al.*, 1973; Brownell, 1974; Goodall and Galeazzi, 1986; and Sigurjónsson *et al.*, 1989, for listings of some mass strandings). Necropsies of animals that have mass stranded do not usually yield evidence of acute infections, whereas singly stranded animals are *usually* characterized by pathological findings (Sergeant, 1982). Brain lesions produced by long-term parasitism may lead to disorientation and stranding of an individual, but whales may also carry a heavy parasite load without obvious symptoms (Medway, 1978).

Hypotheses for causes of mass strandings abound. They include navigational error resulting from reliance on magnetic sensory systems (Kirschvink *et al.*, 1986) (or that induced by the apparent inability of pilot whales to use sonar for simultaneous feeding and navigating) and socially or behaviourally induced stranding. If a whale of extreme social importance or strong filial bond strands due to pathological or navigational problems, others in the pod may strand also and then be unable to remain off the beach once removed due to a secondary social or epimeletic response. This social, or perhaps epimeletic, response was used successfully to keep a pod of long-finned pilot whales from repeated strandings by researchers in New Zealand (Dawson *et al.*, 1985). Because the "distress calls" of the beached young of the pod appeared to evoke a stranding response from the older whales, the younger whales were towed offshore and moored to buoys, an action which lured the older animals back out to sea.

It is likely that a combination of two or more of the possible causes of mass strandings leads to such an event.

Parasites and Disease

The most common disease in a sample of 55 fishery-killed long-finned pilot whales off Newfoundland was infestation with a parasitic worm of the upper respiratory tract. Every whale was parasitized except a calf with no erupted teeth (Cowan, 1966b). Cowan also found that the incidence of lesions of the myocardium and arteriosclerosis increased in frequency with age in this group.

In their examination of fishery-killed pilot whales from the Faroes, Raga and Balbuena (1993) found a different parasite composition, with stomach worms (*Anisakis simplex*) being the most prevalent parasite. They noted that probably this difference is due to different diets between the areas. They found the prevalence and intensity of infection increased with age, and was higher in males.

Common parasites of pilot whales are "whale lice" (*Isocvamus delphini*), cestodes (*Phyllobothrium delphini*), and nematodes (*Anisakis simplex*). Dailey and Brownell (1972) listed known parasites for both species.

Pathogenic bacteria and viruses have been isolated from stranded and captive pilot whales (*Streptococcus*, *Pseudomonas*, *Escherichia*, *Staphylococcus*, and influenza to name a few), but are not necessarily portents of acute infection. As in humans, these and other bacteria and viruses can exist in the system innocuously, becoming acute when the animal is stressed or otherwise compromised (Buck, 1984; Higgins *et al.*, 1980; Hinshaw *et al.*, 1986). Duignan *et al.* (1995) have shown that a high percentage of pilot whales tested have high titres to morbillivirus, suggesting a high rate of infection in western Atlantic animals.

Live Maintenance

Capture of long-finned pilot whales for public display has been neither extensive nor as consistent as the take of short-finned pilot whales. Prior to 1973, a small number of long-finned pilot whales were captured off New England. Since then, two stranded specimens from the US east coast were held in captivity for a short time (Reeves and Leatherwood, 1984).

Short-finned pilot whales have been collected in southern Californian, Hawaiian, and Japanese waters for aquaria and oceanaria. During 1966–72, 33 pilot whales were taken off the coast of southern California (Walker, 1975), and 17 have been taken since 1973 (Reeves and Leatherwood, 1984). Two of these resided at Sea World in San Diego, California. During this same time period (1963–72), 20 whales were captured near Hawaii (Reeves and Leatherwood, 1984). The annual mean number of short-finned pilot whales taken alive (for public display) in the drive fishery at Taiji during 1974–84 was about six (Kasuya *et al.*, 1984).

Pilot whales have a high mortality rate in captivity (Walker, 1975). Between 1966 and 1973, animals held for public display had an average annual survival rate of 0.51 (D. DeMaster, personal communication; data are from Walker, 1975, Table 6). Animals that survived for at least 90 days had an average annual survival rate of 0.73 (D. DeMaster, personal communication), which indicates that acclimatization to captivity was a problem in these years. More recent information on survival in captivity has not been published.

Activity patterns

Shurley *et al.* (1969) monitored the daily activity patterns of a healthy, trained captive 4-year-old female short-finned pilot whale from the North Pacific. The mean activity pattern for this individual was described as 54% active waking, 26% inactive alert, and 22% behavioural sleep per 24-hour period. It was noted both that the above states were distributed nearly parallel with gross human activity in the vicinity and that the animal was raised in captivity. Gilmore (1962) reported observations on another captive short-finned pilot whale during 2 hours of behavioural sleep. The young male was seen to interrupt its sleep 83 times during the 2-hour period, taking one breath during each. Eighty of the intervals between interruptions were between 15 sec and 1 min 14 sec, while the remaining were 1 min 45 sec, 1 min 50 sec, and 2 min 30 sec.

Kritzler (1952) discussed activity patterns of newly captive short-finned pilot whales from Florida waters. Four of 46 animals that stranded at St. Augustine Beach were moved to tanks at Marineland of Florida. Initially, these whales were distinctly nocturnal in activity. After a period of months, however, the sole survivor had adjusted his activity to the chronology of his feeding schedule.

Epimeletic (care-giving) behaviour has been recorded for at least one captive short-finned pilot whale; he was observed interacting with a dead Pacific white-sided dolphin (*L. obliquidens*) for half a day. The whale brought the dolphin to the surface repeatedly and interfered when divers tried to remove it from the tank (Caldwell *et al.*, 1963). Cooperative actions have also been observed in terrestrial mammals with a matricentral or matrilineal structure. In odontocetes, these types of behaviour are exhibited particularly by subadult and adult females (Gaskin, 1982).

Human Effects

Directed fisheries

Long-finned pilot whales have been hunted by man at least since the sixteenth century near the Faroe Islands. The fishery there has continued to the present, with catches in the early 70s at about 1500 per year (Mitchell, 1975a,b) and increasing during the mid 80s to near 2500 (Hoydal, 1986). When Joensen and Zachariassen (1982) examined catch statistics for this fishery for 1709–1978, they found a pulse of 100–120 years between what they called "really good years", that is, catches between several hundred to several thousand animals. Hoydal (1986) suggested that the historical catch levels have not affected the numbers of whales subsequently caught in the Faroe fishery.

Small numbers of pilot whales have been taken elsewhere in the North Atlantic, in Norway, West Greenland (Kapel, 1975), Iceland, the Irish coast (O'Riordan, 1975), and at Cape Cod (during the late nineteenth–early twentieth centuries), where a drive fishery took 2000–3000 per year (Mitchell, 1975b). The fishery for the long-finned pilot whale at Newfoundland took a peak annual catch of nearly 10 000 in 1956 but declined after the early 1960s (Mitchell, 1975a) and is no longer in operation (IWC, 1992). Exploitation of this species in the Southern Hemisphere has been sporadic and very low (Mitchell, 1975a).

The short-finned pilot whale has been exploited for centuries in the western North Pacific. Kasuya and Marsh (1984) present data from whaling operations for 1948–1980, on both northern and southern forms (although prior to 1965 the figures are an underestimate of the take of the southern form). The small-type whaling vessels (harpoon gun-equipped) off Hokkaido and Sanriku took a mean annual catch of 91 (range = 0–415) of the northern form during 1948–1979. Between 1948 and 1980, drive and harpoon fisheries took a mean annual catch of 302 (range = 0–781) of the southern form at Taiji, Izu and Okinawa. Both fisheries peaked in the late 40s–early 50s, and subsequently declined with a few exceptional years (see Kasuya and Marsh, 1984; Kasuya, 1986b). In recent years, the southern form continues to sustain a higher kill than the northern form. In 1982, the drive fishery at Taiji expanded and harpooning of the northern form was resumed off Sanriku and Hokkaido. Between 1982 and 1985, 1755 whales of the southern form were killed (Kasuya, 1986b), and 519 of the northern form were taken during this same time (Kasuya and Tai, 1986). From 1985 to 1989, Japan took a total of 2326 short-finned pilot whales.

Elsewhere, a small, intermittently active fishery takes around 220 pilot whales per year in the Lesser Antilles in the Caribbean at St. Vincent Island (Caldwell and Caldwell, 1975), and there are indications of a small fishery at St. Lucia Island (Gaskin and Smith, 1977). There is evidence that early whalers took substantial numbers of these whales (Mitchell, 1975a). In the 19th century, whalers of the Azores hunted pilot whales to supplement the reduced sperm whale catch in winter, while American sperm whalers used pilot whales for "practice" hunting (Clarke, 1981). There are no records of directed catches of pilot whales in the Lesser Antilles or Azores since 1986 (IWC, 1988).

By-catches

Coastal fish and squid fisheries that utilize drift and set and drift gill nets, seines, unintentionally catch marine mammals along with their target species. This problem is worldwide in occurrence, but not all species of marine mammals are recovered or reported in most countries. Therefore, there are probably more pilot whales taken incidentally than are presently documented. Based on preliminary data (and very few first-hand observations), the squid round-haul fishery in southern California waters is estimated to have taken 30 short-finned pilot whales in one year (Miller *et al.*, 1983; DeMaster *et al.*, 1985). In the California drift gill net fishery between 1993 and 1995, the mean annual take of short-finned pilot whales was 20 (Barlow *et al.*, 1997).

On the east coast of the USA, the foreign Atlantic mackerel fishery was responsible for the take of 141 pilot whales in 1988. This fishery was suspended in early May of that year as a direct result of this anomalously high take (Fairfield *et al.*, 1993).

A 1990 workshop to review mortality of cetaceans in passive nets and traps documented an annual kill of 50–100 *G. melas* off the Atlantic coast of France, and 350–750 *G. macrorhynchus* off Japan (Perrin, 1994).

Pollution

A growing threat to the integrity of the marine environment is the all-pervasive problem of pollution. Of particular concern are the organochlorine contaminants such as DDT and PCB, which usually have terrestrial origins. Based upon the current rate of PCB pollution, Reijnders (1989) predicts that within the next century, all of the coastal-dwelling marine mammal species will have felt its deleterious effects.

Long-finned pilot whales off the Faroes, France, UK and the eastern USA appear to be carrying high levels of DDT and PCB in their tissues (Taruski *et al.*, 1975; Alzieu and Duguy, 1979; Gaskin, 1982; Martin *et al.*, 1986; Borrell and Aguilar, 1993). However, those animals examined off the Newfoundland and Tasmanian coasts had very low levels, at least of DDT (Anonymous, 1983; Muir *et al.*, 1988).

There is as well a wide variation in contaminant loads in short-finned pilot whales. High concentrations of DDT and PCB were found in whales off the Pacific coast of the USA in the mid 70s (O'Shea *et al.*, 1980), while low levels were found in whales from the Antilles and off Japan (Gaskin *et al.*, 1977; O'Shea *et al.*, 1980; Tanabe *et al.*, 1987).

Heavy metals such as cadmium and mercury also have been found in pilot whales from the Faroes (Simmonds *et al.*, 1994). Because these contaminants accumulate in tissues over time, older animals tend to have higher concentrations. Although not healthful for the animals carrying them, these pollutants may one day play a role in stock differentiation (Borrell and Aguilar, 1993; Caurant *et al.*, 1993).

Acknowledgments

Many colleagues and friends contributed their time and expertise to this document. We wish to thank the following people for kindly reviewing and improving the manuscript: I. Barrett, J. Carr, D. DeMaster, T. Kasuya, C. Lockyer, H. Marsh, W.F. Perrin, and F.G. Wood. Also P. Folkens, J. Heyning, C. Lockyer, and C. Potter assisted us in obtaining photographs or drawings.

References

Alagarswami, K., Bensam, P., Rajapandian, M. E. and Bastian Fernando, A. (1973). Mass stranding of pilot whales in the Gulf of Mannar. *Indian J. Fish.* **20**, 269–279.

Alzieu, C. and Duguy, R. (1979). Teneurs in composés organochlorés chez les cétacés et pinnipèdes fréquentant les côtes françaises. *Oceanolo. Acta* **2**, 107–120.

Amos, B., Schlotterer, C. and Tauz, D. (1993). Social structure of pilot whales revealed by analytical DNA profiling. *Science* **260**, 670–672.

Andersen, L. W. and Friedrich, U. (1988). The karyotype of the long-finned pilot whale, *Globicephala melaena. Hereditas* **109**, 245–251.

Anonymous (1983). Australia, progress report on cetacean research, June, 1981 to May, 1982. *Rep. Int. Whal. Commn* **33**, 191–194.

Barlow, J. and Gerrodette, T. (1996). Abundance of cetaceans in California waters based on 1991 and 1993 ship surveys. NOAA Technical Memo NMFS, NOAA-TM-NMFS-SWFSC-223.

Barlow, J., Forney, K. A., Hill, P. S., Brownell Jr., R. L., Caretta, J. V., DeMaster, D. P., Julian, F., Lowry, M. S., Ragen, T. and Reeves, R. R. (1997). U.S. Pacific marine mammal stock assesments: 1996. NOAA Technical Memo NMFS-SWFSC-248, 223 pp.

Barnes, L. G. (1976). Outline of eastern North Pacific fossil cetacean assemblages. *Syst. Zool.* **25**, 321–343.

Barnes, L. G. (1984). Whales, dolphins and porpoises: origin and evolution of cetacea. *In* "Mammals. Notes for a Short Course Organized by P. D. Gingerich and C. E. Badgley" (Ed. T. W. Broadhead) pp. 139–154. University of Tennessee Department of Geological Sciences, 234 pp.

Barnes, L. G., Domning, D. P. and Ray, C. E. (1986). Status of studies on fossil marine mammals. *Mar. Mamm. Sci.* **1**, 15–53.

Bloch, D. (1992). Studies on the long-finned pilot whale in the Faroes, 1976–1986. *Fróòskparrit* **38–39**, 35–61.

Bloch, D. and Lockyer, C. (1988). Killer whales (*Orcinus orca*) in Faroese waters. *Rit. Fiskideildar* **11**, 55–64.

Bloch, D., Lockyer, C. and Zachariassen, M. (1993a). Age and growth parameters of the long-finned pilot whale off the Faroe Islands. *In* "Biology of Northern Hemisphere Pilot Whales" (Eds G. P. Donovan, C. H. Lockyer and A. R. Martin) pp. 163–207. *Rep. Int. Whal. Commn*, Spec. Issue No. 14, Cambridge.

Bloch, D., Zachariassen, M. and Zachariassen, P. (1993b). Some external characters of the long-finned pilot whale off the Faroe Islands and a comparison with the short-finned pilot whale. *In* "Biology of Northern Hemisphere Pilot Whales" (Eds G. P.

Donovan, C. H. Lockyer and A. R. Martin) pp. 117–135. *Rep. Int. Whal. Commn*, Spec. Issue No. 14, Cambridge.

Bloch, D., Desportes, G., Hoydal, K. and Jean, P. (1990). Pilot whaling in the Faroe Islands: July 1986–July 1988. *North Atlantic Studies* **2**, 36–44.

Borrell, A. and Aguilar, A. (1993). DDT and PCB pollution in blubber and muscle of long-finned pilot whales from the Faroe Islands. *In* "Biology of Northern Hemisphere Pilot Whales" (Eds G. P. Donovan, C. H. Lockyer, and A. R. Martin) pp. 351–358. *Rep. Int. Whal. Commn*, Spec. Issue No. 14, Cambridge.

Bowers, C. A. and Henderson, R. S. (1972). Project Deep Ops: deep object recovery with pilot and killer whales. NUC Technical Paper 306, 86 pp.

van Bree, P. J. H. (1971). On *Globicephala seiboldi* Gray, 1846, and other species of pilot whales (Notes on Cetacea, Delphinoidea III). *Beaufortia* **19**, 79–87.

van Bree, P. J. H., Best, P. B. and Ross, G. J. B. (1978). Occurrence of the two species of pilot whales (genus *Globicephala*) on the coast of South Africa. *Mammalia* **42**, 323–328.

Brown, D. H. (1960). Behaviour of a captive Pacific pilot whale. *J. Mamm.* **41**, 342–349.

Brown, D. H. (1962). Further observations on the pilot whale in captivity. *Zoologica* **47**, 59–64.

Brown, D. H. and Norris, R. S. (1956). Observations of captive and wild cetaceans. *J. Mamm.* **37**, 311–326.

Brownell, R. (1974). Small odontocetes of the Antarctic. *In* "Antarctic Mammals" (Ed. V. Bushnell). American Geographic Society, New York, 19 pp.

Buck, J. D. (1984). Microbiological observations on two stranded live whales. *J. Wildl. Dis.* **20**, 148–150.

Buckland, S. T., Bloch, D., Cattanach, K. L., Gunnlaugsson, T., Hoydal, K., Lens, S. and Sigurjónsson, J. (1993). Distribution and abundance of long-finned pilot whales in the North Atlantic, estimated from NASS-87 and NASS-89 data. *In* "Biology of Northern Hemisphere Pilot Whales" (Eds G. P. Donovan, C. H. Lockyer and A. R. Martin) pp. 33–49. *Rep. Int. Whal. Commn*, Spec. Issue No. 14, Cambridge.

Caldwell, D. K. and Caldwell, M. C. (1975). Dolphin and small whale fisheries of the Caribbean and West Indies: occurrence, history and catch statistics with special reference to the Lesser Antillean island of St. Vincent. *J. Fish. Res. Bd Can.* **32**, 1105–1110.

Caldwell, M. C., Brown, D. H. and Caldwell, D. K. (1963). Intergeneric behavior by a captive Pacific pilot whale. *Los Angeles Co. Mus. Contr. Sci.* **70**, 3–12.

Caurant, F., Amiard-Triquet, C. and Amiard, J. -C. (1993). Factors influencing the accumulation of metals in pilot whales (*Globicephala melas*) off the Faroe Islands. *In* "Biology of Northern Hemisphere Pilot Whales" (Eds G. P. Donovan, C. H. Lockyer and A. R. Martin) pp. 369–390. *Rep. Int. Whal. Commn*, Spec. Issue No. 14, Cambridge.

Castello, H. P. and Pinedo, M. C. (1985). Report on some sea sightings of different cetacean species for southern Brazil. *In* Primero Reunion de Trabaio de Expertos de Mamiferos Acuaticos de America del Sur.

Clarke, R. (1981). Whales and dolphins of the Azores and their exploitation. *Rep. Int. Whal. Commn* **31**, 607–615.

Clarke, R., Aguayo, A. L. and Del Campo, S. B. (1978). Whale observation and whale marking off the coast of Chile in 1964. *Sci. Rep. Whales Res. Inst.* **30**, 117–177.

Cope, E. D. (1869). Introductory note (to the cetaceans of the western coast of North America by C. M. Scammon). *Proc. Acad. Nat. Sci. Philadelphia* **21**, 13–32.

Cowan, D. F. (1966a). Observations on the pilot whale *Globicephala melaena*: organ weight and growth. *Anat. Rec.* **155**, 623–628.

Cowan, D. F. (1966b). Pathology of the pilot whale — a comparative survey. *Arch. Pathol.* **82**, 178–189.

Dailey, M. D. and Brownell Jr, R. L. (1972). A checklist of marine mammal parasites. *In* "Mammals of the Sea: Biology and Medicine" (Ed. S. H. Ridgway) pp. 528–589. Charles C. Thomas, Springfield, IL.

Davies, J. L. (1960). The southern form of the pilot whale. *J. Mamm.* **41**, 29–34.

Davies, J. L. (1963). The antitropical factor in cetacean speciation. *Evolution* **17**, 107–116.

Dawson, S. M., Whitehouse, S. and Williscroft, M. (1985). A mass stranding of pilot whales in Tryphena Harbour, Great Barrier Island. *Invest. Cetacea* **17**, 165–173.

de Kock, L. L. (1956). The pilot whale stranding on the Orkney Island of Westray, 1955. *Scottish Naturalist* **68**, 63–70.

de Kock, L. L. (1969). The arterial vessels of the neck in the pilot whale (*Globicephala melaena* Traill) and the porpoise (*Phocoena phocoena*) in relation to the carotid body. *Acta Anat.* **36**, 274–292.

DeMaster, D., Miller, D., Henderson, J. R. and Coe, J. M. (1985). Conflicts between marine mammals and fisheries off the coast of California. *In* "Marine Mammals and Fisheries" (Eds J. R. Beddington, R. J. H. Beverton and D. M. Lavigne) pp. 111–118. Allen & Unwin, London.

Desportes, G. and Mouritsen, R. (1993). Diet of the pilot whale, *Globicephala melas*, around the Faroe Islands. Preliminary results. *In* "Biology of Northern Hemisphere Pilot Whales" (Eds G. P. Donovan, C. H. Lockyer and A. R. Martin) pp. 305–324. *Rep. Int. Whal. Commn*, Spec. Issue No. 14, Cambridge.

Desportes, G. (1994). Symmetry in testicular development in long-finned pilot whales. *Mar. Mamm. Sci.* **10**, 376–380.

Dohl, T. P., Norris, K. S., Guess, R. C., Bryant, J. D. and Honig, M. W. (1980). Final Contract Report AA550-CT7-36, to Minerals Management Survey, Vol. II, 414 pp.

Dohl, T. P., Guess, R. C., Duman, M. L. and Helm, R. C. (1983). Cetaceans of central and northern California, 1980–1983: status, abundance, and distribution. POCS, MMS, US Department of the Interior, Contract Report 14-12-0001-29090, 284 pp.

Donovan, G. P. (1984). Small cetaceans seen during the IWC/IDCR research cruise in the eastern tropical Pacific, and in particular off Peru. *Rep. Int. Whal. Commn* **34**, 561–567.

Donovan, G. P., Lockyer, C. H. and Martin, A. R. (1993). "Biology of Northern Hemisphere Pilot Whales". *Rep. Int. Whal. Commn*, Spec. Issue No. 14, Cambridge, 479 pp.

Dudok van Heel, W. H. (1962). Sound and Cetacea. *Netherlands J. Sea Res.* **1**, 407–507.

Duignan, P. J., House, C., Geraci, J. R., Early, G., Copland, H. G., Walsh, M. T., Bossart, G. D., Cray, C., Sadove, S., St. Aubin, D. J. and Moore, M. (1995). Morbillivirus infection in two species of pilot whales (*Globicephala* sp.) from the western Atlantic. *Mar. Mamm. Sci.* **11**, 150–162.

Elias, H., Lange, W., Huag, H., Schwartz, D. E. and Schlinska, G. (1968). Cerebro-cortical surface areas stereologically determined. *Anat. Rec.* **160**, 344.

Elsner, R., Franklin, D. L., Van Critters, R. L. and Kenney, D. W. (1966). Cardiovascular defense against asphyxia. *Science* **153**, 941–949.

Evans, W. E. (1967). Vocalization among marine mammals. *In* "Marine Bio-Acoustics", Vol. 2, pp. 159–186. Pergamon Press, Oxford.

Evans, W. E., Thomas, J. A. and Kent, D. B. (1986). A study of pilot whales (*Globicephala macrorhynchus*) in the southern California Bight. International Whaling Commission Scientific Committee Meeting Document SC38/SM3, 21 pp.

Fairfield, C. P., Waring, G. T. and Sano, M. H. (1993). Pilot whales incidentally taken during the distant water fleet Atlantic mackerel fishery in the mid-Atlantic bight, 1984–88. *In* "Biology of Northern Hemisphere Pilot Whales" (Eds G. P. Donovan, C. H. Lockyer and A. R. Martin) pp. 107–116. *Rep. Int. Whal. Commn*, Spec. Issue No. 14, Cambridge.

Felts, W. J. L. and Spurrell, F. A. (1965). Structural orientation and density in cetacean humeri. *Am. J. Anat.* **116**, 171–203.
Felts, W. J. L. and Spurrell, F. A. (1966). Some structural and developmental characteristics of cetacean (Odontocete) radii: study of adaptive osteogenesis. *Am. J. Anat.* **118**, 103–134.
Forney, K. A. (1994). Recent information on the status of odontocetes in Californian waters. NOAA Technical Memo NMFS-SWFSC-202, 87 pp.
Fraser, F. C. (1950). Two skulls of *Globicephala macrorhyncha* (Gray) from Dakar. *Atlantide Rep.* **1**, 49–60.
Gales, R. and Pemberton, D. (1992). Stomach contents of long-finned pilot whales (*Globicephala melas*) and bottlenose dolphins (*Tursiops truncatus*) in Tasmania. *Mar. Mamm. Sci.* **8**, 405–413.
Gaskin, D. E. (1968). The New Zealand Cetacea. *N.Z. Mar. Dept Fish. Res. Div. Bull.*, New Ser., **1**, 1–92.
Gaskin, D. E. (1972). "Whales, Dolphins and Seals". St. Martins Press, New York, 200 pp.
Gaskin, D. E. (1982). "The Ecology of Whales and Dolphins". Heinemann, London, 459 pp.
Gaskin, D. E. and Smith, G. J. D. (1977). The small whale fishery of St. Lucia, W. I. *Rep. Int. Whal. Commn* **27**, 493.
Gilmore, R. M. (1962). "Bubbles and Other Pilot Whales". Barley Brae Printers, Del Mar, CA, 16 pp.
Glezer, I. I. and Morgane, P. J. (1990). Ultrastructure of synapses and golgi analysis of neurons in neocortex of the lateral gyrus (visual cortex) of the dolphin and pilot whale. *Brain Res. Bull.* **24**, 401–427.
Glezer, I. I., Hof, P. R., Leranth, C. and Morgane, P. J. (1993). Calcium-binding protein-containing neuronal populations in mammalian visual cortex: A comparative study in whales, insectivores, bats, rodents and primates. *Cerebral Cortex* **3**, 249–272.
Glezer, I. I., Hof, P. R., Istomin, V. V. and Morgane, P. J. (1995). Comparative immunochemistry of calcium-binding protein positive neurons in visual and auditory systems of cetacean and primate brains. *In* "Sensory Systems of Aquatic Mammals" (Eds R. A. Kastelein, J. A. Thomas and P. E. Nachtigall) pp. 477–513. De Spil Publishers, Woerden, The Netherlands.
Goodall, R. N. P. and Galeazzi, A. R. (1986). A review of strandings of pilot whales *Globicephala melaena* south of 30°S, with emphasis on Tierra del Fuego. International Whaling Commission Scientific Meeting Document SC38/SMWP2, 21 pp.
Gray, J. E. (1846). Mammalia: on the cetaceous animals. *In* "The Zoology of the Voyage of H.M.S. Erebus and Terror under Command of Captain Sir James Clark Ross, R.N., F.R.S., during Years 1839 to 1843", Vol. 1 (Eds J. Richardson and J. E. Gray) pp. 13–53. Janson, London.
Hacker, E. S. (1986). Stomach content analysis of short-finned pilot whales (*Globicephala macrorhynchus*) and northern elephant seals (*Mirounqa anqustirostis*) from the southern California Bight. Southwest Fisheries Center Administration Report No. LJ-86-08C, 34 pp.
Hain, J. H. W. and Leatherwood, S. (1982). Two sightings of white pilot whales, *Globicephala melaena*, and summarized records of anomalously white cetaceans. *J. Mamm.* **63**, 338–343.
Harrison, R. S. and Young, B. A. (1969). Stellate cells in the delphinid adenohypophysis (of) *Globicephala scammoni, Delphinus bairdi, Lagenorhynchus obliquidens. J. Endocrinol.* **43**, 323–324.
Hashimoto, S., Dayton, S. and Roberts Jr, J. C. (1967). Aliphate wax alcohols and other lipids in atheromota and arterial tissues of cetaceans. *Comp. Biochem. Physiol.* **20**, 975–986.

Haug, H. (1987). Brain sizes, surfaces, and neuronal sizes of the cortex cerebri: a stereological investigation of man and his variability and a comparison with some mammals (Primates, whales, marsupials, insectivores, and one elephant). *Am. J. Anat.* **180**, 126–142.

Hay, R. (1982). Aerial line-transect estimates of abundance of humpback, fin, and long-finned pilot whales in the Newfoundland–Labrador area. *Rep. Int. Whal. Commn* **32**, 475–480.

Higgins, R., Claveau, R. and Roy, R. (1980). Bronchopneumonia caused by *Streptococcus equi* in a North Atlantic pilot whale (*Globicephala melaena*). *J. Wildl. Dis.* **16**, 319–321.

Hinshaw, V. S., Bean, W. J., Geraci, J., Fiorelli, P., Early, G. and Webster, R. G. (1986). Characterization of two influenza A viruses from a pilot whale. *J. Virol.* **58**, 655–656.

Hof, P. R., Glezer, I. I., Revischin, A. V., Bouras, C., Charnay, Y. and Morgane, P. J. (1995). Distribution of dopaminergic fibers and neurons in visual auditory cortices of the harbor porpoise and pilot whale. *Brain Res. Bull.* **36**, 275–284.

Hoydal, K. (1986). Data on the long-finned pilot whale (*Globicephala melaena*, Traill), in Faroe waters and an attempt to use the 274 years time series of catches to assess the state of the stock. International Whaling Commission Scientific Committee Meeting Document SC/38/SM7, 27 pp.

Hubbs, C. L. (1952). Antitropical distribution of fishes and other organisms. *In* "Proceedings of the Seventh Pacific Scientific Congress of the Pacific Science Association", Vol. III, "Meteorology and Oceanography", pp. 324–329.

Hui, C. A. (1985). Undersea topography and the comparative distributions of two pelagic cetaceans. *Fish. Bull.* **83**, 472–475.

International Whaling Commission (1988). Report of the small cetacean subcommittee. *Rep. Int. Whal. Commn* **38**, 117–125.

International Whaling Commision (1992). Report of the small cetaceans subcommittee. *Rep. Int. Whal. Commn* **42**, 178–234.

Joensen, J. S. and Zachariassen, P. (1982). Grindatol 1584–1640 og 1709–1978. *Serprent Frodskaparrit* **30**, 71–102.

Jones Jr, J. K., Carter, D. C., Genoways, H. H., Hoffmann, R. S., Rice, D. W. and Jones, C. (1986). Revised checklist of North American mammals north of Mexico, 1986. *Occ. Papers Mus. Texas Tech. Univ.* **107**, 5.

Kapel, F. O. (1975). Preliminary notes on the occurrence and exploitation of smaller Cetacea in Greenland. *J. Fish. Res. Bd Can.* **32**, 1079–1082.

Kasamatsu, F. and Joyce, G. G. (1995). Current status of odontocetes in the Antarctic. *Antarct. Sci.* **7**, 365–379.

Kasuya, T. (1975). Past occurrence of *Globicephala melaena* in the western North Pacific. *Sci. Rep. Whales Res. Inst.* **27**, 95–110 + 3 plts.

Kasuya, T. (1986a). Comparison of life history parameters between two stocks of short-finned pilot whales off the Pacific coast of Japan. International Whaling Commission Scientific Committee Meeting Document SC/38/SM10, 32 pp.

Kasuya, T. (1986b). Japanese fisheries exploiting southern form short-finned pilot whales. International Whaling Commission Scientific Committee Meeting Document SC/38/SM19, 12 pp.

Kasuya, T. and Marsh, H. (1984). Life history and reproductive biology of the short-finned pilot whale, *Globicephala macrorhynchus*, off the Pacific coast of Japan. *In* "Reproduction in Whales, Dolphins and Porpoises" (Eds W. F. Perrin, R. L. Brownell and D. P. DeMaster) pp. 259–310. *Rep. Int. Whal. Commn*, Spec. Issue No. 6, Cambridge.

Kasuya, T. and Matsui, S. (1984). Age determination and growth of the short-finned pilot whale off the Pacific coast of Japan. *Sci. Rep. Whales Res. Inst.* **35**, 57–91.

Kasuya, T. and Tai, S. (1986). Operation of Japanese small-type whaling on the short finned pilot whale off northeastern Japan. International Whaling Commission Scientific Committee Meeting Document SC/38/SM16.

Kasuya, T. and Tai, S. (1993). Life history of short-finned pilot whale stocks off Japan and a description of the fishery. *In* "Biology of Northern Hemisphere Pilot Whales" (Eds G. P. Donovan, C. H. Lockyer and A. R. Martin) pp. 439–479. *Rep. Int. Whal. Commn*, Spec. Issue No. 14, Cambridge.

Kasuya, T., Tobayama, T. and Matsui, S. (1984). Review of live-capture of small cetaceans in Japan. *Rep. Int. Whal. Commn* **34**, 597–602.

Kasuya, T., Miyashita, T. and Kasamatsu, F. (1986a). Segregation of two types of short-finned pilot whales off the Pacific coast of Japan. International Whaling Commission Scientific Committee Meeting Document SC/38/SM12, 12 pp.

Kasuya, T., Sergeant, D. E. and Tanaka, K. (1986b). Re-examination of some life history parameters of *Globicephala melaena* in Newfoundland waters. International Whaling Commission Scientific Committee Meeting Document SC/38/SM11, 18 pp.

Kasuya, T., Miyashita, T. and Kasamatsu, F. (1988). Segregation of two forms of short-finned pilot whales off the Pacific Coast of Japan. *Sci. Rep. Whales Res. Inst. Tokyo* **39**, 77–90.

Kirschvink, J. L., Dizon, A. E. and Westphal, J. A. (1986). Evidence from strandings for geomagnetic sensitivity in cetaceans. *J. Exp. Biol.* **120**, 1–24.

Kritzler, H. (1952). Observations on the pilot whale in captivity. *J. Mamm.* **33**, 321–334.

Kulu, D. D. (1972). Evolution and cytogenetics. *In* "Mammals of the Sea: Biology and Medicine" (Ed. S. H. Ridgway) pp. 503–527. Charles C. Thomas, Springfield, IL.

Ketten, D. R. and Wartzok, D. (1990). Three-dimensional reconstructions of the dolphin ear. *In* "Sensory Abilities of Cetaceans: Laboratory and Field Evidence" (Eds J. A. Thomas and R. A. Kastelein) pp. 81–105. Plenum Press, New York.

Lankester, E. R. (1864). On new mammalia from the red crag. *Annals ant. Mag. Nat. Hist.*, Third Series, **14**, 353–360.

Lawrence, B. and Schevill, W. S. (1965). The functional anatomy of the delphinid nose. *Bull. Mus. Comp. Zool. Harvard* **114**, 103–151.

Leatherwood, J. S. (1974). A note on gray whale behavioral interactions with other marine mammals. *Mar. Fish. Rev.* **36**, 50.

Leatherwood, J. S. and Dahlheim, M. E. (1978). Worldwide distribution of pilot whales and killer whales. Naval Ocean Systems Center Technical Note 443, 39 pp.

Leatherwood, J. S., Evans, W. E. and Lingle, G. E. (1973). The Pacific pilot whale, *Globicephala* cf. *G. scammoni*. Evaluation of its effect on high frequency sonar and its utility as an oceanographic monitoring platform. Naval Underseas Center Technical Note 933, 48 pp.

Leatherwood, J. S., Caldwell, D. K. and Winn, H. (1976). Whales, dolphins and porpoises of the western north Atlantic: a guide to their identification. NOAA Technical Report NMFS Circular-396, 176 pp.

Lesson, R. (1828). "Complement des Oeuvres de Buffon ou Histoire Naturelle des Animaux rares decouverts par les Naturalists et les Voyageurs depuis la Mort de Buffon", Vol. 1, "Cetaces". Paris, 442 pp + 12 plts.

Lockyer, C. (1993a). A report on patterns of deposition of dentine and cement in teeth of pilot whales, genus *Globicephala*. *In* "Biology of Northern Hemisphere Pilot Whales" (Eds G. P. Donovan, C. H. Lockyer and A. R. Martin) pp. 137–161. *Rep. Int. Whal. Commn*, Spec. Issue No. 14, Cambridge.

Lockyer, C. (1993b). Seasonal changes in body fat condition of Northeast Atlantic pilot whales, and their biological significance. *In* "Biology of Northern Hemisphere Pilot Whales" (Eds G. P. Donovan, C. H. Lockyer and A. R. Martin) pp. 325–350. *Rep. Int. Whal. Commn*, Spec. Issue No. 14, Cambridge.

Marsh, H. and Kasuya, T. (1984). Changes in the ovaries of the short-finned pilot whale, *Globicephala macrorhynchus*, with age and reproductive activity. *Rep. Int. Whal. Commn*, Spec. Issue No. 6, pp. 311–335.

Marsh, H. and Kasuya, T. (1991). An overview of the changes in the role of a female pilot whale with age. *In* "Dolphin Societies—Discoveries and Puzzles" (Eds K. Pryor and K. S. Norris) pp. 281–285. University of California Press, Berkeley, CA.

Martin, A. R. and Desportes, G. (1987). Preliminary studies of pilot whales from Faroese waters since 1986: reproduction and foetal growth. International Whaling Commission Scientific Meeting Document SC/39/SM15.

Martin, A. R. and Rothery, P. (1993). Reproductive parameters of female long-finned pilot whales (*Globicephala melas*) around the Faroe Islands. *In* "Biology of Northern Hemisphere Pilot Whales" (Eds G. P. Donovan, C. H. Lockyer and A. R. Martin) pp. 263–304. *Rep. Int. Whal. Commn*, Spec. Issue No. 14, Cambridge.

Martin, A. R., Reynolds, P. and Richardson, M. G. (1986). Aspects of the biology of pilot whales (*Globicephala melaena*) in recent mass strandings on the British coast. International Whaling Commission Scientific Meeting Document SC/38/SM6, 29 pp.

Mead, J. G. (1975). Anatomy of the external nasal passages and facial complex in the Delphinidae (Mammalia: Cetacea). *Smithsonian Contrib. Zool.* **207**, 1–72.

Medway, W. (1978). Suppurative meningoencephalitis in a stranded pilot whale (*Globicephala melaena*). *Aquat. Mamm.* **6**, 99–100.

Medway, W. and Moldovan, F. (1966). Blood studies on the north Atlantic pilot (pothead) whale, *Globicephala melaena* (Traill, 1809). *Physiol. Zool.* **39**, 110–116.

Mercer, M. C. (1975). Modified Leslie–DeLury population models of the long-finned pilot whale (*Globicephala melaena*) and annual production of the short-finned squid (*Illex illecebrosus*) based upon their interaction at Newfoundland. *J. Fish. Res. Bd Can.* **32**, 1145–1154.

Mitchell, E. D. (1975a). Porpoise, dolphin and small whale fisheries of the world, status and problems. *Int. Union Conserv. Nat. Resour. Morges Monogr.* **3**, 1–129.

Mitchell, E. D. (1975b). Report of the Meeting on Smaller Cetaceans, Montreal, 1–11 April 1979, Subcommittee on Small Cetaceans, Scientific Committee, International Whaling Commission. *J. Fish. Res. Bd Can.* **32**, 889–983.

Miller, D. J., Herder, M. J. and Scholl, J. P. (1983). California marine mammal–fishery interaction study, 1979–1981. Southwest Fisheries Center Administration Report No. LJ-83-13C, 233 pp.

Miyashita, T. (1986). Abundance of some globicephalid cetaceans in the adjacent waters of Japan. International Whaling Commission Scientific Committee Meeting Document SC/38/SM17, 18 pp.

Morgane, P. J. and Glezer, I. I. (1990). Sensory neocortex in dolphin brain. *In* "Sensory Abilities of Cetaceans: Laboratory and Field Evidence" (Eds J. A. Thomas and R. A. Kastelein) pp. 107–136. Plenum Press, New York.

Muir, D. G. C., Wagemann, R., Grift, N. P., Norstrom, R. J., Simon, M. and Lien, J. (1988). Organochlorine chemical and heavy metal contaminants in white-beaked dolphins (*Lagenorhynchus albirostris*) and pilot whales (*Globicephala melaena*) from the coast of Newfoundland, Canada. *Arch. Environ. Contam. Toxicol.* **17**, 613–629.

Murie, J. (1874). On the organization of the Caa'ing whale, *Globicephalus melas*. *Trans. Zool. Soc. Lond.* **8**, 235–301 + 9 plts.

Nelson, D. and Lien, J. (1996). The status of the long-finned pilot whale, *Globicephala melas*, in Canada. *Can. Field-Naturalist* **110**, 511–524.

Nishiwaki, M. (1972). General biology. *In* "Mammals of the Sea: Biology and Medicine" (Ed. S. H. Ridgway) pp. 3–204. Charles C. Thomas, Springfield, IL.

Norris, K. S. and Prescott, J. H. (1961). Observations on Pacific cetaceans of California and Mexican waters. *Univ. Calif. Berkeley Pub. Zool.* **63**, 291–402.

Oelschlager, H. A. (1990). Evolutionary morphology and acoustics in the dolphin skull. *In* "Sensory Abilities of Cetaceans: Laboratory and Field Evidence" (Eds J. A. Thomas and R. A. Kastelein) pp. 137–162. Plenum Press, New York.

Olsen, C. R., Hale, F. C., Elsner, R. and Kenney, D. W. (1967). Effects of hypercapnia on airways and ventilation in the pilot whale. *Physiologist* **10**, 267.

Olsen, C. R., Hale, F. C., Elsner, R. and Kenney, D. W. (1969). Mechanics of ventilation in the pilot whale. *Respir. Physiol.* **7**, 137–149.

O'Riordan, C. E. (1975). Long-finned pilot whales, *Globicephala melaena*, driven ashore in Iceland, 1800–1973. *J. Fish. Res. Bd Can.* **32**, 1101–1103.

O'Shea, T. J., Brownell, R. L., Clark, D. R., Walker, W. A., Gay, M. L. and Lamont, T. G. (1980). Organochlorine pollutants in small cetaceans from the Pacific and South Atlantic Oceans, November 1968–June 1976. *Pest. Monit. J.* **14**, 35–46.

Payne, P. M. and Heinemann, D. W. (1993). The distribution of pilot whales *Globicephala* spp. in the shelf-edge and slope waters of the northeastern United States, 1978–1988. *In* "Biology of Northern Hemisphere Pilot Whales" (Eds G. P. Donovan, C. H. Lockyer and A. R. Martin) pp. 51–68. *Rep. Int. Whal. Commn*, Spec. Issue No. 14, Cambridge.

Perrin, W. F. (1994). Report of the workshop on mortality of cetaceans in passive fishing nets and traps. *In* "Gillnets and Cetaceans" (Eds W. F. Perrin, G. P. Donovan and J. Barlow) pp. 6–71. *Rep. Int. Whal. Commn*, Spec. Issue No. 15, Cambridge.

Perrin, W. F. and Reilly, S. B. (1984). Reproductive parameters of dolphins and small whales of the family Delphinidae. *In* "Reproduction in Whales, Dolphins and Porpoises" (Eds W. F. Perrin, R. L. Brownell and D. P. DeMaster) pp. 97–133. *Rep. Int. Whal. Commn*, Spec. Issue No. 6, Cambridge.

Pilson, M. E. Q. (1970). Pilot whale tears. *J. Mamm.* **51**, 625–626.

Polisini, J. M. (1980). A comparison of *Globicephala macrorhynchus* (Gray, 1846) with the pilot whale of the north Pacific ocean: an analysis of the skull of the broad-rostrum pilot whales of the genus *Globicephala*. Ph.D. thesis, University of Southern California, 299 pp.

Raga, J. A. and Balbuena, J. A. (1993). Parasites of the long-finned pilot whale, *Globicephala melas* (Traill, 1809), in European waters. *In* "Biology of Northern Hemisphere Pilot Whales" (Eds G. P. Donovan, C. H. Lockyer and A. R. Martin) pp. 391–406. *Rep. Int. Whal. Commn*, Spec. Issue No. 14, Cambridge.

Reeves, R. R. and Leatherwood, S. (1984). Live-capture fisheries for cetaceans in USA and Canadian waters, 1973–1982. *Rep. Int. Whal. Commn* **34**, 497–507.

Reijnders, P. J. H. (1989). The cumulative role of pollution in marine mammal habitat degradation. *In* Abstracts of the 8th Biennial Conference on the Biology of Marine Mammals, Pacific Grove, CA.

Reilly, S. B. (1977). The distribution of pilot whales *Globicephala macrorhynchus* Gray, 1896, in the eastern tropical Pacific. M.S. thesis, California Polytechnic State University, San Luis Obispo, CA, 79 pp.

Rice, D. W. (1977). A list of marine mammals of the world. NOAA Technical Report NMFS SSRF-711, 15 pp.

Rice, D. W. and Scheffer, V. B. (1968). A list of marine mammals of the world. *U.S. Fish Wildl. Serv. Spec. Rep. Fish.* 579.

Ridgway, S. H. (1972). Homeostasis in the aquatic environment. *In* "Mammals of the Sea: Biology and Medicine" (Ed. S. H. Ridgway) pp. 590–748. Charles C. Thomas, Springfield, IL.

Ridgway, S. H. and Patton, G. S. (1971). Dolphin thyroid: some anatomical and physiological findings. *Z. vergl. Physiol.* **71**, 129–141.

Ridgway, S. H., Simpson, J. G., Patton, G. S. and Gilmartin, W. G. (1970). Hematologic findings in certain small cetaceans. *J. Am. Vet. Med. Assoc.* **157**, 566–575.
Ridgway, S. H. and Brownson, R. H. (1984). Relative brain sizes and cortical surface areas in odontocetes. *Acta Zool. Fenn.* **172**, 149–152.
Ross, G. J. B. (1984). The smaller cetaceans of the south east coast of southern Africa. *Ann. Cape Prov. Mus. (Nat. Hist.)* **15**, Part 2, 410 pp.
Sellards, E. H. (1960). Fossil vertebrates from Florida: a new Miocene fauna; new Pliocene species; the Pleistocene fauna. Florida State Geographical Survey Eighth Annual Report, pp. 77–119.
Sergeant, D. E. (1962a). The biology of the pilot or pothead whale *Globicephala melaena* (Traill) in Newfoundland waters. *Bull. Fish. Res. Bd Can.* **132**, 1–84.
Sergeant, D. E. (1962b). On the external characters of the blackfish or pilot whales (Genus *Globicephala*). *J. Mamm.* **43**, 395–413.
Sergeant, D. E. (1982). Mass strandings of toothed whales (Odontoceti) as a population phenomenon. *Sci. Rep. Whales Res. Inst.* **34**, 1–47.
Sergeant, D. E. (1986). Possible connexion of pilot whale (*Globicephala melaena*) populations at Newfoundland and the Faroe Islands. International Whaling Commission Scientific Committee Meeting Document SC/38/SM9, 6 pp.
Shane, S. H. (1986). Detailed observations of a single pod of pilot whales and sightings of other marine mammals at Catalina Island, California in winter, 1984–5. Southwest Fisheries Center Contract No. 40-ABNF 5-702, 49 pp.
Shane, S. H. and Costello, L. (1993). Life-threatening contact between a woman and a pilot whale captured on film. *Mar. Mamm. Sci.* **9**, 331–336.
Shurley, J. T., Serafetinides, E. A., Brooks, R. E., Elsner, R. and Kenney, D. W. (1969). Sleep in cetaceans: I. The pilot whale, *Globicephala scammoni*. *Psychobiology* **6**, 230.
Sigurjónsson, J., Vikingsson, G. and Lockyer, C. (1989). On two recent mass strandings of pilot whales (*Globicephala melas*) on the west and southwest coast of Iceland. International Whaling Commission Scientific Committee Meeting Document SC/41/SM26.
Simmonds, M. P., Johnston, P. A., French, M. C., Reeve, R. and Hutchinson, J. D. (1994). Organochlorines and mercury in pilot whale blubber consumed by Faroe islanders. *Sci. Total Environ.* **149**, 97–111.
Simpson, J. G. and Gardner, M. B. (1972). Comparative microscopic anatomy of selected marine mammals. *In* "Mammals of the Sea: Biology and Medicine" (Ed. S. H. Ridgway) pp. 298–418. Charles C. Thomas, Springfield, IL.
Tanabe, S., Loganathan, B. G., Subramanian, A. and Tatsukawa, R. (1987). Organochlorine residues in short-finned pilot whales. Possible use as tracers of biological parameters. *Mar. Poll. Bull.* **18**, 561–563.
Tarpley, R. J. and Ridgway, S. H. (1994). Corpus callosum size in delphinid cetaceans. *Brain Behav. Evol.* **44**, 156–165.
Taruski, A. G. (1979). The whistle repertoire of the North Atlantic pilot whale (*Globicephala melaena*) and its relationship to behavior and environment. *In* "Behavior of Marine Animals", Vol. 3, "Cetaceans" (Eds H. E. Winn and B. L. Olla) pp. 345–368. Plenum Press, New York.
Tomilin, A. G. (1967). "Mammals of the USSR and Adjacent Countries", Vol. IX, "Cetacea" (Ed. V. G. Heptner). Nauk USSR, Moscow (English translation, Israel Program for Scientific Translations, Jerusalem).
Traill, T. S. (1809). Description of a new species of whale, *Delphinus melas*. *Nicholson's J. Nat. Phil. Chem. Arts* **22**, 81–83.
Van Waerebeek, K. and Reyes, J. C. (1986). The first records of the short-finned pilot whale *Globicephala macrorhynchus* for Peruvian waters. *In* Abstracts of the Fourteenth

Symposium of the European Association for Aquatic Mammals, Barcelona, 16–19 March.

Wada, S. (1986). Genetic differentiation between two forms of short-finned pilot whales off the Pacific coast of Japan. International Whaling Commission Scientific Committee Meeting Document SC/38/SM18, 10 pp.

Wade, P. W. and Gerrodette, T. (1993). Estimates of cetacean abundance and distribution in the eastern tropical Pacific. *Rep. Int. Whal. Commn* **43**, 477–493.

Walen, K. and Modin, S. (1965). Comparative chromosome analysis of the bottlenosed dolphin (*Tursiops truncatus*) and the pilot whale (*Globicephala scammoni*). *Am. Nat.* **99**, 349–354.

Walker, W. A. (1975). Review of the live-capture fishery for smaller cetaceans taken in southern California waters for public display, 1966–73. *J. Fish. Res. Bd Can.* **32**, 1197–1211.

Waring, G. T., Gerrior, P., Payne, P. M., Parry, B. L. and Nicholas, J. R. (1990). Incidental take of marine mammals in foreign fishery activities off the northeast United States, 1977–1988. *Fish. Bull. U.S.* **88**, 347–360.

Wedmid, Y., Litchfield, C., Ackman, R. G., Sipos, J. C., Eaton, C. A. and Mitchell, E. D. (1973). Heterogeneity of lipid composition within the cephalic melon tissue of the pilot whale (*Globicephala melaena*). *Biochim. Biophys. Acta* **326**, 439–447.

Weller, D. W., Würsig, B., Whitehead, H., Norris, J. C., Lynn, S. K., Davis, R. W., Clauss, N. and Brown, P. (1996). Observations of an interaction between sperm whales and short-finned pilot whales in the Gulf of Mexico. *Mar. Mamm. Sci.* **12**, 588–594.

Yonekura, M., Matsui, S. and Kasuya, T. (1980). On the external characters of *Globicephala macrorhynchus* off Taiji, Pacific coast of Japan. *Sci. Rep. Whales Res. Inst.* **32**, 67–95 + 2 plts.

11

Killer Whale
Orcinus orca (Linnaeus, 1758)

Marilyn E. Dahlheim and John E. Heyning

Genus and Species

Taxonomy

Various scientific names have been assigned to the killer whale (Hershkovitz, 1966; Heyning and Dahlheim, 1988). These numerous names can be explained by sexual and age differences in the size of the dorsal fin, individual variations in colour patterns, and the cosmopolitan distribution of the species. Currently the genus *Orcinus* is considered monotypic, with geographical variation noted in size and pigmentation patterns. Two recently described Antarctic species, *Orcinus nanus* (Mikhalev *et al.*, 1981) and *O. glacialis* (Berzin and Vladimirov, 1982, 1983), both appear to refer to the same population of smaller individuals. The data presented to support a new species indicate modal differences in several aspects of morphology and ecology, but are based on extremely small sample sizes. A worldwide review of specimens is needed to document geographical variation.

The killer whale is the largest species within the family Delphinidae. The genus *Orcinus* has been placed in the subfamily Orcininae or Globicephalinae by several taxonomists. This taxon is inconsistent in its membership and may include *Orcinus*, *Pseudorca*, *Globicephala*, *Orcaella*, and *Feresa* (Fraser and Purves, 1960; Mead, 1975; Slijper, 1936) or only *Pseudorca* and *Orcinus* (Kasuya, 1973). A cladistic analysis of the myoglobin amino acid sequence indicates that *O. orca* is

Handbook of Marine Mammals, Volume 6
ISBN 0-12-588506-7

more similar to *Globicephala melas* than to *Stenella attenuata*, *Delphinus delphis*, or *Tursiops truncatus* (Meuth *et al.*, 1981).

Evolution

Numerous large delphinoid teeth, primarily from the Pliocene, have been attributed to species within the genus *Orcinus*. Such teeth have been reported from the Pliocene of Italy (Sarra, 1933) and Japan (Matsumoto, 1937). One of the few fossils represented by a good skull is *O. citoniensis* (Capellini, 1883) from the Pliocene of Italy. It appears to be a smaller species than *O. orca* with an estimated total length of less than 4 m. *O. citoniensis* has a slightly higher tooth count (14/14) and proportionately smaller teeth than the extant species. In these characters, *O. citoniensis* seems to be intermediate between the more typical delphinids and *O. orca* (Pilleri and Pilleri, 1982).

Common names

The name "killer whale" originated from this species' predatory habits; it is known to feed on baleen whales and dolphins. Early whalers referred to this animal as the "whale killer". Through the years, this was transposed to killer whale. "Orca" is Latin for a kind of whale and "Orcinus" is derived from the Latin "Orcus" meaning of the netherworld. Other common names include orca, blackfish (English), kasatka, kosatka (Russian), sakamata, shachi (Japanese), epaulard, orque (French), hahyrna, sverdifscur, huyding (Icelandic), spekkhugger, stourvagn, starhynning, staurhval, vaghund (Norwegian), spächuggare (Swedish), zvarrdwalvis (Dutch), and swordfish (Newfoundland).

Distribution

Range

Killer whales are the world's most widely distributed mammal (Leatherwood and Dahlheim, 1978; Heyning and Dahlheim, 1988; Fig. 1). Although observed in tropical waters and the open sea, they are most abundant in coastal habitats and high latitudes. Killer whales inhabiting coastal areas often enter shallow bays, estuaries, and river mouths (Leatherwood *et al.*, 1976).

The killer whale's occurrence has been documented as far north as the Chukchi and Beaufort seas in the Arctic Ocean (Lowry *et al.*, 1987). In the northeastern Pacific Ocean, killer whales occur in the eastern Bering Sea, near the Aleutian Islands, south of the Alaskan Peninsula, and waters surrounding Kodiak Island (Scammon, 1874; Murie, 1959; Braham and Dahlheim, 1982; Dahlheim and Waite, 1993; Dahlheim, 1994). Their occurrence in Prince William Sound (South central Alaska) is well documented (Hall, 1981; Braham and Dahlheim,

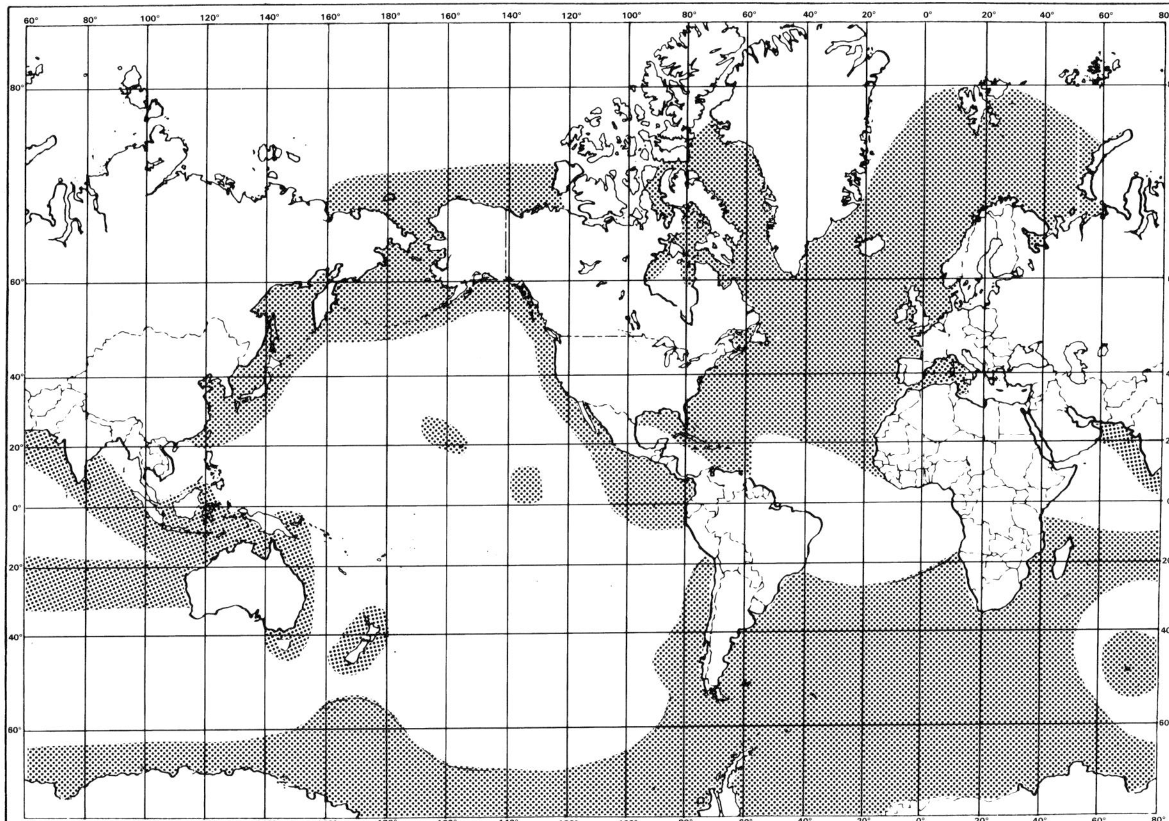

FIG. 1 Geographical distribution of *Orcinus orca* based on documented records. Killer whales most probably occur in all oceans and most seas.

1982; Leatherwood *et al.*, 1990a). They occur frequently in the waters of southeastern Alaska (Scheffer, 1967; Dahlheim *et al.*, 1997) and the intracoastal waterways of British Columbia and Washington State (Balcomb and Goebel, 1976; Bigg *et al.*, 1987; Osborne *et al.*, 1988). There are scattered reports of occurrences along the coasts of Washington, Oregon, and California (Norris and Prescott, 1961; Fiscus and Niggol, 1965; Rice, 1968; Gilmore, 1976; Black *et al.*, 1993), both coasts of Baja California (Dahlheim *et al.*, 1982), the offshore tropical Pacific (Dahlheim *et al.*, 1982), the Gulf of Panama, and the Galapagos Islands. In the western North Pacific, killer whales occur frequently along the Soviet coast in the Bering Sea, the Sea of Okhotsk, the Sea of Japan, and along the eastern side of Sakhalin and the Kuril Islands (Tomilin, 1957). There are numerous accounts of their occurrence off China (Wang, 1985) and Japan (Nishiwaki and Handa, 1958; Kasuya, 1971; Ohsumi, 1975). Data from the central Pacific are scarce. They have been reported off Hawaii, but do not appear to be abundant in these waters (Tomich, 1986).

Killer whales are widely distributed in the North Atlantic Ocean (Reeves and Mitchell, 1988a; Hammond and Lockyer, 1988). Killer whales are noted off Greenland (Heide-Jörgensen, 1988), Iceland (Sigurjónsson *et al.*, 1988), and in the Barents and White Seas (Tomilin, 1957). Many records document their presence off Norway (Christensen, 1988; Lien *et al.*, 1988a; Lyrholm, 1988; and Oien, 1988), the Faroe Islands (Bloch and Lockyer, 1988), the UK and Ireland (Evans, 1988), Denmark, Holland, Belgium, and France. Reports from the Mediterranean are sparse (McBrearty *et al.*, 1986). Records are available from Western Africa (Cadenat, 1959). In the western North Atlantic, their occurrence has been recorded off Baffin Island and in Lancaster Sound (Sergeant and Fisher, 1957; Reeves and Mitchell, 1988b), and south along the eastern coasts of Canada and the United States (Katona *et al.*, 1988; Lien *et al.*, 1988b; Mitchell and Reeves, 1988). They occur in the Gulf of Mexico, but are considered uncommon at least along continental shelf waters (Caldwell *et al.*, 1956; O'Sullivan and Mullin, 1977). Reports from Florida (Moore, 1953), the West Indies, the Bahamas (Backus, 1961), and St. Vincent are noted.

In the southern oceans, killer whales are commonly seen off Australia and New Zealand (Iredale and Troughton, 1969; Baker, 1983). Sporadic occurrence is known off Ecuador and Peru, with greater abundance off the Chilean coast (Aguayo, 1975). In the South Atlantic, they commonly occur among the South Shetland Islands and the South Orkneys (Gaskin, 1968). Their occurrence off Uruguay is uncommon (Ximinez *et al.*, 1972). Killer whales have been observed off South Africa (Ross, 1984), but records become fewer northward along the coastline of Africa. Based on sighting data, killer whales are widely distributed in the South Atlantic and adjacent Antarctic waters (Budylenko, 1980). Records from the Indian Ocean include catch reports from the Natal coast of South Africa (Best and Ross, 1977) and sightings off the Crozet Islands (Voisin, 1972; Guinet, 1988) and Kerguelen Islands (Robineau, 1989). Leatherwood *et al.*

(1990b) summarized sighting data throughout the Indian Ocean. Mid-ocean distribution in the Southern Hemisphere is poorly known.

In Antarctic waters, *Orcinus* has been recorded as far south as the Ross Sea (Brown *et al.*, 1974). It is frequently observed along the edge of the pack ice throughout Antarctic waters (Brown *et al.*, 1974; Berzin and Vladimirov, 1983; Kasamatsu *et al.*, 1987), in sharp contrast to its distribution in the Arctic.

External Characteristics

Colour pattern

Killer whales are one of the most strikingly pigmented of all cetaceans (Fig. 2), making field identification easy. Killer whales are black dorsally and white ventrally. The white region extends from the tip of the lower jaw posteriorly, constricts medially between the flippers, widens slightly posteriorly and ends just caudal of the urogenital region. A lateral white flank patch that extends dorso-posteriorly above the urogenital region is continuous with the ventral white area. The ventral aspect of the flukes is white or light grey, and may be bordered in black. A conspicuous white oval patch is located slightly above and behind the eye. A highly variable grey or white saddle is usually present posterior to the dorsal fin. Saddle shape varies among individuals, pods, and from one side to the other on a single animal (Baird and Stacey, 1988). The saddle patch is indistinct in young individuals, becoming more obvious as the animal matures.

In young animals, white areas are yellowish in colour (Scheffer, 1971). Animals born at Sea World and Marineland had "white" areas that were described as light orange to ochre coloration (B. Andrews, personal communication). The yellowish hue has been reported also on adult whales from the Antarctic (Berzin and Vladimirov, 1982, 1983; Evans *et al.*, 1982) and on some animals in the North Pacific (Scammon, 1874; Scheffer and Slipp, 1948; M. Dahlheim, personal observations). In the case of adults, yellowish coloration may be due to diatom deposits on the skin (Hart, 1935). Melanistic individuals have been seen in the North Pacific (Scammon, 1874; Scheffer and Slipp, 1948), as well as partially albinistic animals (Carl, 1960; Scheffer and Slipp, 1948). Haley (1973) reported one killer whale which was white as a result of Chediak–Higashi Syndrome. The scientific documentation of the syndrome was provided by Taylor and Farrell (1973).

Individual and geographical variations in the pigmentation pattern are well established (Carl, 1946; Evans *et al.*, 1982). Many individuals in the Antarctic have lighter pigmented skin revealing the primitive delphinid "cape" pattern (Fig. 3) defined by Perrin (1972). This cape is present on at least some foetuses from the North Atlantic (Guldberg and Nansen, 1894: Fig. 3, Plate 5). Observations on a captive held North Pacific and North Atlantic whale also

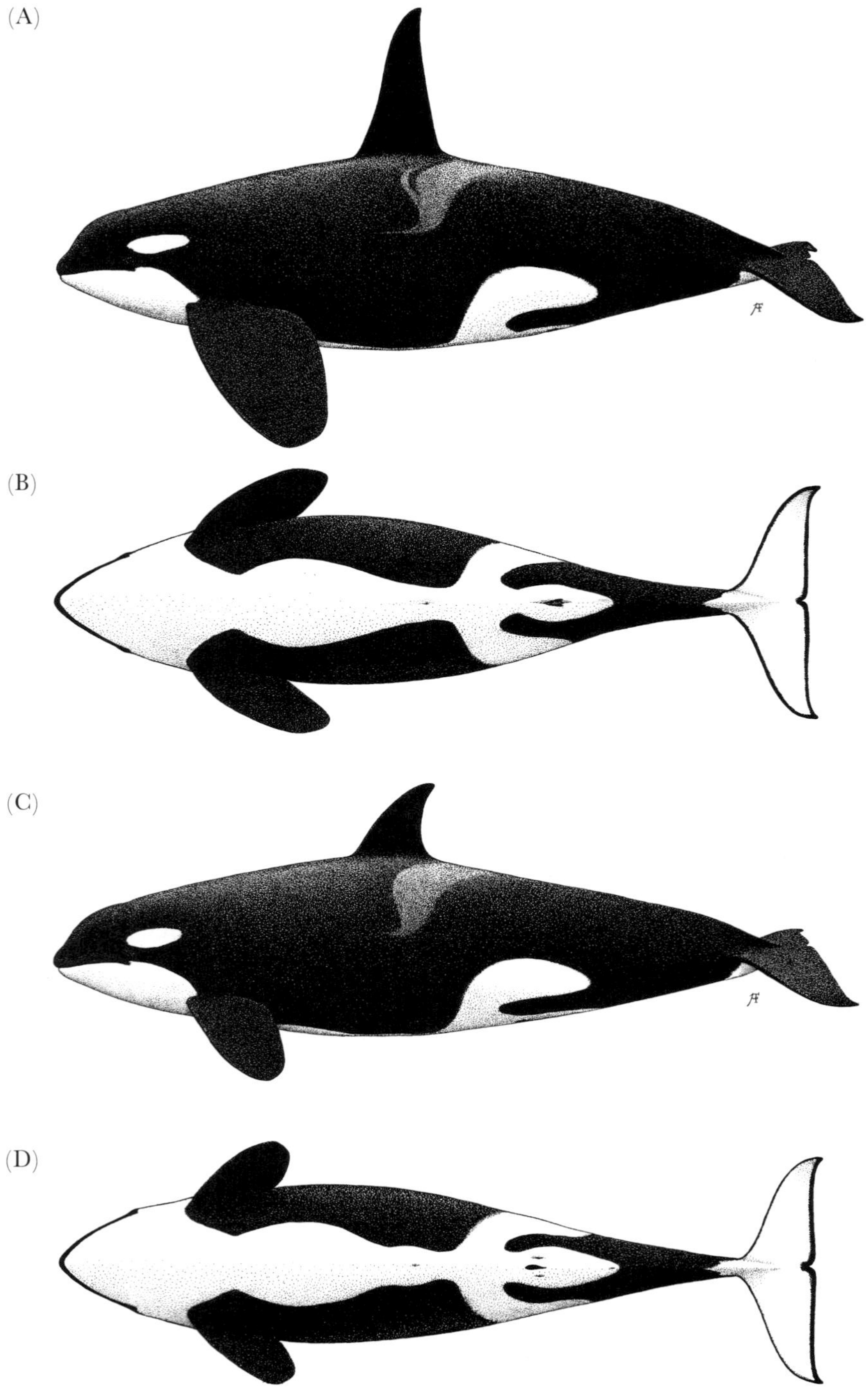

FIG. 2 Lateral and ventral views of a typical adult killer whale male (A, B) and female (C, D). (Illustration by Pieter Folkens)

documented the presence of a cape (D. Bain, personal communication). Many of the Southern Hemisphere animals are also distinctively lighter in colour posterior to the dorsal fin (Fig. 3).

Size and shape

Sexual dimorphism occurs in body size, flipper size, and height of the dorsal fin. Females attain a body length of up to 7.7 m and males 9.0 m (Heyning and Brownell, in prep.). In adult males, the dorsal fin is erect and may be from 1.0 to 1.8 m tall, whereas the dorsal fins of females are less than 0.7 m and distinctly falcated (Heyning and Brownell, in prep.). Height of the dorsal fin is a useful character for identifying adult male composition within pods.

The head of the killer whale is somewhat rounded with a slight demarcation of a beak. The relatively large ovate flippers are positioned about one-fourth of the distance from the snout to flukes. The flipper shape contrasts sharply with the sickle-shaped flippers of most delphinids. Flipper length may attain 20% of the body length in males and 11–13% of the body length in females (Eschricht, 1866). Total spread of the flukes may be over one-fifth of the body length for both sexes (Nishiwaki, 1972).

Weight

Although few animals have been weighed, maximum accurate weights of 3810 kg for a 6.70 m female and 5568 kg for a 6.75 m male have been obtained (B. Andrews, personal communication). A length/weight equation of $W = 0.000208L^{2.577}$, with W in kg and L in cm, has been calculated (Bigg and Wolman, 1975).

Internal Anatomy

Skull

Skulls of adult *O. ocra* typically are distinguished from those of other species of delphinids by their large size (condylobasal length to 100 cm), dental formula, and large teeth. Skulls from subadult killer whales may be confused with those from false killer whales (*Pseudorca crassidens*). In killer whales, the width across the premaxillae is usually less than 50% of the rostral width (measured just anterior to the antorbital notches) and the lateral border of the premaxillae is slightly more sigmoid and wider distally (Fig. 4). The pterygoids are widely separated in killer whales and their teeth are compressed anteroposteriorly at the roots (Glass, 1974). The tympanic and periotic bone complex of *O. orca* is characterized by the lack of a ventral keel, closure of the elliptical foramen, and massive anterior and posterior periotic processes (Kasuya, 1973). The temporal fossa is noticeably

FIG. 3 Photographs of killer whales taken off the Antarctic Peninsula. Note the distinctive cape pattern (A and B) and the lighter dorsum posterior to the saddle patch (B). (Photos by J.E. Heyning, 23 January 1990)

A

B

C

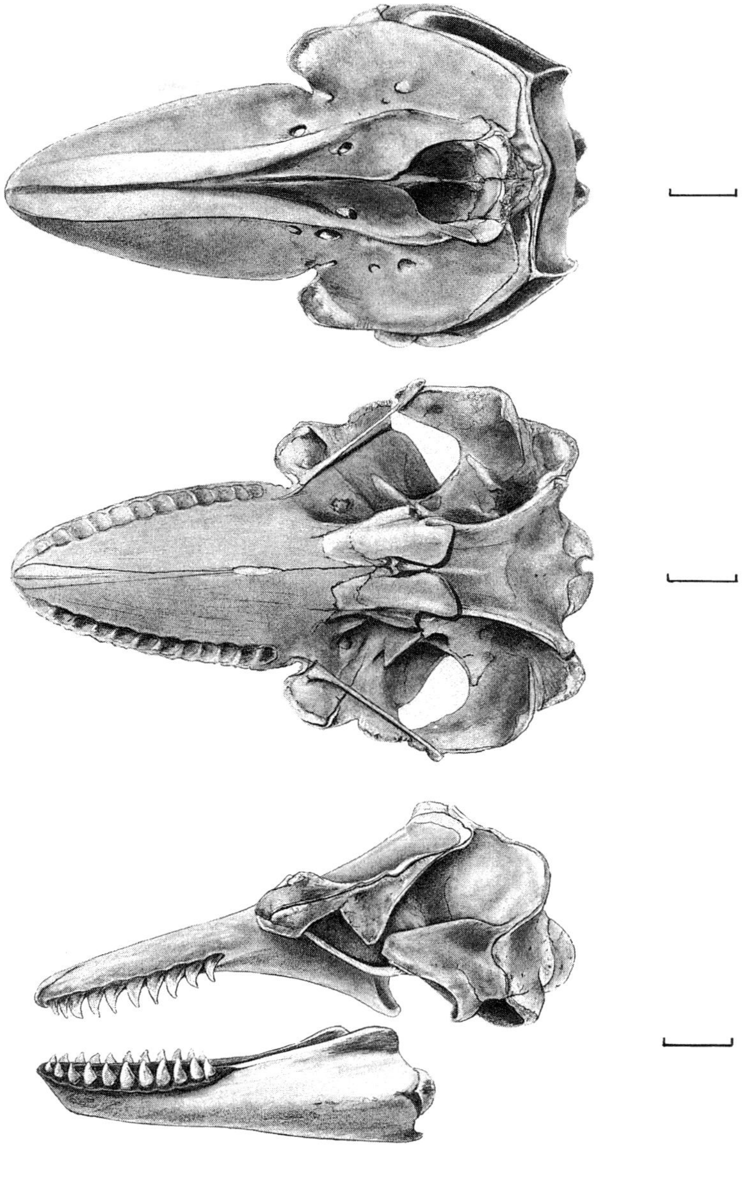

FIG. 4 Skull of *Orcinus orca*. Female 492 cm in length (National Marine Mammal Laboratory #90) from the Bering Sea (North of Akutan Island, 54° 33′N, 165° 54′W). (A) Dorsal view, (B) ventral view. (C) Lateral view of *Orcinus orca* skull and mandible. Male 518 cm in length (National Marine Mammal Laboratory, Seattle, Washington #77) from the North Pacific (east of Kodiak, Alaska, 57° 46′N, 152° 13′W). Calibration mark is 10 cm long. (Illustration by K. Zecca)

large, indicating a large and powerful temporalis muscle for jaw closure. Ness (1967) reported that the degree of skull symmetry in *Orcinus* is high compared with other odontocetes; however, Heyning (1989) noted scaling problems due to rostral lengths with Ness's analysis. Killer whales exhibit similar skull asymmetry as other delphinids. The lower jaw is relatively short.

The dental formula is usually listed as 10–12/10–12; however, up to 14 teeth per row have been counted (Eschricht, 1866; Scammon, 1874). The adult tooth may be as long as 13 cm (Nishiwaki, 1972). Both the mandibular and maxillary alveoli are deep, with approximately two-thirds of the tooth embedded. The tips are pointed and curved inwards and backwards. When the jaws close, the upper and lower teeth interlock. Older animals may exhibit extensive wear on the teeth (Caldwell and Brown, 1964).

Postcranial skeleton

The vertebral count is C_7, T_{11-13}, L_{10-12}, Ca_{20-24}, for a total of 50–54 vertebrae (Eschricht, 1866; Nishiwaki, 1972). Rib counts range from 11 to 13 per side (Eschricht, 1866), with the anterior six or seven ribs attached to the vertebrae by both the capitulum and tuberculum and the remainder attached only by the tuberculum. Ribs 1–6 attach to the sternum.

The phalanges are wider than they are long. The phalangeal formula is listed as I_1, II_{4-6}, III_{3-4}, IV_{2-3}, and V_2 (Heyning and Dahlheim, 1988). The ends of the phalanges and most carpal elements were composed of cartilage for an adult male examined by Eschricht (1866). Harmer (1927) hypothesized that the accelerated secondary growth of the flippers in maturing males was related to the continued growth of these cartilages.

Organs, tissues and physiology

Little information is available on the internal anatomy of killer whales. The tongue of *O. orca* is more protrusible than that of *Tursiops* (Donaldson, 1977). The general plan of the digestive system is similar to that of other delphinoids. The forestomach is large and extremely distensible to accommodate large prey items. The intestines of one individual measured 54.2 m (Eschricht, 1866). The reniculi of the kidneys are arranged in units of four that are contiguous therein. The renicular venous drainage is intrarenicular, with no peripheral venous plexus (Cave, 1977). The nasal diverticulae and associated facial musculature are similar to these features in other delphinids except some structures seemed proportionally smaller. The melon also is relatively small in *O. orca* (Mead, 1975).

The brain of *Orcinus*, like those of other odontocete whales, has highly convoluted cerebral hemispheres and lacks olfactory bulbs or olfactory nerves. The width of the brain exceeds its length and the corpus callosum is small relative to the huge size of the brain (Ridgway, 1986; Tarpley and Ridgway, 1994). Of the cranial nerves, the eighth is the largest, with a particularly well developed audi-

TABLE 1 Organ weights for killer whales, *Orcinus orca*

Specimen number	Sex	Total length (kg)	Total weight (kg)	Pancreas (g)	Spleen (g)	Right kidney (g)	Left kidney (g)	Stomach (g)	Lung (kg)	Intes-tines (kg)	Liver (kg)	Heart (kg)	Brain (g)	Adrenal (g)
LACM 72453	♀	245	239										2800	
LACM 72550	♀	260	213		255	1040	1060	3010	L 3080 R 2590	4380	7.100	2.345	3535	
NMML 83	♂	261	216			2700			5.35		9.0	2.05		
WWU 2255	♂	300	477		450	3180			22.7		12	5.7		
NMML 88	♂	380	1031		770	2950	3100		25	18		8		100
NMML 90	♀	492	1585			10000		44000	25	84	38.5	12		
NMML 77	♂	518	1352	5900	900	5750	4100	33000	19	37	39	10		
UWBM 26183	♀	579	2540			4500	5500		36.5		53	20		
LACM 84291	♀	620	3810		320	16800	17500		L 35 R 53.2		103		5689	L 140 R 230
NMML 84	♂	657							5.35					
LACM 84249	♂	665	3450			16250	11750						7200	L 300 R 350
NMML 80	♂	698	4554									11.5		

tory component. Ridgway and Brownson (1984) estimated an average brain mass of 5617 ± 968 g from killer whales averaging 5.5 m in length. Organ weights are listed in Table 1.

A whale held in Seattle had a surface heart rate of 60 beats/min which declined to 30 beats/min when submerged (Spencer *et al.*, 1967). The oxygen capacity of the blood is reported to be moderate for cetaceans (Lenfant, 1969), suggesting that killer whales may not be deep divers. The whales trained by Bowers and Henderson (1972) did not prove to be deep divers compared with pilot whales. Oxygen carrying capacity of the blood may vary among killer whale pods (D. Duffield, personal communication), and this oxygen capacity may affect diving ability (Ridgway and Johnston, 1966). The well-vascularized fins and flukes probably play an important role in the animal's ability to regulate temperature (Green, 1972). Thermoregulatory variables were measured in captive killer whales by Kasting *et al.* (1989). Thyroid hormone levels are relatively high (Ridgway and Patten, 1971). Ridgway (1972) reported mean blood volume equivalent to 8–10% of body weight, rectal temperature of 36.4 °C, and mean serum osmolality of 337 mOsm. Haematology and blood chemistry values have also been reported (Ridgway *et al.*, 1970; MacNeill, 1975; Cornell, 1983). A lung tidal volume of 46.2 litres and a tidal flow of 129 litres/sec were calculated for a 4.3 m female (Spencer *et al.*, 1967).

In their studies on belugas and killer whales, Kasting *et al.* (1989) proposed that these species had elevated metabolic rates compared with other mammals. Innes and Lavigne (1991) argued that there is no evidence that cetaceans have elevated metabolic rates. Obtaining comparable measures of cetacean metabolic rates is problematic (Kasting, 1991; Gallivan, 1992; Speakman *et al.*, 1993).

Life History and Population Dynamics

Growth and reproduction

Mating and calving seasons often span several months. Breeding cycles do not seem to be synchronized worldwide but may be within more restricted areas. In the northeast Atlantic, mating reportedly occurs from late autumn to midwinter (Jonsgård and Lyshoel, 1970). In the western North Pacific, most mating is suspected to occur between May and July (Nishiwaki, 1972). In Washington State and British Columbia waters, most births occur between October and March (Olesiuk *et al.*, 1990). Killer whales undoubtedly are polygamous.

The length of males at sexual maturity ranges from 5.2 m to 6.2 m (Perrin and Reilly, 1984). One male in captivity became sexually mature at an estimated age of 16 (Duffield and Miller, 1988). Harrison *et al.* (1972) determined that an individual 6.56 m in length had inactive testes with masses of 3632 g (L) and 2270 g (R). A different animal 7.24 m in length with 11 400 g (L) and 12 200 g (R) testes was sexually mature. An investigation of 57 mature males from the

Antarctic indicated an average testis mass of 10 000 g, with a maximum mass of 23 100 g (Mikhalev *et al.*, 1981).

Females attain sexual maturity between 4.6 m and 5.4 m (Perrin and Reilly, 1984). Some of this variation in size at maturity is geographic, with animals from the northeastern Atlantic maturing at the low end of the size range and Antarctic animals the high end (Perrin and Reilly, 1984). Ovaries from mature Antarctic killer whales were reported to average about 10–12 cm in length by 5–7 cm in width (Mikhalev *et al.*, 1981). A female with a 91 cm foetus had a corpus luteum measuring 7.6 cm by 5.1 cm (Turner, 1872). A study conducted by Walker *et al.* (1988) provides a description of reproductive hormone profiles in captive whales during ovarian cycles and pregnancy. Captive female killer whale can be seasonally or non-seasonally polyoestrous or anoestrous (Robeck *et al.*, 1993). Off British Columbia and Washington State, resident female killer whales gave birth to their first viable calves at approximately 15 years of age (Olesiuk *et al.*, 1990). From captive data, females at the birth of their first calf were aged 8–17 years (Duffield *et al.*, 1995). Two second-generation captive births, both to the same female, have been reported (Duffield *et al.*, 1995; D. Odell, personal communication). Based on progesterone levels, gestation is estimated at 517 days (Asper *et al.*, 1988).

Of the mature females taken by Norwegian whalers, 37.3% were pregnant (Christensen, 1984), which calculates to an annual pregnancy rate (APR) of 26.3% based on a 517 day gestation (see Perrin and Reilly, 1984, for terminology). Data from Soviet whaling in the Southern Hemisphere indicates that 27.5% of mature females were pregnant, yielding an APR of 19.2%. Based on an average calving interval of 5.32 years for reproductive females in British Columbia and Washington State (Olesiuk *et al.*, 1990), we calculate an APR of 18.8%, although this figure does not include all mature females.

Observational data from British Columbia and Washington show that of the mature females, only one-half to two-thirds were observed to give birth during the 15-year study period (Olesiuk *et al.*, 1990). The non-reproductive females have been termed either "post-reproductive females" (Olesiuk *et al.*, 1990) or "barren females" (Heimlich-Boran, 1986). We prefer the term non-reproductive females, as the status of these females is unknown at this time. Non-reproductive females seem to exhibit relatively high levels of allomaternal behaviour, which may be related to either a social hierarchy or kinship (Heimlich-Boran, 1986; Haenel, 1986). Some evidence suggests that reproductive rates in some mammals are density dependent (Fowler, 1984, 1987; Kasuya and Marsh, 1984). The above evidence of a higher APR in populations that were heavily exploited, and the presence of non-reproductive females in at least some populations, may suggest some social control over reproduction, and this may be density dependent. Brault and Caswell (1993) further examined the data of Olesiuk *et al.* (1990) and found that the inter-pod variance in growth rates was due to variance in adult reproductive output. However, this variance is not significantly greater than expected based on variation of individual life histories within the population. They concluded that there is no evidence for an effect of social structure on pod-specific population growth rates, and the restriction of growth rates to such a narrow range suggests a possible role for density-dependent processes.

Detailed embryological and placental morphology of *O. orca* has been described for several foetuses (Turner, 1872; Guldberg and Nansen, 1894; Benirschke and Cornell, 1987). The maximum size of foetuses differs regionally and has been documented as 255 cm for the North Atlantic (Perrin and Reilly, 1984), 274 cm for the North Pacific (Nishiwaki and Handa, 1958), and 250 cm for the Antarctic (Mikhalev *et al.*, 1981).

The smallest neonates recorded are 183 cm (North Atlantic), 228 cm (North Pacific, and 220 cm (Antarctic) (Perrin and Reilly, 1984). Average size of north-east Pacific calves is 2.36 m (Heyning, 1988). The nurturing and development of the calf takes place primarily within its mother's pod. Calf dependency in the wild has been reported by Balcomb *et al.* (1979) to last at least 2 years. However, killer whales may begin eating solid food at a younger age than most delphinids (Heyning, 1988). Tooth eruption in killer whale calves begins at approximately 10 weeks with the upper teeth; by 13 weeks the lower teeth are erupting (Asper *et al.*, 1988). Asper *et al.* (1988) summarize nursing frequency and duration of a captive born killer whale, and report the calf eating fish at11 weeks. At 5 months, the calf was eating 6.6 kg smelt. At 15 months, 22 kg of herring and squid were being consumed. Nursing ceased at 18 months. Duffield and Miller (1988) provide a listing of killer whales born in captivity. Sex ratios at birth are probably 1:1.

Myrick *et al.* (1988) examined growth layer groups (GLGs) in tetracycline-labelled and non-labelled teeth of 13 captive animals. Using decalcified and stained thin sections, they found distinct GLGs, but numerous accessory layers make the GLGs difficult to interpret (Perrin and Myrick, 1980; Myrick *et al.*, 1988). The average life span of killer whales, based on sectioned teeth, has been estimated to be 25 years (Jonsgård and Oynes, 1952). The maximum longevity may be 35–40 years (Mitchell and Baker, 1980; Christensen, 1984). Heyning and Hohn (1993) re-evaluated the ageing estimates based on teeth. They found that the growth layer patterns in the tooth sections were very similar to those of other species of delphinids. Using acid-etched thin sections, the maximum age for females was 66 GLGs and the maximum for males was 38 GLGs based on an extremely small sample.

Mortality

Olesiuk *et al.* (1990) studied association patterns in a group of resident whales off the coasts of British Columbia and Washington State. These authors report a mean life expectancy of 50.2 years for female whales and a maximum longevity of about 80–90 years. For males, they report a mean life expectancy of 29.2 years and a maximum longevity of about 50–60 years. Mortality curves were U-shaped for both males and females, but the curve was narrower for males (Olesiuk *et al.*, 1990). The basic model of survivorship based on photo-identification data and the associated strengths and weaknesses of this model are summarized by Buckland (1990) and Brault and Caswell (1993).

Captive animals showed an overall mortality rate of 8.9% per year in one study (Duffield and Miller, 1988), 7% per year in another (DeMaster and Drevenak, 1988), and 6.2% in Small and DeMaster (1995). Killer whales have no significant predators other than man.

Abundance

Abundance estimates are available for some geographical areas. Photo-identification studies conducted in Alaska (eastern Bering Sea to southeastern Alaska) have identified at least 850 individuals (Dahlheim *et al.*, 1997; Dahlheim, 1997; Heise *et al.*, 1991). Off the Queen Charlotte Islands (northern British Columbia) 117 individuals have been identified (Heise *et al.*, 1993). Based on photo-identification studies off eastern and southern Vancouver Island, 260 "resident" whales comprising 19 pods and about 75 "transient" whales comprising 30 pods have been documented (Bigg *et al.*, 1990). Another 75 whales have been discovered recently from this area (Walters *et al.*, 1992). Photo- identification studies have documented 184 individual killer whales off California and another 65 individual whales off Mexico (Black *et al.*, 1996). Population studies using photo-identification techniques result in a minimum count of animals. Due to the highly mobile nature of killer whales, photographic comparisons are required among geographical areas to obtain accurate counts.

In the North Atlantic, information on killer whales in Norwegian coastal waters was solicited through questionnaire surveys. Christensen (1988) reported 483–1507 killer whales for the area. Although more limited in geographical range, photo-identification studies off coastal Norway reported only 26 whales (Lyrholm, 1988). Photo-identification studies conducted off Iceland have identified 143 whales (Sigurjónsson *et al.*, 1988). No matches of photo-identified animals have occurred between Iceland and Norway.

Miyashita (in press) reported preliminary estimates of killer whales off the coast of Japan as 1200 in waters north of 35°N and 700 whales in the waters south of 35°N. An abundance estimate of 70 000 Antarctic killer whales, based on sighting surveys, was reported by Butterworth and De Decker (1990).

Behaviour

Social organization

Killer whales usually occur in small pods, with typically fewer than 40 individuals. Off Japan, Kasuya (1971) estimated the mean pod size to be 6. In Alaskan waters, pod size ranged from 2 to 100 animals, with only 1% of these groups containing 20 or more individuals (Braham and Dahlheim, 1982). In the warm-temperate and tropical eastern Pacific, Dahlheim *et al.* (1982) noted that pods contained a maximum of 75 animals with a mean of 5.3; 91% contained fewer than 10 animals. In the Antarctic, about one-half of the pods had 5–10 animals, but pods of up to 100 whales were observed (Mikhalev *et al.*, 1981). Frequently it is difficult to determine the number of whales in a pod, as a pod can be widely dispersed. Large groups may reflect the temporary merging of a number of pods, possibly related to seasonal peaks in prey availability or reproductive activities. Single whales, usually adult males, occur infrequently (Norris and Prescott, 1961).

Bigg *et al.* (1990) examined the social organization and genealogy of resident killer whales in the coastal waters of British Columbia and Washington State, based on field observations and photographs collected from 1973 to 1987. The authors classified the social organization into communities, pods, subpods, and matrilineal groups. They defined these terms as follows: a community is composed of individuals that share a common range and are associated with one another; a pod is a group of individuals within a community that travel together the majority of time; a subpod is a group of individuals that temporarily fragments from its pod to travel separately; and a matrilineal group consists of individuals within a subpod that travel in very close proximity.

Matrilineal groups are the basic unit of social organization, and consist of whales from 2–3 generations. Membership at each group level is typically stable for resident whales, except for births and deaths. However, there are cases of individuals and matrilineal groups switching to other pods (Heimlich-Boran, 1986; Jacobsen, 1986; Osborne, 1986) and whales from more than one "community" travelling together (Dahlheim *et al.*, 1997). Hoezel *et al.* (1991a) have also used DNA fingerprinting techniques to determine paternity in killer whales.

For the Pacific Northwest (British Columbia and Puget Sound, Washington), approximately 19% of the animals photo-identified within the resident pods are adult males, 31% are adult females, and the remaining 50% are immature whales of both sexes (Olesiuk *et al.*, 1990). For Alaskan waters, approximately 24% of all whales photo-identified are adult males, 47% are either adult females or subadult males, and 29% are either juveniles or calves (Dahlheim, 1997; Dahlheim *et al.*, 1997). For Southern Hemisphere waters, 16% of the animals observed during sighting surveys were adult males and 8% were adult females with accompanying calves. Juveniles, adult females without calves, and subadult males represented 76% (Miyazaki, 1989).

Occasionally, when resident pods encounter each other after being separated for some time, the pods will form two tight lines and approach each other head on. When the groups are within 10–20 m of each other, they stop motionless at the surface and hover, facing each other for 10–30 sec (Osborne, 1986). Intermingling typically follows this greeting ceremony. Intermingling is characterized by slow moving and tight milling concentrations of whales. Many of the whales are in physical contact, rolling and brushing against each other at or near the surface. This greeting ceremony has been described only for the resident pods in Puget Sound, Washington (J, K, and L pods).

Parallel scars caused by the tooth marks of other killer whales indicate that some intraspecific aggression occurs (Greenwood *et al.*, 1974). However, parallel scars could be obtained during other social activities. Many scratches or rake marks may also be the result of interactions with pinnipeds during predation events (M. Dahlheim, personal observations).

Care-giving is well documented (Caldwell and Caldwell, 1966). Waite (1988) describes alloparental care in killer whales. Adolescent females performed most of the alloparenting, although non-reproductive females and males were also observed to alloparent. Adult bulls have been observed to "baby-sit" groups of calves and juveniles (Osborne *et al.*, 1988).

Resident and transient pods

Pods along British Columbia and Washington State have been labelled as "resident" and "transient" (Bigg, 1982; Bigg *et al.*, 1990). While many researchers currently recognize that the terms resident and transient whales are somewhat misleading in their descriptions, the terms are still so entrenched in the popular and scientific literature that they continue to be used. The two sympatric forms are believed to differ in several aspects of morphology, ecology, and behaviour. However, much of what we know about the resident and transient form and/or behaviour of killer whales is derived from research conducted in Pacific Northwest waters and then extrapolated to other areas. It is possible that killer whales occupying other geographical regions or habitats (e.g., pelagic environments) may behave and/or segregate differently.

Bigg *et al.* (1987), based on observational data only, noted that the dorsal fin of resident whales tended to be more rounded than those of transient whales. Baird and Stacey (1988) documented differences in saddle patch pigmentation noting that resident whales exhibited five patterns whereas transient whales exhibited only three of these patterns. Bigg *et al.* (1987) stated that resident pods range in size from 5 to 50 individuals and transient pods range from one to seven. Resident pods are believed to have smaller home ranges than transient whales; however, little is known regarding killer whale winter range and distribution. Differences have been also noted in the vocalization patterns between the two forms (Ford, 1989; Ford and Hubbard-Morton, 1990).

Pacific Northwest resident whales eat mainly fish and rarely consume marine mammals. However, L pod, a resident whale group in Puget Sound, Washington, has been known to prey on marine mammals on at least three occasions (Felleman *et al.*, 1991). Transient whales frequently are seen pursuing marine mammal prey, but there are cases of transient whales eating both fish and marine mammals in southeastern Alaska (M. Dahlheim, personal observations) and Prince William Sound, Alaska (Matkin and Saulitis, 1994). Resident whales have shorter dive times, follow more predictable travel routes, and usually are easier to approach than are transient whales (Bigg *et al.*, 1987; Morton, 1990; Baird *et al.*, 1992). Recent observations, however, suggest that dive times and pod size estimates do overlap between resident and transient whales, and the ease with which a pod can be approached is highly dependent upon the pod's general behaviour. Resident killer whales in Pacific Northwest waters use regions of high relief topography along salmon migration routes, whereas transient whales forage for pinnipeds in shallow protected waters (Heimlich-Boran, 1988). The occurrence and behaviour of transient killer whales observed around southern Vancouver Island was shown to vary seasonally (Baird and Dill, 1995). Some pods foraged in open water while others spent much of their time foraging around pinniped haulouts and along nearshore sites (Baird and Dill, 1995). A comparison of foraging strategies between Pacific Northwest resident and transient whales is presented by Felleman *et al.* (1991).

Resident and transient whales do not intermingle socially (Baird and Dill, 1995). There are a few observations of transient killer whales following resident animals at close range (Matkin and Dahlheim, 1993) and a rare account of an attack by a group of resident whales on a transient pod (D. Ellifrit, personal communication).

In reviewing the discreteness of each character to separate the two forms, Heyning and Dahlheim (1993) concluded that most of the characters used to differentiate between the two forms are behavioural and recommended that more research, especially genetic studies, should be performed to better understand these differences.

Studies on mDNA restriction patterns provide evidence that the resident and transient pods are genetically distinct (Stevens *et al.*, 1989; Hoelzel, 1991a; Hoelzel and Dover, 1991). Genetic materials obtained from killer whales ranging from California to Alaska were compared for genetic variation at three nuclear DNA markers and sequenced for a total of 520 bp from the mitochondrial control region (Hoelzel *et al.*, 1998). Results suggest low levels of genetic dispersal between foraging specialists and a pattern of genetic differentiation consistent with a matrifocal population structure and small effective population size.

Genetic material (i.e. skin) was obtained during biopsy sampling of free-ranging killer whales and from DNA material derived from teeth and bones. Killer whales showed little reaction to biopsy darting (72% of 114 attempts had no response; M. Dahlheim, personal communication). Barrett-Lennard *et al.* (1996), reporting on Canadian killer whale biopsy studies, documented reactions such as momentary shakes or accelerations during 81% of the dart hits and 53% of the misses.

Movements

Although regular sightings and recurrence of the same pods have been documented within specific areas, territoriality among pods has not been documented.

Occupation of well-defined areas may be better described in terms of a home range.

Based on photo-identification studies, numerous individual whales and/or pods have been documented to move between Puget Sound (Washington)/ British Columbia and southeastern Alaska and between southeastern Alaska and Prince William Sound (Ellis, 1984, 1987; Dahlheim *et al.*, 1997). Photographic matches of killer whales also have been documented between Prince William Sound and Kodiak Island (Heise *et al.*, 1991). Movements by individuals/pods of killer whales between Kodiak Island and the southeastern Bering Sea have not as yet been documented. This may be due mostly to limited photo-identification efforts in the area (Dahlheim, 1997). Black *et al.* (1996) documented whale movements from Alaska and British Columbia to California and from California to Mexico. Transient whales appear to cover a more extensive range than residents. A distance of over 2600 km (California to Alaska) has been reported for a transient group (Goley and Straley, 1994). Although transient whales move frequently between Puget Sound, Washington and southeastern Alaska, transient whales have not been documented to move between southeastern Alaska and Prince William Sound. To date, only AG and AF pods (resident groups from southeastern Alaska; Dahlheim *et al.*, 1997) are known to travel between the two areas. While occurring in Prince William Sound, AG and AF pods frequently are seen mixing with Prince William Sound resident whales. Conversely, Prince William Sound resident whales have never been documented in southeastern Alaskan waters.

Feeding

Diet varies considerably from one region to the next and even within specific areas. Killer whales are known to have a variety of prey, including fish, other cetaceans, pinnipeds, mustelids, birds, reptiles, and cephalopods. Jefferson *et al.* (1991) provided a summary list of prey species of killer whales. New prey species have been added recently to Jefferson's list, the moose (*Alces alces*) (M. Opp and J. Barry, personal observations) and the Pacific white-sided dolphin (*Lagenorhynchus obliquidens*) (Dahlheim and Towell, 1994). Fertl *et al.* (1996) provided an extensive listing of elasmobranchs consumed by killer whales. Shevchenko (1976) reported killer whale remains in the stomachs of two male killer whales collected in the warm-temperate zone of the Southern Hemisphere. In Norwegian waters, Jonsgård and Lyshoel (1970) noted that herring constituted the whales' main diet, whereas off Japan, Nishiwaki and Handa (1958) reported that cod, flatfish, and cephalopods are the primary components of the diet. Rice (1968) noted that adult bulls in the northeastern Pacific fed mainly on other marine mammals based on extremely small samples. In Puget Sound, the main food of resident killer whales during the summer and fall is salmon (Balcomb *et al.*, 1979). Of 362 stomachs examined from Antarctic killer whales, 217 contained fish, 75 contained remains of minke whales

Balaenoptera acutorostrata, 35 contained pinnipeds, and 35 contained squid (Ivashin, 1981). Most food items are swallowed whole. However, when whales attack larger prey, they rip away smaller pieces of flesh and then consume them. The tongues, lips, and genital regions of baleen whales seem to be the favoured parts. Food sharing, provisioning, and foraging strategies of killer whale predation on marine mammals off Argentina are discussed in Hoelzel (1991b). In most geographical regions, killer whale movements may be related to movements of their prey (Sergeant and Fisher, 1957; Jonsgård and Lyshoel, 1970). Whales may travel 125–200 km per day while foraging. Although killer whale pod size varies considerably, pod size of foraging transient whales is consistent with the maximization of energy intake hypothesis (Baird and Dill, 1996). Sergeant (1969) reported that killer whales maintained in captivity consumed an equivalent of 4% of their body weight daily. Food consumption of a female killer whale held in captivity for 11 years is summarized in Kastelein and Vaughan (1989).

When attacking large whales, killer whales approach from several angles. The prevalence of tooth scars on the flukes of gray whales (*Eschrichtius robustus*) suggests that the killer whales' most common strategy is to immobilize and drown their prey (Rice and Wolman, 1971). Each killer whale may have a specific role when attacking large prey. When chasing gray whales in the Bering Sea, killer whales approached in a linear formation, then gave pursuit in a loose crescent-shaped formation with individual killer whales spaced approximately 300 m apart (Ljungblad and Moore, 1983). Two interesting accounts of killer whale predation on balaenopterids are given for the Gulf of California. Silber *et al.* (1990) describe the pursuit of a Bryde's whale (*Balaenoptera edeni*) by killer whales, and Vidal and Pechter (1989) report behavioural observations of fin whales (*Balaenoptera physalus*) in the presence of killer whales. Off central California, a pod of eight killer whales was observed to chase nine adult gray whales from a distance of 2 km off the coast into the nearshore kelp beds. While being pursued, the gray whales were tightly grouped together, spending the majority of time close to the surface. For more than 45 minutes following this encounter, all gray whales ($n = 36$) were observed to be within close proximity of the kelp, swimming very slowly and showing only their blowholes (termed snorkling behaviour) when surfacing to breathe (M. Dahlheim, personal observations).

When hunting pinnipeds, killer whales use a variety of strategies, probably dependent upon the prey species being targetted. The particular hunting or feeding strategy, or target prey, probably dictates the concentration and dispersal of killer whales. Killer whales are known to kill seals and sea lions by hitting the prey into the air with their flukes. Reports of killer whales ramming pinnipeds, causing internal damage, have been noted as well. Off Argentina, out of 568 observations of killer whales hunting pinnipeds, 94% involved cooperative hunting. During cooperative hunting, killer whales had 31% success rate resulting in capture or death of pinnipeds. Sixty-four per cent involved killer whales briefly stranding themselves to capture pinnipeds, a technique taught to younger whales by adult males (Lopez and Lopez, 1985). Intentional strandings and predatory behaviour of killer whales around the Crozet Islands is discussed in Guinet (1991, 1992). In icy regions, killer whales have been observed to peer onto ice floes for seals and sometimes hit the ice from underneath to knock pinnipeds or penguins into the water (Fraser, 1949). Smith *et al.* (1981) describe how a pod of whales located a crabeater seal on an ice floe, moved away from the floating ice, then swam as a group close to the floe. The movements of the whales caused a wave, which tipped up the ice floe and broke over it, causing the seal to be thrown in the water. Killer whales are known to follow the open leads created by large ships and icebreakers in pursuit of prey (Thomas *et al.*, 1987).

While chasing salmon in Puget Sound, killer whales often hunt individually. However, Evans and Bastian (1969) reported killer whales circling a school of fish; one or two whales would then break ranks and dash through the middle of the school to feed. Similä and Ugarte (1993) report on surface and underwater observations of cooperatively feeding killer whales in northern Norway. These authors describe how whales herd herring into dense concentrations and then stun their prey by slapping the edge of the fish school with the underside of their flukes, eating stunned fish one by one.

Killer whales consume fish of commercial importance. Troll catches of salmon show a decline when killer whales are in the area (Pike and MacAskie, 1969) and damage to fishing gear has also been reported (Scheffer and Slipp, 1948; Nishiwaki and Handa, 1958). Off Iceland, killer whales are attracted to herring operations. In the 1950s, because of complaints about damage to fishing gear and reduction in fish landed, Iceland reduced local killer whale numbers (Hoyt, 1990). However, the success of this whale reduction effort off Iceland has been questioned. Incidences of killer whales eating longline-caught tuna are available from various areas (e.g. North Atlantic, coast of Japan, and Indian Ocean). Off Tasmania, killer whales were known to prey on trevalla (*Hyperoglyphe porosa*) caught in the dropline fishery. Reports of depredation by killer whales on longline-catches of sablefish (*Anoplopoma fimbria*) in Alaskan waters (Bering Sea and Prince William Sound) have been documented (Dahlheim, 1988; Yano and Dahlheim, 1995a,b). Longline fisheries interactions involving killer whales in southeastern Alaska also have been recently observed (Dahlheim *et al.*, 1997). Killer whales are known to follow fish-processing vessels for many miles feeding off discarded fish. In the Bering Sea, the same pod of whales was reported to follow a vessel for 31 days for approximately 1600 km (National Marine Mammal Laboratory, unpublished).

Association with other species

Being a top predator, the killer whale utilizes the available resources in a complex fashion. Killer whales often associate with other marine mammals (cetaceans and pinnipeds) without attacking them (Jefferson *et al.*, 1991). Limited data suggest that some killer whale pods may selectively predate on a given prey species (Matkin and Dahlheim, 1995). This is reasonable when one considers the learning required by the whales to successfully pursue, catch, and consume different prey items. The complexity of predator–prey interactions is well documented in the following accounts. Dolphin (1987) observed Southeastern Alaskan humpback whales (*Megaptera novaeangliae*), including small individuals and a female with a calf, join a pod of killer whales that were actively attacking a sea lion (*Eumetopias jubatus*). A transient pod of killer whales passed directly by three Steller sea lions and two harbour porpoise. The whales then encountered a single harbour seal, which they immediately attacked and consumed (M. Dahlheim, personal observations). On a separate occasion, two adult male Steller sea lions were observed to chase killer whales away from a rookery area (M. Dahlheim, personal observations).

In the late 1800s and early 1900s, an unusual cooperative relationship between killer whales and Australian shore-based whalers occurred. Killer whale pods alerted whalers to the presence of humpback whales (*Megaptera novaeangliae*) in the area. After following the killer whales to the prey, the whalers killed the humpbacks and allowed the killer whales to feed upon the dead whales' tongues (Wellings, 1964).

Swimming and diving

Norris and Prescott (1961) stated that, in general, killer whales took three to five short dives of 10–35 sec duration followed by a longer dive lasting 1–4 min. Erickson (1978), summarizing data collected from two radio-tagged transient killer whales, reported a mean dive cycle of 5.77 min. The cycle consisted of a long dive followed by three or four surface blows of 3–4 sec spaced a mean 21 sec apart. The longest recorded dive was 17 min. Respiration rates are highly variable and are associated with the behavioural state of the animal. Baird and Goodyear (1993), using recoverable time–depth recorders and VHF radio tags, obtained records of 737 consecutive dive profiles for killer whales off Vancouver Island, British Columbia. The maximum depth of dive was 173 m (bottom depth). Velocity during descents and ascents reached 6 m/sec. The maximum recorded depth of dive of a trained killer whale was 260 m (Bowers and Henderson, 1972). Swimming speeds usually are 6–10 km/hr with a maximum of 40 km/hr (Lang, 1966). *O. orca* has been documented to bowride ships (Dahlheim, 1980).

Aerial and other behaviour

Behaviour such as breaching (Fig. 5), spyhopping (Fig. 6), flipper slapping, and lobtailing is common (Jacobsen, 1986). The significance of such behaviour is not known, and occurrence is within various behavioural contexts. Play behaviour is well documented in killer whales (Osborne *et al.*, 1988). In British Columbia (Hoyt, 1990) and Prince William Sound, Alaska (National Marine Mammal Laboratory, unpublished data), killer whales have been documented to rub along the gravel beds of shallow-water beaches.

FIG. 5 Breaching killer whale in Puget Sound, WA. (Photo by Dave Ellifrit, Center for Whale Research, Friday Harbor, WA)

Sound production

Grieg (1906) contributed one of the earliest published reports on killer whale sounds, noting "flutelike" calls from young animals and "roars" from old bulls during whaling operations. Schevill and Watkins (1966) analysed calls from a young captive male in British Columbia, noting clicks believed to be used in echolocation and screams assumed to be for communication. Killer whales have been proved to use echolocation (Diereks *et al.*, 1971). Killer whales have an extensive repertoire of sounds covering a wide range of frequencies (Dahlheim and Awbrey, 1982). Newborn calves produce call types similar to those of their parents; however, their repertoire is limited (Dahlheim and Awbrey, 1982). As with *Tursiops* (Lilly and Miller, 1961), killer whales are capable of producing two different sounds simultaneously, which suggests either dual sound producing structures or a single control point for two separate pathways. A series of sacs off the nasal passageway is believed to be involved in sound production (Mead, 1975; Heyning, 1989).

Dahlheim and Awbrey (1982) documented individual and group differences in the calls of captive whales. Ford and Fisher (1982), working with free-ranging killer whales, documented dialectal variations in the signals among killer whale pods off British Columbia. The extent of vocal differences may reflect the degree

FIG. 6 Spyhopping killer whale. (Photo by Dave Ellifrit, Center for Whale Research, Friday Harbor, WA)

of associations among pods (Ford and Fisher, 1983) and the differences have been used to help determine genealogies. However, Bain (1988) demonstrated that an Icelandic whale, maintained at the same facility as a British Columbia whale, acquired most of the British Columbia repertoire, while retaining the ability to produce native Icelandic calls. These results are in contrast with other cases of captive whales where the vocal repertoire remained stable for over 15 years (Ford, 1989). Regional differences in frequency, structure, timing, and call patterns of killer whale sounds have been noted in the Antarctic (Thomas *et al.*, 1981), North Atlantic (Jehl *et al.*, 1980), and the North Pacific (Awbrey *et al.*, 1982). Ford (1989) analysed vocalizations for each of five activity states, including foraging, travelling, group-resting, socializing, and beach-rubbing. High rates of vocal activity occurred during all behavioural categories except group-resting.

Popper (1980) noted that, among the delphinids thus far studied, killer whales have the best absolute acoustical sensitivity; however, their audible frequency range is limited. Hall and Johnson (1972) presented an audiogram of *Orcinus* with a sensitivity range from 500 Hz to 31 kHz, with greatest sensitivity at 15 kHz. Audiograms recently obtained on two captive female whales noted responses to tones produced at 120 kHz; greatest sensitivity occurred at 20 kHz (Bain *et al.*, in prep.). These authors also demonstrated that absolute hearing sensitivity of killer whales was similar to that reported for other cetaceans. Hall and Johnson (1972) demonstrated that a blindfolded killer whale navigated easily around a 15 m pool; locating and retrieving an air-filled ring. Evans (1973) noted that at a 3 m distance a killer whale could detect an air-filled 10 mm diameter plastic ring whereas, at the same distance, a Pacific white-sided dolphin (*Lagenorhynchus obliquidens*) was capable of detecting a 2 mm diameter sized target. Popper (1980) suggested that killer whales have a diminished use of echolocation and rely more on passive listening.

For many years, it was assumed that underwater acoustics played a major role in coordinating attacks of killer whales on marine mammals. Ford and Hubbard-Morton (1990) noted that transient whales typically forage in silence, probably to avoid detection by their marine mammal prey, but calls are usually given during or following a kill.

Because killer whales are vociferous predators, experiments have been designed to play back killer whale sounds to determine prey response. Experiments have included the following prey: belugas, *Delphinapterus leucas* (Fish and Vania, 1971), gray whales (Cummings and Thompson, 1971; Dahlheim, 1987), right whales, *Eubalaena glacialis* (Cummings *et al.*, 1972), dolphins (*Delphinus delphis* and *Stenella* spp.), sea otters, *Enhydra lutris* (Davis *et al.*, 1988), sharks (Klimley and Myrberg, 1979), and jackass penguins, *Spheniscus demersus* (Frost *et al.*, 1975). Prey responses ranged from avoidance of the sound source to no overt response. During the projection of killer whale sounds, a complete cessation of all gray whale vocalizations occurred (Dahlheim, 1987). Sperm whales (*Physeter macrocephalus*) and narwhals (*Monodon monoceros*) also cease sound production in the presence of killer whales (Arnbom *et al.*, 1987 and Campbell *et al.*, 1988, respectively). The behavioural significance of the responses to projected killer whale sounds used during these experiments is unknown.

Vision

Visual acuity is apparently well developed in killer whales (White *et al.*, 1971).

Parasites and Disease

Killer whales are known to have the following endoparasites: Trematoda, *Fasciola skriabini*; Cestoda, *Trigonocotyle spasskyi*, *Phyllobothrium* sp.; and Nematoda, *Anasakis simplex* (Daily and Brownell, 1972). Killer whales are relatively free of external parasites, but barnacles *Xenobalanus globicipitis* (Heyning and Dahlheim, 1988) and *Cryptolepas rhachianecti* (Samaras, 1989) have been observed on the rostrum and trailing edge of the flukes of a calf. The attachment of *Cryptolepas*, considered to be host-specific to gray whales, is considered a rare phenomenon. The presence of *Xenobalanus* barnacles, attached to the trailing edge of the dorsal fin of older animals, also has been noted (B. Tershey, personal observations). Lockyer (1979) photographed a remora (Echeneididae) attached to a killer whale. Diatoms on the skin of killer whales found in high latitudes have been noted (Hart, 1935; Nemoto *et al.*, 1980). Two species of whale lice (*Cyamus orcini* and *C. antarcticensis*) have been found on killer whales (Leung, 1970; Berzin and Vlasova, 1982).

Killer whale teeth often show extensive wear and are worn flush with the gumline, resulting in exposure of the pulp cavity. Infection may penetrate through the pulp cavity, causing jaw abscesses. Simpson and Gardner (1972) reported frequent abscessed follicles of the vestigal hair on the rostrum of captive animals that eventually spread over the entire skin surface. Greenwood and Taylor (1985) reported the causes of death for captive killer whales ($n = 32$) to be pneumonia (25%), systemic mycosis (22%), other bacterial infections (15.6%), and mediastinal abscesses (9.4%), with 28% remaining undiagnosed. Severe atherosclerosis was found in the coronary arteries of one old stranded female (Roberts *et al.*, 1965). Hodgkin's disease has been reported in killer whales (Yonezawa *et al.*, 1989).

Individual strandings of killer whales are considered rare. In Alaska, there are a few cases of killer whales found stranded live, probably as a result of whales chasing fish or other prey into shallow water (e.g. Bristol Bay and Cook Inlet areas in Alaska). Whales inadvertently become stranded as a result of an outgoing tide (National Marine Mammal Laboratory, unpublished data). In 1994, a group of killer whales was trapped in a saltwater lake (Barnes Lake) in southeastern Alaska for approximately 60 days. Researchers intervened and successfully escorted 7 of the 9 whales to open water (D. Bain, personal communication). Mass strandings of killer whales are not common, with only about one dozen records worldwide (Mitchell and Reeves, 1988).

Live Maintenance

In the past, killer whales were considered ruthless killers to be feared and destroyed. Subsequent to the capture of a killer whale in the Pacific Northwest, it was found that, after capture, these large predators were usually gentle, intelligent, and remarkably adaptable. A live-capture fishery was active from 1962 to 1976 in the local waters of Washington State and British Columbia. By 1976, 55 whales had been taken for display purposes (Asper and Cornell, 1977). Bigg and Wolman (1975) provided data on catch statistics for the Pacific Northwest with slightly different numbers. Since 1976, no live-capture attempts have taken place in this area, with the exception of one collection in August 1977, when an immature female was found wounded and was taken for rehabilitation. A live-capture fishery has been operational since 1976 in the North Atlantic off Iceland (Sigurjónsson and Leatherwood, 1988). Demographics for killer whales in oceanaria in the United States and Canada are given in Duffield and Miller (1988) and Duffield *et al.* (1995). Ridgway (1979) has listed some causes of death.

In the United States, live-capture permits are required and basic standards must be met on pool size, water quality, nutrition, and health care. Millions of people have been given the opportunity to view killer whales at close range in aquariums and zoos throughout the world, making them more aware of the whales' presence in nature and providing researchers with rare opportunities to study these animals.

Human Effects

Directed fisheries

Killer whales have been exploited at low levels in several regions worldwide and more heavily exploited in three areas as summarized by Hoyt (1990). Norwegian whalers in the eastern North Atlantic took an average of 56 whales per year from 1938 to 1981. The Japanese took an average of 43 whales per year along their coastal waters from 1946 to 1981. The Soviets, whaling primarily in the Antarctic, took an average of 26 animals annually from 1935 to 1979, but took 916 animals in the 1979/80 Antarctic season (Dahlheim, 1981).

Tomilin (1957) listed the oil yield for killer whales as 750–950 kg per animal and Eschricht (1866) noted that 205 litres was yielded by a 6.4 m animal. Killer whale oil in Japan was equal in price to the oil of sperm whales (*Physeter macrocephalus*). *Orcinus* oil has an acid value of 0.63, saponification value of 211.9, and an iodine value of 86.4 (Nishiwaki and Handa, 1958). The fresh meat is used for human consumption in Japan while the aged meat and viscera are utilized as fertilizer or bait (Nishiwaki and Handa, 1958). In Norway, killer whale meat was used only for animal consumption (Jonsgård and Lyshoel, 1970).

By-catches

Incidental takes during fishing operations occur, but are considered rare (National Marine Mammal Laboratory, unpublished; Heyning, 1988; Leatherwood and Reeves, 1989).

Pollution

Effects of environmental pollutants are not known for cetaceans. Calambokidis *et al.* (1984) found high levels of PCBs and DDT (250 ppm and 640 ppm, respectively) in the blubber of an adult male transient killer whale in Washington State and 38 ppm PCB and 59 ppm DDE wet weight levels in a resident male killer whale. The levels in the transient whale are much higher than any reported for all other examined marine mammals in Washington State. Ono *et al.* (1987)

analysed blubber specimens collected from three killer whales taken off the coast of Japan for polychlorinated dibenzo-*p*-dioxins (PCDDs) and polychlorinated dibenzofurans (PCDFs). Several PCDF highly toxic congeners were identified in the samples.

Research following the 24 March 1989 Exxon Valdez oil spill in Prince William Sound, Alaska, documented killer whales transiting through the areas covered with a light sheen of oil (Harvey and Dahlheim, 1994; Matkin *et al.*, 1994). These authors noted that no attempt was made by the whales to avoid contaminated waters. Between 1989 and 1991, the resident AB pod in Prince William Sound lost 14 individuals (Matkin *et al.*, 1994). In late 1996, the integrity of AB pod was still in question. There is a strong correlation between the loss of the 14 whales and the Exxon Valdez oil spill, but there is no clear cause and effect relationship (Dahlheim and Matkin, 1994).

Habitat encroachment

Much concern has been raised about the effect of increasing human maritime activities and associated noise levels on whales. Phillips and Baird (1993) noted that increasing boat traffic had no adverse impact on killer whales. Acoustic investigations on captive killer whales concluded that (1) vessel noise can reduce the whale's ability to detect pure tones, (2) noise has the strongest effect when it is from the side or behind the whale, and (3) the whale's ability to detect broadband signals is not substantially affected by low levels of vessel noise (Bain and Dahlheim, 1994).

Directed attacks by free-ranging killer whales on humans resulting in death have not been recorded. However, Stephens (1962) lists several close approaches by killer whales to divers. In the early 1990s, a marine mammal trainer working at Sealand of the Pacific (Victoria, British Columbia) fell into a tank containing three captive killer whales. The woman drowned as a result of being pulled underwater by the adult male whale. Aggressive behaviour towards boats also has been noted (Notarbartolo di Sciara, 1978); however, these attacks may have been provoked by attempts to capture, harpoon, or harass the whales.

Acknowledgments

B. Andrews, D. Bain, R. Baird, K.C. Balcomb III, M. Bigg, R.L. Brownell, Jr, D. Odell, and F.G. Wood provided critical reviews of this manuscript. The National Marine Mammal Laboratory's Publication Review Committee and NMFS's Publication Unit also provided valuable reviews. Sea World kindly provided unpublished data. Artwork was completed by the Alaska Fisheries Science Center graphics unit (Kathy Zecca) and Pieter Folkens, Oceanic Society, San Francisco, California.

References

Aguayo L., A. (1975). Progress report on small cetacean research in Chile. *J. Fish. Res. Bd Can.* **32**, 1123–1143.

Arnbom, T., Papastavrou, V., Weilgart, L. S. and Whitehead, H. (1987). Sperm whales react to an attack by killer whales. *J. Mamm.* **66**, 450–453.

Asper, E. D. and Cornell, L. H. (1977). Live capture statistics for the killer whale (*Orcinus orca*) 1961–1976, in California, Washington, and British Columbia. *Aquat. Mamm.* **5**, 21–26.

Asper, E. D., Young, G. W. and Walsh, M. T. (1988). Observations on the birth and development of a captive-born killer whale. *Int. Zoo. Yearbook* **27**, 295–304.

Awbrey, F. T., Thomas, J. A., Evans, W. E. and Leatherwood, S. (1982). Ross Sea killer whale vocalizations: preliminary description and comparison with those of some northern hemisphere killer whales. *Rep. Int. Whal. Commn* **32**, 667–670.

Backus, R. H. (1961). Stranded killer whale in the Bahamas. *J. Mammal.* **42**, 418–419.

Bain, D. E. (1988). An evaluation of evolutionary processes: studies of natural selection, dispersal, and cultural evolution in killer whales (*Orcinus orca*). Ph.D. dissertation, University of California at Santa Cruz, 257 pp.

Bain, D. E. and Dahlheim, M. E. (1994). Effects of masking noise on detection thresholds of killer whales. *In* "Impacts of the Exxon Valdez Oil Spill on Marine Mammals" (Ed. T. R. Loughlin) pp. 243–256. Academic Press, San Diego.

Bain, D. E., Dahlheim, M. E., Kriete, B., Mullen, J., Marrin-Cooney, D. and Povey, P. (in prep.). Auditory thresholds and critical ratios of killer whales (*Orcinus orca*).

Baird, R. W. and Dill, L. W. (1995). Occurrence and behavior of transient killer whales: seasonal and pod specific variability, foraging behavior, and prey handling. *Can. J. Zool.* **73**, 1300–1311.

Baird, R. W. and Dill, L. W. (1996). Ecological and social determinants of group size in transient killer whales. *Behav. Ecol.* **7**, 408–416.

Baird, R. W. and Goodyear, J. D. (1993). An examination of killer whale diving behavior using a recoverable, suction-cup attached TDR/VHF tag. *In* Abstracts of the Tenth Biennial Conference on the Biology of Marine Mammals, 11–15 November, Galveston, TX.

Baird, R. W. and Stacey, P. J. (1988). Variation in saddle patch pigmentation in populations of killer whales (*Orcinus orca*) from British Columbia, Alaska, and Washington State. *Can. J. Zool.* **66**, 2582–2585.

Baird, R. W., Abrams, P. A. and Dill, L. M. (1992). Possible indirect interactions between transient and resident killer whales: implications for the evolution of foraging specializations in the genus *Orcinus*. *Oecologia* **89**, 125–132.

Baker, A. N. (1983). "Whales and Dolphins of New Zealand and Australia: an Identification Guide". Victoria University Press, Wellington, 133 pp.

Balcomb, K. C., III, and Goebel, C. A. (1976). A killer whale study in Puget Sound. Final Report to the Marine Mammal Division, National Marine Fisheries Service, Contract No. NASO-6-35330, Seattle, WA, 11 pp.

Balcomb, K. C., III, Boran, J. R., Osborne, R. W. and Haenel, N. J. (1979). Observations of killer whales (*Orcinus orca*) in greater Puget Sound, State of Washington. Report to the Marine Mammal Commission, Washington, DC.

Barrett-Lennard, L. G., Smith, T. G. and Ellis, G. M. (1996). A cetacean biopsy system using lightweight pneumatic darts, and its effect on the behavior of killer whales. *Mar. Mamm. Sci.* **12**, 14–27.

Benirschke, K. and Cornell, L. H. (1987). The placenta of the killer whale (*Orcinus orca*). *Mar. Mamm. Sci.* **3**, 82–86.

Berzin, A. A. and Vladimirov, V. L. (1982). A new species of killer whale from the Antarctic. *Priroda*, No. 6, 31 (Translated by S. Pearson, National Marine Mammal Laboratory, Seattle, WA).

Berzin, A. A. and Vladimirov, V. L. (1983). A new species of killer whale (Cetacea Delphinidae) from Antarctic waters. *Zool. Zh.* **62**, 287–295 (Translated by S. Pearson, National Marine Mammal Laboratory, Seattle, WA).

Berzin, A. A. and Vlasova, L. P. (1982). Fauna of the cetacean Cyamidae (Amphipoda) of the world ocean. *Invest. Cetacea* **13**, 149–164.

Best, P. B. and Ross, G. J. B. (1977). Exploitation of small cetaceans off Southern Africa. *Rep. Int. Whal. Commn* **27**, 494–497.

Bigg, M. A. (1982). An assessment of killer whale (*Orcinus orca*) stocks off Vancouver Island, British Columbia. *Rep. Int. Whal. Commn* **32**, 655–666.

Bigg, M. A. and Wolman, A. A. (1975). Live-capture killer whale (*Orcinus orca*) fishery, British Columbia and Washington, 1962–1973. *J. Fish. Res. Bd Can.* **32**, 1213–1221.

Bigg, M. A., Ellis, G. M., Ford, J. K. B. and Balcomb, K. C. (1987). "Killer whales — A Study of their Identification, Genealogy and Natural History in British Columbia and Washington State". Phantom Press and Publishers, Nanaimo, British Columbia, 79 pp.

Bigg, M. A., Olesiuk, P. F., Ellis, G. M., Ford, J. K. B. and Balcomb, K. C. (1990). Social organization and genealogy of resident killer whales (*Orcinus orca*) in the coastal waters of British Columbia and Washington State. *In* "Individual Recognition of Cetaceans: Use of Photo-identification and other Techniques to Estimate Population Parameters" (Eds P. S. Hammond, S. A. Mizroch and G. P. Donovan) pp. 383–405. *Rep. Int. Whal. Commn*, Spec. Issue No. 12, Cambridge.

Black, N. A., Shulman-Janiger, A., Ternullo, R. L. and Guerrero-Ruiz, M. (1996). A killer whale catalog for California and western Mexico. Report for National Marine Fisheries Service, Southwest Fisheries Science Center/NOAA, Contract No. 40JGNF500356.

Black, N. A., Schulman, A., Ellifrit, D., Shearwater, D., Baldridge, A., Ternullo, R. L., Goley, D., Calambokidis, J. and Webber, M. (1993). Photo-identification of killer whales off California. *In* Abstracts of the Tenth Biennial Conference on the Biology of Marine Mammals, 11–15 November, Galveston, TX.

Bloch, D. and Lockyer, C. (1988). Killer whales (*Orcinus orca*) in Faroese waters. *In* "North Atlantic Killer Whales" (Eds J. Sigurjónsson and S. Leatherwood) pp. 55–64. *Rit Fiskideildar (Journal of the Marine Institute Reykjavik)*, Vol. XI.

Bowers, C. A. and Henderson, R. S. (1972). Project deep ops: deep object recovery with pilot and killer whales. Naval Undersea Center Technical Paper 306, 86 pp.

Braham, H. W. and Dahlheim, M. E. (1982). Killer whales in Alaska documented in the Platforms of Opportunity Program. *Rep. Int. Whal. Commn* **32**, 643–646.

Brault, S. and Caswell, H. (1993). Pod-specific demography of killer whales (*Orcinus orca*). *Ecology* **74**, 1444–1454.

Brown, S. G., Brownell, J. R., Jr, Erickson, A. W., Hofman, R. J., Llano, G. A. and Mackintosh, N. A. (1974). "Antarctic Mammals". American Geographic Society Folio 18.

Buckland, S. T. (1990). Estimation of survival rates from sightings of individually identifiable whales. *In* "Individual Recognition of Cetaceans: Use of Photo-identification and other Techniques to Estimate Population Parameters" (Eds P. S. Hammond, S. A. Mizroch and G. P. Donovan) pp. 149–153. *Rep. Int. Whal. Commn*, Spec. Issue No. 12, Cambridge.

Budylenko, G. A. (1981). Distribution and some aspects of biology of killer whales in the South Atlantic. *Rep. Int. Whal. Commn* **31**, 523–525.

Butterworth, D. S. and De Decker, J. G. (1990). Estimates of abundance for Antarctic blue, fin, sei, sperm, humpback, killer and pilot whales from the 1978/79 to 1985/86 IWC/IDCR sighting survey cruises. Report submitted to International Whaling Commission, SC/41/O20, 598–599.

Cadenat, J. (1959). Rapport sur les petits Cetaces ouest-africains. *Bull. Inst. Fr. Afr. Noire* (A)**21**, 1367–1409.

Calambokidis, J., Peard, J., Steiger, G., Cubbage, J. and DeLong, R. L. (1984). Chemical contaminants in marine mammals from Washington State. US Department of Commerce, NOAA Technical Memorandum F/NWC-6, 1–167.

Caldwell, D. K. and Brown, D. H. (1964). Tooth wear as a correlate of described feeding behavior by the killer whale, with notes on a captive specimen. *Bull. South. Calif. Acad. Sci.* **63**, 128–140.

Caldwell, M. C. and Caldwell, D. K. (1966). Epimeletic (care-giving) behaviour in Cetacea. *In* "Whales, Porpoises and Dolphins" (Ed. K. S. Norris) pp. 755–789. University of California Press, Berkeley, CA.

Caldwell, D. K., Layne, J. N. and Siebenaler, J. B. (1956). Notes on a killer whale (*Orcinus orca*) from the northeastern Gulf of Mexico. *Q. J. Florida Acad. Sci.* **19**, 189–196.

Campbell, R. R., Yurick, D. B. and Snow, N. B. (1988). Predation on narwhals, *Monodon monoceros*, by killer whales, *Orcinus orca*, in the Eastern Canadian Arctic. *Can. Field-Naturalist* **102**, 689–696.

Capellini, G. (1883). Di un'orca fossile scoperta a Cetona in Toscana. *Mem. Accad. Sci. Int. Bologna*, Ser. 4, 665–687.

Carl, G. C. (1946). A school of killer whales stranded at Estevan Point, Vancouver Island. Report for the Provincial Museum of Natural History and Anthropology 1945, pp. 821–828.

Carl, G. C. (1960). Albinistic killer whales in British Columbia. Report for the Provincial Museum of Natural History and Anthropology, pp. 29–36.

Cave, A. J. E. (1977). The reniculus in *Hyperoodon* and *Orcinus*. *Invest. Cetacea* **8**, 103–120 + 3 plts.

Christensen, I. (1984). Growth and reproduction of killer whales, *Orcinus orca*, in Norwegian coastal waters. *In* "Reproduction in Whales, Dolphins, and Porpoises" (Eds W. F. Perrin, R. L. Brownell, Jr and D. P. DeMaster) pp. 253–258. *Rep. Int. Whal. Commn*, Spec. Issue No. 6, Cambridge.

Christensen, I. (1988). Distribution, movements and abundance of killer whales (*Orcinus orca*) in Norwegian coastal waters, 1982–1987, based on questionnaire surveys. *In* "North Atlantic Killer Whales" (Eds J. Sigurjónsson and S. Leatherwood) pp. 79–88. *Rit Fiskideildar (Journal of the Marine Institute Reykjavik)*, Vol. XI.

Cornell, L. H. (1983). Hematology and clinical chemistry values in the killer whales, *Orcinus orca* L. *J. Wildl. Dis.* **19**, 259–264.

Cummings, W. C. and Thompson, P. O. (1971). Gray whales, *Eschrichtius robustus*, avoid the underwater sounds of killer whales, *Orcinus orca*. *Fish. Bull.* **69**, 525–530.

Cummings, W. C., Fish, J. F. and Thompson, P. O. (1972). Sound production and other behaviour of southern right whales, *Eubalaena glacialis*. *Trans. San Diego Soc. Nat. Hist.* **17**, 1–13.

Dahlheim, M. E. (1980). Killer whales observed bowriding. *Murrelet* **61**, 78.

Dahlheim, M. E. (1981). A review of the biology and exploitation of the killer whale, *Orcinus orca*, with comments on recent sightings from Antarctic. *Rep. Int. Whal. Commn* **31**, 541–546.

Dahlheim, M. E. (1987). The Bio-acoustics of the gray whale (*Eschrichtius robustus*). Ph.D. dissertation, University of British Columbia, 317 pp.

Dahlheim, M. E. (1988). Killer whale (*Orcinus orca*) depredation on longline catches of sablefish (*Anoplopoma fimbria*) in Alaskan waters. Northwest and Alaska Fisheries Center, National Marine Fisheries Service, NWAFC Processed Report 88–14, 31 pp.
Dahlheim, M. E. (1994). Abundance and distribution of killer whales (*Orcinus orca*) in Alaska, 1993. Annual Report to the MMPA Assessment Program, Office of Protected Resources, NMFS, NOAA, Silver Spring, MD.
Dahlheim, M. E. (1997). A photographic catalog of killer whales (*Orcinus orca*) from the central Gulf of Alaska to the southeastern Bering Sea. NOAA Technical Report, NMFS 131, November 1997, 54 pp..
Dahlheim, M. E. and Awbrey, F. (1982). A classification and comparison of vocalizations of captive killer whales (*Orcinus orca*). *J. Acoust. Soc. Am.* **72**, 661–670.
Dahlheim, M. E. and Towell, R. T. (1994). Occurrence and distribution of Pacific white-sided dolphins (*Lagenorhynchus obliquidens*) in southeastern Alaska, with notes on an attack by killer whales (*Orcinus orca*). *Mar. Mamm. Sci.* **10**, 458–464.
Dahlheim, M. E., Ellifrit, D. K. and Swenson, J. D. (1997). "Killer Whales of Southeast Alaska: A Catalogue of Photo-identified Individuals". Day Moon Press, Seattle, WA, 79 pp.
Dahlheim, M. E., Leatherwood, J. S. and Perrin, W. E. (1982). Distribution of killer whales in the warm temperate and tropical eastern Pacific. *Rep. Int. Whal. Commn* **32**, 647–653.
Dahlheim, M. E. and Matkin, C. O. (1994). Assessment of injuries to Prince William Sound killer whales. *In* "Impacts of the Exxon Valdez Oil Spill on Marine Mammals" (Ed. T. R. Loughlin) pp. 163–171. Academic Press, San Diego.
Dahlheim, M. E. and Waite, J. M. (1993). Abundance and distribution of killer whales (*Orcinus orca*) in Alaska, 1992. Annual Report to the MMPA Assessment Program, Office of Protected Resources, NMFS, NOAA, Silver Spring, MD.
Dailey, M. D. and Brownell, R. L., Jr. (1972). A checklist of marine mammal parasites. *In* "Mammals of the Sea: Biology and Medicine" (Ed. S. H. Ridgway) pp. 528–589. Charles C. Thomas, Springfield, IL.
Davis, R. W., Williams, T. M. and Awbrey, F. (1988). Sea otter oil spill avoidance study. US Department of the Interior, Mineral Management Service, MMS-88-0051, 65 pp.
DeMaster, D. P. and Drevenak, J. K. (1988). Survivorship patterns in three species of captive cetaceans. *Mar. Mamm. Sci.* **4**, 297–311.
Diereks, K. J., Troehta, R. T., Greenlaw, C. F. and Evans, W. E. (1971). Recordings and analysis of dolphin echolocation sightings. *J. Acoust. Soc. Am.* **49**, 1729–1732.
Dolphin, W. F. (1987). Observations of humpback whale, *Megaptera novaeangliae* — killer whale, *Orcinus orca*, interactions in Alaska: comparison with terrestrial predatory-prey relationships. *Can. Field-Naturalist* **101**, 70–75.
Donaldson, B. J. (1977). The tongue of bottlenose dolphin (*Tursiops truncatus*). *In* "Functional Anatomy of Marine Mammals", Vol. 3 (Ed. R. J. Harrison) pp. 175–197. Academic Press, New York.
Duffield, D. A. and Miller, K. W. (1988). Demographic features of killer whales in oceanaria in the United States and Canada, 1965–1987. *In* "North Atlantic Killer Whales" (Eds J. Sigurjónsson and S. Leatherwood) pp. 297–306. *Rit Fiskideildar (Journal of the Marine Institute Reykjavik)*, Vol. XI.
Duffield, D. A., Odell, D. K., McBain, J. F. and Andrews, B. (1995). Killer whale (*Orcinus orca*) reproduction at Sea World. *Zoo. Biol.* **14**, 417–430.
Ellis, G. (1984). Killer whales of southern Alaska: a catalogue of individuals photo-identified in 1984. Hubbs Sea World Research Institute Technical Report No. 84-176, San Diego, 73 pp.

Ellis, G. (1987). Killer whales of Prince William Sound and Southeast Alaska: a catalogue of individuals photoidentified, 1976–1986. Hubbs Sea World Research Institute Technical Report No. 87–200, San Diego, 76 pp.

Erickson, A. W. (1978). Population studies of killer whales (*Orcinus orca*) in the Pacific Northwest: a radio-marking and tracking study of killer whales. Report to the Marine Mammal Commission, Contract PB-285 615, 31 pp.

Eschricht, D. F. (1866). On the species of the genus *Orca* inhabiting the northern seas. *In* "Recent Memoirs on the Cetacea" (Ed. W. H. Flower) pp. 151–188. Ray Society, London, 312 pp. + 4 plts.

Evans, P. G. H. (1988). Killer whales (*Orcinus orca*) in British and Irish waters. *In* "North Atlantic Killer Whales" (Eds J. Sigurjónsson and S. Leatherwood) pp. 42–54. *Rit Fiskideildar (Journal of the Marine Institute Reykjavik)*, Vol. XI.

Evans, W. E. (1973). Echolocation by marine delphinids and one species of freshwater dolphin. *J. Acoust. Soc. Am.* **54**, 191–199.

Evans, W. E. and Bastian, J. (1969). Marine mammal communication: social and ecological factors. *In* "The Biology of Marine Mammals" (Ed. H. T. Andersen) pp. 425–475. Academic Press, New York.

Evans, W. E., Yablokov, A. V. and Bowles, A. E. (1982). Geographic variation in the color pattern of killer whales. *Rep. Int. Whal. Commn* **32**, 687–694.

Felleman, F. L., Heimlich-Boran, J. R. and Osborne, R. W. (1991). The feeding ecology of killer whales (*Orcinus orca*) in the Pacific Northwest. *In* "Dolphin Societies: Discoveries and Puzzles" (Eds K. Pryor and K. Norris) pp. 113–147. University of California Press, Berkeley, CA.

Fertl, D., Acevedo-Gutierrez, A. and Darby, F. L. (1996). A report of killer whales (*Orcinus orca*) feeding on a carcharhinid shark in Costa Rica. *Mar. Mamm. Sci.* **12**, 606–611.

Fiscus, C. H. and Niggol, K. (1965). Observations of cetaceans off California, Oregon and Washington. US Fish and Wildlife Service, Special Scientific Report Fisheries No. 498, 27 pp.

Fish, J. F. and Vania, J. S. (1971). Killer whale, *Orcinus orca*, sounds repel white whales, *Delphinapterus leucas*. *Fish. Bull.* **69**, 531–535.

Ford, J. K. B. (1989). Acoustic behavior of resident killer whales (*Orcinus orca*) off Vancouver Island, British Columbia. *Can. J. Zool.* **67**, 727–745.

Ford, J. K. B. and Fisher, H. D. (1982). Killer whale (*Orcinus orca*) dialects as an indicator of stocks in British Columbia. *Rep. Int. Whal. Commn* **32**, 671–679.

Ford, J. K. B. and Fisher, H. D. (1983). Group-specific dialects of killer whales (*Orcinus orca*) in British Columbia. *In* "Communication and Behavior of Whales" (Ed. R. Payne) pp. 129–161. AAAS Selected Symposia Series, Westview Press, Boulder, CO.

Ford, J. K. B. and Hubbard-Morton, A. B. (1990). Vocal behavior and dialects of transient whales in coastal waters of British Columbia, California, and Southeast Alaska. *In* Abstracts of the Third International Orca Symposium, Victoria, B.C., Canada, 9–12 March, 1990.

Fowler, C. W. (1984). Density dependence in cetacean populations. *In* "Reproduction in Whales, Dolphins, and Porpoises" (Eds W. F. Perrin, R. L. Brownell, Jr and D. P. DeMaster) pp. 373–380. *Rep. Int. Whal. Commn*, Spec. Issue No. 6, Cambridge.

Fowler, C. W. (1987). A review of density dependence in populations of large mammals. *In* "Current Mammalogy" (Ed. H. H. Genoways) pp. 401–441. Plenum Press, New York.

Fraser, F. C. (1949). Whales and dolphins. *In* "Field Book of Giant Fishes" (Eds J. R. Norman and F. C. Fraser) pp. 201–349. G. Putnam's Sons, New York.

Fraser, F. C. and Purves, P. E. (1960). Hearing in cetaceans. *Bull. Brit. Mus. (Nat. Hist.) Zool.* **7**, 1–140.

Frost, D. H. H., Shaughnessy, P. D., Semmelink, A., Sketch, M. and Siegfried, W. R. (1975). The response of jackass penguins to killer whale vocalizations. *S. Afr. J. Sci.* **71**, 157–158.
Gallivan, G. J. (1992). What are the metabolic rates of cetaceans? *Physiol. Zool.* **65**, 1285–1297.
Gaskin, D. E. (1968). Distribution of Delphinidae (Cetacea) in relation to sea surface temperatures off eastern and southern New Zealand. *NZ Mar. J. Freshwater Resources* **2**, 527–534.
Gilmore, R. (1976). Killer whales in the San Diego Area, Del Mar to the Coronado Islands. American Cetacean Society Newsletter, San Diego, CA, 2 pp.
Glass, B. P. (1974). "A Key to Skulls of North American Mammals". Oklahoma State University, Stillwater, 59 pp.
Goley, P. D. and Straley, J. M. (1994). Attack on gray whales (*Eschrichtius robustus*) in Monterey Bay, California, by killer whales (*Orcinus orca*) previously identified in Glacier Bay, Alaska. *Can. J. Zool.* **72**, 1528–1530.
Green, R. F. (1972). Observations on the anatomy of some cetaceans and pinnipeds. *In* "Mammals of the Sea: Biology and Medicine" (Ed. S. H. Ridgway) pp. 247–297. Charles C. Thomas, Springfield, IL.
Greenwood, A. G. and Taylor, D. C. (1985). Captive killer whales in Europe. *Aquat. Mamm.* **1**, 10–12.
Greenwood, A. G., Harrison, R. J. and Whitting, H. W. (1974). Functional and pathological aspect of the skin of marine mammals. *In* "Functional Anatomy of Marine Mammals" (Ed. R. J. Harrison) pp. 73–110. Academic Press, New York.
Grieg, J. A. (1906). Nogle notiser fra et spaekhuggerstaeng ved bildostrommen, 1 Jan. 1904. Bergens Mus. Arbok., Naturvitensk. *Rekke* **2**, 1–28.
Guinet, C. (1988). Historique de la presence d'orques autour de l'ile de la Possession, archipel Crozet: photo-identification 1964–1986. *Mammalia* **52**, 285–289 (English translation by A. V. Murray, Marine Mammal Research Group, Victoria).
Guinet, C. (1991). Intentional stranding apprenticeship and social play in killer whales (*Orcinus orca*). *Can. J. Zool.* **69**, 2712–2716.
Guinet, C. (1992). Predatory behavior of killer whales (*Orcinus orca*) around the Crozet Islands. *Can. J. Zool.* **70**, 1656–1667.
Guldberg, G. and Nansen, F. (1894). "On the Development and Structure of the Whales", Part I, "On the Development of the Dolphin". Bergen Museum, 70 pp. + 7 plts.
Haenel, N. J. (1986). General notes on the behavioral ontogeny of Puget Sound killer whales and the occurrence of allomaternal behavior. *In* "Behavioral Biology of Killer Whales" (Eds B. Kirkevold and J. S. Lockard) pp. 285–300. Alan R. Liss, New York.
Haley, D. (1973). Albino killer whale. *Sea Frontiers* **19**, 66–71.
Hall, J. D. (1981). Aspects of the natural history of cetaceans of Prince William Sound, Alaska. Ph.D. dissertation, University of California at Santa Cruz, 148 pp.
Hall, J. D. and Johnson, C. S. (1972). Auditory threshold of a killer whale, *Orcinus orca* (Linnaeus). *J. Acoust. Soc. Am.* **51**, 515–517.
Hammond, P. S. and Lockyer, C. (1988). Distribution of killer whales in the eastern North Atlantic. *In* "North Atlantic Killer Whales" (Eds J. Sigurjónsson and S. Leatherwood) pp. 24–41. *Rit Fiskideildar (Journal of the Marine Institute Reykjavik)*, Vol. XI.
Harmer, S. F. (1927). Report on Cetacea stranded on the British coasts from 1913 to 1926. *Brit. Mus. (Nat. Hist.)* **10**, 1–91 + 7 maps.

Harrison, R. J., Brownell, R. L., Jr, and Boice, R. L. (1972). Reproduction and gonadal appearances in some odontocetes. *In* "Functional Anatomy of Marine Mammals", Vol. 1 (Ed. R. J. Harrison) pp. 362–429. Academic Press, New York.

Hart, T. J. (1935). On the diatoms of the skin film of whales, and their possible bearing on problems of whale movements. *Discov. Rep.* **10**, 247–282, + 1 plt.

Harvey, J. T. and Dahlheim, M. E. (1994). Cetaceans in oil. *In* "Impacts of the *Exxon Valdez* Oil Spill on Marine Mammals" (Ed. T. R. Loughlin) pp. 257–264. Academic Press, San Diego.

Heide-Jörgensen, M. P. (1988). Occurrence and hunting of killer whales in Greenland. *In* "North Atlantic Killer Whales" (Eds J. Sigurjónsson and S. Leatherwood) pp. 115–135. *Rit Fiskideildar (Journal of the Marine Institute Reykjavik)*, Vol. XI.

Heimlich-Boran, S. L. (1986). Cohesive relationships among Puget Sound killer whales. *In* "Behavioral Biology of Killer Whales" (Eds B. Kirkevold and J. S. Lockard) pp. 261–284. Alan R. Liss, New York.

Heimlich-Boran, J. R. (1988). Behavioral ecology of killer whales (*Orcinus orca*) in the Pacific Northwest. *Can. J. Zool.* **66**, 565–578.

Heise, K., Ellis, G. and Matkin, C. (1991). "A Catalogue of Prince William Sound Killer Whales". North Gulf Oceanic Society, Homer, AK, 51 pp.

Heise, K., Barrett-Lennard, L., Ford, J. K. B. and Ellis, G. (1993). Killer whales of the Queen Charlotte Islands. Report prepared for the South Moresby/Gwaii Haanas National Park Reserve, Canadian Parks Service, Queen Charlotte City, B.C., 22 pp.

Hershkovitz, P. (1966). Catalog of living whales. *U.S. Natl Mus. Bull.* **246**, 81–86.

Heyning, J. E. (1988). Presence of solid food in a young calf killer whale (*Orcinus orca*). *Mar. Mamm. Sci.* **4**, 68–71.

Heyning, J. E. (1989). Comparative facial anatomy of beaked whales (*Ziphiidae*) and a systematic revision among the families of extant Odontoceti. Contributions in Science, Natural History Museum of Los Angeles County 405, 64 pp.

Heyning, J. E. and Brownell, R. L., Jr (in prep.). Variation in external morphology of killer whales.

Heyning, J. E. and Dahlheim, M. E. (1988). "*Orcinus orca*". Mammalian Species Account, American Society of Mammalogists, No. 304, 9 pp. + 4 figs.

Heyning, J. E. and Dahlheim, M. E. (1993). How distinct are resident and transient whales? Abstracts of the Tenth Biennial Conference on the Biology of Marine Mammals, Galveston, TX, 11–15 November.

Heyning, J. E. and Hohn, A. (1993). Refining age estimation techniques for killer whales. Contract report to the National Marine Mammal Laboratory, Seattle, WA, 8 pp.

Hoelzel, A. R. (1991a). Analysis of regional mitochondrial DNA variation in the killer whale; implications for cetacean conservation. *In* "Genetic Ecology of Whales and Dolphins" (Ed. A. R. Hoelzel) pp. 225–233. *Rep. Int. Whal. Commn*, Spec. Issue No. 13, Cambridge.

Hoelzel, A. R. (1991b). Killer whale predation on marine mammals in Punta Norte, Argentina; food sharing, provisioning, and foraging strategy. *Behav. Ecol. Sociobiol.* **29**, 297–304.

Hoelzel, A. R. and Dover, G. A. (1991). Genetic differentiation between sympatric killer whale populations. *J. Heredity* **66**, 191–195.

Hoelzel, A. R., Dahlheim, M. E. and Stern, S. J. (1998). Low genetic variation among killer whales (*Orcinus orca*) in the eastern North Pacific, and differentiation between foraging specialists. *J. Heredity* **89**, 121–128.

Hoelzel, A. R., Ford, J. K. B. and Dover, G. A. (1991). A paternity test case for the killer whale (*Orcinus orca*) by DNA fingerprinting. *Mar. Mamm. Sci.* **7**, 35–43.

Hoyt, E. (1990). "The Whale Called Killer". E.P. Dutton, New York, 291 pp.

Innes, S. and Lavigne, D. M. (1991). Do cetaceans really have elevated metabolic rates? *Physiol. Zool.* **84**, 1130–1134.

Iredale, T. and Troughton, E. Le G. (1969). A checklist of the mammals of Australia. *Rec. Aust. Mus. Sydney* **6**, 63–68.

Ivashin, M. V. (1981). USSR progress report on cetacean research June 1979–May 1980. *Rep. Int. Whal. Commn* **32**, 221–226.

Jacobsen, J. J. (1986). The behavior of *Orcinus orca* in the Johnstone Strait, British Columbia. *In* "Behavioral Biology of Killer Whales" (Eds B. Kirkevold and J. S. Lockard) pp. 135–185. Alan R. Liss, New York.

Jefferson, T. A., Stacey, P. J. and Baird, R. W. (1991). A review of killer whale interactions with other marine mammals: predation to co-existence. *Mamm. Rev.* **21**, 151–180.

Jehl, J. R., Jr, Evans, W. E., Awbrey, F. T. and Drieschman, W. S. (1980). Distribution and geographic variation in the killer whale (*Orcinus orca*) populations of the Antarctic and adjacent waters. *Antarct. J. U.S.* **15**, 161–163.

Jonsgård, Å. and Lyshoel, P. B. (1970). A contribution to the knowledge of the biology of the killer whale, *Orcinus orca* (L.). *Nytt. Mag. Zool.* **18**, 41–48.

Jonsgård, Å. and Oynes, P. (1952). Om bottlenosen (*Hyperoodon rostratus*) og spekkoggeren (*Orcinus orca*). *Fauna* **1**, 1–17.

Kasamatsu, F., Hembree, D., Joyce, G., Tsunoda, L., Rowlett, R. and Nakano, T. (1987). Distribution of cetacean sightings in the Antarctic: Results obtained from the IWC/ICDR Minke Whale Assessment Cruises, 1978/79 to 1983/84. Working manuscript SC/39/O 10 submitted to the International Whaling Commission, 77 pp.

Kastelein, R. A. and Vaughan, N. (1989). Food consumption, body measurements and weight changes of a female killer whale (*Orcinus orca*). *Aquat. Mamm.* **15**, 18–21.

Kasting, N. (1991). Reply to technical comment. *Physiol. Zool.* **64**, 1135–1136.

Kasting, N. W., Adderley, S. A. L., Safford, T. and Hewlett, K. G. (1989). Thermoregulation in beluga (*Delphinapterus leucas*) and killer (*Orcinus orca*) whales. *Physiol. Zool.* **62**, 687–701.

Kasuya, T. (1971). Consideration of distribution and migration of toothed whales off the Pacific coast of Japan based on aerial sighting records. *Sci. Rep. Whales Res. Inst.* **23**, 37–60.

Kasuya, T. (1973). Systematic consideration of recent toothed whales based on the morphology of tympano-periotic bone. *Sci. Rep. Whales Res. Inst.* **25**, 1–103.

Kasuya, T. and Marsh, H. (1984). Life history and reproductive biology of the short-finned pilot whale, *Globicephala macrorhynchus*, off the Pacific coast of Japan. *In* "Reproduction in Whales, Dolphins, and Porpoises" (Eds W. F. Perrin, R. L. Brownell, Jr and D. P. DeMaster) pp. 259–310. *Rep. Int. Whal. Commn*, Spec. Issue No. 6, Cambridge.

Katona, S. K., Beard, J. A., Girton, P. E. and Wenzel, F. (1988). Killer whales (*Orcinus orca*) from the Bay of Fundy to the Equator, including the Gulf of Mexico. *In* "North Atlantic Killer Whales" (Eds J. Sigurjónsson and S. Leatherwood) pp. 205–224. *Rit Fiskideildar (Journal of the Marine Institute of Reykjavik)*, Vol. XI.

Klimley, A. P. and Myrberg, A. A., Jr (1979). Acoustic stimuli underlying withdrawal from a sound source by adult lemon sharks, *Negaprion brevirostris* (Poey). *Bull. Mar. Sci.* **29**, 447–458.

Lang, T. G. (1966). Hydrodynamic analysis of cetacean performance. *In* "Whales, Porpoises and Dolphins" (Ed. K. S. Norris) pp. 410–432. University of California Press, Berkeley, CA.

Leatherwood, J. S. and Dahlheim, M. E. (1978). Worldwide distribution of pilot whales and killer whales. Naval Ocean Systems Center Technical Report 443, 39 pp.

Leatherwood, S. and Reeves, R. R. (1989). "Marine Mammal Research and Conservation in Sri Lanka 1985–1986". United Nations Environment Programme, Nairobi, Kenya, 138 pp.
Leatherwood, S., Caldwell, D. K. and Winn, H. E. (1976). Whales, dolphins and porpoises of the western North Atlantic: a guide to their identification. NOAA Technical Report NMFS Circular 396, 176 pp.
Leatherwood, S., Matkin, C. O., Hall, J. D. and Ellis, G. M. (1990a). Killer whales, *Orcinus orca*, photo-identified in Prince William Sound, Alaska, 1976 through 1987. *Can. Field-Naturalist* **104**, 362–371.
Leatherwood, J. S., McDonald, D., Prematunga, W. P., Girton, P., McBrearty, D. and Ilangakoon, A. (1990b). Records of the "blackfish" (killer, false killer, pilot, pygmy killer, and melon-headed whales) in the Indian Ocean Sanctuary, 1772–1976. *In* "Cetaceans and Cetacean Research in the Indian Ocean Sanctuary" (Eds S. Leatherwood and G. P. Donovan) pp. 32–65. United Nations Environment Programme, Marine Mammal Technical Publication No. 3, Nairobi, Kenya.
Lenfant, C. (1969). Physiological properties of blood of marine mammals. *In* "The Biology of Marine Mammals" (Ed. H. T. Andersen) pp. 95–116. Academic Press, New York.
Leung, Y. M. (1970). *Cyamus orcina*, a new species of whale louse (Cyamidae, Amphipoda) from a killer whale. *Bull. Inst. Fr. Afr. noire* (A) **32**, 669–675.
Lien, J., Christensen, I., Lien, M. and Jones, P. W. (1988a). A note on killer whales (*Orcinus orca*) near Svolvaer, Norway in November–December 1984. *In* "North Atlantic Killer Whales" (Eds J. Sigurjónsson and S. Leatherwood) pp. 95–98. *Rit Fiskideildar (Journal of the Marine Institute Reykjavik)*, Vol. XI.
Lien, J., Stenson, G. B. and Jones, P. W. (1988b). Killer whales (*Orcinus orca*) in waters of Newfoundland and Labrador, 1978–1986. *In* "North Atlantic Killer Whales" (Eds J. Sigurjónsson and S. Leatherwood) pp. 194–201. *Rit Fiskideildar (Journal of the Marine Institute Reykjavik)*, Vol. XI.
Lilly, J. C. and Miller, A. M. (1961). Sounds emitted by the bottlenose dolphin. *Science* **133**, 1689–1693.
Ljungblad, D. K. and Moore, S. E. (1983). Killer whales (*Orcinus orca*) chasing gray whales (*Eschrichtius robustus*) in the northern Bering Sea. *Arctic* **36**, 361–364.
Lockyer, C. (1979). Response of orcas to tagging. *Carnivore* **2**, 19–21.
Lopez, J. D. and Lopez, D. (1985). Killer whales (*Orcinus orca*) of Patagonia, and their behavior of intentional stranding while hunting nearshore. *J. Mammal.* **66**, 181–183.
Lowry, L. F., Nelson, R. R. and Frost, K. J. (1987). Observations of killer whales, *Orcinus orca*, in Western Alaska: sightings, strandings, and predation on other marine mammals. *Can. Field-Naturalist* **101**, 6–12.
Lyrholm, T. (1988). Photoidentification of individual killer whales (*Orcinus orca*) off the coast of Norway, 1983–1986. *In* "North Atlantic Killer Whales" (Eds J. Sigurjónsson and S. Leatherwood) pp. 89–94. *Rit Fiskideildar (Journal of the Marine Institute Reykjavik)*, Vol. XI.
MacNeill, A. C. (1975). Blood values for some captive cetaceans *Can. Vet. J.* **16**, 187–193.
Matkin, C. O. and Saulitis, E. (1994). Killer whale (*Orcinus orca*) biology and management in Alaska. Contract No. T75135023 to the Marine Mammal Commission, Washington, DC.
Matkin, C. O., Ellis, G. M., Dahlheim, M. E. and Zeh, J. (1994). Status of killer whales in Prince William Sound (1985–1992). *In* "Impacts of the *Exxon Valdez* Oil Spill on Marine Mammals" (Ed. T. R. Loughlin) pp. 141–162. Academic Press, San Diego.
Matkin, D. R. and Dahlheim, M. E. (1995). Feeding behaviors of killer whales in northern southeastern Alaska. *In* "Proceedings of the Third Glacier Bay Science Symposium 1993" (Ed. D. R. Engstrom) pp. 246–253. National Park Service, Anchorage, AK.

Matsumoto, H. (1937). A new species of orca from the Basal Calabrian at Naganuma, Minato Town, Province of Kazusa, Japan. *Zool. Mag. Jap.* **49**, 191–193.

McBrearty, D. A., Message, M. A. and King, G. A. (1986). Observations on small cetaceans in the north-east Atlantic Ocean and Mediterranean Sea: 1978–1982. *In* "Research on Dolphins" (Eds M. M. Bryden and R. Harrison) pp. 225–249. Clarendon Press, Oxford.

Mead, J. G. (1975). Anatomy of the external nasal passages and facial complex in the Delphinidae (Mammalia: Cetacea). *Smithsonian Contrib. Zool.* **207**, 1–72.

Meuth, J. L., Jones, B. N. and Gurd, F. R. N. (1981). Reassignment of residue 122 in the myoglobin from the killer whale, *Orcinus orca*. *J. Molec. Evol.* **17**, 163–166.

Mikhalev, Y. A., Ivashin, M. V., Savusin, V.P. and Zelenaya, F. E. (1981). The distribution and biology of killer whales in the Southern Hemisphere. *Rep. Int. Whal. Commn* **1**, 551–566.

Mitchell, E. D. and Baker, A. N. (1980). Age of reputedly old killer whale, *Orcinus orca*, from Eden, Twofold Bay, Australia. *In* "Age Determination of Toothed Whales and Sirenians" (Eds W. F. Perrin and A. C. Myrick) pp. 143–154. *Rep. Int. Whal. Commn*, Spec. Issue No. 3, Cambridge.

Mitchell, E. and Reeves, R. R. (1988). Records of killer whales in the western North Atlantic, with emphasis on eastern Canadian waters. *In* "North Atlantic Killer Whales" (Eds J. Sigurjónsson and S. Leatherwood) pp. 161–193. *Rit Fiskideildar (Journal of the Marine Institute Reykjavik)*, Vol. XI.

Miyashita, T. (in press). Distribution and abundance of some globicephalid cetaceans in the adjacent waters of Japan. *Rep. Int. Whal. Commn*, Spec. Issue No. 14, Cambridge.

Miyazaki, N. (1989). Notes on the school composition of killer whales in the Southern Hemisphere. *Bull. Natl Mus. Tokyo*, Ser. A, **15**, 53–59.

Moore, J. C. (1953). Distribution of marine mammals in Florida waters. *Am. Midl. Naturalist* **49**, 138–140.

Morton, A. B. (1990). A quantitative comparison of behavior in resident and transient killer whales of the central British Columbia coast. *In* "Individual Recognition of Cetaceans: Use of Photo-Identification and other Techniques to Estimate Population Parameters" (Eds P. S. Hammond, S. A. Mizroch and G. P. Donovan) pp. 245–248. *Rep. Int. Whal. Commn*, Spec. Issue No. 12, Cambridge.

Murie, O. J. (1959). "Fauna of the Aleutian Islands and Alaska Peninsula". North American Fauna No. 61, US Department of the Interior, Fish and Wildlife Service, 406 pp.

Myrick, A., Yochem, P. and Cornell, L. (1988). Toward calibrating dentinal layers in captive killer whales by use of tetracycline labels. *In* "North Atlantic Killer Whales" (Eds J. Sigurjónsson and S. Leatherwood) pp. 285–296. *Rit Fiskideildar (Journal of the Marine Research Institute Reykjavik)*, Vol. XI.

Nemoto, T., Best, P. B., Ishimaru, K. and Takano, H. (1980). Diatom films on whales in South African waters. *Sci. Rep. Whales Res. Inst.* **32**, 97–103.

Ness, A. R. (1967). A measure of asymmetry of the skulls of odontocete whales. *J. Zool.* **153**, 209–221.

Nishiwaki, M. (1972). General biology. *In* "Mammals of the Sea: Biology and Medicine" (Ed. S. H. Ridgway) pp. 3–204. Charles C. Thomas, Springfield, IL.

Nishiwaki, M. and Handa, C. (1958). Killer whales caught in coastal waters off Japan. *Sci. Rep. Whales Res. Inst.* **13**, 85–96.

Norris, K. S. and Prescott, J. H. (1961). "Observations of Pacific Cetaceans of Californian and Mexican Waters". University of California, Publications in Zoology 63, pp. 291–402.

Notarbartolo di Sciara, G. (1978). A killer whale *Orcinus orca* attacks and sinks a sailing boat. *Natura* **68**, 218–220.

O'Sullivan, S. and Mullin, K. D. (1997). Killer whales (*Orcinus orca*) in the northern Gulf of Mexico. *Mar. Mamm. Sci.* **13**, 141–147.

Ohsumi, S. (1975). Review of Japanese small-type whaling. *J. Fish. Res. Bd Can.* **32**, 1111–1121.

Oien, N. (1988). The distribution of killer whales (*Orcinus orca*) in the North Atlantic based on Norwegian catches, 1938–1981, and incidental sightings, 1967–1987. *In* "North Atlantic Killer Whales" (Eds J. Sigurjónsson and S. Leatherwood) pp. 65–78. *Rit Fiskideildar (Journal of the Marine Institute Reykjavik)*, Vol. XI.

Olesiuk, P. F., Bigg, M. A. and Ellis, G. M. (1990). Life history and population dynamics of resident killer whales (*Orcinus orca*) in the coastal waters of British Columbia and Washington State. *In* "Individual Recognition of Cetaceans: Use of Photo-Identification and other Techniques to Estimate Population Parameters" (Eds P. S. Hammond, S. A. Mizroch and G. P. Donovan) pp. 209–243. *Rep. Int. Whal. Commn*, Spec. Issue No. 12, Cambridge.

Ono, M., Kanan, N., Wakimoto, T. and Tatsukawa, R. (1987). Dibenzofurans a greater global pollutant than dioxins? Evidence from analyses of open ocean killer whale. *Mar. Pollut. Bull.* **18**, 640–643.

Osborne, R. W. (1986). A behavioral budget of Puget Sound killer whales. *In* "Behavioral Biology of Killer Whales" (Eds B. Kirkevold and J. S. Lockard) pp. 211–249. Alan R. Liss, New York.

Osborne, R., Calambokidis, J. and Dorsey, E. M. (1988). "A Guide to Marine Mammals of Greater Puget Sound". Island Publishers, Friday Harbor, WA, 191 pp.

Perrin, W. F. (1972). Color patterns of spinner porpoises (*Stenella* cf. *S. longirostris*) of the eastern Pacific and Hawaii, with comments on delphinid pigmentation. *Fish. Bull.* **70**, 983–1003.

Perrin, W. F. and Myrick, A. C., Jr (1980). "Age Determination of Toothed Whales and Sirenians". *Rep. Int. Whal. Commn*, Spec. Issue No. 3, Cambridge, 229 pp.

Perrin, W. F. and Reilly, S. B. (1984). Reproductive parameters of dolphins and small whales of the family Delphinidae. *In* "Reproduction in Whales, Dolphins, and Porpoises" (Eds W. F. Perrin, R. L. Brownell Jr and D. P. DeMaster) pp. 97–134. *Rep. Int. Whal. Commn*, Spec. Issue No. 6, Cambridge.

Phillips, N. E. and Baird, R. W. (1993). Are killer whales harassed by boats? *Victoria Naturalist* **50**, 10–11.

Pike, G. C. and MacAskie, D. B. (1969). Marine mammals of British Columbia. *J. Fish. Res. Bd Can.* **171**, 19–23.

Pilleri, G. and Pilleri, O. (1982). Catalogue of the fossil odontocetes (Cetacea) in the Bologna Giovanni Cappellini Museum of Paleontology with a description of a new species of *Hoplocetus* (Physeteridae). *Memorie de Scienzo Geologiche* **35**, 292–317 + 18 plts.

Popper, A. N. (1980). Sound emission and detection by delphinids. *In* "Cetacean Behaviour: Mechanisms and Functions" (Ed. L. M. Herman) pp. 1–52. Wiley, New York.

Reeves, R. R. and Mitchell, E. (1988a). Killer whale sightings and takes by American pelagic whalers in the North Atlantic. *In* "North Atlantic Killer Whales" (Eds J. Sigurjónsson and S. Leatherwood) pp. 7–23. *Rit Fiskideildar (Journal of the Marine Institute, Reykjavik)*, Vol. XI.

Reeves, R. R. and Mitchell, E. (1988b). Distribution and seasonality of killer whales in the eastern Canadian Arctic. *In* "North Atlantic Killer Whales" (Eds J. Sigurjónsson and S. Leatherwood) pp. 136–160. *Rit Fiskideildar (Journal of the Marine Institute, Reykjavik)*, Vol. XI.

Rice, D. W. (1968). Stomach contents and feeding behaviour of killer whales in the eastern North Pacific. *Nor. Hvalfangst-Tid.* **57**, 35–38.

Rice, D. W. and Wolman, A. A. (1971). "Life History and Ecology of the Gray Whale (*Eschrichtius robustus*)". American Society of Mammalogists Special Publication No. 3, 142 pp.

Ridgway, S. H. (1972). Homeostasis in the aquatic environment. *In* "Mammals of the Sea: Biology and Medicine" (Ed. S. H. Ridgway) pp. 590–747. Charles C. Thomas, Springfield, IL.

Ridgway, S. H. (1979). Reported causes of death of captive killer whales (*Orcinus orca*). *J. Wildl. Dis.* **15**, 99–104.

Ridgway, S. H. (1986). Physiological observations on dolphin brains. *In* "Dolphin Cognition and Behavior" (Eds R. Schusterman, J. Thomas and F. Wood) pp. 31–59. Lawrence Erlbaum, Hillsdale, NJ.

Ridgway, S. H. and Brownson, R. H. (1984). Relative brain sizes and cortical surface areas in odontocetes. *Acta Zool. Fenn.* **172**, 149–152.

Ridgway, S. H. and Johnston, D. G. (1966). Blood oxygen and ecology of porpoises of three genera. *Science* **151**, 456–458.

Ridgway, S. H. and Patton, G. S. (1971). Dolphin thyroid: some anatomical and physiological findings. *Z. vergl. Physiol.* **71**, 129–141.

Ridgway, S. H., Simpson, J. G., Patton, G. S. and Gilmartin, W. G. (1970). Hematologic findings in certain small cetaceans. *J. Am. Vet. Med. Assn* **157**, 566–575.

Robeck, T. R., Schneyer, A. L., McBain, J. F., Dalton, L. M., Walsh, M. T., Czekala, N. M. and Kraemer, D. C. (1993). Analysis of urinary immunoreactive steroid metabolites and gonadotropins for characterization of the estrous cycle, breeding period, and seasonal estrous activity of captive killer whales (*Orcinus orca*). *Zoo Biol.* **12**, 173–187.

Roberts, J. C., Jr, Boice, R. C., Brownell, R. L., Jr and Brown, D. H. (1965). Spontaneous atherosclerosis in Pacific toothed and baleen whales. *In* "Comparative Atherosclerosis: the Morphology of Spontaneous and Induced Atherosclerotic Lesions in Animals and its Relation to Human Disease" (Eds J. C. Roberts, Jr and R. Straus) pp. 151–155. Harper and Row, New York.

Robineau, D. (1989). Les cetaces des iles Kerguelen. *Mammalia* **33**, 265–278.

Ross, G. J. B. (1984). The smaller cetaceans of the south east coast of southern Africa. *Ann. Cape Provinc. Mus. (Nat. Hist.)* **15**, 173–410.

Samaras, W. F. (1989). New host record for the barnacle *Cryptolepas rhachianecti* Dall, 1872 (Balanomorpha: Coronulidae). *Mar. Mamm. Sci.* **5**, 84–87.

Sarra, R. (1933). Denti di pesci del Cretaceo e di mammiferi del Pliocene rinvenuti in Basilicata. *Riv. Ital. Paleontol.* **39**, 29–34.

Scammon, C. M. (1874). "The Marine Mammals of the Northwestern Coast of North America". J. H. Carmany, San Francisco, 319 pp.

Scheffer, V. B. (1967). The killer whale. *Pacif. Search* **1**, 1–2.

Scheffer, V. B. (1971). Fat-chopper: killer whales. *In* "Toothed Whales in the Eastern North Pacific and Arctic Waters" (Ed. Alice Seed) pp. 12–15. Pacific Search Books, Seattle, WA.

Scheffer, V. B. and Slipp, J. W. (1948). The whales and dolphins of Washington State with a key to the cetaceans of the west coast of North America. *Am. Midl. Naturalist* **39**, 274–287.

Schevill, W. E. and Watkins, W. A. (1966). Sound structure and directionality in *Orcinus* (killer whale). *Zoologica* **51**, 71–76.

Sergeant, D. E. (1969). Feeding rates of cetacea. *FiskDir. Skr. Ser. HavUnders* **15**, 246–258.

Sergeant, D. E. and Fisher, H. D. (1957). The smaller Cetacea of eastern Canadian waters. *J. Fish. Res. Bd Can.* **14**, 83–115.

Shevchenko, V. I. (1976). The nature of the interrelationships between killer whales and other cetaceans. *Morsk. mlekopitayuschchie Chust'* **2**, 173–174 (1975) (Translated by the Translation Bureau (WRI), St. Anne de Bellevue, Multilingual Services Division, Department of the Secretary of the State of Canada, Translation Series 3839).

Sigurjónsson, J. and Leatherwood, S. (1988). The Icelandic live-capture fishery for killer whales, 1976–1988. *In* "North Atlantic Killer Whales" (Eds J. Sigurjónsson and S. Leatherwood), pp. 307–316. *Rit Fiskideildar (Journal of the Marine Institute, Reykjavik)*, Vol. XI.

Sigurjónsson, J., Lyrholm, T., Leatherwood, S., Jónsson, E. and Vikingsson, G. (1988). Photoidentification of killer whales (*Orcinus orca*) off Iceland, 1981–1986. *In* "North Atlantic Killer Whales" (Eds J. Sigurjónsson and S. Leatherwood) pp. 99–114. *Rit Fiskideildar (Journal of the Marine Institute, Reykjavik)*, Vol. XI.

Silber, G. K., Newcomer, M. W. and Perez-Cortes, H. (1990). Killer whales (*Orcinus orca*) attack and kill a Bryde's whale (*Balaenoptera edeni*). *Can. J. Zool.* **68**, 1603–1606.

Similä, T. and Ugarte, F. (1993). Surface and underwater observations of cooperatively feeding killer whales in northern Norway. *Can. J. Zool.* **71**, 1494–1499.

Simpson, J. C. and Gardner, M. D. (1972). Comparative microscopic anatomy of selected marine mammals. *In* "Mammals of the Sea: Biology and Medicine" (Ed. S. H. Ridgway) pp. 298–418. Charles C. Thomas, Springfield, IL.

Slijper, E. J. (1936). "Die Cetaceen: Vergleichend-anatomisch und systematisch". Reprinted 1973, A. Asher, Amsterdam, 590 pp. (in German).

Small, R. J. and DeMaster, D. P. (1995). Survival of five species of marine mammals. *Mar. Mamm. Sci.* **11**, 209–226.

Smith, T. G., Siniff, D. B., Reichle, R. and Stone, S. (1981). Coordinated behavior of killer whales, *Orcinus orca*, hunting a crabeater seal, *Lobodon carcinophagus*. *Can. J. Zool.* **59**, 1185–1189.

Speakman, J. R., McDevitt, R. M. and Cole, K. R. (1993). Measurement of basal metabolic rates: don't lose sight of reality in the quest for comparability. *Physiol. Zool.* **66**, 1045–1049.

Spencer, M. P., Gornall, T. A. and Poulter, T. C. (1967). Respiratory and cardiac activity of killer whales. *J. Appl. Physiol.* **22**, 974–981.

Stephens, W. M. (1963). The killer (*Orcinus orca*). *Sea Frontiers* **9**, 262–273.

Stevens, T. A., Duffield, D. A., Asper, E. D., Hewlett, K. G., Bolz, A., Gage, L. J. and Bossart, G. D. (1989). Preliminary findings of restriction fragment differences in mitochondrial DNA among killer whales (*Orcinus orca*). *Can. J. Zool.* **67**, 2592–2595.

Tarpley, R. J. and Ridgway, S. H. (1994). Corpus callosum size in delphinid cetaceans. *Brain Behav. Evol.* **44**, 156–165.

Taylor, R. F. and Farrel, R. K. (1973). Light and electron microscopy of peripheral blood neutrophils in a killer whale affected with Chediak–Higashi syndrome. *Fed. Proc.*, 822.

Thomas, J. A., Ferm, L. M. and Kuechle, V. B. (1987). Silence as an antipredation strategy by Weddell seals. *Antarct. J. U.S.* **22**, 232–234.

Thomas, J. A., Leatherwood, S., Evans, W. E., Jehl, J. R. and Awbrey, F. (1981). Ross Sea killer whale distribution, behavior, color pattern, and vocalizations. *Antarct. J. U.S.* **15**, 157–158.

Tomich, P. Q. (1986). "Mammals in Hawaii". Bishop Museum Press, Honolulu, HI, 375 pp.

Tomilin, A. G. (1957). "Mammals of the U.S.S.R. and Adjacent Countries", Vol. IX, "Cetacea" (Ed. V. G. Heptner). Nauk USSR, Moscow (English translation, 1967, Israel Program for Scientific Translations, Jerusalem).

Turner, W. (1872). On the gravid uterus and on the arrangement of the foetal membranes in the Cetacea. *Trans. Roy. Soc. Edinburgh* **26**, 467–504 + 2 plts

Vidal, O. and Pechter, G. (1989). Behavioral observations of fin whale, *Balaenoptera physalus*, in the presence of killer whale, *Orcinus orca*. *Fish. Bull.* **87**, 370–373.
Voisin, J. F. (1972). Notes on the behavior of the killer whale, *Orcinus orca* (L.). *Norwegian J. Zool.* **20**, 93–96.
Waite, J. M. (1988). Alloparental care in killer whales (*Orcinus orca*). M.Sc. thesis, University of California at Santa Cruz, 65 pp.
Walker, L. A., Cornell, L., Dahl, K. D., Czekala, N. M., Dargen, C. M., Joseph, B., Hsueh, A. J. W. and Lasley, B. L. (1988). Urinary concentrations of ovarian steroid hormone metabolites and bioactive follicle-stimulating hormone in killer whales (*Orcinus orchus*) during ovarian cycles and pregnancy. *Biol. Reproduct.* **39**, 1013–1020.
Walters, E. L., Baird, R. W. and Guenther, T. J. (1992). New killer whale "pod" discovered near Victoria. *Victoria Naturalist* **49**, 7.
Wang, P. (1985). Distribution of cetaceans in Chinese waters. Southwest Fisheries Center, National Marine Fisheries Service, Administrative Report LJ-85-24, 11 pp.
Wellings, H. P. (1964). Shore whaling at Twofold Bay. *Eden* 1–15.
White, D., Cameron, N., Spong, P. and Bradford, J. (1971). Visual acuity of the killer whale (*Orcinus orca*). *Exp. Neurol.* **32**, 230–236.
Ximinez, A., Langguth, A. and Praderi, R. (1972). Lista sistematica de los mamiferos del Uruguay. *Anales Mus. Nac. Hist. Nat. Montevideo* **7**, 1–49.
Yano, K. and Dahlheim, M. E. (1995a). Killer whale (*Orcinus orca*) depredation on longline catches of bottomfish in the southeastern Bering Sea and adjacent waters. *Fish. Bull.* **93**, 355–373.
Yano, K. and Dahlheim, M. E. (1995b). Behavior of killer whales (*Orcinus orca*) during longline fishery interactions in the southeastern Bering Sea and adjacent waters. *Fish. Sci.* **61**, 584–589.
Yonezawa, M., Nakamine, H., Tanaka, T. and Miyaji, T. (1989). Hodgkin's disease in a killer whale (*Orcinus orca*). *J. Comp. Pathol.* **100**, 203–207.

12

Harbour Porpoise

Phocoena phocoena (Linnaeus, 1758)

Andrew J. Read

Genus and Species

Taxonomy

The harbour porpoise was described as *Delphinus phocoena* by Linnaeus in 1758 from "Oceano, Europaeo & Balthico" and assigned to the genus *Phocoena* by Cuvier in 1816. As noted by Schevill *et al.* (1969), both the genus and species names have been misspelled occasionally as "Phocaena phocaena" from the Latin spelling of porpoise, "phocaena".

The harbour porpoise is one of three species currently recognized in the genus *Phocoena* and one of six members of the Family Phocoenidae. The other members of the genus are the Burmeister's porpoise *Phocoena spinipinnis* Burmeister 1865 and the vaquita *Phocoena sinus* Norris and McFarland 1958. The family also includes the Dall's porpoise *Phocoenoides dalli* (True, 1885), the finless porpoise *Neophocaena phocaenoides* (Cuvier 1829) and the spectacled porpoise *Phocoena dioptrica* (Lahille 1912). There is controversy as to the generic identity of the spectacled porpoise. Barnes (1985) proposed that the spectacled porpoise be placed in the genus *Australophocaena* based on morphological grounds, but a recent molecular analysis supports the retention of this species in the genus *Phocoena*

Handbook of Marine Mammals, Volume 6
ISBN 0-12-588506-7

(Rosel, Haygood and Perrin, 1995). The taxonomy and evolutionary history of the family is described by Gaskin (1982) and Barnes (1985).

Subspecies

Several subspecies of harbour porpoises have been proposed (Tomilin, 1957; Gaskin *et al.*, 1974). Recent analyses of geographic variation in mitochondrial DNA (mtDNA) support the existence of three subspecies: *P. p. phocoena* in the North Atlantic, *P. p. vomerina* in the North Pacific and *P. p. relicta* in the Black Sea and Sea of Azov (Rosel, Dizon and Haygood, 1995; Wang *et al.*, 1996). This view is consistent with morphometric analyses of porpoise skulls from the North Atlantic and North Pacific (Yurick and Gaskin, 1987; Amano and Miyazaki, 1992).

Common names

The harbour porpoise is given many vernacular names throughout its range, including: bruinvisch (The Netherlands), common porpoise (United Kingdom), marsouin commun (France), marsvin (Denmark), pourcil (Quebec), puffing pig (Atlantic Canada), schweinswal (Germany), toninha (Portugal), tumbler (Scotland), and tumlare (Sweden). A complete list of common names is given by Schulze (1987).

Distribution

Range

Harbour porpoises occur throughout temperate, shelf waters of the northern hemisphere (Fig. 1). The distribution of the species has been reviewed in detail by Gaskin (1984) and, more recently, by the Scientific Committee of the International Whaling Commission (IWC, 1996).

The southern limit of harbour porpoises in the western North Atlantic is delimited by two stranding records from northern Florida (Polacheck *et al.*, 1995). These two records are believed to be extralimital, however, as very few porpoises occur south of Cape Hatteras, North Carolina (Polacheck *et al.*, 1995). The northernmost record in the western North Atlantic is from Upernavik, West Greenland at approximately 72°N (Gaskin, 1984).

Four discrete populations are believed to exist in the western North Atlantic: Bay of Fundy–Gulf of Maine, Gulf of St. Lawrence, Newfoundland and Labrador, and Greenland (Gaskin, 1984; Palka *et al.*, 1995; IWC, 1996; Wang *et al.*, 1996). In each of these populations, porpoises move into coastal waters during summer. In some areas, harbour porpoises move offshore to avoid advancing ice cover during winter. In general, however, our understanding of the winter distribution of porpoises is limited by a lack of survey data.

In the Northeast Atlantic, harbour porpoises are distributed discontinuously from Novaja Zemlja (73°N), in the Barents Sea to Senegal on the northwestern coast of Africa (IWC, 1996). The northernmost record is a sighting at 77°N, off the coast of Spitsbergen, Svalbard (Bjorge and Oien, 1995). This distribution

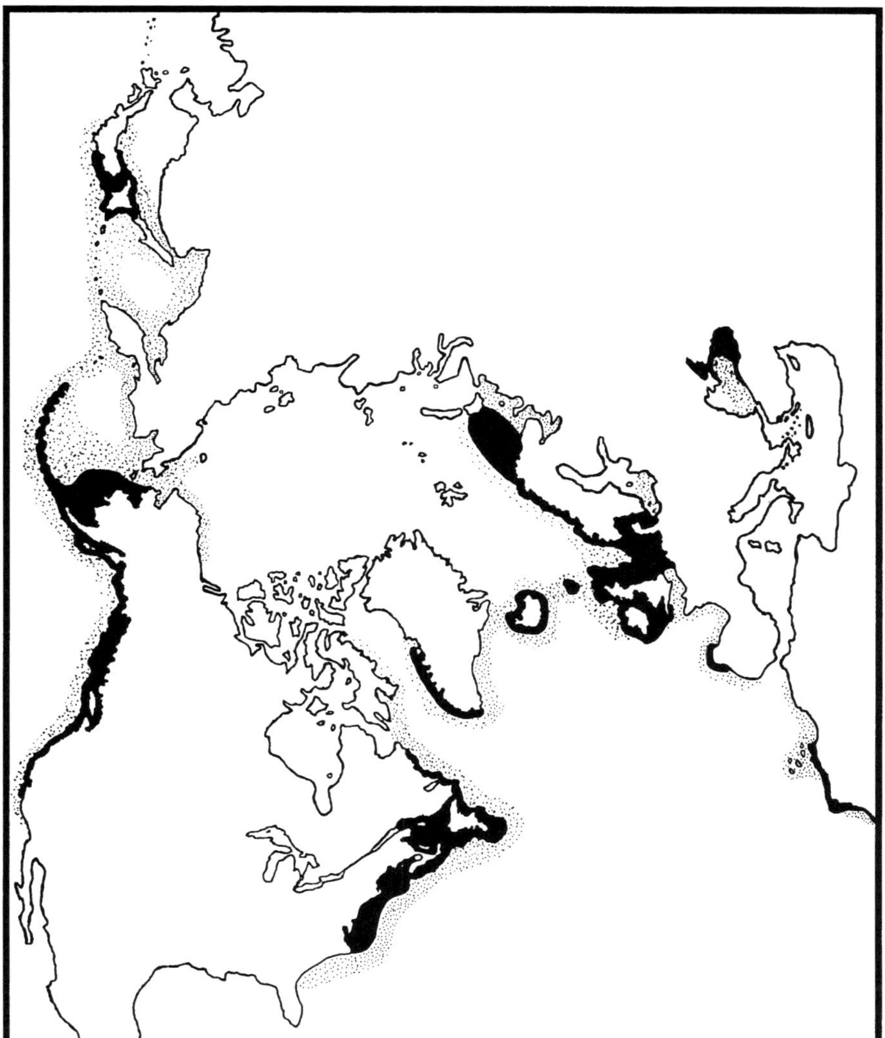

FIG. 1 Distribution of the harbour porpoise, after Gaskin (1984) and IWC (1996). (Figure prepared by A. J. Westgate)

includes the Irish Sea, North Sea, and Kattegat and Skagerrak Seas (Berggren and Arrhenius, 1995; IWC, 1996). Harbour porpoises are found seasonally also around the coasts of Iceland (Saemundsson, 1939) and the Faroe Islands (IWC, 1996). Harbour porpoises were, at one time, common in the Baltic Sea, but their distribution and abundance there has been greatly reduced in recent decades (IWC, 1996). A small resident population may still exist in ice-free areas of the eastern Baltic (Skora *et al.*, 1988). The seasonal movements of these animals are poorly understood; in the past, porpoises were taken in a drive fishery as they moved through the Danish belt waters in autumn, presumably on their way to the North Sea (Kinze, 1995). A hiatus in distribution occurs in the southern part of the North Sea, extending to the English Channel and Bay of Biscay, separating porpoises from the Celtic Shelf (Rogan and Berrow, 1996) from those along the northwestern coast of Iberia. Harbour porpoises are present year-round in Portuguese waters (Sequira, 1996). The population structure of harbour porpoises in the northeastern Atlantic is complex and not well understood (IWC, 1996).

There have been no records of the harbour porpoises in the Mediterranean Sea for over 15 years (IWC, 1996), so porpoises in the Black Sea and Sea of Azov are effectively isolated from the remainder of the species' range.

In the North Pacific, harbour porpoises are found as far east as the Mackenzie delta (van Bree *et al.*, 1977), and extend into the Bering Sea, along the Aleutian Island chain and west to Sakhalin Island (Gaskin, 1984). In these northern waters, porpoises are limited seasonally by ice cover. It is not known whether the range of the harbour porpoise extends into the Chukchi Sea. The distribution extends, more or less continuously, south along the western coast of North America from Alaska to Monterey Bay, California. Gaskin *et al.* (1993) noted that the confirmed southern limit of the species in the western Pacific is at Wakayama, Japan (34°N). As is the case in the northeastern Atlantic, the population structure of harbour porpoises in the North Pacific is not well understood (see Gaskin, 1984; Calambokidis and Barlow, 1991; Barlow *et al.*, 1995).

Habitat

Throughout its range, the species is limited to the waters of the continental shelf by its demersal foraging behaviour and diving capacity (see below). The species is seldom found in waters warmer than 17°C (Gaskin *et al.*, 1993). It is also possible that, in some areas, agonistic interactions with bottlenose dolphins (*Tursiops* sp.) may also play a role in determining the limits of the range (Ross and Wilson, 1996). Analyses of distribution over several spatial scales indicate that porpoises are found most frequently in cool waters, where aggregations of prey are concentrated (Watts and Gaskin, 1985; Palka, 1995).

External Characteristics

Colour pattern

The pigmentation pattern of harbour porpoises has been described in detail by Mercer (1973) and Koopman and Gaskin (1994). The species exhibits a basic pattern of countershading common in many small odontocetes (Fig. 2). The dorsal cape is charcoal grey to black and extends laterally to a variable extent. Some reports suggest that the dorsal surface exhibits a brown hue, but these are in error, perhaps due to lighting conditions at sea or possibly diatoms on the skin. The flanks are greyish white and fade ventrally to white in subtle and complex patterns. Many individuals possess a dark eye patch and flipper stripes, which vary in number, shape and extent. Koopman and Gaskin (1994) reported that all individuals exhibit some degree of bilateral asymmetry in pigmentation. Pigmentation patterns are unique to each animal, but are not gender-dependent and are not useful in discriminating individuals among populations (Koopman and Gaskin, 1994).

Size and shape

The harbour porpoise is one of the smallest cetaceans, and females are consistently larger than males in all populations (Gaskin *et al.*, 1984). In the Bay of Fundy, mature female porpoises reach an asymptotic length of 158 cm and an asymptotic mass of 65 kg; the corresponding values for males are 143 cm and 50 kg (Read and Tolley, 1997). Length distributions from European samples may be found in Møhl-Hansen (1954) and van Bree (1973). Selected morphometric measurements from physically mature porpoises from the Bay of Fundy are provided in Table 1. Similar data are presented by Lockyer (1995a) for British porpoises and by Stuart and Morejohn (1980) for porpoises from California. Harbour porpoises vary regionally in size; porpoises in Mauritania (Smeenk *et al.*, 1992) and Portugal (Sequira, 1996), for example, are known to exceed 2 m in length.

Harbour porpoises are robust animals, which helps to minimize heat loss to a cold environment (Slijper, 1958). Maximum girth occurs just anterior to the dorsal fin and corresponds to approximately 64% of total length in adults, similar to corresponding values from other phocoenids (e.g. Brownell *et al.*, 1987). Like all phocoenids, the head lacks an external rostrum, and the tip of the lower jaw does not extend past the tip of the upper jaw. The eyes are situated just caudal and dorsal to the margin of the gape. The tiny external auditory meatus is visible in live and in freshly dead specimens. The flippers insert at approximately 20% of the total length of the body and are more rounded than those of *P. sinus* or *P. spinipinnis*. The dorsal fin is triangular in shape and situated at the mid-portion

(A)
(B)

FIG. 2 (A) Pigmentation patterns and external morphology of a harbour porpoise killed in a herring weir in the Bay of Fundy. Note the robust shape, charcoal grey cape and complex pattern of grey on the flanks, fading to white ventrally. (Photo by A. J. Read) (B) Pigmentation patterns of the head of a harbour porpoise killed in a herring weir in the Bay of Fundy. Note the subtle eye ring, distinctive flipper stripe and chin patch. The external auditory meatus is clearly visible. (Photo by J. Skeleteon)

TABLE 1 External morphometric measurements from physically mature female ($n = 32$) and male ($n = 62$) harbour porpoises from the Bay of Fundy (lengths are in cm, mass is in kg). Data are taken from Read and Tolley (1997)

	Females			Males		
Code	Mean	SD	Range	Mean	SD	Range
Mass	62	7.0	43–76	49	5.2	36–61
Total length	157	7.3	131–168	144	5.5	129–157
Length of gape	10	1.1	7–12.5	9	1.2	6.5–14
Rostrum to eye	14	1.1	12.5–16.5	14	1.6	10–18.5
Rostrum to blowhole	13	2.0	10.5–18	13	1.9	10–18
Rostrum to tip of dorsal fin	20	1.4	17–23	19	1.9	15.5–24
Rostrum to ear	87	4.7	71.5–96	78	3.2	71–85
Rostrum to axilla	30	1.9	26–37	28	1.9	24–34
Rostrum to umbilicus	61	4.0	52.5–72	55	3.4	45–61.5
Rostrum to anterior genital slit	93	5.2	81.5–107	61	3.6	52.5–70
Rostrum to anus	107	5.6	90–116	100	4.5	80–108
Girth to axilla	90	4.3	80–101	83	3.3	75–90
Maximum girth	100	6.0	87–107	92	3.6	84–100.5
Girth to anus	56	3.9	45.5–65	52	4.5	33–61
Length of dorsal fin base	25	3.5	19.5–31	23	3.4	16.5–31
Height of dorsal fin	11	1.0	9–13	10	1.1	8–12
Maximum width of flipper	9	0.60	7–10	8	0.63	6.5–9.5
Width of flipper at base	10	0.85	8–11	9	0.73	6.5–10.5
Fluke span	37	2.2	32.5–42	34	3.0	24–40.5

of the dorsal surface. In all but the youngest specimens (Van Deinse, 1931), several rows of epidermal tubercles are present on the leading edge of the dorsal fin, although the function of these structures is unknown. The genital aperture of males is found closer to the umbilicus than to the anus, similar to the anatomical arrangement of other phocoenids. Small pre-anal pores of unknown function are present in male specimens. The caudal peduncle is laterally compressed and a slight keel arises from the presence of dorsal and ventral blubber ridges along the tail stock (Koopman, 1998).

Internal Anatomy

Skull

The morphology of the skull of the harbour porpoise (Fig. 3) is well known. As noted by Barnes (1985), the skulls of all living phocoenids, including the harbour porpoise, exhibit paedomorphic features including: relatively large and rounded brain cases, short rostra, delayed suture fusion in ontogeny, and thin-walled

(A)
(B)
(C)

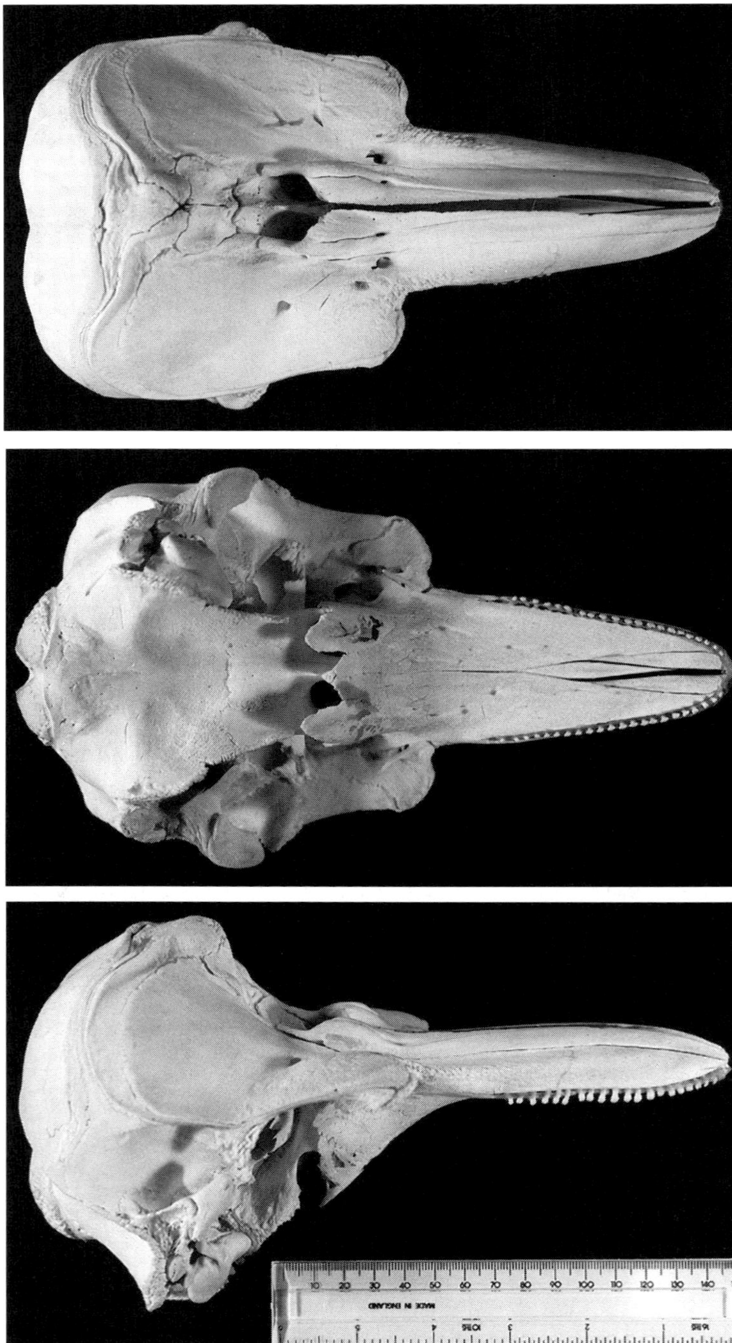

FIG. 3 (A) Dorsal view, (B) ventral view and (C) lateral view of the cranium of a harbour porpoise.

(D)

(E)

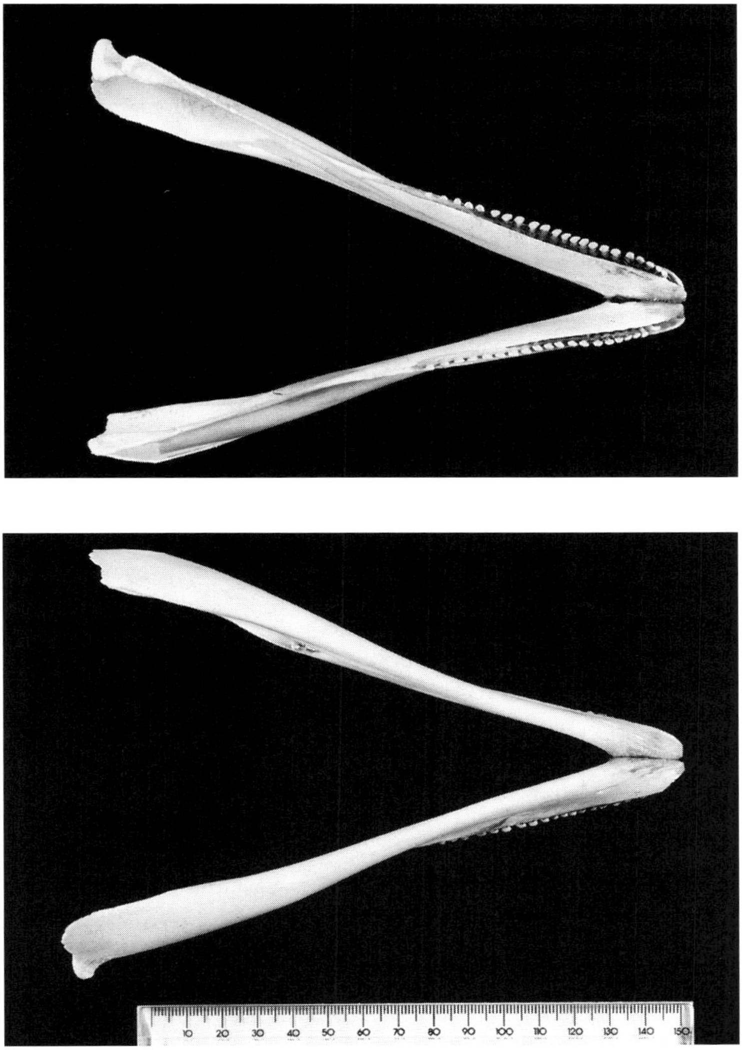

FIG.3 *Continued* (D) Dorsal view and (E) ventral view of the mandibles of a harbour porpoise. (Photos courtesy M. Klinowska)

occipital and basioccipital bones. Compared with *P. sinus*, the skull of the harbour porpoise is larger, has a longer and narrower rostrum, a smaller foramen magnum, and the posterior edge of the palate forms a W-shaped, rather than U-shaped, indentation (Norris and McFarland, 1958). The skull of *P. spinipinnis* has a less compressed brain case, with more developed jugals and shorter tooth rows than that of the harbour porpoise (Allen, 1925).

Condylobasal lengths of 108 immature and mature specimens from the North Atlantic and North Pacific varied from 221 mm to 314 mm (Amano and Miyazaki, 1992); female skulls are consistently larger than those of males in all populations (Yurick and Gaskin, 1987). As in other odontocetes, the harbour porpoise skull exhibits sinistral asymmetry, believed to be associated with sound production (Yurick and Gaskin, 1988). Harbour porpoise teeth, like those of other phocoenids, are relatively small and spatulate. Like all odontocetes, harbour porpoises are homodont and monodont; the number of teeth varies from 21 to 29 in each maxillary row and from 20 to 29 in each mandibular row (Yurick and Gaskin, 1987). The teeth begin to erupt in the middle portion of the tooth rows when calves attain a body mass of approximately 25 kg (Møhl-Hansen, 1954). Observations on the development and intraspecific variation of cranial morphology can be found in Stuart and Morejohn (1980), Kinze (1985), Miyazaki *et al.* (1987), Yurick and Gaskin (1987) and Amano and Miyazaki (1992).

Postcranial skeleton

The vertebral column of the harbour porpoise consists of 62–67 vertebrae, including seven cervical vertebrae (fused and compressed in adults), 12–14 thoracic vertebrae, 13–18 lumbar vertebrae and 27–32 caudal vertebrae (Gaskin *et al.*, 1974; Smith *et al.*, 1976). The fusion of cervical vertebrae restricts mobility of the head to some extent. The large spinous and transverse processes on thoracic and lumbar vertebrae provide extensive surface areas for muscle attachment and increase the mechanical efficiency of the axial musculature (Smith *et al.*, 1976). There are usually six or seven double-headed ribs and six or seven sternal ribs present. The bones of the forelimbs are described in Smith *et al.* (1976). Comparisons between the postcranial skeleton of the harbour porpoise and those of other phocoenids can be found in True (1889), Allen (1925), Norris and McFarland (1958) and Noble and Fraser (1971). Interested readers should also consult the classic account of Slijper (1936).

Blubber thickness

In comparison with many other small odontocetes, harbour porpoises possess a relatively thick layer of lipid-rich blubber to insulate them from their cold environment (Worthy and Edwards, 1990). The thickness of the blubber varies over the body and also with age, reproductive status, and nutritive condition of individuals (Parry, 1949; Read, 1990a; Lockyer, 1995a; Koopman, 1998). Koopman (1998) examined blubber thickness at 48 sites on 114 porpoises from the Bay of Fundy and Gulf of Maine and found blubber to be thickest over the thoracic and abdominal surfaces of the animal and thinnest laterally on the caudal. Dorsal and ventral blubber are present along the caudal peduncle. Calves possess the thickest and lactating females the thinnest blubber (Read, 1990a;

Koopman, 1998), reflecting the different energetic and thermoregulatory requirements of these animals. Lockyer (1995a) found that the blubber of stranded neonates in the United Kingdom was relatively thin, suggesting that porpoises are born with a thin blubber layer which increases substantially during early lactation. The blubber of older calves can represent more than 45% of total body mass, but this proportion decreases during ontogeny (Slijper, 1958; Kanwisher and Sundnes, 1965; Read, 1990a). In general, blubber thickness exhibits a negative correlation with body size, an unusual relationship for a marine mammal (Ryg *et al.*, 1993; Koopman, 1998).

The chemical composition of the blubber of harbour porpoises is similar to that of other marine mammals (Lovern, 1934), but composition varies over the body surface (Koopman *et al.*, 1996). Blubber over the thorax and abdomen is divided into inner and outer layers; the inner layer is likely to be more active metabolically and may represent the site of energy deposition and mobilization. Blubber in the caudal peduncle is less stratified and may function in shaping body form rather than as a thermoregulatory or insulative layer (Koopman *et al.*, 1996; Koopman, 1998). Harbour porpoise blubber contains high levels of isovaleric acid (2–27%), a compound which has been observed also in delphinids, monodontids and other phocoenids, but not in ziphiids, physeterids or platanistids (Litchfield *et al.*, 1975). In general, the lipid content of harbour porpoise blubber is extremely high, reaching levels of 88% (Worthy and Edwards, 1990) and 93% (Lockyer, 1995a) wet weight, reflecting the relatively high insulative value of this tissue.

Organs and soft tissues

The internal soft anatomy of the harbour porpoise has been the subject of many descriptions, and it is not possible to provide a full review of this body of work here. Interested readers should consult Slijper (1936) or Schulze (1987) for more detailed descriptions.

The musculature of the harbour porpoise is similar to that of other odontocetes (e.g. Pabst, 1990). The epaxial musculature is dominated by the large longissimus, which has its origin on the caudal portion of the skull and inserts on the caudal vertebrae. The dominant ventral muscle is the hypaxialis, which originates ventrad on the transverse processes of thoracic and lumbar vertebrae and inserts on the caudal vertebrae and chevron bones (Smith *et al.*, 1976). The muscles of the head, shoulder and forelimbs are complex and are described in detail by Smith *et al.* (1976). In adult porpoises from the United Kingdom, muscle comprises 31% of total mass of males and 23% of females (Lockyer, 1995a). These proportions increase significantly during ontogeny.

A striking feature of the vascular system of the harbour porpoise is the extensive retia in the thoracic region. The heart is dorsoventrally flattened, cone-shaped (Halina and Gaskin, 1978), and relatively large, comprising approximately 0.5% of total body mass (Rowlatt and Gaskin, 1975). These proportional values of heart mass are less than those of *P. dalli*, which average 1.3 (Ridgway and Johnston, 1966). Hilton and Gaskin (1978) describe the vascular supply to the liver. The anatomy and functional morphology of the heart are described in detail by Rowlatt and Gaskin (1975) and Halina and Gaskin (1978).

The respiratory system of the harbour porpoise is described by Wislocki (1942). The lungs of this species have many of the adaptations associated with diving that are found in other odontocetes, including sphincters in the distal portions of the bronchioles and incomplete cartilaginous rings supporting the trachea, bronchi and bronchioles (Gaskin *et al.*, 1974). Amundin and Cranford (1990) describe the anatomy of the upper nasal passages in relation to sound production in this species.

The digestive system of the harbour porpoise is described by Smith (1972) and compared with that of other cetaceans by Gaskin (1978). The stomach consists of three compartments: the forestomach, the main stomach, and the pyloric stomach (Smith, 1972). The forestomach is non-glandular, lined with keratinized stratified squamous epithelium and is essentially a bag-like extension of the esophagus. The lining of the forestomach is white and characterized by a complex system of folds or rugae. In this compartment, solid prey is reduced to a liquid state by mechanical and chemical digestion, with digestive juices presumably originating in the main stomach (Smith, 1972). Most digestion occurs in the main stomach, which is homologous to the fundic stomach of terrestrial mammals and contains many secretory gland cells (Smith, 1972). The pyloric stomach is attached to the main stomach by a connecting channel. The liver is large, contributing between 1.6% and 5.6% of total body mass (Slijper, 1958). Like all cetaceans, harbour porpoises lack a gall bladder. The pancreas is flattened and located between the liver and the stomach complex. The spleen is small and consists of a main node (approximately 10 g) and several discrete nodules (< 1 g).

The anatomy of the harbour porpoise urogenital system is similar to that of other small odontocetes. The kidneys are large, lobulate and are comprised of many reniculi, which are bound together by connective tissue. Each reniculus may contain one or more renal papillae, which act as the functional units of the kidney. The mean number of papillae per kidney was 293 (range 260–361) in six porpoises examined by Hedges *et al.* (1979). The kidneys represent approximately 0.84% (range 0.46–1.76%) of total body mass (Hedges *et al.*, 1979). Further description of the kidneys of the harbour porpoise may be found in Hedges *et al.* (1979) and Van der Spoel (1963).

The brain and nervous system of the harbour porpoise are described by Fraser (1952) and Jansen and Jansen (1969). In a sample of 109 harbour porpoises from the Bay of Fundy and Gulf of Maine, brain mass varied from 342 g to 546 g (W. A. McLellan and D. A. Pabst, unpublished data).

The anatomy of the male and female reproductive systems are described by Meek (1918) and Slijper (1949). Perhaps the most striking aspect of the reproductive anatomy is the relatively large size of testes during the breeding season (see below), which may exceed more than 4% of body mass (Read, 1990b).

Ovulation is almost exclusively sinistral in this species, and the right ovary maintains an immature appearance throughout life in most individuals (Slijper, 1949; Fisher and Harrison, 1970; Sorensen and Kinze, 1994). The placenta is described by Wislocki (1933), and detailed observations on foetal development may be found in Guldberg and Nansen (1894), Slijper (1949) and Møhl-Hansen (1954).

Physiology

The metabolic rate of the harbour porpoise has been the subject of some controversy (Yasui and Gaskin, 1986). Some researchers (e.g. Kanwisher and Sundnes, 1965) asserted that these small cetaceans must exhibit an elevated metabolic rate to maintain their body temperature in a cold aquatic environment. More recent measurements of metabolic rate (Worthy *et al.*, 1987) do not support the hypothesis that harbour porpoises possess an elevated rate of metabolism.

Blood chemistry values for wild harbour porpoises from the Bay of Fundy are presented by Koopman *et al.* (1995). Comparable data from captive or rehabilitated stranded porpoises may be found in Andersen (1966), Nielsen and Andersen (1982), and Kastelein *et al.* (1990). In general, these values are similar to those of other small coastal odontocetes.

Life History and Population Dynamics

Growth and maturation

Length and mass at birth are approximately 70 cm and 5 kg, respectively (Sorensen and Kinze, 1994; Lockyer, 1995b). Growth curves for male and female harbour porpoises from the Bay of Fundy are presented in Fig. 4. Similar growth curves for the population of harbour porpoises in the United Kingdom may be found in Lockyer (1995b). Females grow at a faster rate and for a longer period than males before reaching physical maturity. In the Bay of Fundy, females reach physical maturity at approximately 7 years of age and males reach maturity at approximately 5 years of age; when physically mature, females are 9% longer and 27% heavier than males (Read and Tolley, 1997).

In the Bay of Fundy, female harbour porpoises reach sexual maturity in their second, third or fourth year of life. Mean age and length at sexual maturation for females in this population is 3.44 years and 143 cm (Read, 1990c; Read and Gaskin, 1990). Mean age at sexual maturation for females in the Danish population is 3.64 years (Sorensen and Kinze, 1994). Age at sexual maturation in females is known to vary temporally within populations (Read and Gaskin,

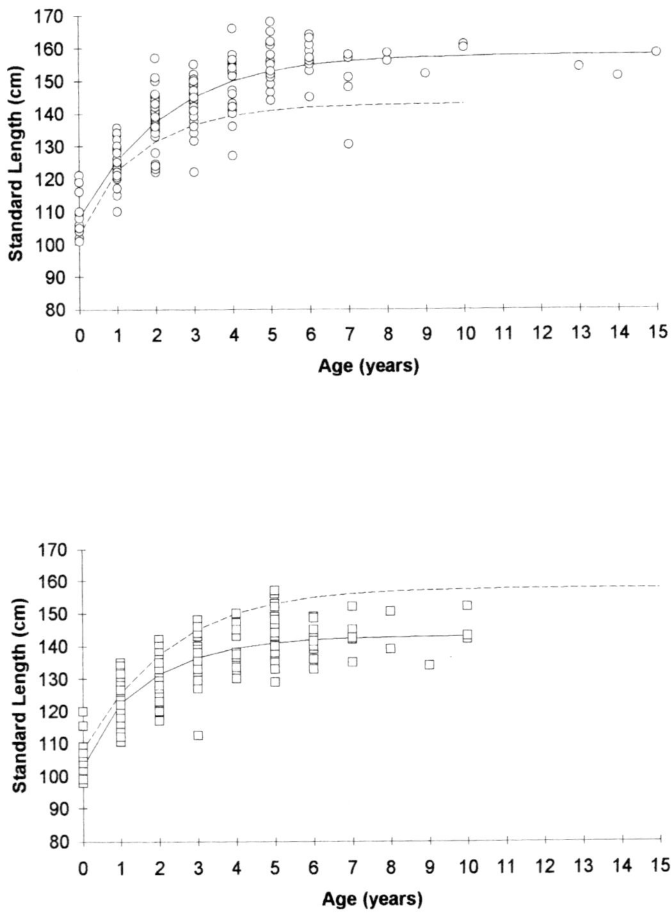

FIG. 4 Growth curves for standard length of harbour porpoise females (top) and males (bottom) from the Bay of Fundy. The Gompertz curve for each sex is indicated by the solid line in each figure, and the corresponding curve for the other sex is indicated by the broken line. Data from Read and Tolley (1997).

1990) and also likely varies between populations, depending on porpoise density and prey availability.

Maximum longevity is reported as 24 years in both the United Kingdom (Lockyer, 1995b) and California (A. A. Hohn and R. L. Brownell, unpublished data). The maximum age reported from the Bay of Fundy and Gulf of Maine is 17 years (Read and Hohn, 1995). Porpoises clearly are capable of living past 20 years of age, but very few individuals survive for such long periods, particularly in populations that are subjected to heavy rates of mortality in commercial fisheries (see below).

Reproduction

Reproduction is extremely seasonal in all populations of harbour porpoises studied to date (Gilpin, 1875; Møhl-Hansen, 1954; Fisher and Harrison, 1970; Gaskin *et al.*, 1984; Simons, 1984; Kinze, 1989; Read, 1990b; Sorensen and Kinze, 1994; Addink *et al.*, 1995; Lockyer, 1995b; Read and Hohn, 1995). Ovulation and conception occur during a relatively short period in spring or summer, resulting in a synchronized period of foetal growth, which last for approximately 10.6 months (Read, 1990b; Sorensen and Kinze, 1994). In Danish, Dutch and UK waters, parturition occurs primarily in June and July (Sorensen and Kinze, 1994; Addink *et al.*, 1995; Lockyer, 1995b). In the Bay of Fundy and Gulf of Maine, parturition occurs in May (Gaskin *et al.*, 1984; Read, 1990b; Read and Hohn, 1995). Lactation is the most poorly understood portion of the reproductive cycle, but females are believed to nurse their calves for between 8 months and 12 months (Møhl-Hansen, 1954; Read, 1990b). Weaning is a gradual process and calves start to take solid food when they are still nursing (see below).

Most female harbour porpoises in Denmark and the Bay of Fundy become pregnant each year and are simultaneously lactating and pregnant for much of their adult lives (Møhl-Hansen, 1954; Read, 1990c; Sorensen and Kinze, 1994; Read and Hohn, 1995). In contrast, female porpoises in California do not appear to reproduce each year (A. A. Hohn and R. L. Brownell, unpublished data). Studies in other populations would be particularly useful to help determine the factors responsible for this intraspecific variation in reproductive interval.

Abundance

Growing concern over the impact of mortality of harbour porpoises in commercial fisheries has led to recent improvements in the estimation of abundance for

TABLE 2 Abundance estimates for selected areas of the range of the harbour porpoise (CV is coefficient of variation)

Area	Year	Estimate	CV	Source
Gulf of Maine	1991	37 500	0.29	Blaylock *et al.* (1995)
Gulf of Maine	1992	67 500	0.23	Blaylock *et al.* (1995)
Central California	1988–1993	4 120	0.22	Barlow *et al.* (1995)
Northern California	1988–1993	9 250	0.23	Barlow *et al.* (1995)
Oregon and Washington	1990–1991	26 175	0.21	Barlow *et al.* (1995)
Skaggerak and Belt Seas	1994	36 046	0.30	IWC (1996)
North Sea	1994	279 367	0.18	IWC (1996)
Ireland and western UK	1994	36 280	0.57	IWC (1996)

this species (IWC, 1996). Developments in the techniques used specifically for harbour porpoises (e.g. Polacheck, 1995) have allowed researchers to develop reliable estimates of abundance for the first time. Large scale surveys have been conducted in both the North Pacific and North Atlantic in recent years (Table 2). Notable among these endeavours was the SCANS survey, conducted during the summer of 1994, which covered most of the range of the harbour porpoise in the eastern North Atlantic (IWC, 1996). Abundance estimates are lacking for large parts of the range of this species, particularly the western Pacific and the Black Sea.

Abundance estimates of harbour porpoises are currently generated with some precision, generally with coefficients of variation varying between 0.2 and 0.3 (see Table 2). Rather unsettling, however, is the large discrepancy between abundance estimates made for the Gulf of Maine in 1991 and 1992; both surveys covered the known range of this population, but produced dramatically different estimates of abundance in different years. The factors responsible for this inter-annual variation may be related to variation in water temperature on a regional scale, but are not fully understood (Palka, 1995).

Behaviour

Social organization

Harbour porpoises are often described as exhibiting little social behaviour, but this sentiment is more a reflection of the limitations of our observation techniques rather than of the true nature of the species. In the wild, porpoises are difficult to approach and follow, do not approach moving boats, and it is difficult to identify individuals from natural markings (Fig. 5). When researchers have overcome these difficulties, they have discovered that harbour porpoises exhibit a rich social life, much like that of other small odontocetes (Amundin and Amundin, 1974). For example, in the coastal waters of the Bay of Fundy, Watson (1976) was able to identify thirteen individual porpoises from scars or patterns of pigmentation over three summers. Several adult females returned to the same small cove in successive years, accompanied by new calves each season. The porpoises were found in a number of different temporary social groupings. Observations of captive harbour porpoises also indicate that porpoises exhibit complex patterns of social behaviour, mediated by frequent sexual and agonistic interactions (Andersen and Dziedzic, 1964; Andersen, 1969; Amundin and Amundin, 1971).

In many areas, particularly during the spring and early summer, harbour porpoises are found in small groups consisting of a few individuals. For example, mean group size of harbour porpoises in aerial surveys off the eastern coast of the USA was 2.9, with a mode of 1 (CETAP, 1982). Large aggregations of tens

FIG. 5 Harbour porpoise surfacing in the Bay of Fundy. (Photo by A. J. Read)

to hundreds of individuals have been recorded in late summer and fall in the western North Atlantic (Hoek, 1992; Katona *et al.*, 1993). In the Bay of Fundy, there is a distinct increase in mean group size over the course of the summer, perhaps reflecting changes in social behaviour after the cessation of reproductive activity in late spring and early summer.

Females and calves are found together for the duration of lactation, but it is not known whether individuals of other age and sex classes exhibit long term social bonds. Segregation by age and sex appears to be a common feature of harbour porpoise populations, and has been documented from sightings (Smith and Gaskin, 1983), strandings (A. J. Read, unpublished data), and directed catches (Møhl-Hansen, 1954). The mechanisms responsible for this segregation are not understood.

Movements

With the exception of limited observations of naturally marked individuals (Watson, 1976; Gaskin and Watson, 1985), all information on harbour porpoise movements has come from telemetry studies. Early VHF telemetry studies in the Bay of Fundy (Gaskin *et al.*, 1975; Read and Gaskin, 1985) suggested that porpoises exhibit limited ranges during the summer months. These studies were plagued by short periods of radio contact with tagged porpoises, which limited the inferences that could be drawn from the movement data. More recent long term studies using satellite-linked radio telemetry (Read and Westgate, 1997) indicate that porpoises are extremely mobile and are capable of covering large distances in relatively short periods. The mean daily distance travelled by eight porpoises equipped with satellite-linked transmitters in the Bay of Fundy, for example, varied between 14 km and 58 km. Tagged individuals made rapid point-to-point excursions lasting from hours to days that were interspersed with longer periods of residency in restricted areas. These animals moved throughout the Bay of Fundy and Gulf of Maine, utilizing home ranges that encompassed tens of thousands of km^2 (Read and Westgate, 1997).

Feeding

Harbour porpoises forage independently and seldom use co-operative strategies for concentrating or obtaining prey, at least not in the manner reported for many delphinids (Würsig, 1986). A record of harbour porpoises feeding co-operatively in California (Fink, 1959) probably was an error in species identification (K. S. Norris, personal communication).

Harbour porpoises feed on small, schooling clupeoid and gadid fishes. Individual prey are generally less than 40 cm in length and typically range from 10 cm to 30 cm in length. In some but not all areas, the prey of harbour porpoises are found near the sea floor.

In the Bay of Fundy, harbour porpoises feed primarily on juvenile Atlantic herring *Clupea harengus* and secondarily on silver hake *Merluccius bilinearis* (Smith and Gaskin, 1974; Recchia and Read, 1989). Significant variation due to season (A. J. Read and J. E. Craddock, unpublished data), year (Recchia and Read, 1989), reproductive status (Recchia and Read, 1989) and ontogeny (Smith and Read, 1992) exists in the diet of porpoises from this population. In the Bay of Fundy, calves begin to take solid food in their first summer, often by ingesting small crustaceans before progressing to small fish (Smith and Read, 1992).

In most other areas, the diet of harbour porpoises is also dominated by clupeoid fishes, although the species composition varies by area. For example, in the southern Gulf of St. Lawrence porpoises feed primarily on herring, but in the northern Gulf, where herring is replaced ecologically by capelin, *Mallotus villosus*, the diet of porpoises is dominated by capelin (Fontaine *et al.*, 1994). Similarly, in northern Scandinavia porpoises feed on capelin, while in Danish and Swedish waters the diet is primarily composed of herring (Aarefjord *et al.*, 1995). Capelin is also consumed by harbour porpoises in West Greenland (Kinze, 1989). Herring and sprat *Clupea sprattus* are important prey items in the Baltic and North Seas (Lindroth, 1962; Rae, 1965, 1973). In most areas where clupeoids are the dominant prey, the diet is supplemented to a significant extent by gadids or other demersal fishes.

The daily ration of porpoises of various age and sex classes has not been well documented. Andersen (1965) fed a daily average of 4.3 kg of herring and whiting, *Gadus merlangus*, to eight captive harbour porpoises, ranging from 129 cm to 162 cm in length. This daily ration varied between 5% and 14% of body mass. Using a variety of feeding regimes, Koga (1991) estimated that immature porpoises required a daily ration of 7–8% of body mass to maintain their weight. Obviously, higher rations are required to support the energetic costs of growth and reproduction. The ration will also vary with the energetic content of prey, which can vary ontogenetically and seasonally, particularly in clupeoid fishes (Yasui and Gaskin, 1986).

Association with other species

Harbour porpoises are not generally found in close association with other species of cetaceans. For example, CETAP (1982) aerial surveys off the eastern coast of the United States found harbour porpoises in association with other species in only 3% of sightings, the lowest frequency of multispecies sightings for any cetacean species. Approximately 70% of these multi-species sightings were with baleen whales (*Balaenoptera acutorostrata*, *B. physalus*, and *Megaptera novaeangliae*).

An unusual case of inter-specific association was documented by Ross and Wilson (1996), who described violent interactions between harbour porpoises and bottlenose dolphins around the Moray Firth, Scotland. In these interactions, bottlenose dolphins pursued, attacked and killed harbour porpoises for reasons that are unclear. Such lethal, non-predatory interactions are extremely rare in mammals and challenge our perceptions of the benign nature of these animals.

Predation

White sharks (*Carcharodon carcharias*) and killer whales (*Orcinus orca*) are the two major predators of harbour porpoises. White sharks are known to prey on harbour porpoises in the Bay of Fundy and Gulf of Maine (Arnold, 1972). The tail stocks of most porpoise carcasses found in the stomachs of white sharks have been severed, suggesting that attack may come from behind and possibly below (Arnold, 1972). Unlike larger odontocetes, such as bottlenose dolphins, live harbour porpoises are seldom seen carrying healed scars from shark bites, suggesting that most encounters with these predators are fatal. A neonatal porpoise (79 cm) recovered from the stomach of a Greenland shark *Somniosus microcephalus* off Newfoundland in 1962 (Williamson, 1963), was likely scavenged by this sluggish shark. Killer whales are known to prey on harbour porpoises in many areas where their ranges overlap (Jefferson *et al.*, 1991). The effects of predation on harbour porpoise demography are not known.

Diving and surfacing behaviour

Harbour porpoises make brief dives, generally lasting for less than 5 min (Watson and Gaskin, 1983; Read and Gaskin, 1985; Westgate *et al.*, 1995). Most ventilation sequences consist of several breaths interspersed with short submergences. These sequences separate longer dives, during which porpoises may descend to depth. Detailed descriptions of surfacing behaviour are given by Amundin (1974) and Smith *et al.* (1976). On calm days, porpoises often lie motionless at the surface for periods varying from several seconds to over a minute (Watson and Gaskin, 1983); during these periods porpoises may be either resting or scanning the water below them.

Westgate *et al.* (1995) studied the diving behaviour of harbour porpoises by placing time–depth recorders on animals released from herring weirs in the Bay of Fundy. The maximum depth of dive made by these porpoises was 226 m, close to the maximum depth of the study area. In 8167 recorded dives, mean depths and durations for individuals ranged from 14 m to 41 m and 44 s to 103 s, respectively (Westgate *et al.*, 1995). Individual rates of descent and ascent were similar and ranged from 0.9 m/sec to 2.3 m/sec. Individual porpoises spent 33–60% of their monitoring periods in the upper 2 m of the water column. Recent data from satellite-linked radio tags, also deployed on harbour porpoises in the Bay of Fundy, indicate that porpoises spend 3–7% of the time at the water's surface (Read and Westgate, 1997).

Sound production and hearing

In general, our knowledge of the echolocation abilities of harbour porpoises is better than our understanding of their use of sound in a social context. Harbour porpoises produce click trains that are believed to be used for both communication and echolocation and do not produce frequency-modulated whistles (Schevill *et al.*, 1969). Click trains are produced in two major frequency ranges: one near 2 kHz (Schevill *et al.*, 1969) and one around 130 kHz (Mohl and Andersen, 1973). Recent observations of sound production made by captive porpoises are described by Verboom and Kastelein (1995). Both experimental (Amundin and Andersen, 1983) and anatomical (Amundin and Cranford, 1990) evidence supports the hypothesis that sound is produced in the area of the nasal plugs. The sounds produced by harbour porpoises are strikingly similar to those produced by dolphins of the genus *Cephalorhynchus*, leading Watkins *et al.* (1977) to comment on the possibility that these two genera have experienced convergent evolution in sound production in similar coastal habitats.

Harbour porpoises have excellent auditory capabilities. Andersen (1970) presents a behavioural audiogram made with a young captive porpoise, which indicated that the range of best sensitivity was between 8 kHz and 32 kHz.

Parasites and Disease

The parasite fauna of the harbour porpoise are relatively numerous and individual animals often host heavy loads of parasites (Møhl-Hansen, 1954; Clausen and Andersen, 1988; van Nie, 1989; Baker and Martin, 1992; Larsen, 1995). Twenty-two helminth species have been reported as parasites of harbour porpoises (Raga *et al.*, 1989). Several parasites are typical components of the helminth fauna of the harbour porpoise: the nematodes *Pseudalius*, *Stenurus* and *Halocercus* and *Tonyurus* in the air passages and sinuses (Arnold and Gaskin, 1975; Kinze, 1989; Balbuena *et al.*, 1994); the trematode *Campula* in the liver and pancreas (Botero-Anug, 1995); and the nematode *Anisakis* in the stomach (Smith, 1989). *Crassicauda* is occasionally present in the blubber and mammary gland (Botero-Anug, 1995). Most adult harbour porpoises carry heavy loads of parasites, even in the absence of disease or health problems. For example, Botero-Anug (1995) found that *Campula oblonga* occurred frequently in the liver and pancreas of porpoises killed in gill nets in the Bay of Fundy and Gulf of Maine, but that neither the flukes nor the associated fibrosis affected the health of these individuals. These heavy parasite loads may exacerbate other health problems, but are unlikely to pose significant risk to robust animals.

Several authors have described pathological findings from stranded porpoises (van Nie, 1989; Siebert, 1995)or those killed incidentally to commercial fishing operations (Andersen, 1974, 1978; Clausen and Andersen, 1988). Common pathological features found even in apparently healthy animals are skin lesions caused by pox virus (van Utrecht, 1959) and fibrosis associated with parasitic infestations of the lung, liver, pancreas and stomach (Smith, 1989; Baker and Martin, 1992; Botero-Anug, 1995).

Some harbour porpoises in the Bay of Fundy carry antibodies to morbillivirus, indicating past infection in this population (Duignan *et al.*, 1995). A few harbour porpoises died in the morbillivirus epizootic that claimed thousands of harbour seals in Europe during 1988 (Kennedy *et al.*, 1988; Welsh *et al.*, 1992). The significance of morbillivirus as an agent of mortality in harbour porpoise populations is not known (Duignan *et al.*, 1995). It is also unclear whether any link exists between such epizootics and the presence of heavy loads of anthropogenic contaminants in the tissues of these animals (see below).

Live Maintenance

Sporadic attempts have been made to maintain harbour porpoises for over a century (Collet, 1984), generally with poor results. The first concerted programme to obtain porpoises for captivity occurred in Denmark during the winter of 1957–58, when Dudok van Heel (1962) obtained several porpoises and transported them to Holland. All 19 porpoises died within six months, many within minutes or hours of capture. A second series of porpoises was obtained by Soren Andersen (1978) from Danish pound nets and maintained in Denmark. Andersen collected 67 porpoises between 1962 and 1974; these animals also experienced very high mortality in the initial months of captivity, but he was able to keep one specimen alive for 39 months. These collective experiences, together with other unsuccessful attempts to maintain the species in captivity for research or display, led to the belief that harbour porpoises are difficult, if not impossible, to maintain in a captive environment.

Recent rehabilitation efforts with live stranded have changed our perception of the maintenance of this species in captivity. Notable among these efforts has been the work of Kastelein *et al.* (1990) in The Netherlands, where several animals have been released back into the wild and others have been the focus of intensive research on the behaviour of harbour porpoises around fishing gear (Nachtigall *et al.*, 1995). These rehabilitated porpoises represent a unique resource to researchers interested in understanding the mechanisms responsible for entanglement in fishing gear and in finding ways of mitigating this problem (also see the recent book on the biology of the harbour porpoise by Read *et al.* (Eds.), 1997).

Human Effects

Directed fisheries

Harbour porpoises have been taken in directed fisheries throughout much of their range at one time or another. Readers should refer to the detailed reviews of Mitchell (1975), Gaskin (1984) and Klinowska (1991) for further information on this exploitation.

In the Bay of Fundy, large numbers of porpoises were shot by Micmac and Passamaquoddy hunters in the 19th and early 20th centuries (Leighton, 1937). Porpoises were taken on calm days by hunters working from canoes (Gilpin, 1875). The total number of porpoises taken in this fashion each year is unknown, but probably numbered in the hundreds to low thousands (see Mitchell, 1975). A similar hunt occurred in the Pacific Northwest (Scammon, 1874). The primary products from this hunt were oil, rendered from the blubber, and a high quality lubricant produced from mandibular fat. Also the meat was used for human consumption. There are no records of porpoises being hunted in the Bay of Fundy prior to the arrival of Europeans (Gilpin, 1875). The directed hunt of porpoises by aboriginal Americans dwindled in the 20th century and disappeared altogether in the early 1980s (Gaskin, 1984).

A large drive fishery operated in Danish belt waters during autumn as porpoises moved out of the Baltic Sea (Kinze, 1995). In this area, porpoises were driven into small coves by fishermen beating the surface of the water with long, slender sticks. Once porpoises entered a cove, their exit was blocked by a net, and the animals were hauled ashore and killed by ex-sanguination (Kinze, 1995). The main product of the hunt was oil. Records of this hunt extend back as far as 1357, with an average of 1122 porpoises taken between 1834 and 1874 (Andersen, 1982; Kinze, 1995). Detailed statistics on this hunt are presented by Kinze (1995). The magnitude of the hunt declined in the 20th century, perhaps due to a decrease in porpoise numbers or to economic competition with other oil products. This fishery was resumed in both World Wars, when oil products were scarce (Møhl-Hansen, 1954; Andersen, 1982; Kinze, 1995).

Large numbers of harbour porpoises were taken by purse seine and rifle in the Black Sea for more than a century (Smith, 1982). The main products of this exploitation were oil, meat and meal. The hunt has continued in Turkish waters into the 1990s (Klinowska, 1991). Detailed statistics from this fishery are provided by the Scientific Committee of the International Whaling Commission (IWC, 1983).

A directed hunt for harbour porpoises also operates in the waters of West Greenland during the spring, summer and autumn months. The number of porpoises taken by this fishery was roughly estimated as 1000 animals per year between 1950 and 1970 (Kapel, 1975). This hunt continues today, although there are no estimates of its current magnitude.

By-catches

Due to their habitat in productive coastal waters, harbour porpoises are captured incidentally in commercial fisheries throughout their range (Gaskin, 1984). Porpoises are taken in a variety of gear types including weirs, pound nets, cod traps and surface gill nets, but the vast majority of this mortality occurs in bottom-set gill nets (Read and Gaskin, 1988; Jefferson and Curry, 1994). Particularly large by-catches have occurred in gill net fisheries in the Bay of Fundy and Gulf of Maine, the Gulf of St. Lawrence, Newfoundland, western Greenland, Iceland and the North Sea (Table 3). Smaller by-catches occur in many other areas and may have significant effects on local populations. In the past, and in some areas still today, fishermen and their families used the meat of harbour porpoises to supplement their diet. Interested readers should consult the reviews of Gaskin (1984), Klinowska (1991), Jefferson and Curry (1994), and the Scientific Committee of the International Whaling Commission (1996) for detailed statistics on these by-catches.

In general, with the possible exception of the Black and Baltic Seas, incidental mortality in commercial fisheries poses the greatest current threat to harbour porpoises. In its recent review of the status of this species in the North Atlantic, the Scientific Committee of the International Whaling Commission (1996) noted that, in the areas where adequate data on abundance and by-catch levels exist, incidental mortality exceeds sustainable levels. The Committee "perceives

TABLE 3 Estimates of annual incidental mortality in commercial fisheries for selected areas of the range of the harbour porpoise. Only estimates generated from independent observer programmes have been included

Area	Years	Estimates	Source
Gulf of Maine	1990–1993	1200–2900	Bravington and Bisack (1996)
Bay of Fundy	1993–1994	101–424	Trippel *et al.* (1996)
West Greenland	1990–1993	134–1531	IWC (1996)
Skagerrak	1995	17	IWC (1996)
North Sea[a]	1993–1994	4629	IWC (1996)
Celtic Shelf	1993–1994	2049	IWC (1996)
Central California	1989–1992	44–92	Barlow *et al.* (1995)

[a] This estimate includes only mortality in Danish gill net fisheries. Additional mortality occurs in the fisheries of other North Sea countries.

a broad-scale risk to harbour porpoise populations in the North Atlantic due to problems related to fishery by-catch". The World Conservation Union has also recognized this threat, listing the species as vulnerable throughout its range.

Pollution

A considerable body of literature exists describing the levels of various pollutants in tissues of the harbour porpoise (Aguilar and Borell, 1995; Westgate *et al.*, 1996). In many industrialized areas, harbour porpoises carry high levels of organochlorines, although the levels of some compounds, particularly PCBs and DDTs, have decreased recently in some areas (Granby and Kinze, 1991; Westgate *et al.*, 1996). Porpoises also carry high levels of trace metals originating from human activity and natural sources, such as mercury, cadmium, copper and zinc (Falconer *et al.*, 1983; Johnston, 1995). Contaminant levels in harbour porpoises often vary geographically and may serve as useful markers in studies of population structure (Calambokidis and Barlow, 1991; Palka *et al.*, 1996). The dynamics of contaminants are believed to be similar to those in other small cetaceans; levels of organochlorines increase with age in males, for example, but decrease in mature females, who pass them to their offspring through placental and lactational transfer (Duinker and Hilldebrand, 1979; Granby and Kinze, 1991).

A fair amount is known regarding the levels of contaminants in the tissues of harbour porpoises, but very little about the potential effects of these anthropogenic chemicals. Kuiken *et al.* (1994) attempted to link pollutant burdens with health status in stranded porpoises in the United Kingdom and found no correlation between PCB levels, body condition and cause of death. Likewise, Kuiken *et al.* (1993) found no relationship between levels of organochlorines and adrenocortical hyperplasia in stranded porpoises. The levels of these compounds in the tissues of harbour porpoises are high enough to warrant concern (Aguilar and

Borell, 1995), but our understanding of their effects must await more sophisticated studies of ecotoxicological risk assessment (Reijnders, 1994).

Acknowledgments

This chapter is dedicated to the memory of William E. Schevill, a pioneer in the study of porpoises, dolphins and whales, and an inspiration to all of us who study the natural history of marine mammals. I would like to thank Tara Cox, Heather Koopman and Andrew Westgate for their helpful comments on the manuscript and for help with references. I also thank David Gaskin for introducing me to the harbour porpoise and for the many ways in which he has facilitated my work over the past two decades. My research on harbour porpoises has been supported by the US National Marine Fisheries Service, US Office of Naval Research, US Marine Mammal Commission, World Wildlife Fund Canada, Fisheries and Oceans Canada, Whale and Dolphin Conservation Society and Greenpeace International. I thank Margaret Klinowska for the photographs of the harbour porpoise skull.

References

Aarefjord, H., Bjorge, A. J., Kinze, C. C. and Lindstedt, I. (1995). Diet of the harbour porpoise (*Phocoena phocoena*) in Scandinavian waters. *Rep. Int. Whal. Commn*, Spec. Issue No. 16, Cambridge, pp. 211–222.

Addink, M. J., Sorensen, T. B. and Garcia Hartmann, M. (1995). Aspects of reproduction and seasonality in the harbour porpoise from Dutch waters. *In* "Whales, Seals, Fish and Man" (Eds A. S. Blix, L. Walloe and O. Ulltang) pp. 459–464. Elsevier, Amsterdam.

Aguilar, A. and Borell, A. (1995). Pollution and harbour porpoises in the eastern North Atlantic. *Rep. Int. Whal. Commn*, Spec. Issue No. 16, Cambridge, pp. 231–242.

Allen, G. M. (1925). Burmeister's Porpoise (*Phocaena spinipinnis*). *Bull. Mus. Comp. Zool.* **67.**

Amano, M. and Miyazaki, N. (1992). Geographic variation in skulls of the harbor porpoise, *Phocoena phocoena. Mammalia* **56**, 133–144.

Amundin, B. and Amundin, M. (1971). Några etologiska iakttagelser över tumlaren, *Phocoena phocoena* (L.), i. fångenskap. *Zool. Revy* **33**, 51–59.

Amundin, M. (1974). Functional analysis of the surfacing behaviour in the harbour porpoise, *Phocoena phocoena* (L.). *Z. Saugetierkunde* **39**, 313–318.

Amundin, M. and Amundin, B. (1974). On the behaviour and study of the harbour porpoise, *Phocoena phocoena*, in the wild. *Invest. Cetacea* **5**, 317–328.

Amundin, M. and Andersen, S. H. (1983). Bony nares air pressure and nasal plug muscle activity during click production in the harbour porpoise, *Phocoena phocoena*, and bottlenosed dolphin, *Tursiops truncatus*. *J. Exp. Biol.* **105**, 275–282.

Amundin, M. and Cranford, T. (1990). Forehead anatomy of *Phocoena phocoena* and *Cephalorhynchus commersoni*: 3-dimensional computer generated reconstructions with emphasis on the nasal diverticula. *In* "Sensory Abilities of Cetacea" (Eds J. Thomas and R. Kastelein) pp. 1–18. Plenum Press, New York.

Andersen, S. (1965). L'alimentation du marsouin (*Phocaena phocoaena* L.) en captivité. *Vie et Milieu* **16**, 799–810.

Andersen, S. (1966). The physiological range of the formed elements in the peripheral blood of the harbour porpoise, *Phocaena phocaena* (L.) in captivity. *Nord. Vet.-Med.* **18**, 51–65.

Andersen, S. (1969). Epimeletic behaviour in captive harbour porpoise, *Phocaena phocaena* (L.). *Invest. Cetacea* **1**, 203–205.

Andersen, S. (1970). Auditory sensitivity of the harbour porpoise *Phocoena phocoena*. *Invest. Cetacea* **2**, 255–259.

Andersen, S. H. (1974). A typical case of netcaught harbour porpoise, *Phocoena phocoena*, from Danish waters. *Aquat. Mamm.* **2**, 1–5.

Andersen, S. H. (1978). Experiences with harbour porpoises, *Phocoena phocoena*, in captivity: Mortality, autopsy findings and influence of the captive environment. *Aquat. Mamm.* **6**, 39–49.

Andersen, S. H. (1982). Change in occurrence of the harbour porpoise, *Phocoena phocoena*, in Danish waters as illustrated by catch statistics from 1834 to 1970. *In* "Mammals in the Seas", FAO Fisheries Series Vol. 5, pp. 131–133.

Andersen, S. and Dziedzic, A. (1964). Behaviour patterns of captive harbour porpoise *Phocaena phocaena* (L.). *Bull. Inst. Océanogr. Monaco* **63**, 1–20.

Arnold, P. W. (1972). Predation on harbour porpoise, *Phocoena phocoena*, by a white shark, *Carcharodon carcharias*. *J. Fish. Res. Bd Can.* **29**, 1213–1214.

Arnold, P. W. and Gaskin, D. E. (1975). Lungworms (Metastrongyloidea: Pseudaliidae) of harbor porpoise *Phocoena phocoena* (L. 1758). *Can. J. Zool.* **53**, 713–735.

Baker, J. R. and Martin, A. R. (1992). Causes of mortality and parasites and incidental lesions in harbour porpoises (*Phocoena phocoena*) from British waters. *Vet. Rec.* **130**, 554–558.

Balbuena, J. A., Aspholm, P. E., Andersen, K. I. and Bjorge, A. (1994). Lung-worms (Nematoda: Pseudaliidae) of harbour porpoises (*Phocoena phocoena*) in Norwegian waters: patterns of colonization. *Parasitology* **108**, 343–349.

Barlow, J., Brownell Jr, R. L., DeMaster, D. P., Forney, K. A., Lowry, M. S., Osmek, S., Ragen, T. J., Reeves, R. R. and Small, R. J. (1995). U.S. Pacific marine mammal stock assessments. NOAA Technical Memorandum NMFS-SWFSC-219.

Barnes, L. G. (1985). Evolution, taxonomy and antitropical distributions of the porpoises (Phocoenidae, Mammalia). *Mar. Mamm. Sci.* **1**, 149–165.

Berggren, P. and Arrhenius, F. (1995). Densities and seasonal distribution of harbour porpoises (*Phocoena phocoena*) in the Swedish Skagerrak, Kattegat and Baltic Seas. *Rep. Int. Whal. Commn,* Spec. Issue No. 16, Cambridge, pp. 109–121.

Bjorge, A. and Oien, N. (1995). Distribution and abundance of harbour porpoise, *Phocoena phocoena*, in Norwegian waters. *Rep. Int. Whal. Commn*, Spec. Issue No. 16, Cambridge, pp. 89–98.

Blaylock, R. A., Hain, J. E., Hansen, L. J., Palka, D. L. and Waring, G. T. (1995). U.S. Atlantic and Gulf of Mexico marine mammal stock assessments. NOAA Technical Memorandum NMFS-SEFSC 363.

Botero-Anug, A. M. (1995). Trematode infection as a determinant of fitness in harbor porpoise (*Phocoena phocoena*). D.V.Sc. thesis, University of Guelph, Canada.

Bravington, M. V. and Bisack, K. D. (1996). Estimates of harbour porpoise bycatch in the Gulf of Maine sink gillnet fishery, 1990–1993. *Rep. Int. Whal. Commn* **46**, 567–574.

van Bree, P. J. H. (1973). On the length distribution of harbour porpoises, *Phocoena phocoena* (Linnaeus, 1758), from west European and Baltic waters. *Mammalia* **37**, 359–360.

van Bree, P. J. H., Sergeant, D. E. and Hoek, W. (1977). A harbour porpoise, *Phocoena phocoena* (Linnaeus, 1758), from the Mackenzie River delta, Northwest Territories, Canada. *Beaufortia* **26**, 99–105.

Brownell Jr, R. L., Findley, L. T., Vidal, O., Robles, A. and Manzanilla, S. (1987). External morphology and pigmentation of the vaquita, *Phocoena sinus* (Cetacea: Mammalia). *Mar. Mamm. Sci.* **3**, 22–30.

Calambokidis, J. and Barlow, J. (1991). Chlorinated hydrocarbons and their use for describing population discreteness in harbor porpoises from Washington, Oregon, and California. NOAA Technical Report NMFS 98, pp. 101–110.

CETAP. (1982). A characterization of marine mammals and turtles in the Mid- and North Atlantic areas of the U.S. outer continental shelf. Final Report to the US Bureau of Land Management.

Clausen, B. and Andersen, S. (1988). Evaluation of bycatch and health status of the harbour porpoise (*Phocoena phocoena*) in Danish waters. *Danish Rev. Game Biol.* **13**, 1–20.

Collet, A. (1984). Live-capture of cetaceans for European institutions. *Rep. Int. Whal. Commn* **34**, 603–607.

Dudok van Heel, W. H. (1962). Sound and Cetacea. *Netherlands J. Sea Res.* **1**, 407–507.

Duignan, P. J., House, C., Geraci, J. R., Duffy, N., Rima, B. K., Walsh, M. T., Early, G., St. Aubin, D. J., Sadove, S., Koopman, H. and Rhinehart, H. (1995). Morbillivirus infection in cetaceans of the western North Atlantic. *Vet. Microbiol.* **44**, 241–249.

Duinker, J. C. and Hilldebrand, M. T. J. (1979). Mobilisation of organochlorines from female lipid tissue and transplacental transfer to fetus in a harbour porpoise (*Phocoena phocoena*) in a contaminated area. *Bull Environ. Contam. Toxicol.* **23**, 728–732.

Falconer, C. R., Davies, I. M. and Topping, G. (1983). Trace metals in the common porpoise, *Phocoena phocoena*. *Mar. Environ. Res.* **8**, 119–127.

Fink, B. D. (1959). Observation of porpoise predation on a school of Pacific sardines. *Calif. Fish Game* **45**, 216–217.

Fisher, H. D. and Harrison, R. J. (1970). Reproduction in the common porpoise (*Phocoena phocoena*) of the North Atlantic. *J. Zool. Lond.* **161**, 471–486.

Fontaine, P., Hammill, M. O., Barrette, C. and Kingsley, M. C. (1994). Summer diet of the harbour porpoise (*Phocoena phocoena*) in the estuary and the northern Gulf of St. Lawrence. *Can. J. Fish. Aquat. Sci.* **51**, 172–178.

Fraser, F. C. (1952). "Handbook of R. H. Burne's Cetacean Dissections". British Museum (Natural History), London.

Gaskin, D. E. (1978). Form and function in the digestive tract and associated organs in Cetacea, with a consideration of metabolic rates and specific energy budgets. *Oceanogr. Mar. Biol. Ann. Rev.* **16**, 313–345.

Gaskin, D. E. (1982). "The Ecology of Whales and Dolphins". Heinemann, London.

Gaskin, D. E. (1984). The harbour porpoise *Phocoena phocoena* (L.): regional populations, status, and information on direct and indirect catches. *Rep. Int. Whal. Commn* **34**, 569–586.

Gaskin, D. E., Arnold, P. W. and Blair, B. A. (1974). *Phocoena phocoena*. *Mamm. Species* **42**, 1–8.

Gaskin, D. E, Smith, G. J. D. and Watson, A. P. (1975). Preliminary study of the movements of harbour porpoises (*Phocoena phocoena*) in the Bay of Fundy using radiotelemetry. *Can. J. Zool.* **53**, 1466–1471.

Gaskin, D. E., Smith, G. J. D., Watson, A. P., Yasui, W. Y. and Yurick, D. B. (1984). Reproduction in the porpoises (Phocoeneidae): implications for management. *Rep. Int. Whal. Commn*, Spec. Issue No. 6, Cambridge, pp. 135–148.

Gaskin, D. E.. and Watson, A. P. (1985). The harbor porpoise, *Phocoena phocoena*, in Fish Harbour, New Brunswick, Canada: occupancy, distribution, and movements. *Fish. Bull.* **83**, 427–442.

Gaskin, D. E., Yamamoto, S. and Kawamura, A. (1993). Harbor porpoise, *Phocoena phocoena* (L.), in the coastal waters of northern Japan. *Fish. Bull.* **91**, 440–454.

Gilpin, J. B. (1875). On the smaller cetaceans inhabiting the Bay of Fundy and shores of Nova Scotia. *Proc. Nova Scotia Inst. Nat. Sci.* **4**, 21–34.

Granby, K. and Kinze, C. C. (1991). Organochlorines in Danish and West Greenland harbour porpoises (*Phocoena phocoena*). *Mar. Pollut. Bull.* **22**, 458–462.

Guldberg, G. and Nansen, F. (1894). "On the Development and Structure of the Whale", Part 1, "On the Development of the Dolphin". Bergens Museum, Bergen, 70 pp.

Halina, W. G. and Gaskin, D. E. (1978). Functional anatomy of the coronary system of the harbour porpoise *Phocoena phocoena* (L.). *Can. J. Zool.* **56**, 1643–1653.

Hedges, N. A., Gaskin, D. E. and Smith, G. J. D. (1979). Renicular morphology and renal vascular system of the harbour porpoise *Phocoena phocoena* (L.). *Can. J. Zool.* **57**, 868–875.

Hilton, J. W. and Gaskin, D. E. (1978). Comparative volumes and vascular microanatomy of the intrahepatic venous system of the harbour porpoise, *Phocoena phocoena* (L.). *Can. J. Zool.* **56**, 2292–2298.

Hoek, W. (1992). An unusual aggregation of harbor porpoises (*Phocoena phocoena*). *Mar. Mamm. Sci.* **8**, 152–155.

IWC (1983). Report of the sub-committee on small cetaceans. *Rep. Int. Whal. Commn* **33**, 152–170.

IWC (1996). Report of the sub-committee on small cetaceans. *Rep. Int. Whal. Commn* **46**, 160–179.

Jansen, J. and Jansen, J. K. S. (1969). The nervous system of Cetacea. *In* "The Biology of Marine Mammals" (Ed. H. T. Anderson) pp. 175–252. Academic Press, New York.

Jefferson, T. A. and Curry, B. E. (1994). A global review of porpoise (*Cetacea: Phocoenidae*) mortality in gillnets. *Biol. Conserv.* **67**, 167–183.

Jefferson, T. A., Stacey, P. J. and Baird, R. W. (1991). A review of killer whale interactions with other mammals: predation to co-existence. *Mamm. Rev.* **21**, 151–180.

Johnston, D. W. (1995). Spatial and temporal differences in heavy metal concentrations in the tissues of harbour porpoises (*Phocoena phocoena* L.) from the western North Atlantic. M.Sc. thesis, University of Guelph, Canada.

Kanwisher, J. and Sundnes, G. (1965). Physiology of a small cetacean. *Hvalradets Skrifter* **48**, 45–53.

Kapel, F. O. (1975). Preliminary notes on the occurrence and exploitation of smaller Cetacea in Greenland. *J. Fish. Res. Bd Can.* **32**, 1079–1082.

Kastelein, R. A., Bakker, M. J. and Dokter, T. (1990). The medical treatment of 3 stranded harbour porpoises (*Phocoena phocoena*). *Aquat. Mamm.* **15**, 181–202.

Katona, S. K., Rough, V. and Richardson, D. T. (1993). "A Field Guide to Whales, Porpoises, and Seals from Cape Cod to Newfoundland". Smithsonian Institution Press, Washington, DC.

Kennedy, S., Smyth, J. A., Cush, P. F., McCullogh, S. J., Allen, G. M. and McQuaid, S. (1988). Viral distemper now found in porpoises. *Nature* **336**, 21.

Kinze, C. C. (1985). Intraspecific variation in Baltic and North Sea harbour porpoises (*Phocoena phocoena* (L., 1758)). *Viddensk Meddr dansk naturh. Foren.* **146**, 63–74.

Kinze, C. C. (1989). On the reproduction, diet and parasitic burden of the harbour porpoise *Phocoena phocoena* in West Greenlandic waters. *Eur. Res. Cetaceans* **3**, 91–95.

Kinze, C. C. (1995). Exploitation of harbour porpoises (*Phocoena phocoena*) in Danish waters: a historical review. *Rep. Int. Whal. Commn*, Spec. Issue No. 16, Cambridge, pp. 141–153.

Klinowska, M. (1991). "Dolphins, Porpoises and Whales of the World". The IUCN Red Data Book, IUCN, Gland, Switzerland.

Koga, T. (1991). Relationship between food consumption and growth of harbour porpoises *Phocoena phocoena* in captivity. *Int. Marine Biol. Res. Inst. Rep.* **2**, 71–73.

Koopman, H. N. (1998). Topographical distribution of the blubber of harbor porpoises (*Phocoena phocoena*). *J. Mamm.* **79**, 260–270.

Koopman, H. N. and Gaskin, D. E. (1993). Individual and geographical variation in pigmentation patterns of the harbour porpoise, *Phocoena phocoena* (L.). *Can. J. Zool.* **72**, 135–143.

Koopman, H. N., Iverson, S. J. and Gaskin, D. E. (1996). Stratification and age-related differences in blubber fatty acids of the male harbour porpoise (*Phocoena phocoena*). *J. Comp. Physiol.* **165**, 628–639.

Koopman, H. N., Westgate, A. J., Read, A. J. and Gaskin, D. E. (1995). Blood chemistry of wild harbor porpoises *Phocoena phocoena* (L.). *Mar. Mamm. Sci.* **11**, 123–135.

Kuiken, T., Bennett, P. M., Allchin, C. R., Kirkwood, J. K., Baker, J. R., Lockyer, C. H., Walton, M. J. and Sheldrick, M. C. (1994). PCBs, cause of death and body condition on harbour porpoises (*Phocoena phocoena*) from British waters. *Aquat. Toxicol.* **28**, 13–28.

Kuiken, T., Höfle, U., Bennett, P. M., Allchin, C. R., Kirkwood, J. K., Baker, J. R., Appleby, E. C., Lockyer, C. H., Walton, M. J. and Sheldrick, M. C. (1993). Adrenocortical hyperplasia, disease, and chlorinated hydrocarbons in the harbour porpoise (*Phocoena phocoena*). *Mar. Pollut. Bull.* **26**, 440–446.

Larsen, B. H. (1995). Parasites and pollutants in seven harbour porpoises (*Phocoena phocoena*) from the Faroe Islands, 1987–1988. *Rep. Int. Whal. Commn*, Spec. Issue No. 16, Cambridge, pp. 223–230.

Leighton, A. H. (1937). The twilight of the Indian porpoise hunter. *Nat. Hist.* **40**, 410–416.

Lindroth, A. (1962). Baltic salmon fluctuations 2: Porpoise and salmon. *Rep. Inst. Freshwater Res. Drottningholm* **44**, 105–112.

Litchfield, C., Greenberg, A. J., Caldwell, D. K., Caldwell, M. C., Sipos, J. C. and Ackman, R. G. (1975). Comparative lipid patterns in acoustical and nonacoustical fatty tissues of dolphins, porpoises and toothed whales. *Comp. Biochem. Physiol.* **50B**, 591–597.

Lockyer, C. (1995a). Aspects of the morphology, body fat condition and biology of the harbour porpoise, *Phocoena phocoena*, in British waters. *Rep. Int. Whal. Commn*, Spec. Issue No. 16, Cambridge, pp. 199–209.

Lockyer, C. (1995b). Investigation of aspects of the life history of the harbour porpoise, *Phocoena phocoena*, in British waters. *Rep. Int. Whal. Commn*, Spec. Issue No. 16, Cambridge, pp. 189–197.

Lovern, J. A. (1934). Liver fat metabolism in fishes. III. Selective formation of fat deposits. *Biochem. J.* **28**, 394–402.

Meek, A. (1918). The reproductive organs of Cetacea. *J. Anat.* **52**, 186–210.

Mercer, M. C. (1973). Observations on distribution and intraspecific variation in pigmentation patterns of odontocete Cetacea in the western North Atlantic. *J. Fish. Res. Bd Can.* **30**, 1111–1130.

Mitchell, E. (1975). "Porpoise, dolphin and small whale fisheries of the world. Status and problems". IUCN Monograph No. 3, IUCN, Morges, Switzerland.

Miyazaki, N., Amano, M. and Fujise, Y. (1987). Growth and skull morphology of harbour porpoises in Japanese waters. *Mem. Nat. Sci. Mus.* **20**, 137–146.

Mohl, B. and Andersen, S. H. (1973). Echolocation: high frequency component in the click of the harbour porpoise (*Phocoena ph.* L.). *J. Acoust. Soc. Am.* **54**, 1368–1372.

Møhl-Hansen, U. (1954). Investigations on reproduction and growth of the porpoise (*Phocaena phocaena* (L.)) from the Baltic. *Viddensk Medd Dan Naturhist Foren* **116**, 369–396.

Nachtigall, P. E., Lien, J., Au, W. W. L. and Read, A. J., Eds. (1995). "Harbour Porpoises—Laboratory Studies to Reduce Bycatch". De Spil, Woerden, The Netherlands.

van Nie, C. J. (1989). Postmortem findings in stranded harbour porpoises (*Phocoena phocoena* L. 1758) in the Netherlands from 23 March 1983 to 25 June 1986. *Aquat. Mamm.* **15**, 80–83.

Nielsen, E. and Andersen, S. (1982). Clinical chemistry and hematological findings in the harbour porpoise (*Phocoena phocoena*) from Danish waters. *Aquat. Mamm.* **9**, 1–3.

Noble, B. A. and Fraser, F. C. (1971). Description of a skeleton and supplementary notes on the skull of a rare porpoise *Phocoena sinus* Norris & McFarland 1958. *J. Nat. Hist.* **5**, 447–464.

Norris, K. S. and McFarland, W. N. (1958). A new harbor porpoise of the genus *Phocoena* from the Gulf of California. *J. Mamm.* **39**, 22–39.

Pabst, D. A. (1990). Axial muscles and connective tissues of the bottlenose dolphin. *In* "The Bottlenose Dolphin" (Eds S. Leatherwood and R. R. Reeves) pp. 51–67. Academic Press, San Diego.

Palka, D. (1995). Influences on spatial patterns of Gulf of Maine harbor porpoises. *In* "Whales, Seals, Fish and Man" (Eds A. S. Blix, L. Walloe and O. Ulltang) pp. 69–75. Elsevier, Amsterdam.

Palka, D. L., Read, A. J., Westgate, A. J. and Johnston, D. W. (1995). Summary of current knowledge of harbour porpoises in US and Canadian Atlantic waters. *Rep. Int. Whal. Commn* **46**, 559–565.

Parry, D. A. (1949). The structure of whale blubber, and a discussion of its thermal properties. *Quart. J. Microsc. Sci.* **90** 13–25.

Polacheck, T. (1995). Double team field tests of line transect methods for shipboard sighting surveys for harbor porpoises. *Rep. Int. Whal. Commn*, Spec. Issue No. 16, Cambridge, pp. 51–68.

Rae, B. B. (1965). The food of the common porpoise (*Phocaena phocaena*). *J. Zool. Lond.* **146**, 114–122.

Rae, B. B. (1973). Additional notes on the food of the common porpoise (*Phocoena phocoena*). *J. Zool. Lond.* **169**, 127–131.

Raga, J. A., Kinze, C. C., Balbuena, J. A., Ortiz, T. and Fernández, M. (1989). New data on helminth parasites of the harbour porpoise *Phocoena phocoena* in Danish waters. *In* "Proceedings of the Third Annual Conference of the European Cetacean Society" (Eds P. G. H. Evans and C. Smeenk).

Read, A. J. (1990a). Estimation of body condition in harbour porpoises. *Phocoena phocoena*. *Can. J. Zool.* **68**, 1962–1966.

Read, A. J. (1990b). Reproductive seasonality in harbour porpoises, *Phocoena phocoena*, from the Bay of Fundy. *Can. J. Zool.* **68**, 284–288.

Read, A. J. (1990c). Age at sexual maturity and pregnancy rates of harbour porpoises *Phocoena phocoena* from the Bay of Fundy. *Can. J. Fish. Aquat. Sci.* **47**, 561–565.

Read, A. J. and Gaskin, D. E. (1985). Radio tracking the movements and activities of harbor porpoises, *Phocoena phocoena* (L.), in the Bay of Fundy, Canada. *Fish. Bull.* **83**, 543–552.

Read, A. J. and Gaskin, D. E. (1988). Incidental catch of harbor porpoises by gill nets. *J. Wildl. Manage.* **52**, 517–523.

Read, A. J. and Gaskin, D. E. (1990). Changes in growth and reproduction of harbour porpoises, *Phocoena phocoena*, from the Bay of Fundy. *Can. J. Fish. Aquat. Sci.* **47**, 2158–2163.

Read, A. J. and Hohn, A. A. (1995). Life in the fast lane: the life history of harbor porpoises from the Gulf of Maine. *Mar. Mamm. Sci.* **11**, 423–440.

Read, A. J., Nicolas, J. R. and Craddock, J. E. (1996). Winter capture of a harbor porpoise in a pelagic drift net off North Carolina. *Fish. Bull.* **94**, 381–383.

Read, A. J., Wiepkema, P. R. and Nachtigall, P. E. (1997). "The Biology of The Harbour Porpoise". De Spil Publishers, Woerden, The Netherlands, 409 pp.

Read, A. J. and Tolley, K. A. (1997). Postnatal growth and allometry of harbour porpoises from the Bay of Fundy. *Can. J. Zool.* **75**, 122–130.

Read, A. J. and Westgate, A. J. (1997). Monitoring the movements of harbour porpoises (*Phocoena phocoena*) with satellite telemetry. *Mar. Biol.* **130**, 315–322.

Recchia, C. A. and Read, A. J. (1989). Stomach contents of harbour porpoises, *Phocoena phocoena* (L.), from the Bay of Fundy. *Can. J. Zool.* **67**, 2140–2146.

Reijnders, P. J. H. (1994). Toxicokinetics of chlorobiphenyls and associated physiological responses in marine mammals, with particular reference to their potential for ecotoxicological risk assessment. *Sci. Total Environ.* **154**, 229–236.

Ridgway, S. H. and Johnston, D. G. (1966). Blood oxygen and ecology of porpoises of three genera. *Science* **151**, 456–458.

Rogan, E. and Berrow, S. D. (1996). A review of harbour porpoises, *Phocoena phocoena*, in Irish waters. *Rep. Int. Whal. Commn* **46**, 595–605.

Rosel, P. E., Dizon, A. E. and Haygood, M. G. (1995). Variability of the mitochondrial control region in populations of the harbour porpoise, *Phocoena phocoena*, on inter-oceanic and regional scales. *Can. J. Fish. Aquat. Sci.* **52**, 1210–1219.

Rosel, P. E., Haywood, M. G. and Perrin, W. F. (1995). Phylogenetic relationships among the true porpoises (Cetacea: Phocoenidae). *Molec. Phylogen. Evol.* **4**, 463–474.

Ross, H. M. and Wilson, B. (1996). Violent interactions between bottlenose dolphins and harbour porpoises. *Proc. R. Soc. Lond.* **263**, 283–286.

Rowlatt, U. and Gaskin, D. E. (1975). Functional anatomy of the heart of the harbor porpoise, *Phocaena phocaena*. *J. Morphol.* **146**, 479–494.

Ryg, M., Lydersen, L., Knutsen, L. O., Bjorge, A., Smith, T. G. and Oritsland, N. A. (1993). Scaling of insulation in seals and whales. *J. Zool. Lond.* **230**, 193–206.

Saemundsson, B. (1939). Mammalia. *Zool. Iceland* **4**, Part 76, 1–38.

Scammon, C. M. (1874). "The Marine Mammals of the North-western Coast of North America, Described and Illustrated Together with an Account of the American Whale-Fishery". John H. Carman, San Francisco.

Schevill, W. E., Watkins, W. A. and Ray, C. (1969). Click structure in the porpoise, *Phocoena phocoena*. *J. Mamm.* **50**, 721–728.

Schulze, G. (1996). "Die Schweinswale". Ziemsen Verlag, Wittenberg.

Sequira, M. (1996). Harbour porpoises, *Phocoena phocoena*, in Portuguese waters. *Rep. Int. Whal. Commn* **46**, 583–586.

Siebert, U. (1995). Analyse des Quecksilbergehaltes in Organen von Kleinwalen aus deutschen Gewassern der Nord- und Ostsee und seine moglichen Beziehungen zu Erkrankungs- und Todesursachen. Inaugural-Dissertation zur Erlangung des Doktorgrades beim Fachbereich Veterinarmedizin, Justus-Leibigs-Universitat, Gieben, Germany.

Simons, L. S. (1984). Seasonality of reproduction and dentinal structures in the harbor porpoise (*Phocoena phocoena*) of the North Pacific. *J. Mamm.* **65**, 491–495.

Skora, K. E., Pawliczka, I. and Klinowska, M. (1988). Observations of the harbour porpoise (*Phocoena phocoena*) on the Polish Baltic coast. *Aquat. Mamm.* **14**, 113–119.

Slijper, E. J. (1936). Die Cetaceen, vergleichend-anatomisch und systematisch. *Capita Zoologica* **7**, 1–600.

Slijper, E. J. (1949). On some phenomena concerning pregnancy and parturition of the Cetacea. *Bijdragen tot de Dierkunde* **28**, 416–448.

Slijper, E. J. (1958). Organ weights and symmetry problems in seals and porpoises. *Arch. Neerl. Zool.* **13**, 97–113.

Smeenk, G., Leopold, M. F. and Addink, M. J. (1992). Note on the harbour porpoise *Phocoena phocoena* in Mauritania, West Africa. *Lutra* **35**, 98–104.
Smith, G. J. D. (1972). The stomach of the harbour porpoise *Phocoena phocoena* (L.). *Can. J. Zool.* **50**, 1611–1616.
Smith, G. J. D., Browne, K. W. and Gaskin, D. E. (1976). Functional myology of the harbour porpoise, *Phocoena phocoena* (L.). *Can. J. Zool.* **54**, 716–729.
Smith, G. J. D. and Gaskin, D. E. (1974). The diet of harbour porpoises (*Phocoena phocoena* (L.)) in coastal waters of Eastern Canada, with special reference to the Bay of Fundy. *Can. J. Zool.* **52**, 777–782.
Smith, G. J. D. and Gaskin, D. E. (1983). An environmental index for habitat utilization by female harbour porpoises with calves near Deer Island, Bay of Fundy. *Ophelia* **22**, 1–13.
Smith, J. W. (1989). Ulcers associated with larval *Anisakis simplex* B (Nematoda: Ascaridoidea) in the forestomach of harbour porpoises, *Phocoena phocoena* (L.). *Can. J. Zool.* **67**, 2270–2276.
Smith, R. J. and Read, A. J. (1992). Consumption of euphausiids by harbour porpoise (*Phocoena phocoena*) calves in the Bay of Fundy. *Can. J. Zool.* **70**, 1629–1632.
Smith, T. D. (1982). Current understanding of the status of small cetacean populations in the Black Sea. *In* "Mammals in the Seas", FAO Fisheries Series No. 5, Vol. 4, pp. 121–130.
Sorensen, T. B. and Kinze, C. C. (1994). Reproduction and reproductive seasonality in Danish harbour porpoises, *Phocoena phocoena*. *Ophelia* **39**, 159–176.
Stuart, L. J. and Morejohn, G. V. (1980). Developmental patterns in osteology and external morphology in *Phocoena phocoena*. *Rep. Int. Whal. Commn*, Spec. Issue No. 3, Cambridge, pp. 133–142.
Tomilin, A. G. (1957). "Mammals of the U.S.S.R. and Adjacent Countries", Volume IX, "Cetacea" (Ed. V. G. Heptner). Nauk USSR, Moscow (English translation, Israel Program for Scientific Translation, Jerusalem).
Trippel, E. A., Wang, J. Y., Strong, M. B., Carter, L. S. and Conway, J. D. (1996). Incidental mortality of harbour porpoise (*Phocoena phocoena*) by the gill-net fishery in the lower Bay of Fundy. *Can. J. Fish. Aquat. Sci.* **53**, 1294–1300.
True, F. W. (1889). Contributions to the natural history of the cetaceans. A review of the Delphinidae. *Bull. U.S. Nat. Mus.* **36**, 1–149.
van Utrecht, W. L. (1959). Wounds and scars in the skin of the common porpoise, *Phocaena phocaena* (L.). *Mammalia* **23**, 100–122.
Van Deinse, A. B. (1931). De fossile en recente Cetacea van Nederland. Dissertation, University of Utrecht.
Van der Spoel, S. (1963). The vascular system in the kidneys of the common porpoise (*Phocoena phocoena* (L.)). *Bijdr. Dierk.* **33**, 71–82.
Verboom, W. C. and Kastelein, R. A. (1995). Acoustic signals by harbour porpoises (*Phocoena phocoena*). *In* "Harbour Porpoises — Laboratory Studies to Reduce Bycatch" (Eds P. E. Nachtigall, J. Lien, W. W. L. Au and A. J. Read) pp. 1–39. De Spil, Woerden, The Netherlands.
Wang, J. Y., Gaskin, D. E. and White, B. N. (1996). Mitochondrial DNA analysis of harbour porpoise, *Phocoena phocoena*, subpopulations in North America waters. *Can. J. Fish. Aquat. Sci.* **53**, 1632–1645.
Watkins, W. A., Schevill, W. E. and Best, P. B. (1977). Underwater sounds of *Cephalorhynchus heavisidii* (Mammalia: Cetacea). *J. Mamm.* **58**, 316–320.
Watson, A. P. (1976). The diurnal behaviour of the harbour porpoise (*Phocoena phocoena* L.) in the coastal waters of the western Bay of Fundy. M.Sc. thesis, University of Guelph, Canada.

Watson, A. P. and Gaskin, D. E. (1983). Observations on the ventilation cycle of the harbour porpoise *Phocoena phocoena* (L.) in coastal waters of the Bay of Fundy. *Can. J. Zool.* **61**, 126–132.

Watts, P. and Gaskin, D. E. (1985). Habitat index analysis of the harbor porpoise (*Phocoena phocoena*) in the southern coastal Bay of Fundy, Canada. *J. Mamm.* **66**, 733–744.

Welsh, M. J., Lyons, C., Trudgett, A., Rima, B. K., McCullogh, S. J. and Orvell, C. (1992). Characterization of a cetacean morbillivirus isolated from a porpoise (*Phocoena phocoena*). *Arch. Virol.* **74**, 631–641.

Westgate, A. J., Muir, D. C. G., Gaskin, D. E. and Kingsley, M. C. S. (1996). Concentrations and accumulation patterns of organochlorine contaminants in the blubber of harbour porpoises, *Phocoena phocoena*, from the coast of Newfoundland, the Gulf of St. Lawrence and the Bay of Fundy/Gulf of Maine. *Environ. Pollut.* **90**, 1–14.

Westgate, A. J., Read, A. J., Berggen, P., Koopman, H. N. and Gaskin, D. E. (1995). Diving behaviour of harbour porpoises, *Phocoena phocoena*. *Can J. Fish. Aquat. Sci.* **52**, 1064–1073.

Williamson, G. R. (1963). Common porpoise from the stomach of a Greenland shark. *J. Fish. Res. Bd Can.* **20**, 1085–1086.

Wislocki, G. B. (1933). On the placentation of the harbour porpoise (*Phocaena phocaena* Linnaeus). *Biol. Bull.* **65**, 80–98.

Wislocki, G. B. (1942). The lungs of Cetacea, with special reference to the harbour porpoise (*Phocaena phocaena* Linnaeus). *Anat. Rec.* **84**, 117–123.

Worthy, G. A. J. and Edwards, E. F. (1990). Morphometric and biochemical factors affecting heat loss in a small temperate cetacean (*Phocoena phocoena*) and a small tropical cetacean (*Stenella attenuata*). *Physiol. Zool.* **63**, 432–442.

Worthy, G. A. J., Innes, S., Braune, B. M. and Stewart, R. E. A. (1987). Rapid acclimation of cetaceans to an open-system respirometer. *In* "Marine Mammal Energetics" (Eds A. C. Huntley, D. P. Costa, G. A. J. Worthy and M. A. Castellini) pp. 115–126. Society for Marine Mammalogy Special Publication No. 1, Allen Press, Kansas.

Würsig, B. (1986). Delphinid foraging strategies. *In* "Dolphin Cognition and Behavior: A Comparative Approach" (Eds R. J. Schusterman, J. A. Thomas and F. G. Wood) pp. 347–359. Lawrence Erlbaum Associates, Hillsdale, NJ.

Yasui, W. Y. and Gaskin, D. E. (1986). Energy budget of a small cetacean, the harbour porpoise, *Phocoena phocoena* (L.). *Ophelia* **25**, 183–197.

Yurick, D. B. and Gaskin, D. E. (1987). Morphometric and meristic comparisons of skulls and harbour porpoise *Phocoena phocoena* (L.) from the North Atlantic and North Pacific. *Ophelia* **27**, 53–75.

Yurick, D. B. and Gaskin, D. E. (1988). Asymmetry in the skull of the harbour porpoise *Phocoena phocoena* (L.) and its relationship to sound production and echolocation. *Can. J. Zool.* **66**, 399–402.

13

Vaquita
Phocoena sinus
Norris and McFarland, 1958

Omar Vidal, Robert L. Brownell, Jr and Lloyd T. Findley

Genus and Species

Taxonomy

Phocoena sinus was described by Norris and McFarland in 1958 from three skulls collected along the northeast shore of Punta San Felipe and at Bahia San Felipe, Baja California Norte (BCN), in the Gulf of California, Mexico. Two of these specimens (holotype and a paratype) are deposited in the Museum of Vertebrate Zoology, University of California at Berkeley. The other, a paratype, is in the National Museum of Natural History, Smithsonian Institution, Washington, DC. The late Carl L. Hubbs suggested the specific name *sinus* to Norris and McFarland. In Latin it means "bay", referring to the occurrence of the species in the Gulf of California (Sea of Cortez).

The six extant members of the family Phocoenidae are *Phocoena phocoena*, *P. spinipinnis*, *P. sinus*, *Neophocaena phocaenoides*, *Phocoena dioptrica* and *Phocoenoides dalli*. The first four have been placed in the subfamily Phocoeninae and the others in the subfamily Phocoenoidinae (Barnes, 1985). The geologically earliest undisputed fossil phocoenid is *Salumiphocaena stocktoni* (Wilson, 1973) from the late Miocene deposits in southern California (Barnes, 1985). Three other fossil phocoenids have been described: *Piscolithax tedfordi* Barnes, 1984 from latest

Handbook of Marine Mammals, Volume 6
ISBN 0-12-588506-7

Miocene deposits in Mexico; *P. longirostris* de Muizon, 1983 from Early Pliocene deposits in Peru; and *P. boreios* Barnes, 1984 from latest Miocene deposits in Mexico (Barnes, 1985). All these fossil phocoenids belong in the Phocoenoidinae. Several periotics of an undetermined fossil species of Phocoeninae were collected recently from the Pliocene at Santa Rosalía, Baja California Sur (BCS), Mexico (Vidal, 1991).

Common names

Vaquita (= "little cow") is the name used by local fishermen in the northern Gulf of California, Mexico, where for many years these porpoises have been taken incidental to fishing operations (Brownell *et al.*, 1987; Vidal, 1995). Cochito (= "little pig") had been a commonly used vernacular name for *P. sinus* because it was adopted by the International Whaling Commission in 1975 (Mitchell, 1975). However, in the Gulf of California, "cochito" is used for more than one species of small cetacean. The English name used for this animal is the Gulf of California harbour porpoise.

Evolution

Phocoenids have been hypothesized to represent a paedomorphic evolutionary lineage (i.e. some or several foetal or juvenile characters persist in adults). Although examples of paedomorphosis may be found in some species in other cetacean families, in no other family group is this phenomenon as universally presented and as pronounced as it is in the living phocoenids (Barnes, 1985). *P. sinus* appears to be strongly paedomorphic (L. G. Barnes, personal communication; Torre-Cosío, 1995).

Norris and McFarland (1958) discussed the possible origin of *P. sinus* from the Northern Hemisphere *P. phocoena* or from *P. spinipinnis*, which lives off South America. Although *P. sinus* is in some respects intermediate between both species, they favoured *P. spinipinnis* as the more likely ancestor because the cranium of *P. sinus* resembles that of *P. spinipinnis* more than that of *P. phocoena*. Noble and Fraser (1971) noted that the axial skeleton of *P. sinus* suggests a closer affinity to *P. spinipinnis* than to *P. phocoena*. All of the axial skeleton material that we and a colleague (Torre-Cosío, 1995) have examined agrees generally with the specimen described by Noble and Fraser (1971).

Externally, *P. sinus* lacks the peculiarly shaped and heavily tuberculated dorsal fin of *P. spinipinnis*. A few low tubercles do appear along the leading edge of the fin and are more pronounced in older individuals. The relative sizes of the flippers and several other external measurements of *P. sinus* are closer to those of *P. spinipinnis* than they are to *P. phocoena* (Brownell *et al.*, 1987). Thus, *P. sinus* and *P. spinipinnis* appear to be more closely related. We believe, as Norris and McFarland (1958) suggested, that *P. sinus* probably evolved as a result of a northward movement of an ancestral *P. spinipinnis*-like population into the Gulf of California during one of the Pleistocene glacial ages.

Distribution

Range

This porpoise is restricted to the relatively shallow waters of the northern Gulf of California, with most of the records from near San Felipe, Rocas Consag, and El Golfo de Santa Clara (Fig. 1). This distribution is based only on confirmed records, and other reputed records of the species in the southern Gulf of California are considered unreliable (Brownell, 1986). Ninety confirmed records of *P. sinus* have been collected, all north of Puertecitos (BCN) and Puerto Peñasco, Sonora (Vidal, 1991, 1995; Brownell, 1986).

The distribution of this porpoise in the upper Gulf of California appears to be highly localized, with the highest densities offshore of San Felipe and Rocas Consag, and offshore of El Golfo de Santa Clara (Silber, 1990a; Silber and Norris, 1991; Gerrodette *et al.*, 1995; Vidal, 1995). An analysis of all available sightings led Silber (1990a) and Silber and Norris (1991) to suggest that vaquitas occupy the northern Gulf year-round.

Two unconfirmed sightings in 1983 near Isla Cerralvo, in the southern Gulf of California (about 850 km south of the southernmost confirmed sighting), were reported by Silber (1990a). A strong El Niño (ENSO) event occurred during 1982–1983, and water temperatures were unusually high at that time in the southern Gulf (Cane, 1983). Extralimital range records have been reported for most species of cetaceans around the world for various reasons, including El Niño events. Therefore, these southern Gulf records, if valid, may be extralimital and thus do not necessarily imply a wider geographic range for this porpoise. In the absence of confirmed records (i.e. voucher specimens or photographs) from the southern Gulf, we conclude that this species' normal range is confined to the northernmost Gulf of California and is the most restricted range of any marine cetacean.

Habitat

Vaquitas are known only from relatively shallow waters. Silber (1990a) reported 51 sightings in water depths of 13.5–37 m, and most of these sightings were 11–25 km from shore. All but two of the sightings were less than 40 km from San Felipe (mainly between this locality and Rocas Consag). Water visibility ranged from 0.9 m to 12 m. Both of the sightings by Wells *et al.* (1981) were in water depths of around 19 m, and were about 18 km from shore. All known incidental entanglements of this species in gill nets have occurred in water depths of from

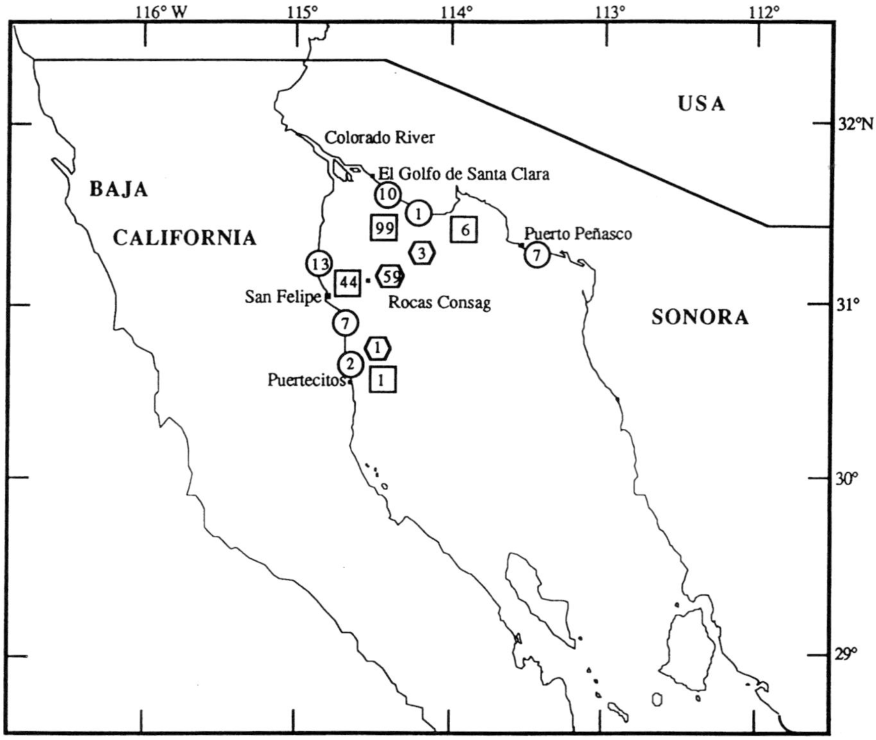

FIG. 1 Range of the vaquita, *Phocoena sinus*: circles indicate beached specimens, squares indicate incidental mortality in gill nets, and hexagons indicate sightings.

4 m to 36 m (estimated by the fishermen operating the nets or by reference to nautical charts) and between 3 km and 33 km from shore (Vidal, 1995).

External Characteristics

Colour pattern

Generally the pigmentation pattern (Figs 2 and 3) can be divided into three parts: dark-grey cape, pale-grey lateral field, and white ventral field (terminology of Perrin, 1972; Perrin *et al.*, 1981). The most conspicuous features of the coloration are the relatively large black eye and lip patches, which contrast sharply with the surrounding light grey of the side of the head and thoracic regions. The flipper stripe is also a dominant feature of the coloration. The shapes of the eye patch, lip patches, and flipper stripe all vary slightly among individuals. Additional details of the colour pattern are given by Brownell *et al.* (1987).

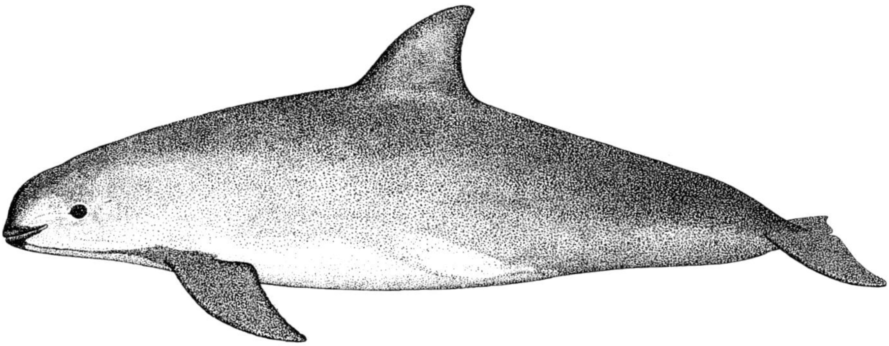

FIG. 2 Lateral view of vaquita, *Phocoena sinus*. (Drawing by Pieter A. Folkens)

Size and shape

This species is the shortest of all delphinoids. Two recently studied series of 13 (Brownell *et al.*, 1987) and 30 (Vidal, 1995) specimens ranged from 70.3 cm to 148.2 cm in total length. A near-term foetus measured 71.5 cm. Some selected external measurements are given in Table 1. No significant sexual dimorphism was noted in any of these measurements.

As with other phocoenids, the vaquita is small but robust, with a mean maximum girth (measured just in front of the anterior insertion of the dorsal fin) of about 70% of total length. In lateral view, the head appears as a truncated cone with the posterior part of the melon sloping downward towards the blowhole, especially in adult males (Fig. 2). Anteriorly, the melon slopes abruptly to the snout tip. There is no "beak" present, but the snout protrudes very slightly from the base of the melon. The lower jaw does not project beyond the snout tip as it does in most delphinoids. The flippers are attenuated and apically pointed (Figs 2 and 3), much like those of *P. spinipinnis* and unlike those of *P. phocoena*. The shape and height of the dorsal fin (Fig. 2) is the most striking external feature of *P. sinus*. It is proportionally much higher than in other species of *Phocoena*. The shape is roughly triangular, with the anterior edge slightly convex and longer than the posterior edge, which is nearly straight or slightly concave. In the mature males we have examined, the dorsal fin is taller and less falcate (i.e. more erect). The genital aperture in males, as in some other phocoenids, is closer to the umbilicus than to the anus, which contrasts with the condition typical of many other delphinoids.

Small tubercles of variable development were present along the upper 1/2 to 3/4 of the anterior edge of the dorsal fin. These were more prominent and more numerous in some of the largest individuals. Although the tubercles were prominent to the touch in the larger individuals, none of our specimens had the more prominent horny tubercles found in *P. spinipinnis* (Allen, 1925; Brownell and Praderi, 1984). All of the 17 males we examined had a small, blind opening (not

FIG. 3 External views of *Phocoena sinus* incidentally caught near El Golfo de Santa Clara, Sonora, Gulf of California, Mexico. (Photos by J. M. Nava)

TABLE 1 Ranges of selected external measurements (cm) of vaquitas, *Phocoena sinus*, from Brownell *et al.* (1987), based on a sample size of 12

Measurement	Range	Mean	Percentage of mean
Total length	90.3–143.5	113.7	—
Tip of upper jaw to:			
angle of gape	4.9–11.8	6.9	6.1
centre of blowhole	8.0–13.0	10.3	9.1
centre of eye	8.5–12.0	10.6	9.3
auditory meatus	12.6–17.6	15.5	13.6
ant. insertion flipper	18.0–25.3	21.9	19.3
umbilical scar	42.5–61.0	50.2	44.1
tip of dorsal fin	54.5–85.0	68.8	60.5
centre of genital slit	52.0–95.0	70.0	61.6
centre of anus	66.1–100.1	79.1	69.6
Ant. length of flipper	17.5–29.1	23.5	20.7
Axilla length of flipper	13.2–22.5	17.7	15.6
Max. width of flipper	6.4–10.1	8.5	7.5
Height of dorsal fin	9.6–15.0	12.3	10.8
Length of dorsal fin base	12.7–21.0	16.3	14.3
Width of flukes	23.8–44.7	34.6	30.4
Width of fluke, from notch to nearest point on anterior border	7.2–10.9	9.3	8.2
Depth of notch	1.2–2.6	1.9	0.2
Width of blowhole	1.7–2.8	2.3	0.2
Girth, at axilla	51.5–75.5	65.7	57.8
Girth, maximum	63.0–93.0	77.0	67.7
Girth, at anus	38.5–58.0	47.2	41.5

yet described in detail) just anterior to the anus that appeared to be much like that described for *Neophocaena phocaenoides* by Nishiwaki and Kureha (1975). One of the males had two of these openings (parallel to each other).

The four largest males (134.5, 135, 142.2 and 144 cm) and the nine largest females (135, 135, 135, 139.7, 139.7, 140.5, 140.9, 143.5 and 148.2 cm) examined by Brownell *et al.* (1987) and Vidal (1995) were all physically mature. Three additional large specimens include two females (both dehydrated and decomposed), 139 cm and 150 cm; and one male (decomposed), 145 cm long (Brownell, 1982; Silber and Norris, 1991). Although the sample is small, the maximum female body length is greater than that of males, as in the harbour porpoise, *P. phocoena*.

Weight

Whole body weights of the above-mentioned 43 fresh specimens ranged from 7.5 kg to 47.5 kg. The total lengths of these specimens ranged from 70.3 cm to 148.2 cm.

Internal Anatomy

Skull

Norris and McFarland (1958) described the first skulls, and additional details were reported by Orr (1969), Noble and Fraser (1971), Brownell (1983), Magatagan *et al.* (1984), Barnes (1985), G. Carvallo and L. T. Findley (unpublished data), and Torre-Cosío (1995). The adult skulls are smaller, and have relatively much broader and shorter rostra than those of other phocoenids. The condylobasal lengths of 36 skulls (of immature and mature individuals) range from 180 mm to 243 mm. Ranges of additional selected skull measurements (in mm) are zygomatic width 123–157 ($n = 21$), mandible length 158–184 ($n = 21$), and rostral length 79–93 ($n = 17$) (Brownell, 1983 and unpublished data; Torre-Cosío, 1995). The skull is illustrated in Fig. 4. The tympanic bullae and periotics were illustrated by Brownell (1983).

The teeth have spatulate crowns and essentially are uniform throughout the mouth, as in other species of *Phocoena*. In 24 specimens, the number of erupted teeth in the maxillary tooth rows was 16–22 and in the mandibular tooth rows 17–20 (Brownell *et al.*, 1987; O. Vidal, unpublished data).

Postcranial skeleton

In most individuals examined by us the first three of the seven cervical vertebrae were fused, as in *P. spinipinnis*. In three specimens the first four vertebrae were fused. In another specimen the fourth and fifth vertebrae were also fused but separated from the first three. There are 12 or 13 thoracic, 11–15 lumbar, and 29–31 caudal vertebrae, with 19–23 chevron bones (Noble and Fraser, 1971; Brownell, 1983; Magatagan *et al.*, 1984; L. T. Findley, G. Carvallo, and J. Torre-Cosío, unpublished data). The total number of vertebrae ranges between 58 and 64. The first six or seven ribs have both capitular and tubercular attachments to the vertebrae. Six or seven pairs of sternal ribs are present of which the first three are attached to the sternum. Based on counts from radiographs or prepared flipper skeletons of nine individuals, the phalangeal formula is I_2, II_{7-10}, III_{7-9}, IV_{4-5}, V_{1-2} (L. T. Findley and J. Torre-Cosío, unpublished data).

(A)

(B)

(C)

(D)

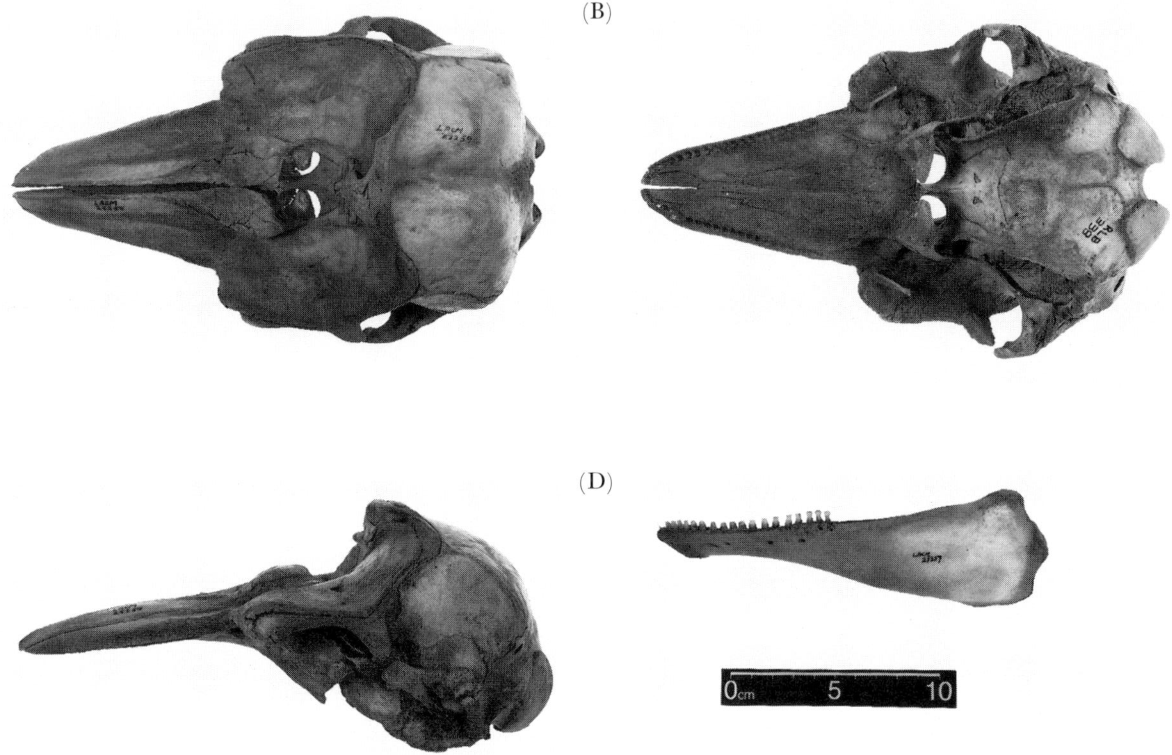

FIG. 4 Ventral (A), dorsal (B) and lateral (C) views of the skull and the mandible (D) from a 150 cm *Phocoena sinus* (LACM 28259) from San Felipe, Baja California, Mexico.

Blubber thickness

The range of blubber thickness in 19 vaquitas that varied in total length between 72.8 cm and 148.2 cm was: 8–33 mm mid-dorsal (adjacent to the anterior edge of dorsal fin), 8–19 mm mid-lateral, and 8–21 mm mid-ventral (O. Vidal, unpublished data). The relative blubber weight in one of the smallest porpoises (93.5 cm) was 33.3% but only 26.5% in one of the largest specimens (148.2 cm).

Internal organs

Ranges of selected organ weights from 19 specimens we examined are presented in Table 2.

Microanatomy

G. E. Michel (unpublished data) provides histological descriptions for tissues of 13 organs from a total of nine vaquitas. In general, the microanatomy of *P. sinus* was found to resemble that of other delphinoids, such as the bottlenose dolphin, *Tursiops truncatus*.

TABLE 2 Organ weights (g) of the largest and smallest male and female vaquitas, *Phocoena sinus*

Organ	Male 93.5 cm	Female 105.9 cm	Male 142.2 cm	Female 148.2 cm
Heart	82.2	121.7	198.1	233.7
Lungs				
left	180.7	149.9	204.7	436.4
right	168.6	176.3	284.9	472.6
Stomach	118.5	390.0	—	—
Intestine	367.8	940.0	610.6	—
Liver	300.0	502.0	806.3	1235.0
Kidneys				
left	64.7	86.4	153.5	180.8
right	65.3	92.4	142.0	177.3
Pancreas	15.4	43.2	75.3	77.2
Spleen	4.0	6.8	7.0	4.6
Thyroid	5.0	—	—	—
Adrenals				
left	1.0	1.7	3.5	6.3
right	1.1	2.9	4.1	6.7
Gonads				
left	4.0	—	284.3	2.5
right	3.7	—	325.6	1.2
Blubber	5.5 kg	9.5 kg	7.36 kg	11.65 kg
Total weight	16.5 kg	22.0 kg	42.50 kg	44.00 kg

Physiology and biochemistry

Nothing has been published on the physiology or biochemistry of this porpoise. Recent genetic studies based on samples of vaquitas taken at different times and locations in the upper Gulf found no sequence polymorphisms (i.e. no genetic variability in mitochondrial DNA) which may represent the results of a population bottleneck, founder effect, or severe inbreeding in the species (Rosel, 1992; Rosel and Rojas-Bracho, 1993).

Life History and Population Dynamics

Growth and reproduction

The life history of the vaquita appears, in many ways, to be similar to the harbour porpoise from the Bay of Fundy, Canada (Read and Gaskin, 1990). Nothing is known about the gestation period, but it is probably similar to that of *P. phocoena* (about 11 months) (see also Hohn *et al.*, 1996). Neonates ($n = 11$) are approximately 70–78 cm in total length and weigh at least 7.5 kg; they are born mainly in March and April (Brownell, 1983; Brownell *et al.*, 1987; Silber and Norris, 1991; Vidal, 1995; Hohn *et al.*, 1996). Vidal (1995) also reported six other possible neonates (two referred to by fishermen as "very small" and the others estimated by them at around 50, 60, 70, and 80 cm) that died during fishing operations in March 1989 and February and April of 1990. Silber (1988, 1990a) reported seven calves (8.2% of all individuals) sighted during his surveys, six between 25 and 27 March 1986 and one "very young, probably less than two days old" on 9 April 1987. Based on the limited data, parturition occurs in the late winter and early spring, between February and April, with the peak of calving in late March and early April (Vidal, 1995; Hohn *et al.*, 1996). The age structure of a recently studied sample of vaquitas killing in fishing nets was bimodal, with 62% of the specimens of 0–2 years and 31% of the specimens of 11–16 years (Hohn *et al.*, 1996). The oldest individual was estimated to be 21 years (Hohn *et al.*, 1996).

We examined a total of 43 fresh individuals of known sex (19 males and 24 females). The smallest adult male and female were 128.3 cm and 135 cm, respectively. The weights of these two porpoises were 39 kg and 41 kg, respectively. Also, both of these porpoises were physically mature. The largest immature male and female were 127 cm and 128.7 cm long, respectively.

Based on the minimum testes weights (i.e. the weight of the smaller of the pair, with the epididymis removed) of six males (284.3, 359.4, 480, 620, 642.1 and 643.8 g), vaquitas appear to conform to the general delphinoid pattern of large testis size relative to body size noted by Kenagy and Trombulak (1986). Such relatively large testes suggest some type of multi-male breeding system. The largest immature testis weight (i.e. the weight of the larger of the pair, with the epididymis removed) was 74.3 g.

Abundance

Little is known about the abundance of the vaquita. Although various sightings have been reported, many were opportunistic and do not allow reliable estimates of the population size of this porpoise. Prior to 1986, only one cruise was specifically planned to survey for *P. sinus* (Wells *et al.*, 1981). These authors surveyed a 1959 km transect but recorded only two sightings. Since then, the only other surveys wholly dedicated to *P. sinus* have been those of G. K. Silber and co-workers (Silber, 1988, 1990a; Silber *et al.*, 1988) and of T. Gerrodette and his co-workers (Gerrodette *et al.*, 1994, 1995). A total of 4216 km of boat and aircraft surveys conducted during 77 days in 1986–1989 resulted in only 58 sightings, representing a total of 110 individuals (Silber, 1990a,b). Forty-three of these vaquitas (19 sightings) were recorded during 1715 km of vessel transects, a sighting rate of 2.51 individuals per 100 km surveyed (Silber, 1990a). The remainder of the sightings occurred during non-transect time (Silber, 1990b). In an aerial survey for vaquitas in the northern Gulf, there was only one certain sighting of two vaquitas and a possible sighting of another single vaquita (Barlow *et al.*, 1993). Noting the difficulty of observing vaquitas in the turbid waters of the upper Gulf of California, it was recommended that ship surveys would be more effective for future surveys (Barlow *et al.*, 1993).

Only three serious attempts have been made to estimate the total size of the vaquita population. Based on Silber's censuses for 1986 ("30" individuals in "11" sightings; these figures were later amended by Silber (1988) to "27" and "12" respectively) and on 14 specimens caught incidentally in gill nets and examined during 1985 and 1986 (Findley and Vidal, 1985; Brownell *et al.*, 1987; Robles *et al.*, 1987), Barlow (1986) estimated 50–100 individuals as an approximate lower limit for the population. He noted that it was not possible to estimate any upper limit based on the limited data. It appears that Barlow's minimum estimate was (understandably) too low. It is now known that a minimum of 143 vaquitas were killed in various fishing operations between March 1985 and January 1994 and an annual incidental mortality of 35 vaquitas was estimated (D'Agrosa *et al.*, 1995; Vidal, 1995). Based on a largely separate data set, Turk-Boyer (1989) and Turk-Boyer and Silber (1990) estimated that the annual fishing-related mortality of vaquitas is 32. These recorded kills must represent only a fraction of the total number of vaquitas that are killed in the various regional fishing operations. A population significantly larger than Barlow's (1986) estimate would be necessary to permit this level of mortality.

Recently, the abundance of vaquitas was estimated from four surveys conducted between 1986 and 1993, using a variety of methods (Barlow *et al.*, 1997).

A line-transect method was applied, using parameters from the harbour porpoise. Vaquita abundance was estimated at 503 (CV = 0.63) from 1986–1993 boat surveys, 885 (CV = 0.50) from 1988–1989 aerial surveys, 572 (CV = 1.43) from a 1991 aerial survey, and 224 (CV = 0.39) from a 1993 ship survey. All of these abundance figures indicate that the species is at a critically low level. A weighted log–linear regression indicated a rate of population decline of about 18% (95% CI = −43.2% to +19.3%) between 1986 and 1993.

Based on four years of seasonal surveys, Silber (1990c) presented, but did not discuss quantitative details for, his estimate of 200–500 individuals for the entire population. Silber and co-workers documented 16 sightings with 27 individuals for 1986, 22 sightings with 46 individuals for 1987, 13 sightings with 23 individuals for 1988, and seven sightings with 14 individuals for 1989. The total number of vaquitas observed is only 110. Given the brevity of most sightings (because of the low surfacing profile and relatively short time that this porpoise spends at the surface; e.g. see Silber *et al.*, 1988; Silber, 1990b), the relatively restricted area searched (simply because of the restricted geographical distribution of the vaquita), and the inherent difficulty of differentiating one individual from another, Silber and co-workers probably observed some of the same individuals during different days, months, and years. Thus, considering the scarcity of sightings relative to survey effort, the few individuals per sighting, and the limited geographical range of the vaquita, there can be no doubt that the population of this species is very small.

Behaviour

Social organization

Like other phocoenids, *P. sinus* occurs singly or in small groups. In 58 sightings, 91% comprised from one to three individuals, with a mean group size of 1.9 and a range of 1–7 (Silber, 1990a,b). Loose aggregations of vaquitas in which they were dispersed as single individuals or as small subgroups (from two to four members, greatest number eight to ten) throughout several hundred square metres were observed by Silber (1988). All previous sightings comprised from one to three vaquitas (Norris and McFarland, 1958; Norris and Prescott, 1961; Wells *et al.*, 1981; Vidal *et al.*, 1987).

Wells *et al.* (1981) suggested that these porpoises typically avoid boats, but Silber *et al.* (1988) noted no apparent response to their small boat when following two different female/calf pairs for several hours at distances of 40–200 m, nor to the presence of several nearby fishing skiffs. In each case, however, abbreviated surfacing sequences were noted as a possible reaction to the boats.

Movements

Nothing is available on local or regional movements. Most sightings of these porpoises have been made in a relatively small area during winter and spring months (the seasons when most surveys have been carried out).

Feeding

The remains of bronzestriped grunt, *Orthopristis reddingi* (Haemulidae), and Gulf croaker, *Bardiella icistia* (Sciaenidae), were recovered from the stomach of one *P. sinus* carcass north of San Felipe in 1965 (Fitch and Brownell, 1968). Both fishes are found throughout the demersal zone of the northern Gulf of California. Another beached specimen, collected near El Golfo de Santa Clara in 1984, had remains of sciaenid fishes ("croakers") in its stomach (R. L. Brownell Jr., unpublished data). Two vaquitas collected in 1988 contained several whole fish ("probably *Anchoa nasus* or *Sardinops* spp."), numerous unidentified fish otoliths, and squid beaks (Silber, 1990b). If generic identification of the *Sardinops* is correct, then it could only be *S. sagax caerulea*, the Pacific sardine. Squid beaks, *Lolliguncula panamensis*, were also found in the stomachs of the porpoises collected in 1965 and 1984. Stomach contents of six vaquitas necropsied by us in 1985 yielded upper and lower beaks of the squids *L. panamensis* (103 total beaks and one whole squid body) and *Loliolopsis diomedeae* (25 total beaks) (F. G. Hochberg, personal communication). The former species was present in all six stomachs analysed, and the latter species was found in two of the six stomachs. These squids appear to be abundant in coastal waters throughout the Gulf of California and southward. A recent analysis of stomach contents from 24 vaquita showed the presence of the same two squids plus 17 species of teleost fishes. The sample did not include *Sardinops* or *Anchoa nasus*, although five other species of anchovies were present. Also, the other 12 species of fishes encountered did not include the two species reported by Fitch and Brownell (1968), but we have no doubts about their identifications. All of the 17 fish species can be classified as demersal and/or benthic species inhabiting relatively shallow water in the upper Gulf of California, and it appears that the vaquita is a rather non-selective feeder on small fishes and squids in this zone (Findley *et al.*, in prep.).

Association with other species

No mixed-species aggregations have been encountered, but Silber (1990a) reported two sightings of Bryde's whales, *Balaenoptera edeni*, and three sightings of common dolphins, *Delphinus delphis*, within < 1 km and < 1.5 km of groups of *P. sinus*, respectively. Numerous black storm petrels, *Oceanodroma melania*, and Bonaparte's gulls, *Larus philadelphia*, were observed dipping into the wake of surfacing vaquitas (Silber, 1990a).

Predation

Several shark fishermen of El Golfo de Santa Clara, who regularly capture vaquitas incidentally, reported to one of us (O.V.) that in February to May 1990 and 1991 they found whole or chewed parts of vaquitas in the stomachs of several species of large sharks. The sharks were identified from photographs and/or jaws provided by the informants, and they included at least six species: great white (*Carcharodon carcharias*), shortfin mako (*Isurus oxyrhynchus*), lemon (*Negaprion brevirostris*), black-tipped (*Carcharhinus limbatus*), bigeye thresher (*Alopias superciliosus*) and broad-snout seven-gill (*Notorhynchus cepedianus*) (Vidal, 1995). Both the great white and shortfin mako are known locally as "tiburones tonina" or "dolphin sharks" because of the nearly equal sizes of their caudal-fin lobes. On 18 February 1990, the stomach from a great white shark about 3 m long and 160 kg in weight was said to contain a vaquita that was bitten into three pieces. Two more "tiburones tonina", caught in March 1985 and in early February 1990, reportedly each had a vaquita in their stomachs. These sharks may attack free-swimming vaquitas, or perhaps feed on individuals trapped in gill nets, or both. Other large sharks that may be considered as potential predators of vaquitas include the tiger shark (*Galeocerdo cuvier*) and the scalloped hammerhead (*Sphyrna lewini*).

Killer whales, *Orcinus orca*, are potential predators of *P. sinus*. Killer whales are not uncommon in the Gulf of California (Vidal *et al.*, 1993), and they have been observed harassing and attacking large cetaceans (Vidal and Pechter, 1989; Silber *et al.*, 1990), but no information is available regarding possible predation by them upon vaquitas.

Swimming and diving

Silber *et al.* (1988) provided the only available information on the behaviour and ventilation cycles of the vaquita. Two different female/calf pairs were observed for periods of about 3 hr each. Dive characteristics were similar to those reported for *P. phocoena*. However, mean dive times (see definition for this and the following terms in Watson and Gaskin, 1983), roll intervals, surface times, and rolls per surfacing reported for *P. phocoena* by Watson and Gaskin (1983) were of somewhat greater length than those of *P. sinus*.

Sound production

High frequency sounds were recorded near free-ranging vaquitas in 1986 and 1987 (Silber, 1991). Clicks were sharp, intense, and narrow band ($\bar{x} = 17.1 \pm$ s.d. 5.38 kHz) signals in a frequency range of 122.2–146.9 kHz. The duration of these clicks was 79–193 μsec. The acoustic signals produced by *P. sinus* are similar to those reported for other phocoenids and for one genus of dolphins (*Cephalorhynchus* spp.) (Evans *et al.*, 1988).

Parasites and Diseases

Commensal pseudo-stalked barnacles, *Xenobalanus globicipitis*, were found attached on or near the posterior edges of the dorsal fins, flippers and/or flukes of 14 of 43 fresh vaquitas examined by us. Neither of the larval cestodes (*Phyllobotrium* sp. and *Monorygma* sp.) that have been found in most genera of marine odontocetes was present in the individuals we examined.

Three parasitic trematodes from the intestine of a vaquita were identified as *Synthesium tursionis* and utilized in a redescription of this species by Lamothe-Argumedo (1988). Although this author noted several characters that differed from "typical" specimens of this species, he considered them to be minor and not worthy of recognition as a new species. However, examination of a separate specimen of *Synthesium* from a different vaquita necropsied by us has led M. D. Dailey (California State University at Long Beach, personal communication) to believe that recognition of a new species may be warranted. More specimens are needed to clarify the taxonomic status of this trematode in *P. sinus*. In materials resulting from necropsies of eight vaquitas in 1985, M. D. Dailey (personal communication) identified two specimens of the nematode *Crassicauda* sp., one from muscle connective tissue in the area of a mammary gland of one vaquita, and the other located in the blubber layer 1 cm below the skin and 5 cm anterior to the anus of the other vaquita. Unidentified larval nematodes were found in the stomachs of three vaquitas from the same sample and from muscle tissue in the area adjacent to a mammary gland (one vaquita) and the wall of the pleural cavity (one vaquita).

Live Maintenance

No vaquitas have been maintained in captivity.

Human Effects

By-catches

The major human-induced problem affecting this species is incidental mortality in fishing gear. Vaquitas frequently die in illegal and sporadically permitted "survey-sampling" gill nets set for the endemic and endangered large corvina-like fish called the "totoaba" (*Totoaba macdonaldi*); in legal gill nets set for sharks, rays, mackerels (*Scomberomorus sierra* and *S. concolor*), chano (*Micropogonias megalops*) (a "croaker"), and shrimp (*Penaeus* spp.); and occasionally in commercial shrimp trawls.

The commercial fishery for totoaba in the upper Gulf of California, initially using primitive gill nets, had started by at least the mid-1930s, and unidentified small cetaceans were taken at that time (Brownell, 1982). More efficient nylon gill nets were used during (and probably before) the 1960s and 1970s. After a peak catch of 2261 tons of totoaba meat in 1942, and despite intensified fishing efforts, the annual catches declined to a minimum of approximately 59 tons in 1975. As a consequence of this dramatic decline, the Mexican government declared an indefinite closed season that prohibits any capture of these fish by commercial and sport fisheries (Flanagan and Hendrickson, 1976). Despite the closure, illegal fishing for totoaba continued off El Golfo de Santa Clara and San Felipe until at least January 1994 (Román Rodríguez, 1990; Lagomarsino, 1991; D'Agrosa *et al.*, 1995; Vidal, 1995). Although lacking any quantitative data, some fishing cooperatives in the northern Gulf have claimed that the totoaba stock has recovered and that the fishery should be legally reopened. This pressure has resulted in the issuing of some temporary permits (since 1983) for gill net operations by the Secretariat of Fisheries (SEPESCA) to assess the population status of totoaba or to study various aspects of its biology (Findley and Vidal, 1985; Rosales and Ramírez, 1987; Barrera-Guevara, 1992; Vidal, 1995). To date, however, efforts to assess the totoaba population have proved unsuccessful. Between March 1985 and January 1994, 76 vaquitas were confirmed to have been killed incidentally in totoaba gill nets (Vidal, 1995).

In addition to gill net fishing for totoaba (mesh size 20–30.5 cm), a gill net fishery for several species of large sharks and rays (mesh size 10–15 cm) has been growing rapidly in the upper Gulf of California since the early 1940s, probably together with the totoaba fishery. The shark fishery continues to operate without control (Vidal *et al.*, 1994; D'Agrosa *et al.*, 1995). In recent years, a gill net fishery for mackerel (*Scomberomorus* spp.), chano (*Micropogonias megalops*) and various species of small sharks (mesh size 8.5 cm) has started in the upper Gulf. Between 1985 and 1994, the shark and ray, mackerel, and chano fisheries have incidentally captured at least 45 and three vaquitas, respectively (D'Agrosa *et al.*, 1995; Vidal, 1995). Also, four vaquitas were killed during 1993–1994 in gill nets ("chinchorro de línea") (D'Agrosa *et al.*, 1995).

Norris and Prescott (1961) briefly mentioned a report by a fisherman who had accidentally captured vaquitas in a shrimp trawl. The deaths of nine vaquitas in shrimp trawls (two in 1985, one in 1988, two between 1984 and 1989, three in 1990, and one in 1993) were reported to O. Vidal, A. Robles, or C. D'Agrosa by residents of El Golfo de Santa Clara or San Felipe. All the porpoises taken were referred to as "very small", probably calves or juveniles. Considering the large number (ca. 500) of shrimp boats operating in the upper Gulf of California at the beginning of each typical shrimping season, this fishery poses an additional threat to vaquitas, particularly younger ones.

Pollution

Concerns have been expressed about organochlorine pollutants in the food web containing the vaquita in the upper Gulf of California (Barlow, 1986; Klinowska, 1991). Based on high concentrations of DDT in bivalve molluscs near the mouth of the Colorado River, Guardado (1975) concluded that the adjacent Mexicali Valley, with all of its agricultural activities, is an important source of pollutants in the upper Gulf of California. In the past, pollutants could also have been carried to the upper Gulf by Colorado River water used for irrigating agricultural areas in the southwestern United States. Today, in years with normal precipitation, essentially all of the surface flow of the Colorado River is diverted and utilized before it reaches the Gulf (Flanagan and Hendrickson, 1976). When samples of blubber from eight vaquitas, incidentally killed in fishing activities during 1985 (see Brownell *et al.*, 1987) were analysed for chlorinated hydrocarbons (Calambokidis *et al.*, 1993), relatively low concentrations of total DDT, alpha-BHC, and PCBs were found. These low concentrations were generally much lower than those reported for various odontocetes and marine birds from most other areas. Henny and Anderson (1979) reported pesticide residues (DDE and PCBs) in osprey, *Pandion haliaetus*, eggs from the Gulf of California that were among the lowest for this fishing bird in North America. Therefore, it appears that chlorinated hydrocarbons do not pose as great a threat to the vaquita population in the Gulf of California as they may for some populations of small cetaceans in other areas.

Barlow (1986) reported that exploratory drilling for oil and natural gas started in the uppermost Gulf in the early 1980s. However, the two drilling platforms that were erected near Puerto Peñasco and El Golfo de Santa Clara have been removed (Barlow, 1986). We know of no other planned exploration for fossil hydrocarbons in the upper Gulf at this time. Future development, however, could pose a serious problem for the vaquita and other species of marine organisms if a large oil spill occurred in the upper Gulf of California.

Acknowledgments

We thank L. G. Barnes, R. R. Reeves, G. K. Silber, and T. Jefferson for reading earlier drafts of this paper and for providing useful comments and discussion about vaquitas over the years. J. Torre-Cosío, M. Tahuilán, E. Arizmendi, M. Guerrero, S. Fernández and G. Carvallo assisted with cranial meristic and morphometric data and osteology of recently collected material. O.V. thanks P. Cendón for her help and support during the 1990–1992 field season. Special thanks are due to A. Robles, Conservation International Mexican Program (Mexico, D.F.) for providing unpublished data on vaquita incidental mortality during 1985 and 1986 and for much help over the years in studying these animals.

The vaquita drawing was prepared by Pieter A. Folkens and Fig. 1 was prepared by J. Torre-Cosío. The skulls shown in Figs 5 and 6 were provided courtesy of the San Diego Natural History Museum. Susan Wright and Karen Handschuh typed and corrected various drafts of this chapter.

References

Allen, A. (1925). Burmeister's porpoise (*Phocoena spinipinnis*). *Bull. Mus. Comp. Zool. Harvard College* **67**, 251–261 + 4 plts.

Barlow, J. (1986). Factors affecting the recovery of *Phocoena sinus*, the vaquita or Gulf of California harbor porpoise. NOAA, NMFS, Southwest Fisheries Center Administrative Report LJ-86-37.

Barlow, J., Fleischer, L., Forney, K. and Maravilla, C. O. (1993). An experimental aerial survey for vaquita (*Phocoena sinus*) in the northern Gulf of California, Mexico. *Mar. Mamm. Sci.* **9**, 89–94.

Barlow, J., Gerrodette, T. and Silber, G. (1997). First estimates of vaquita abundance. *Mar. Mamm. Sci.* **13**, 44–58.

Barnes, L. G. (1985). Evolution, taxonomy and antitropical distribution of the porpoises (Phocoenidae: Mammalia). *Mar. Mamm. Sci.* **1**, 149–165.

Barrera-Guevara, J. C. (1992). Biología reproductiva de la totoaba *Totoaba macdonaldi* (Gilbert, 1890) (Teleostei: Sciaenidae). Tesis de Maestría, Inst. Tecn. y Est. Sup. de Monterrey, Campus Guaymas.

Brownell Jr., R. L. (1982). Status of the cochito, *Phocoena sinus*, in the Gulf of California. (FAO Advisory Committee on Marine Resources Research. Mammals in the Seas: Small Cetaceans, Seals, Sirenians, and Otters), *FAO Fisheries Ser. 5*, Vol. 4, pp. 85–90.

Brownell Jr., R. L. (1983). "*Phocoena sinus*". Mammalian Species No. 198, 3 pp.

Brownell Jr., R. L. (1986). Distribution of the vaquita, *Phocoena sinus*, in Mexican waters. *Mar. Mamm. Sci.* **2**, 299–305.

Brownell Jr., R. L., Findley, L. T., Vidal, O., Robles, A. and Manzanilla, N. S. (1987). External morphology and pigmentation of the vaquita, *Phocoena sinus* (Cetacea: Mammalia). *Mar. Mamm. Sci.* **3**, 22–30.

Brownell Jr., R. L. and Praderi, R. (1984). "*Phocoena spinipinnis*". Mammalian Species No. 217, 4 pp.

Calambokidis J., Findley, L. T., Brownell Jr., R. L. and Barlow, J. (1993). Chlorinated hydrocarbon concentrations in the Gulf of California harbor porpoise (*Phocoena sinus*). *In* Abstracts of the Tenth Biennial Conference on the Biology of Marine Mammals, November, Galveston, TX, p. 13.

Cane, A. M. (1983). Oceanographic events during El Niño. *Science* **222**, 1189–1195.

D'Agrosa, C., Vidal, O. and Graham, W. C. (1995). Mortality of the vaquita (*Phocoena sinus*) in gillnet fisheries during 1993–94. *Rep. Int. Whal. Commn*, Spec. Issue No. 16, Cambridge.

Evans, W. E., Awbrey, F. T. and Hackbarth, H. (1988). High frequency pulses produced by free-ranging Commerson's dolphin (*Cephalorhynchus commersonii*) compared to those of phocoenids. *Rep. Int. Whal. Commn*, Spec. Issue No. 9, Cambridge.

Findley, L. T. and Vidal, O. (1985). La marsopa del Golfo de California. *Información Científica y Tecnológica (México)* **7** (105), 15.

Findley, L. T., Nava, J. M. and Torre-Cosío, J. (in prep.). Food habits of the vaquita.

Fitch, J. E. and Brownell Jr., R. L. (1968). Fish otoliths in cetacean stomachs and their importance in interpreting feeding habits. *J. Fish. Res. Bd Can.* **25**, 2561–2574.

Flanagan, C. A. and Hendrickson, J. R. (1976). Observations on the commercial fishery and reproductive biology of the totoaba, *Cynoscion macdonaldi*, in the northern Gulf of California. *Fish. Bull. (U.S.)* **74**, 531–544.

Gerrodette, T., Barlow, J., Taylor, B. L. and Silber, G. S. (1994). First population estimates of the endangered porpoise, *Phocoena sinus*, with implications for monitoring small populations. *In* Abstracts of the 1994 International Meeting of the Society for Conservation Biology, June, Guadalajara, Mexico, p. 90.

Gerrodette, T., Fleischer, L. A., Perez-Cortes, H. and Villa-Ramirez, B. (1995). Distribution of the vaquita, *Phocoena sinus*, based on sightings from systematic surveys. *Rep. Int. Whal. Commn*, Spec. Issue No. 16, Cambridge.

Guardado, P. J. (1975). Concentración de DDT y sus metabolitos en especies filtraoalimentadores y sedimentos en el Valle de Mexicali y alto Golfo de California. *Calif. Coop. Ocean. Fish. Invest. Rep.* **18**, 73–80.

Henny, C. J. and Anderson, D. W. (1979). Osprey distribution, abundance, and status in western North America: III. The Baja California and Gulf of California population. *Bull. So. Calif. Acad. Sci.* **78**, 89–106.

Hohn, A. A., Read, A. J., Fernandez, S., Vidal, O. and Findley, L. T. (1996). Life history of the vaquita, *Phocoena sinus* (Phocoenidae, Cetacea). *J. Zool. Lond.* **39**, 235–251.

Kenagy, G. J. and Trombulak, S. C. (1986). Size and function of mammalian testes in relation to body size. *J. Mammal.* **67**, 1–22.

Klinowska, M. (1991). Vaquita. *In* "Dolphins, Porpoises and Whales of the World", The IUCN Red Data Book, pp. 105–108. IUCN, Gland, Switzerland.

Lagomarsino, I. V. (1991). Endangered species status review: *Totoaba macdonaldi*. NOAA, NMFS Southwest Region Administrative Report SWR-91-01.

Lamothe-Argumedo, R. (1988). Trematodos de mamíferos III. Hallazgo de *Synthesium tursionis* (Marchi, 1873) Stunkard y Alvey 1930 en *Phocoena sinus* (Phocoenidae) en el Golfo de California, Mexico. *An. Inst. Biol. Univ. Natl. Auton. Mexico, Ser. Zool.* **58**, 11–19.

Magatagan, M. D., Boyer, E. H. and Villa-Ramirez, B. (1984). Revision del estado que guarda *Phocoena sinus* Norris and McFarland y descripcion de tres nuevos ejemplares. *An. Inst. Biol. Univ. Natl. Auton. Mexico, Ser. Zool.* **55**, 271–293.

Mitchell, E. D. (1975). Review of the biology and fisheries for smaller cetaceans. *J. Fish. Res. Bd Can.* **32**, 875–1240.

Nishiwaki, M. and Kureha, K. (1975). Strange organ in the anal region of the finless porpoise. *Sci. Rep. Whales Res. Inst.* **27**, 139–140 + 1 plt.

Noble, B. A. and Fraser, F. C. (1971). Description of a skeleton and supplementary notes on the skull of a rare porpoise *Phocoena sinus* Norris and McFarland, 1958. *J. Nat. Hist. Lond.* **5**, 447–464.

Norris, K. S. and McFarland, W. N. (1958). A new harbor porpoise of the genus *Phocoena* from the Gulf of California. *J. Mammal.* **42**, 22–39.

Norris, K. S. and Prescott, J. H. (1961). Observations on Pacific cetaceans of Californian and Mexican waters. *Univ. Calif. Publ. Zool.* **63**, 291–402.

Orr, R. T. (1969). An additional record of *Phocoena sinus*. *J. Mammal.* **50**, 384.

Perrin, W. F. (1972). Color patterns of spinner porpoises (*Stenella* cf. *S. longirostris*) of the eastern Pacific and Hawaii, with comments on delphinid pigmentation. *Fish. Bull. (U.S.)* **70**, 983–1003.

Perrin, W. F., Mitchell, E. D., Mead, J. G., Caldwell, D. K. and van Bree, P. J. H. (1981). *Stenella clymene*, a rediscovered tropical dolphin of the Atlantic. *J. Mammal.* **62**, 583–598.

Read, A. J. and Gaskin, D. E. (1990). Changes in growth and reproduction of harbour porpoises, *Phocoena phocoena*, from the Bay of Fundy. *Can. J. Fish. Aquat. Sci.* **47**, 2158–2163.

Robles, A., Vidal, O. and Findley, L. T. (1987). La totoaba y la vaquita. *Informacion Científica y Tecnologica (Mexico)* **9**(124), 4–6.
Román-Rodríguez, M. J. (1990). Alimentación de *Totoaba macdonaldi* (Gilbert) (Pisces: Sciaenidae) en la parte norte del alto Golfo de California. *Ecologica* (Centro Ecologico de Sonora, Hermosillo, Mexico) **1**, 1–9.
Rosales, J. F. and Ramírez, G. E. (1987). "Estado Actual del Conocimiento Sobre la Totoaba (*Cynoscion macdonaldi*) Gilbert, 1890 [*sic* 1891]". Secretaría de Pesca de Mexico.
Rosel, P. E. (1992). Genetic population structure and systematic relationships of some small cetaceans inferred from mitochondrial DNA sequence variation. Ph.D. dissertation, University of California at San Diego, 191 pp.
Rosel, P. E. and Rojas-Bracho, L. (1993). Genetic variation, or lack thereof, in the vaquita. *In* Abstracts of the Tenth Biennial Conference on the Biology of Marine Mammals, November, Galveston, TX, p. 13.
Silber, G. K. (1988). Recent sightings of the Gulf of California harbor porpoise (*Phocoena sinus*). *J. Mammal.* **69**, 430–433.
Silber, G. K. (1990a). Occurrence and distribution of the vaquita (*Phocoena sinus*) in the northern Gulf of California. *Fish. Bull. (U.S.)* **88**, 339–346.
Silber, G. K. (1990b). Distributional relations of cetaceans in the northern Gulf of California, with special reference to the vaquita, *Phocoena sinus*. Ph.D. dissertation, University of California at Santa Cruz.
Silber, G. K. (1990c). Summary of field study of the vaquita, *Phocoena sinus*. Presented to the Environment Committee, the Commission of the Californias, 18–20 January, Tijuana, Baja California, Mexico.
Silber, G. K. (1991). Acoustic signals of the vaquita (*Phocoena sinus*). *Aquat. Mamm.* **17**, 130–133.
Silber, G. K., Newcomer, M. W. and Barros, G. J. (1988). Observations on the behavior and ventilation cycles of the vaquita (*Phocoena sinus*). *Mar. Mamm. Sci.* **4**, 62–67.
Silber, G. K., Newcomer, M. V. and Perez-Cortes, M. H. (1990). Killer whales (*Orcinus orca*) attack and kill a Bryde's whale (*Balaenoptera edeni*). *Can. J. Zool.* **68**, 1603–1606.
Silber, G. K. and Norris, K. S. (1991). Geographic and seasonal distribution of the vaquita, *Phocoena sinus*. *An. Inst. Biol. Univ. Natl. Auton. Mexico, Ser. Zool.* **62**, 263–268.
Torre-Cosío, J. (1995). Descripcíon del esqueleto, dimorfismo sexual y crecimiento alométrico cráneal de la vaquita, *Phocoena sinus* (Cetacea: Phocoenidae). Master's thesis, Instituto Tecnológico y de Estudios Superiores de Monterrey-Campus Guaymas, Sonora, Mexico.
Turk-Boyer, P. J. (1989). Assessment of the potential impact of fishing activities in the northern Gulf of California on the *Phocoena sinus* population. Report submitted to the Center for Marine Conservation, Washington, DC.
Turk-Boyer, P. J. and Silber, G. K. (1990). Estimate of vaquita, *Phocoena sinus*, mortality in gillnet fisheries in the northern Gulf of California, Mexico. *In* Abstracts of the International Whaling Commission Symposium on Mortality of Cetaceans in Passive Fishing Nets and Traps, La Jolla, CA, p. 49.
Vidal, O. (1991). Catalog of osteological collections of aquatic mammals from Mexico. NOAA Technical Report NMFS 97, 36 pp.
Vidal, O. (1995). Population biology and incidental mortality of the vaquita, *Phocoena sinus*. *Rep. Int. Whal. Commn*, Spec. Issue No. 16, Cambridge.
Vidal, O., Aguayo, L. A., Findley, L., Robles, A., Bourillon, L., Vomend, I., Turk, P., Garate, K., Maronas, L. and Rosas, J. (1987). Avistamientos de mamiferos marinos durante el crucero "Guaymas I" en la region superior del Golfo de California, primavera de 1984. *In* "Memoria, X Reunion Internacional sobre Mamiferos Marinos, 24–27 Marzo 1985, La Paz, Baja California Sur, Mexico", pp. 7–35.

Vidal, O. and Pechter, G. (1989). Behavioral observations on fin whale, *Balaenoptera physalus*, in the presence of killer whale, *Orcinus orca*. *Fish. Bull. (U.S.)* **87**, 370–373.

Vidal, O., Findley, L. T. and Leatherwood, S. (1993). Annotated checklist of the marine mammals of the Gulf of California. *Proc. San Diego Soc. Nat. Hist.* **28**, 1–16.

Vidal, O., Van Waerebeek, K. and Findley, L. T. (1994). Cetaceans and gillnet fisheries in Mexico, Central America and the wider Caribbean: A preliminary review. *Rep. Int. Whal. Commn*, Spec. Issue No. 15, Cambridge.

Watson, A. P. and Gaskin, D. E. (1983). Observations on the ventilation cycle of the harbour porpoise, *Phocoena phocoena* (L.) in coastal waters of the Bay of Fundy. *Can. J. Zool.* **61**, 126–132.

Wells, R. S., Würsig, B. G. and Norris, K. S. (1981). A survey of the marine mammals of the upper Gulf of California, Mexico, with an assessment of the status of *Phocoena sinus*. Report PB 81-1678792, NTIS, Arlington, VA.

Wilson, L. E. (1973). A delphinid (Mammalia, Cetacea) from the Miocene of Palos Verdes Hills, California. *Univ. Calif. Publ. Geol. Sci.* **103**, 1–34

14

Spectacled Porpoise *Phocoena dioptrica* Lahille, 1912

Robert L. Brownell, Jr and Phillip J. Clapham

Genus and Species

Taxonomy

Lahille (1912) first described this porpoise as *Phocaena dioptrica* from a pregnant subadult female collected by Jose de Gaetano after stranding live at Punta Colares, south of Buenos Aires, Argentina. The skeleton of the adult now appears to be lost (Brownell, 1975), but the associated foetus is preserved in alcohol (specimen number MACN 12-14) at the Museo Argentino de Ciencias Naturales. The only junior synonym for this species is *Phocoena storni* (Marelli, 1922); this was based upon a cranium found on Tierra del Fuego and later shown to be *P. dioptrica* (Norris and McFarland, 1958). Several authors (Allen, 1925; Hershkovitz, 1966; Donoso-Barros, 1975) have suggested that the oddly coloured cetacean from Chile, *Phocaena obtusata* Philippi 1893 was a senior synonym and should thus replace *dioptrica*. True (1903) and Brownell (1975) disagreed, and Goodall *et al.* (1988) have demonstrated both the present authors' belief that *P. obtusata* is a specimen of *Cephalorhynchus* (probably *C. eutropia*).

The specific name "dioptrica", from the Greek "diopter" (spectacled), was used by Lahille to describe the appearance of the type specimen's eyes, which were surrounded by a black patch outlined in white.

Handbook of Marine Mammals, Volume 6
ISBN 0-12-588506-7

Barnes (1985) revised the taxonomy of the Phocoenidae and reclassified this species in a new monotypic genus, *Australophocaena*. However, this has been strongly contradicted by recent phylogenetic analysis of phocoenid mitochondrial DNA sequences by Rosel *et al.* (1995). We concur with these authors that the spectacled porpoise should be returned to the genus *Phocoena*.

There are no known fossil remains of this species. The certain geologic range of the Phocoenidae extends from the late Miocene to Recent (Barnes, 1976).

Common names

There are no other common names for the spectacled porpoise in English. In South America, it is known as marsopa de anteojos, or sometimes as tonina.

Distribution

Range

The distribution of the spectacled porpoise (Fig. 1) appears to be circumpolar. The documented range extends from temperate latitudes south into the sub- or low Antarctic, including areas just south of the Antarctic Convergence. It has been recorded from around offshore islands of the West Wind Drift, including Heard, Kerguelen, Auckland, Macquarie, South Georgia, the Falklands and Tierra del Fuego. In 1997, a dead calf washed ashore at Bruny Island, Tasmania; this represents the first record from Australian waters (K. Evans, personal communication). However, sightings, while widely distributed, are few in number: the sixteen reliable or probable observations at sea are summarized by Goodall and Schiavini (1995, their Table 3). The southernmost sighting was at 58° 02′S, 62° 19′ W in Drake Passage (Kasamatsu *et al.*, 1990). The northernmost extent of the known range is represented by a stranded adult male collected by M. C. Pinedo at 32°S, 52°W at Rio Grande do Sul in Brazil (Goodall and Schiavini, 1995). Only five sightings were recorded during ten years of extensive Antarctic surveys for minke whales, *Balaenoptera acutorostrata* (Kasamatsu *et al.*, 1990), and there have been no reliable sightings from other research vessels or cruise ships in this region.

Habitat

This porpoise appears to occur only in cold temperate waters. Although it has been observed or incidentally caught in coastal waters (Goodall and Schiavini, 1995), its habitat is thought to be primarily oceanic in nature. Where recorded, water temperatures associated with sightings have ranged from 5.5°C to 9.5°C (Kasamatsu *et al.*, 1990).

External Characteristics

Colour pattern

Illustrations of the colour pattern of the spectacled porpoise have been published for relatively few specimens. Goodall and Schiavini (1995) illustrate six adults, including two from Bruch (1916) and the Lahille (1912) holototype; they also provide illustrations or descriptions of several calves.

The most distinctive feature of this porpoise is its pigmentation (Fig. 2). The dorsal and dorso-lateral surfaces are black and are separated from the white ventral and ventro-lateral surfaces by a sharply defined line. It is unclear whether the dorsal keel of the caudal peduncle is white in some specimens, as indicated by both Lahille (1912) and Bruch (1916). Goodall and Schiavini (1995) found this area to be black in the specimens that they examined. The upper jaw is black, as are the lips. The eye is surrounded by a black patch, which is sometimes outlined in white. In calves and young animals, the body pigmentation is dark grey dorsally and light grey ventrally; this becomes black and white, respectively, with age.

Dorsally the flukes are black. The ventral side is lighter, being either all white, white with a grey or black border, or mottled. The dorsal pectoral fin pigmentation is variable, ranging from all black to all white, or white with a light grey border. Some individuals have a faint grey stripe extending from the angle of the gape to the anterior insertion of the flipper. Goodall and Schiavini (1995) suggest that the flipper stripe fades or disappears as the animal ages.

Size and shape

To date, published information on length and external measurements are available for only seven females, 11 males and nine animals of unknown sex; these are summarized by Goodall and Schiavini (1995). The females of this subset ranged from 124.6 cm to 203.5 cm in total length. The males ranged from 109 cm+ (a neonate) to 224 cm; the latter animal, the largest specimen of this species so far recorded, was collected in Tierra del Fuego in 1978.

These porpoises are small and robust like other phocoenids; their maximum girth, in front of the dorsal fin, is approximately 50–60% of total length. The snout is rounded and lacks a distinct beak. The flippers are proportionally smaller than those of the vaquita (*Phocoena sinus*) and Burmeister's porpoise (*P. spinipinnis*). The dorsal fin is triangular and shows considerable sexual dimorphism, being greatly enlarged in adult males (Bruch, 1916; Pagnoni and Saba, 1989). Selected external measurements are given in Table 1.

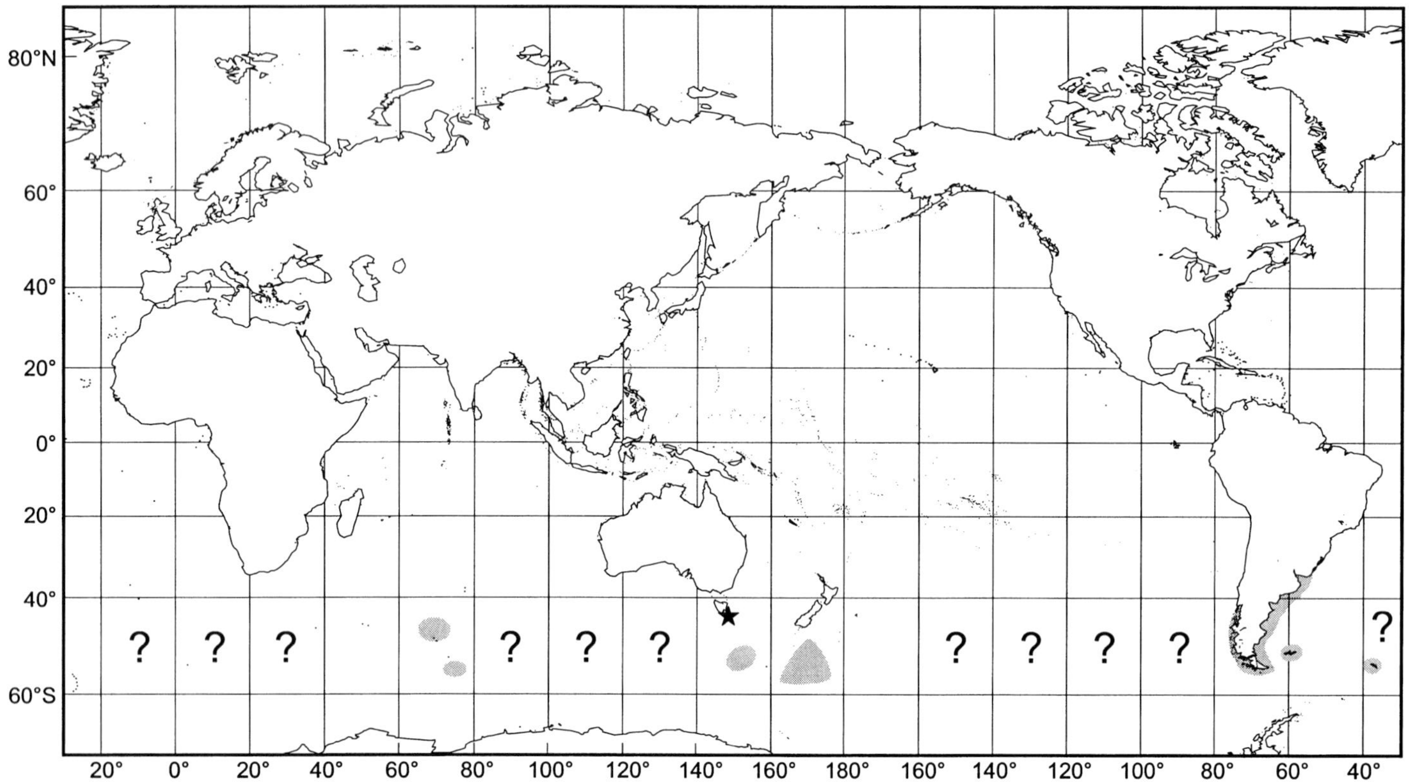

FIG. 1. Known distribution of the spectacled porpoise from strandings and sightings.

FIG. 2. The spectacled porpoise: for variations in skin pigment patterns see Goodall and Schiavini (1995). (Drawing by P. Folkens)

Weight

Data on weight are available for only five specimens. These include the 186 cm subadult female (85 kg) and 48.4 cm foetus (1.6 kg) described by Lahille (1912), and three animals from Goodall and Schiavini (1995): an eviscerated 109 cm calf (14.8 kg), a 124.6 cm calf (24.2 kg) and a 205 cm subadult male (115 kg).

Internal Anatomy

Skull

The skull of the spectacled porpoise is illustrated in Fig. 3. Marelli (1922) was the first to describe and illustrate the skull of this species, based upon the holotype of *Phocaena storni*. Additional crania have been illustrated by Hamilton (1941), Norris and McFarland (1958), Brownell (1975), Baker (1977), Goodall and Cameron (1979), Fordyce *et al.* (1984), Barnes (1985) and Guiler *et al.* (1987). The condylobasal length of 46 adult skulls ranged from 279 mm to 324 mm (Brownell *et al.*, 1989). The periotic and tympanic bullae of this species are illustrated by Brownell (1975).

Barnes (1985) listed several features that distinguish the skull of the spectacled porpoise from that of Burmeister's porpoise. These include: a relatively small, oval-shaped temporal fossa; an elevated, high-vaulted braincase that slopes abruptly onto the narial region; relatively large, high and convex premaxillary bosses; a dorso-ventrally expanded zygomatic process of the squamosal; a short and anteroposteriorly expanded postorbital process of the frontals; and maxillae extending nearly to the dorsal margin of the supraoccipital on top of the skull.

TABLE 1 Selected external measurements (cm) for spectacled porpoises (adapted from Goodall and Schiavini, 1995)

	Neonates and juveniles, both sexes					Females					Males				
Measurement	*n*	Min	Max	Mean	SD	*n*	Min	Max	Mean	SD	*n*	Min	Max	Mean	SD
Total length	5	94.0	135.9	119.5	17.8	5	174.0	203.5	186.5	10.7	8	189.0	224.0	202.4	10.1
Snout to angle of mouth	4	6.0	7.6	6.9	0.7	3	8.9	9.5	9.1	0.3	1	7.5	7.5	7.5	–
Snout to centre of eye	4	12.9	16.5	14.4	1.5	2	15.7	21.5	18.6	4.1	1	17.9	17.9	17.9	—
Snout to tip of dorsal fin	3	58.0	77.0	68.1	9.6	2	99.7	110.3	105.0	7.5	2	117.5	123.0	120.3	3.9
Snout to umbilicus	1	59.7	59.7	59.7	—	2	85.0	91.0	88.0	4.2	—	—	—	—	—
Anterior length of flipper	3	13.0	20.0	16.1	3.6	2	15.1	18.9	17.0	2.7	2	20.0	21.0	20.5	0.7
Max. width of flipper	4	5.6	8.0	6.8	1.0	4	6.9	8.4	7.3	0.7	3	9.8	11.0	10.3	0.6
Height of dorsal fin	5	6.5	10.2	9.0	1.6	4	7.1	12.5	9.2	2.3	9	17.0	32.0	23.0	4.7
Span of flukes	4	22.8	33.0	29.2	4.6	1	32.0	32.0	32.0	—	5	33.5	51.0	42.6	6.8
Girth at axilla	1	61.5	61.5	61.5	—	—	—	—	—	—	1	99.0	99.0	99.0	—

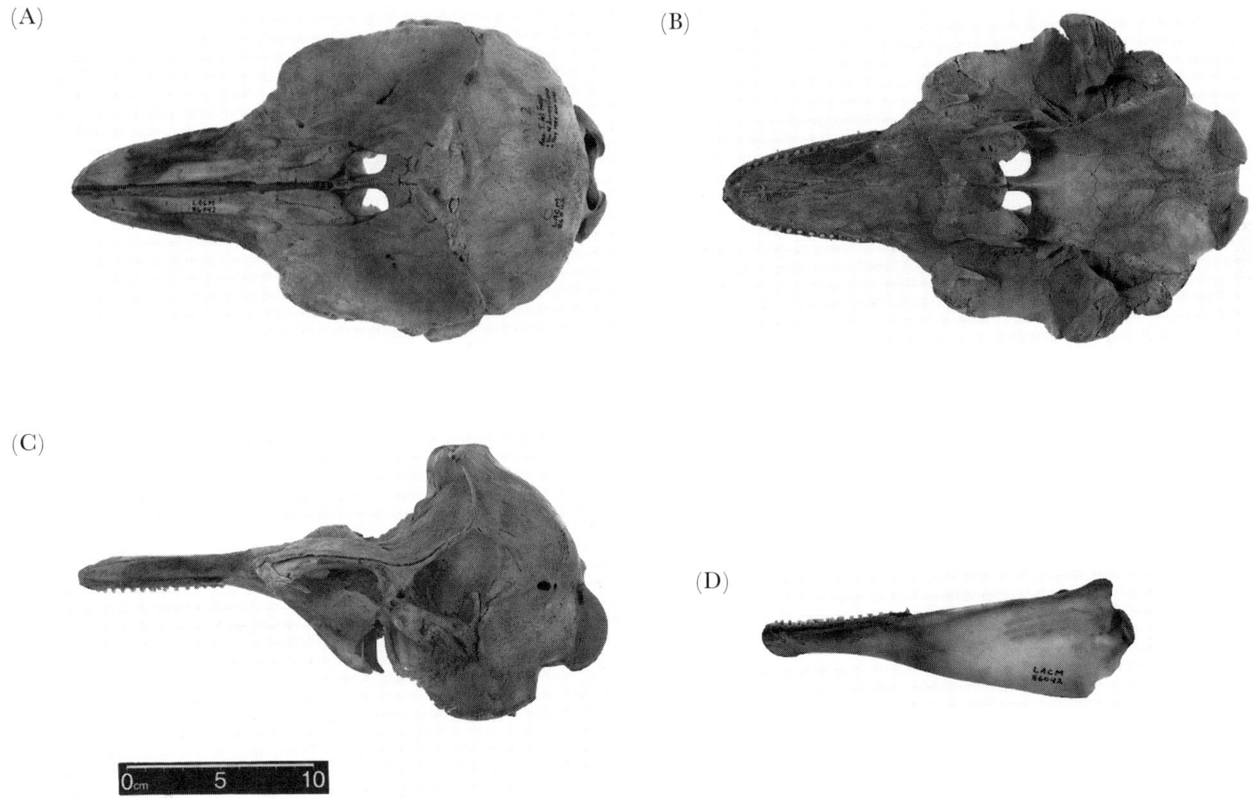

FIG. 3 Skull of a spectacled porpoise (LACM 86042): dorsal (A) view, ventral (B) view, lateral (C) view and mandible (D).

The teeth, like those of other phocoenids, are small and spade-shaped. They are not uniform but rather show variation in both size and shape (Goodall and Schiavini, 1995). Published tooth counts are 16–25 upper (each side, ten specimens; Brownell *et al.*, 1989) and 16–19 lower (each side, six specimens; Brownell, 1975). However, Goodall and Schiavini (1995) give probable counts of 19–26 upper and 19–22 lower. Cabrera and Yepes (1940) erroneously reported the tooth count as 27 in each row. The posterior-most upper teeth in the spectacled porpoise are seated in a shallow aveolar groove and are easily lost; thus their numbers are very difficult to determine from dry specimens.

Postcranial skeleton

A complete skeleton is illustrated by Brownell (1975). The vertebral count based upon only a few specimens is C_7, $T_{13 \text{ or } 14}$, L_{14-16}, $Ca_{32 \text{ or } 33}$. There are at least 19 chevron bones. The first four or five cervical vertebrae are fused. Thirteen ribs are present, with the first nine ribs attached to the vertebrae by both capitulum and tuberculum; the remainder are attached by the tuberculum only. The posterior ribs are markedly flattened. Eight pairs of sternal ribs are present, with the anterior three attached directly to the sternum. The phalangeal count (including the metacarpals) is I_2, II_7, III_4, IV_3, V_4 (Lahille, 1912). Additional details on the osteology of this species are found in Hamilton (1941) and Noble and Fraser (1971).

Organs, tissues and physiology

Very few data exist on the internal anatomy of this species. Goodall and Schiavini (1995) give a weight of 410 g for the heart of a 109 cm neonate, and weights of 86 g and 76 g for its kidneys. They also give weights of 537 g and 387 g for the lungs of a 183 cm female. The facial complex was studied by Purgue (1987), and the blowhole, larynx and hyoids of the type specimen were described by Gallardo (1913).

Life History and Population Dynamics

Growth and reproduction

There are few data available with which to assess age at sexual maturity in the spectacled porpoise. Goodall and Schiavini (1995) report a 183 cm female caught in a fishing net together with a calf; the female had 2–3 growth layer groups (GLGs) in the teeth. This would suggest an age of 2–3 years, although this seems young for an animal of this length. Two other females, both 186 cm long and including the holotype, were pregnant (Lahille, 1912; Bruch, 1916). No length was reported for the Bruch foetus. The foetus from the Lahille specimen was a female 48.4 cm in length, which Lahille believed was close to term. Goodall and Schiavini (1995) reexamined this foetus and concluded from its length and weight relative to the mother that it was not near birth. Reviewing the scant data on lengths of calves and neonates from Tierra del Fuego, Goodall and Schiavini (1995) suggest that this species calves in late spring or summer (November–February).

Based upon a small sample size and GLG readings, Goodall and Schiavini (1995) estimated that females become subadult and perhaps sexually mature by the age of 2 years, while at least some males are physically mature by 4 years. However, an animal of unknown sex that was not yet physically mature was estimated from GLGs at 8 years of age; this stands in contrast to three other specimens of unknown sex which reputedly had attained physical maturity by the age of 3. However, these conclusions must be considered tentative in view of the small sample size. Nothing is known of pregnancy rates, interbirth intervals or duration of lactation in this species.

Mortality

Spectacled porpoises have no known predators. However, it is not unlikely that this species is taken by killer whales (*Orcinus orca*) in portions of its range.

Abundance

Nothing is known of the abundance of this porpoise. It was the most commonly encountered species during preliminary beach surveys undertaken on Tierra del Fuego by R. N. P. Goodall, but once the beaches had been cleared it was exceeded in frequency of occurrence by Commerson's dolphin, *Cephalorhynchus commersonii* (Goodall and Schiavini, 1995).

Behaviour

Social organization

Goodall and Schiavini (1995) note that all of the few confirmed sightings of this species have involved from one to three animals. No mass strandings have been recorded, and nothing is known of the social organization or mating system. There are no observations of aerial behaviour, no recordings of sounds, and no reports of this porpoise associating with other species. However, this is hardly surprising in view of the small number of recorded sightings.

Movements

Nothing is known of the seasonal movements, if any, of this species.

Feeding

Virtually nothing is known of the diet of the spectacled porpoise. Based upon its dentition, it is likely that, like other phocoenids, this species feeds upon fish and squid. The sole record of prey remains are of anchovy (*Engraulis* sp.) and small crustaceans (possibly stomatopods) found in the stomach of a 6 year old male stranded in Chubut, Argentina (Goodall and Schiavini, 1995). There have been no observations of feeding behaviour.

Swimming and diving

The sole description of locomotion and breathing in this species is given by Goodall and Schiavini (1995), who refloated and followed a calf that had stranded in the Beagle Channel. They report a respiration rate of 4–7 (mean 5.6) breaths per minute, as well as "slow, unobtrusive" swimming with a pronounced arch on each dive. Because this animal later died, it is unclear whether its behaviour was representative of the species' normal behaviour.

Parasites and Disease

Lahille (1912) presumably found parasites in the type specimen, because he stated that these would be the subject of later papers; unfortunately, no such work was ever forthcoming. However, nematodes were subsequently found in the blowhole and nasal sacs of this animal (Gallardo, 1913). Goodall and Schiavini (1995) recorded nematodes in the stomachs of three animals from Tierra del Fuego, and "many small worms" in each ear sinus of another spectacled porpoise from this region. No ectoparasites or commensals have been recorded for this species.

Live Maintenance

The spectacled porpoise has never been kept in captivity.

Human Effects

Directed fisheries

Little is known of anthropogenic impacts on this species. Fraser (1968) reported that a specimen made available to him by Wilkins came ashore alive near a whaling station on South Georgia and was immediately "killed by the cook and cut open" (and presumably eaten). Wilkins was told by several whaling captains that these porpoises were "sometimes shot . . . with rifles for food". According to Praderi (1971), fishermen on the coast of Uruguay sometimes hunt these porpoises and could easily recognize them from their pigmentation. Goodall and Schiavini (1995) state that there is no documented direct take of this species, but that it may be taken for crab bait in southern Chile.

By-catches

Goodall *et al.* (1994) reported that at least 34 animals were killed incidentally in coastal gill nets set in Tierra del Fuego between 1975 and 1990. Goodall and Schiavini (1995) noted a co-occurrence of strandings and fishing activity in southeastern Chile, suggesting additional undocumented mortalities from this source. Crespo *et al.* (1994) reported some mortality of spectacled porpoises in bottom and mid-water trawls off the coast of Chubut, Argentina.

Pollution

There have been no studies of pollutants in this species.

Acknowledgments

We acknowledge a debt to R. N. P. Goodall and A. Schiavini, whose thorough review of this species (Goodall and Schiavini, 1995) was the basis for the preparation of much of this chapter. A. Read, W. F. Perrin, L. Barnes and R. N. P. Goodall reviewed the manuscript and made useful suggestions for its improvement. J. E. Heyning generously provided the skull photographs.

References

Allen, G. M. (1925). Burmeister's porpoise (*Phocoena spinipinnis*). *Bull. Mus. Comp. Zool.* **67**, 251–261.

Baker, A. N. (1977). Spectacled porpoise, *Phocoena dioptrica*, new to the subantarctic Pacific Ocean. *NZ J. Mar. Freshwater Res.* **11**, 401–406.

Barnes, L. G. (1976). Outlines of eastern North Pacific fossil cetacean assemblages. *Syst. Zool.* **25**, 321–343.

Barnes, L. G. (1985). Evolution, taxonomy and antitropical distributions of the porpoises (Phocoenidae, Mammalia). *Mar. Mamm. Sci.* **1**, 149–165.

Brownell Jr., R. L. (1975). "*Phocoena dioptrica*". Mammalian Species No. 66, 3 pp.

Brownell Jr., R. L., Heyning, J. E. and Perrin, W. F. (1989). A porpoise, *Australophocaena dioptrica*, previously identified as *Phocoena spinipinnis*, from Heard Island. *Mar. Mamm. Sci.* **5**, 193–195.

Bruch, C. (1916). El macho de *Phocaena dioptrica* Lah. *Physis (B. Aires)* **2**, 461–462 (in Spanish).

Cabrera, A. and Yepes, J. (1940). "Mamiferos Sud-Americanos". Historia Natural Ediar. Co., Argentina Edit., Buenos Aires, 370 pp.

Crespo, E. A., Corcuera, J. F. and López Cazorla, A. (1994). Interactions between marine mammals and fisheries in some coastal fishing areas of Argentina. *Rep. Int. Whal. Commn*, Spec. Issue No. 15, Cambridge, pp. 295–304.

Donoso-Barros, R. (1975). Contribución al conocimiento de los cetáceos vivientes y fósiles del Territorio de Chile. *Gayana (Zool.)* **36**, 1–127 (in Spanish).

Fordyce, R. E., Mattlin, R. H. and Dixon, J. M. (1984). Second record of spectacled porpoise from subantarctic Southwest Pacific. *Sci. Rep. Whales Res. Inst. Tokyo* **35**, 159–164.

Fraser, F. C. (1968). Notes on a specimen of *Phocoena dioptrica* from South Georgia. *Brit. Antarct. Surv. Bull.* **16**, 51–56.

Gallardo, A. (1913). Notas sobre la anatomia del aparato espiricular, laringe y biloides de los delfines: *Phocoena dioptrica* Lahille y *Lagenorhynchus fitzroyi* (Waterhouse) Flower. *An. Mus. Hist. Nat. Buenos Aires* **24**, 235–245.

Goodall, R. N. and Cameron, I. S. (1979). *Phocoena dioptrica*, una nueva especie para aguas chilenas. *Rev. Mus. Argent. Cienc. Nat. 'Bernardino Rivadavia' Inst. Nac. Invest. Cienc. Nat. Zool.* **12**, 143–152.

Goodall, R. N. P. and Schiavini, A. C. M. (1995). On the biology of the spectacled porpoise, *Australophocaena dioptrica*. *Rep. Int. Whal. Commn*, Spec. Issue No. 16, Cambridge, pp. 411–453.

Goodall, R. N. P., Schiavini, A. C. M. and Fermani, C. (1994). Net fisheries and net mortality of small cetaceans off Tierra del Fuego, Argentina. *Rep. Int. Whal. Commn*, Spec. Issue No. 15, Cambridge, pp. 295–304.

Goodall, R. N. P., Sobral, A. P., Labal de Vinuesa, M. and Cameron, I. S. (1988). On *Phocoena obtusata*, with a comparison of certain diagnostic characters in South American phocoenids and the genus *Cephalorhynchus*. *Rep. Int. Whal. Commn*, Spec. Issue No. 9, Cambridge, pp. 259–276.

Guiler, E. R., Burton, H. R. and Gales, N. J. (1987). On three odontocete skulls from Heard Island. *Sci. Rep. Whales Res. Inst. Tokyo* **38**, 117–124.

Hamilton, J. E. (1941). A rare porpoise of the South Atlantic, *Phocaena dioptrica* (Lahille, 1912). *Discovery Rep.* **21**, 229–234.

Hershkovitz, P. (1966). Catalog of living whales. *U.S. Nat. Mus. Bull.* **246**, 1–259.

Kasamatsu, F., Joyce, G. G., Ensor, P. and Mermoz, J. (1990). Current occurrence of cetacea in the Southern Hemisphere: results from the IWC/IDCR Southern Hemisphere minke whale assessment cruises, 1978/79–1987/88. Paper SC/42/O 15 presented to the International Whaling Commission Scientific Committee, 77 pp.

Lahille, F. (1912). Nota preliminar sobre una nueva especie de marsopa del Río de la Plata (*Phocaena dioptrica*). *An. Mus. Nac. Buenos Aires* **23**, 269–278 (in Spanish).

Marelli, C. (1922). *Phocaena storni* sp.n., una nueva especie de marsopa del mar austral argentino. *An. Soc. Cient. Argentina* **94**, 229–240.

Noble, B. A. and Fraser, F. C. (1971). Description of a skeleton and supplementary notes on the skull of a rare porpoise *Phocoena sinus* Norris and McFarland 1958. *J. Nat. Hist.* **5**, 447–464.

Norris, K. S. and McFarland, W. N. (1958). A new harbour porpoise of the genus *Phocoena* from the Gulf of California. *J. Mamm.* **39**, 22–39.

Pagnoni, G. and Saba, S. (1989). New record of the spectacled porpoise. *Mar. Mamm. Sci.* **5**, 201–202.

Philippi, R. A. (1893). Los delfines de la punta austral de la América del Sur. *An. Mus. Nac. Chile*, Sec. I, *Zool.* **6**, 1–18 (in Spanish).

Praderi, R. (1971). Contribución al conocimiento del género *Phocoena* (Cetacea, Phocoenidae). *Rev. Mus. Argent. Cienc. Nat. "Bernardino Rivadavia" Inst. Nac. Invest. Cienc. Nat. Zool.* **7**, 251–266 (in Spanish).

Purgue, A. (1987). Facial complex description of the spectacled porpoise *Phocoena dioptrica* Lahille 1912. Mammalia: Cetacea. Available from author, 19 pp.

Rosel, P. E., Haygood, M. G. and Perrin, W. F. (1995). Phylogenetic relationships among the true porpoises (Cetacea: Phocoenidae). *Molec. Phylogen. Evol.* **4**, 463–474.

True, F. W. (1903). On species of South American Delphinidae described by Dr R. A. Philippi in 1893 and 1896. *Proc. Biol. Soc. Washington* **16**, 133–144.

15

Burmeister's Porpoise *Phocoena spinipinnis* Burmeister, 1865

Robert L. Brownell, Jr and Phillip J. Clapham

Genus and Species

Taxonomy

A new species of porpoise from the "mouth of the River Plata" was described by Burmeister (1865) as *Phocaena spinipinnis*. The dried and mounted skin of the holotype illustrated by Gallardo (1917) still exists (MACNBA 27-2); however, all of the osteological material from this specimen appears to be lost (Brownell and Praderi, 1984). The only junior synonym for this species is *Phocoena philippii* (Philippi, 1893; see True, 1903). The specific name *spinipinnis*, deriving from the Latin "spina", a thorn, and "pinna", a fin, was used by Burmeister in reference to the tubercles on this animal's uniquely shaped dorsal fin. No subspecies are recognized.

Common names

Burmeister's porpoise is known by the Spanish names marsopa espinosa ("thorny porpoise"), chancho marino ("sea pig"), and as tonino or tonina. In Brazil, how-

Handbook of Marine Mammals, Volume 6
ISBN 0-12-588506-7

ever, this species is called the boto de dorsal espinhosa (Portuguese, "thorny dolphin") (Simões-Lopes and Ximénez, 1989).

Evolution

There is only a single known fossil of this species, an incomplete cranium of unknown precedence from Chile (Yañez *et al.*, 1994). The certain geological range of the Phocoenidae extends from the late Miocene to Recent (Barnes, 1976).

Distribution

Range

Burmeister's porpoise occurs along both coasts of South America, in a range which extends from southern Brazil to northern Peru (Fig. 1). A report of this species from Heard Island (53°S, 73° 30′E) in the southern Indian Ocean (Guiler *et al.*, 1987) has been reidentified as the spectacled porpoise, *Phocoena dioptrica* (Brownell *et al.*, 1989). In the Atlantic, the northernmost extent of the known range is represented by a stranding near the Rio Urucanga in Santa Catarina state, Brazil (latitude 28° 48′S; Simões-Lopes and Ximénez, 1989). On the Pacific side, the northernmost record is from Bahia de Paita, Peru (05° 01′S; Allen, 1925; Aguaya, 1975). No sightings have been recorded north of this latter location despite systematic surveys, which Reyes and Van Waerebeek (1995) attribute to the fact that the cool Humboldt current veers away from the coast at this point. The southernmost records of Burmeister's consist of both sightings and incidental catches in the region of the Beagle Channel and Ushuaia (54° 50′S) off the southern coast of Tierra del Fuego (Goodall, 1978; Goodall *et al.*, 1995a), as well as a report of this species taken intentionally for use as crab bait at 55° 10′S near Cape Horn (Cárdenas *et al.*, 1987). Whether Burmeister's porpoise has a continuous distribution throughout its range is unclear. There are numerous gaps in the known distribution along both Atlantic and Pacific coasts, but it is likely that many or most of these simply reflect lack of survey effort in the areas concerned.

Habitat

It has generally been assumed that Burmeister's porpoise is a coastal species (Brownell and Praderi, 1984). It sometimes frequents rivers and estuaries (Aguayo, 1975) and, off Tierra del Fuego, is occasionally observed inside the kelp line (Goodall *et al.*, 1995a). As noted by Goodall *et al.* (1995a), its habitat preferences closely resemble those of the harbour porpoise (*Phocoena phocoena*), which typically is found shoreward of the 100 fathom isobath. However, in recent

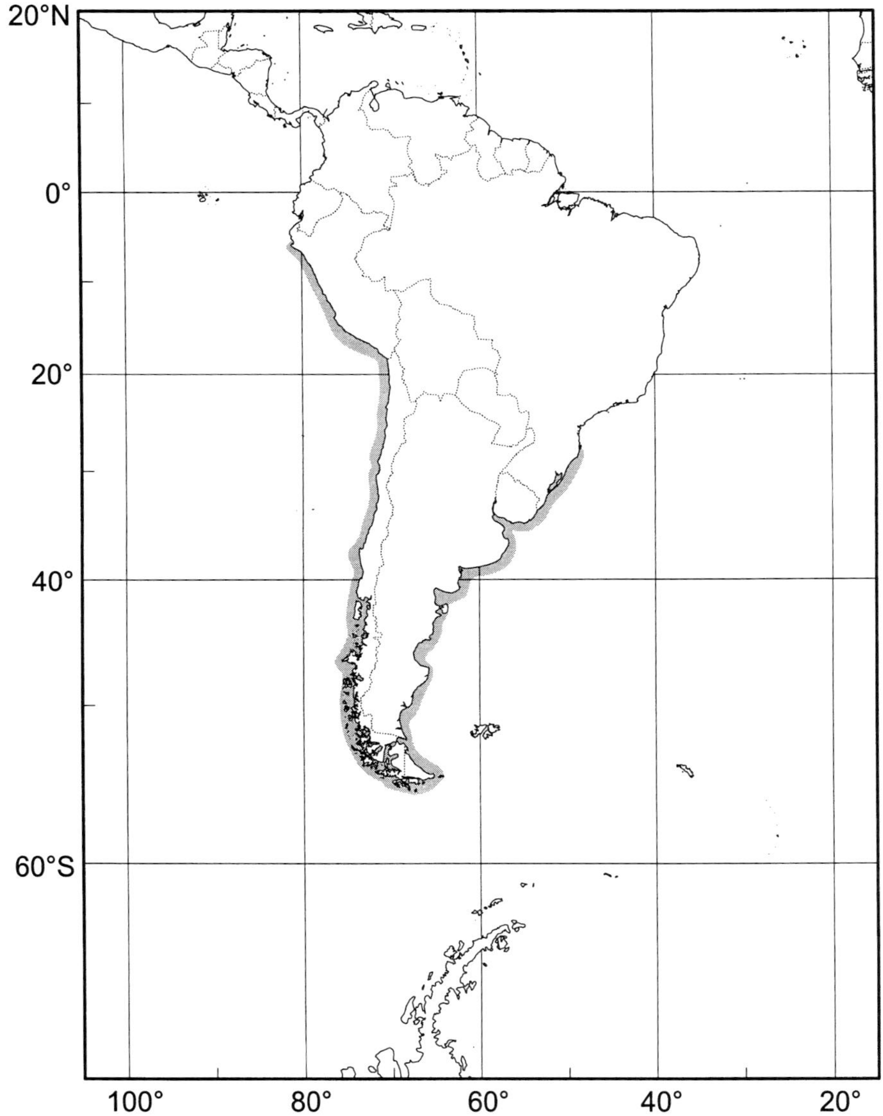

FIG. 1 Known distribution of Burmeister's porpoise, *Phocoena spinipinnis*.

years there has been an increasing number of records of this species from more offshore waters, including out to 50 km from the coast of Argentina (Corcuera, 1991). Incidental captures have been recorded from fisheries whose nets may be set as much as 55 km offshore, but associated depths are largely unknown (Goodall *et al.*, 1995a). These porpoises are associated regularly with water depths ranging from 6 m to 60 m, but occasionally they have been recorded offshore in up to 1000 m of water (Goodall *et al.*, 1995a).

Burmeister's porpoise is found associated with a rather broad range of water temperatures. At the southern limit of its distribution near Cape Horn and Tierra del Fuego, water temperatures range from 3°C in June to about 9°C in the summer months (Goodall *et al.*, 1995a). To the north, the species appears to be associated with warm or temperate waters associated with the two major northward flowing currents of South America, the Humboldt and Falklands currents. The highest recorded temperature associated with a Burmeister's sighting was 19.5°C in Golfo San José, Argentina (Würsig and Würsig, 1980).

External Characteristics

Colour pattern

Overall, the colour pattern of Burmeister's porpoise (Fig. 2) lacks the complexity found in some delphinids, with the visible (above-surface) portion of swimming animals appearing largely uniform in coloration. Some previous descriptions of dead specimens of this species have proved contradictory or misleading, since the skin colour darkens considerably *post-mortem*. Goodall *et al.* (1995b) and Reyes and Van Waerebeek (1995) examined living or recently deceased animals, and presented descriptions that are in broad agreement. These authors described dorsal coloration varying from light brown to grey or greyish-black. There is a light grey or whitish abdominal field extending to the uro-genital region; a pair of stripes are found on this field (Reyes and Van Waerebeek, 1995). Like the vaquita (*Phocoena sinus*) and the spectacled porpoise, Burmeister's has a well defined eye patch, as well as a dark grey flipper stripe that extends from the axilla of the flipper anteriorly along the throat to the lip patch on the lower jaw. The flipper stripe varies among individuals and is asymmetrical, being lower and narrower on the right side of the animal (Reyes and Van Waerebeek, 1995).

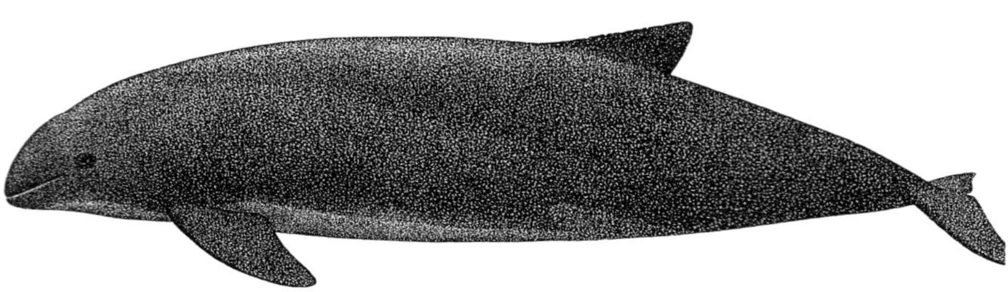

FIG. 2 Burmeister's porpoise: for variations in skin pigmentation patterns see Goodall *et al.* (1995b). (Drawing by P. Folkens)

Size and shape

The largest sample sizes of lengths and body measurements for Burmeister's porpoise come from animals caught in gill nets off Peru ($n = 402$; Reyes and Van Waerebeek, 1995). In addition, Goodall *et al.* (1995b) reviewed published and new data on 39 animals from various locations in southern South America. The longest recorded specimen was a Uruguayan animal of undetermined sex at 200 cm (Brownell and Praderi, 1984). In Reyes and Van Waerebeek's (1995) Peruvian sample, the largest female and male measured 183 cm and 182 cm, respectively. The same sample set showed evidence of slight sexual size dimorphism: the mean length of sexually mature males (170.3 cm, $n = 63$) was significantly greater than that of mature females (165.5 cm, $n = 56$). However, we would note that a comparison of growth curves or of physically mature specimens to assess the existence of dimorphism has yet to be made. Data on selected external measurements taken from Goodall *et al.* (1995b) are given in Table 1.

As is typical of phocoenids, the head is blunt and there is no beak. Measurements reviewed by Goodall *et al.* (1995b) give a maximum girth of 55–66% of total length; thus this species is more robust than the spectacled porpoise, and the flippers are proportionately larger. The most striking feature of Burmeister's porpoise is the dorsal fin, which is sufficiently different from that of all congenerics to be diagnostic. The fin is seated far back and has a rounded, blunt tip. In addition, the anterior margin possesses two to four rows of small blunt tubercles (Brownell and Praderi, 1984; Goodall *et al.*, 1995b).

Weight

Again, by far the best data on weights from this species come from the Peruvian sample described by Reyes and Van Waerebeek (1995). Among 70 males, the heaviest weighed 72 kg (length 170 cm). A 173 cm animal weighing 79 kg was the heaviest of 60 females. Corcuera *et al.* (1995) reported weights of four females and four males taken in gill nets off northern Argentina; the heaviest animal was a 191 cm female weighing 105 kg. The heaviest male weighed 78 kg.

Internal Anatomy

Skull

As noted by Brownell and Praderi (1984), the skull of Burmeister's porpoise is similar to that of the harbour porpoise but differs by having: (1) the brain case

TABLE 1 Selected external measurements (cm) for Burmeister's porpoises from Peru, Chile and Argentina (adapted from Goodall *et al.*, 1995b)

	Juveniles, both sexes					Adult females					Adult males				
Measurement	*n*	Min	Max	Mean	SD	*n*	Min	Max	Mean	SD	*n*	Min	Max	Mean	SD
Total length	5	131.1	146.8	139.6	6.7	7	152.7	185.0	167.3	14.4	4	151.0	175.0	162.7	12.6
Snout to angle of mouth	5	6.0	11.0	9.2	2.1	6	6.4	12.0	9.7	1.9	4	10.4	12.0	11.2	0.8
Snout to centre of eye	5	11.8	17.5	14.6	2.5	6	9.2	20.0	14.9	3.7	3	9.2	18.5	14.8	4.9
Snout to tip of dorsal fin	5	76.0	110.0	94.9	12.5	6	104.5	129.0	116.0	9.2	3	109.0	135.0	118.4	14.4
Snout to umbilicus	4	61.6	69.4	65.3	3.6	4	70.4	77.5	73.1	3.1	2	70.6	80.0	75.3	6.6
Anterior length of flipper	5	21.3	27.0	23.9	2.1	6	25.0	32.0	28.9	2.5	4	24.5	31.0	28.6	2.9
Max. width of flipper	5	7.5	10.0	8.8	0.9	6	9.5	11.5	10.3	0.8	4	9.7	11.0	10.5	0.7
Height of dorsal fin	4	7.0	8.5	7.9	0.6	6	7.4	11.5	9.2	1.6	4	9.0	18.0	11.5	4.3
Span of flukes	4	30.0	40.0	34.2	4.6	6	36.1	45.0	40.6	3.3	3	33.0	41.4	37.5	4.2
Girth at axilla	5	33.0	80.0	66.3	19.3	5	77.6	101.0	87.0	8.8	3	76.0	97.0	88.4	11.0

much less compressed from front to back; (2) a dorsal profile of the supraoccipital bone in line with the dorsal profile of the rostrum instead of tilted at an angle of slightly over 20°; (3) a larger temporal fossa; and (4) fewer teeth. The skull is illustrated in Fig. 3.

Reyes and Van Waerebeek (1995) give data on selected cranial measurements for 13–14 female and 19–22 male Burmeister's porpoises from Peru; some of this information is reproduced in Table 2. An additional ten specimens examined by Brownell and Praderi (1982) provided mean and range values for a condylobasal length of 273 mm and 224–290 mm, respectively.

As in all phocoenids, essentially the teeth are uniform throughout and have spade-shaped crowns (Brownell and Praderi, 1984). Tooth count ranges derived from the combined data ($n =$ approximately 160 animals) of Goodall *et al.* (1995b) and Reyes and Van Waerebeek (1995) are as follows: upper left 11–22, lower left 16–22, upper right 10–20, lower right 16–25. Brownell and Praderi (1984) examined ten specimens and give a minimum count of 15 per side in the lower jaw.

Postcranial skeleton

The vertebral formula for Burmeister's porpoise has been reported as C_7, T_{14}, L_{15}, Ca_{32-35} (Allen, 1925; Corcuera *et al.*, 1995). Pinedo (1989) gave a caudal count of only 31 from a male stranded in Brazil; however, the smaller caudal vertebrae can be easily missed in the absence of X-ray analysis. The first three cervical vertebrae are fused, and the first eight ribs have capitular and tubercular attachments. Seven pairs of sternal ribs are present, with the anterior four attached directly to the sternum. The phalangeal formula given for a single specimen by Allen (1925) was I_2, II_8, III_7, IV_4, V_2. Tympanic bullae and periotics of this species are illustrated by Brownell and Praderi (1984).

Organs, tissues and physiology

The only substantial data on internal organs in Burmeister's porpoise are given by Reyes and Van Waerebeek (1995) from Peru. These data, and information from a single individual examined by Goodall *et al.* (1995b) from Tierra del Fuego, are summarized in Table 3. Goodall *et al.* (1995b) give the intestine length of a 151 cm male as 19 m.

Life History and Population Dynamics

Growth and reproduction

Data on age at sexual maturity in Burmeister's porpoise come principally from the large Peruvian sample of by-caught animals analysed by Reyes and Van

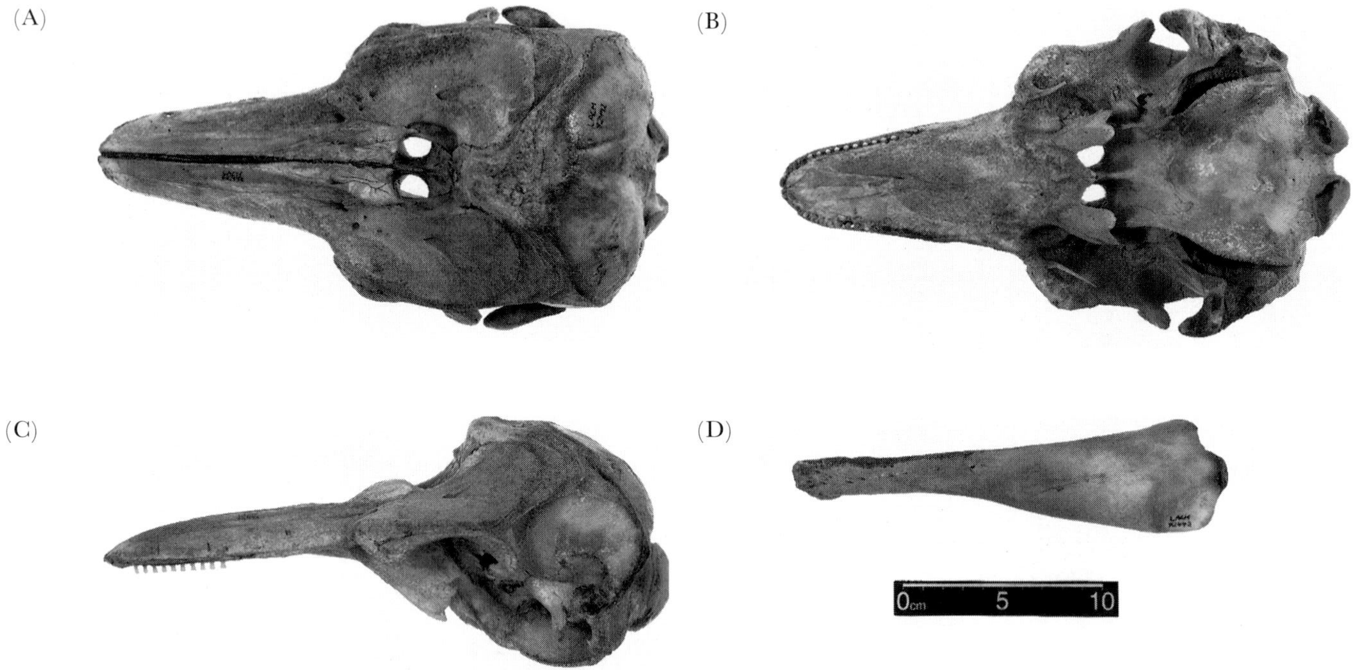

FIG. 3 Skull of Burmeister's porpoise (LACM 72442): dorsal (A) view, ventral (B) view, lateral (C) view and mandible (D).

TABLE 2 Cranial measurements (mm) and meristics of sexually mature Burmeister's porpoises from Peruvian waters. All data from Reyes and Van Waerebeek (1995). Significant intersexual differences were found in the variables LRBA, POOW, ZYW, PARW, HBR, LTV and WTY

	Females				Males			
Variable[a]	*n*	Range	Mean	SD	*n*	Range	Mean	SD
CBL	14	279–292	284.64	3.95	22	270–297	283.68	7.61
LRBA	14	121–131	127.07	2.79	22	116–132	124.64	3.74
LRPT	13	149–165	155.46	4.88	20	141–162	153.45	5.18
WRBA	13	78–87	83.08	2.29	13	77–87	83.00	3.04
WR1/4	14	58–66	61.93	2.62	20	55–64	60.85	2.30
WR1/2	14	48–54	51.71	2.13	21	48–56	51.38	2.50
WP1/2	14	25–29	26.93	1.44	22	23–29	26.68	1.49
WR3/4	14	36–41	39.00	1.75	21	35–43	38.95	2.38
PROW	14	124–139	131.07	4.34	21	127–140	133.38	3.47
POOW	14	147–156	152.21	2.69	21	150–162	154.81	2.82
ZYW	14	150–160	156.00	2.88	21	155–166	159.10	2.90
PARW	14	128–137	132.21	3.19	22	127–144	135.27	4.07
HBR	14	79–90	87.07	2.70	22	86–95	90.05	2.61
LBR	14	109–119	113.00	3.37	22	110–121	114.64	3.36
MWPR	14	32–43	39.36	2.98	22	36–45	40.14	2.32
WEN	14	27–34	30.36	2.06	22	27–35	30.73	2.21
TFL	13	55–63	59.85	2.15	21	53–65	59.76	3.74
TFW	13	45–54	47.62	3.31	21	40–45	48.48	3.84
ORL	13	45–55	49.23	2.83	20	46–59	50.75	3.31
LAP	14	24–30	26.86	1.70	21	19–32	28.10	2.72
SEPT	14	22–27	23.57	1.40	21	18–27	22.19	2.11
WIN	14	44–51	47.71	2.02	22	43–52	47.68	2.28
UTRL	14	72–103	84.86	10.6	22	69–99	77.86	6.86
LTRL	14	86–97	90.50	2.77	22	82–98	89.77	4.00
HRA	14	53–59	57.14	1.92	22	53–60	56.68	2.23
LRA	14	211–229	219.57	4.36	22	209–225	218.14	4.69
WFM	14	28–34	31.29	1.73	21	28–34	31.00	1.64
WINA	14	25–31	28.07	2.09	22	24–33	29.10	2.69
WIPA	14	47–54	49.86	1.96	21	47–54	51.14	1.98
LTV	14	29–32	30.36	0.74	19	30–33	31.47	0.70
WTY	14	18–20	19.50	0.65	19	18–21	20.05	0.78
LPE	14	27–30	28.86	1.03	20	26–31	29.10	1.29

[a] Condylobasal length (CBL), length rostrum from base (LRBA), length rostrum from pterygoid (LRPT), width rostrum at base (WRBA), width rostrum at quarter length (WR1/4), width rostrum at half length (WR1/2), width premaxillaries at half length (WP1/2), width rostrum at three-quarter length (WR3/4), pre-orbital width (PROW), postorbital width (POOW), zygomatic width (ZYW), parietal width (PARW), height braincase (HBR), length braincase (LBR), maximum width of premaxillaries (MWPR), width external nares (WEN), temporal fossa length (TFL), temporal fossa width (TFW), orbital length (ORL), length of antorbital processus (LAP), separation pterygoids (SEPT), width of internal nares (WIN), height of mandibular ramus (HRA), length of mandibular ramus (LRA), width of foramen magnum (WFM), maximum width of nasals (WINA), maximum width of palatines (WIPA), length of tympanic bulla (LTY), width of tympanic bulla (WTY), and length of periotic (LPE).

TABLE 3 Weights of selected internal organs (g) of adult Burmeister's porpoises. The first listed male came from Tierra del Fuego (Goodall *et al.*, 1995b); all other specimens were from Peru (Reyes and Van Waerebeek, 1995, from which this table is adapted). Additional organ weights for numbered animals: [1]liver 2450 g, heart 400 g, left lung 1130 g, right lung 1150 g; [2]liver 2500 g, heart 520 g, each lung 767 g; [3]liver 1600 g, heart 290 g, left lung 860 g, right lung 740 g

	Length (cm)	Total wt. (kg)	Left gonad	Right gonad	Left kidney	Right kidney	Pancreas	Spleen
Females	153	55	11	1	150	181	90	21
	159	58			214	248	93	22
	160	59	26	8	210	225	130	26
	164	62	17	2	270	270	160	
	164	62	7	2	200	210	84	25
	164	60	12	8	305	270	165	24
	165	65	26	1	278	291	190	14
	166[1]	71	17	2	290	300	160	30
	169	68	12	1	228	224	132	20
	173	63	7	1	300	325	100	52
	173	79	14	1	280	275	150	21
Males	151[2]	59	132	135	266	259		48
	160	62	240	240	180	185	110	
	162	56	183	182	204	186		52
	162	52	100	85	215	215	150	
	163	56	256	240	197	210	95	16
	169	56	228	216	205	223		31
	170[3]	62	300	300	170	190	80	10
	170	72	360	320	280	280		60
	171	67	296	247	160	170	154	
	174	69	290	270	285	270	160	54

Waerebeek (1995); their results concur with a much smaller sample from various locations in southern South America reviewed by Goodall *et al.* (1995b). In the Peruvian animals, minimum and average length at attainment of sexual maturity was found to be 153 cm and 154.8 cm, respectively, for females ($n = 56$), and 158 cm and 159.9 cm, respectively, for males ($n = 63$). This intersexual difference is significant.

The pregnancy rate in the Peruvian sample was high: 60% of mature females were pregnant, and 34% of these were simultaneously pregnant and lactating, a reproductive rate which is similar to that of the harbour porpoise (Read and Hohn, 1995). Gestation was estimated from foetal length at 11–12 months, with the peak of conception and parturition occurring in summer (February–March). The sex ratio among 31 foetuses was at parity. Average length at birth was estimated at 86 cm; thus it seems clear that a 44 cm foetus taken off Uruguay and called "full-term" by Brownell and Praderi (1984) was not near birth. Corcuera *et al.* (1995) suggest (based on what they acknowledge are scant data) that reproduction in Burmeister's porpoise is not highly seasonal or that peaks of calving are not unimodal.

Mortality

There is little information on survival rates or life expectancy in this species. Counts of growth layer groups (GLGs) in teeth have been published for only about thirty animals (Corcuera *et al.*, 1995; Goodall *et al.*, 1995b; Reyes and Van Waerebeek, 1995). Of these, the oldest was a 196 cm female from northern Argentina who was at least 12 years old (Corcuera *et al.*, 1995). Nothing is known concerning natural mortality or rates of increase.

Abundance

There are no quantitative data on abundance, although this species is assumed to be numerous in southern South American coastal waters.

Behaviour

Social organization

These porpoises usually occur in small groups. Aguayo (1975) observed a group of eight at the mouth of the Loa River, Chile (21° 25′S), and the largest group seen by Goodall *et al.* (1995a) in the Golfo San José, Argentina, was also eight. Goodall *et al.* (1995a) reviewed group size frequencies for southern South American sightings; they found that most sightings involved between one and six animals, with the modal group size being two. A "loose" group of 22 was reported off Peru by Van Waerebeek and Reyes (1990), and Guerra *et al.* (1987) saw 70 animals, supposedly associated, off northern Chile. Nothing is known of either the social organization or the mating system of this species.

Movements

Data on seasonal movements of Burmeister's porpoise are sparse and come largely from entanglements and incidental sightings. At Golfo San José, Argentina, they are observed almost exclusively in spring and summer (Würsig *et al.*, 1977). This suggests that seasonal movement (either north–south or inshore–offshore) does occur, although whether this is correlated with water temperature or abundance of prey is unknown. A seasonal movement in porpoise captures by the "corvina" fishery occurs off Valdivia, Chile, with animals caught inshore (up to 18.5 km from the coast) in summer and offshore (18–37 km) in winter (Oporto and Brieva, 1994). However, the fishermen move their nets offshore in winter, so it is unclear whether the by-catch truly reflects movements by the porpoises or simply seasonal sampling of a population that is distributed throughout both areas all year. If the former, it is possible that Burmeister's porpoise migrates offshore to follow seasonal movements of sardines (Goodall *et al.*, 1995a). However, if seasonal movements occur, they are not universal in this species. A year-round population of Burmeister's porpoise appears to exist in the Beagle Channel; sightings have been made in every month except August and September (Goodall *et al.*, 1995a).

Feeding

Published data on stomach contents of a total of 168 Burmeister's porpoises from various locations in both eastern and western South America are reviewed by Goodall *et al.* (1995b). These data are given in Tables 4 and 5; the latter summarizes quantitative data compiled by Reyes and Van Waerebeek (1995) and McKinnon (1988) on incidence of prey items in the stomachs of animals from Peruvian waters. The diet from all areas was predominantly fish, notably anchovy (*Engraulis* spp.) and hake (*Merluccius gayi*). Squid, mysid shrimp and euphausiids are also taken, and the stomachs of some Chilean animals also contained small snails, crustaceans and mollusc egg capsules.

Nothing is known of the foraging behaviour of this species, and there are no confirmed observations of feeding, although Goodall *et al.* (1995b) report an observation of six or seven animals which may have been feeding on Fuegian sardines (*Sprattus fuegensis*) off Tierra del Fuego.

Association with other species

Burmeister's porpoise has been observed frequently in the same areas as other cetaceans, but there is no record of active association with other animals. This topic is briefly reviewed by Goodall *et al.* (1995a); they suggest also that, in Golfo San José, Argentina, spatial habitat separation may be maintained between Burmeister's porpoise and two other species found there, the dusky dolphin (*Lagenorhynchus obscurus*) and the bottlenose dolphin (*Tursiops truncatus*). A similar pattern is evident in Peruvian waters (McKinnon, 1988).

Swimming and diving

Burmeister's porpoise is very difficult to detect in any but calm conditions, a fact that may explain the discrepancy between the assumed abundance of this animal in coastal waters on the one hand and the relative rarity of field observations on the other. The animal's respiratory and diving behaviour does not lend itself to easy observation: swimming is highly unobtrusive, surfacing is quiescent, and

TABLE 4 Summary of prey species found in stomach contents of Burmeister's porpoises from southern South America. Adapted from Goodall *et al.* (1995b). Data from Peru are given separately in Table 5

Country	Common name	Scientific name
Chile ($n = 37$)	Anchovy	*Engraulis ringens*
	Hake	*Merluccius gayi*
	Jack mackerel	*Trachurus murphyi*
	Sardine	*Sardinops sagax*
	Sardina común	*Clupea bentincki*
	Congrio colorado	*Genypterus chilensis*
	Róbalo	*Eleginops maclovinus*
	Silverside	*Odontestes regia*
	Patagonian squid	*Loligo gahi*
	Small snails	—
	Small crustaceans	—
	Mollusc egg capsules	—
Argentina ($n = 5$)	Unidentified shrimp	—
	Mysid shrimp	—
	Euphausiids	—
	Tailed hake	*Macruronus magellanicus*
	Anchovy, anchoita	*Engraulis anchoita*
	Calamarete	*Loligo sanpaulensis*
	Unidentified shrimp	—
Uruguay ($n = 1$)	Patagonian hake	*Merluccius hubbsi*
	Porgy	*Pagrus sedecim*
	Unidentified squid	—
Brazil ($n = 1$)	Pescadilla	*Cynoscion striatus*
	Molluscs	*Mactra* sp.
	—	*Pelecytallas*

relatively prolonged dives of 1–3 min are common (Würsig *et al.*, 1977; Goodall *et al.*, 1995a). The porpoise makes a slow roll at the surface, exposing the blowhole and the dorsal fin but barely breaking the water surface (Goodall *et al.*, 1995a). The location and shape of the dorsal fin also help to make this a cryptic species. The sole measurement of underwater swimming speed, made by theodolite at Golfo San José, returned a value of 4 km/hr (Würsig *et al.*, 1977).

Aerial behaviour

Aerial activity appears to be almost absent in this species. Goodall *et al.* (1995a) report only two cases of leaps by Burmeister's porpoises, both associated with surfing behaviour.

Sound production

Nothing is known of the vocalizations or mode of sound production in this species.

TABLE 5 Prey composition from stomach content analysis of Burmeister's porpoise from Peruvian waters (adapted from Reyes and Van Waerebeek, 1995). Total $n = 123$ animals

Common name	Scientific name	Incidence of items (%)	
		Reyes and Van Waerebeek (1995)	McKinnon (1988)
Anchovy	*Engraulis ringens*	96.3	94.2
Hake	*Merluccius gayi*	13.0	3.8
Drum	*Sciaena deliciosa*	7.4	17.3
Jack mackerel	*Trachurus murphyi*	7.4	—
Silverside	*Odontestes regia*	5.5	—
Sardine	*Sardinops sagax*	3.7	5.8
Patagonian squid	*Loligo gahi*	3.7	9.6
Weakfish	*Cynoscion analis*	—	9.6
Blackruff	*Seriolella violacea*	—	1.9
Unidentified/traces		5.5	—

Parasites and Disease

Until recently, the only data on parasites in Burmeister's porpoises came from a single Uruguayan individual examined by Brownell and Praderi (1982); the stomach contained nematodes. More comprehensive information has since been reported from Peru by Reyes and Van Waerebeek (1995), which we reproduce here as Table 6. Corcuera *et al.* (1995) provide additional information from northern Argentina but do not quantify the occurrence of each parasite. The great majority of examined individuals in Peru were infected with the trematode *Synthesium tursionis*, as well as nematodes of the genus *Stenurus* and the nematode *Pseudalius inflexus* (Reyes and Van Waerebeek, 1995). The two nematodes were also highly abundant in the Argentinian animals, which overall contained a lower diversity of parasites than those from Peru (possibly due to small sample size; Corcuera *et al.*, 1995). Two individuals from Argentina had the Acanthocephalan *Polymorphus cetaceum* in their stomachs; this represents the first record of such infection for Burmeister's porpoise, although *Polymorphus* is the most abundant parasite in the sympatric franciscana (*Pontoporia blainvillei*).

The only recorded ectoparasites are cyamids of the genus *Isocyamus* (species undetermined), found on only seven of 116 Peruvian porpoises; none was found among the Argentinian animals. Although not a parasite, *Xenobalanus globicipitis* has been observed attached to the fins of Burmeister's porpoises off Peru (Huaman and Reyes, 1986, cited in Reyes and Van Waerebeek, 1995).

Very little is known about diseases or other pathological conditions in this species. Goodall *et al.* (1995b) describe gastric ulcers in the forestomach, and

TABLE 6 Parasites found in Burmeister's porpoises from Peru (data from Reyes and Van Waerebeek, 1995)

Parasites	Infection site	Incidence (%)
Trematoda		
Synthesium tursionis	Pyloric stomach, middle ear	74.6
Pholeter gastrophilus	Pyloric stomach, small intestine	13.5
Nasitrema globicephalae	Air sinuses, small intestine	9.3
Braunina cordiformis	Pyloric stomach	1.8
Nematoda		
Stenurus sp.	Air sinuses, middle ear	93.4
Pseudalius inflexus	Lungs	84.4
Anisakis typica	Stomachs	28.1
Halocerus sp.	Lungs	16.7

haemorrhages in the submucosa of the pyloric stomach of a mature female from Argentina. Extensive necrosis of soft tissue and associated osteolysis of the underlying bones on the left side of the head were noted in one animal by Reyes and Van Waerebeek (1995). Diseased, deformed or rehealed bones have also been found in some other specimens (Philippi, 1896; Goodall *et al.*, 1995b; Reyes and Van Waerebeek, 1995). Van Bressem *et al.* (1993) described pox virus in Burmeister's porpoise from Peru.

Live Maintenance

Only a single Burmeister's porpoise has been taken into captivity: a 126 cm, emaciated young male that live-stranded south of Buenos Aires. It was held in captivity for eight days before it died, apparently of bronchial pneumonia (Loureiro, 1986).

Human Effects

Directed fisheries

It is widely suspected that Burmeister's porpoises are shot or harpooned for use as crab bait in southern Chile (Cárdenas *et al.*, 1987; Lescrauwaet and Gibbons, 1994). However, because quantitative data are lacking, the extent of this problem is unknown.

By-catches

The first documentation of incidental take in Burmeister's porpoise was made by Clarke (1962), who discovered a specimen in a fish market in Chimbote, Peru. Today, it is known that by-catch occurs in various areas within the species' range, including Peru, Chile, Argentina, Uruguay and Brazil. The most extensive known takes occur in Peruvian waters, where Burmeister's porpoise is caught primarily in net fisheries, and where it has been used extensively for human consumption (Perrin *et al.*, 1994; Goodall *et al.*, 1995b; Reyes and Van Waerebeek, 1995). In such areas, there is little distinction between directed takes and by-catches. Mortality in Peru was recently estimated as "> 450 per annum" (IWC, 1994). Net captures also occur off the Chilean coast (Aguayo, 1975; Oporto and Brieva, 1994; Perrin *et al.*, 1994). Coastal or shark gill net fisheries are also responsible for mortality in Burmeister's porpoise in Argentina ("> 12 per annum", IWC, 1994), Tierra del Fuego, and, to a lesser extent, Uruguay (Brownell and Praderi, 1984; Corcuera *et al.*, 1994; Crespo *et al.*, 1994; Goodall *et al.*, 1994). Takes are poorly documented in all areas, but the high mortality of this species in Peru is cause for considerable concern (Perrin *et al.*, 1994).

Pollution

There has been only one study of pollutants in this species, that conducted by Corcuera *et al.* (1995) on eight animals caught in gill nets off northern Argentina. Organochlorine levels in all animals were low, a finding which is consistent with the relatively low degree of pollution known from local waters. There are no data on contaminant burdens of Burmeister's porpoises from areas known to have significant industrial activity, which is not currently the case in northern Argentina.

Acknowledgments

Helpful reviews of earlier versions of this manuscript were given by R. N. P. Goodall, T. Jefferson, A. Read, K. Van Waerebeek and B. Würsig. J. E. Heyning generously provided the skull photographs.

References

Aguayo, L. A. (1975). Progress report on small cetacean research in Chile. *J. Fish. Res. Bd Can.* **37**, 1123–1143.

Allen, G. M. (1925). Burmeister's porpoise (*Phocoena spinipinnis*). *Bull. Mus. Comp. Zool.* **67**, 251–261.

Barnes, L. G. (1976). Outlines of eastern North Pacific fossil cetacean assemblages. *Syst. Zool.* **25**, 321–343.

Brownell Jr., R. L. and Praderi, R. (1982). Status of Burmeister's porpoise, *Phocoena spinipinnis*, in southern South American waters. FAO Fisheries Ser. Vol. 5, pp. 91–96.

Brownell Jr., R. L. and Praderi, R. (1984). "*Phocoena spinipinnis*". Mammalian Species No. 217, 4 pp.

Brownell Jr., R. L., Heyning, J. E. and Perrin, W. F. (1989). A porpoise, *Australophocaena dioptrica*, previously identified as *Phocoena spinipinnis*, from Heard Island. *Mar. Mamm. Sci.* **5**, 193–195.

Burmeister, H. (1865). Description of a new species of porpoise in the Museum of Buenos Aires. *Proc. Zool. Soc. Lond.* 228–231.

Cárdenas, J. C., Oporto, J., Stutzin, M. and Gibbons, J. (1987). Impacto de la pesquiría de centolla (*Lithodes antarctica*) y centollón (*Paralomis granulosa*) sobre las poblaciones de cetáceos y pinnípedos de Magallanes, Chile. *In* "Anais da 2a Reuniao de Trabalho de Especialistas em Mamiferos Acuáticos da América do Sul (Eds S. Siciliano and B. Fiori). FBCN, Rio de Janeiro, Brazil, pp. 57–59 (in Portuguese).

Clarke, R. (1962). Whale observation and whale marking off the coast of Chile in 1958 and from Ecuador towards and beyond the Galápagos Islands in 1959. *Norsk Hvalfangsttid.* **51**, 265–287.

Corcuera, J. (1991). Marsopa espinosa, *Phocoena spinipinnis* (Burmeister, 1865). *In* "Mares regionales. Estado de Conservación de los Mamíferos Marinos del Atlántico Sudoccidental", No. 138 (Eds H. L. Capozzo and M. Junín), pp. 27–30. Informes y Estudios del Programa de Mares regionales del PNUMA (in Spanish).

Corcuera, J., Monzón, F., Crespo, E. A., Aguilar, A. and Raga, J. A. (1994). Interactions between marine mammals and the coastal fisheries of Necochea and Claromecó (Buenos Aires Province, Argentina). *Rep. Int. Whal. Commn*, Spec. Issue No. 15, Cambridge, pp. 283–290.

Corcuera, J., Monzón, F., Aguilar, A., Borrell, A. and Raga, J. A. (1995). Life history data, organochlorine pollutants and parasites from eight Burmeister's porpoises, *Phocoena spinipinnis*, caught in northern Argentine waters. *Rep. Int. Whal. Commn*, Spec. Issue No. 16, Cambridge, pp. 365–372.

Crespo, E. A., Schiavini, A. C. M., Perez Macri, G., Reyes, L. and Dans, S. (1994). Estudios sobre determinación de edad en mamíferos marinos del Atlántico sudoccidental. *An. Reun. Trab. Esp. Mamif. Acuat.* **IV**, 31–39 (in Spanish).

Gallardo, A. (1917). Sobre el tipo de la *Phocoena spinipinnis* Burmeister. *Physis Buenos Aires* **3**, 83–84 (in Spanish).

Goodall, R. N. P. (1978). Report on the small cetaceans stranded on the coasts of Tierra del Fuego. *Sci. Rep. Whales Res. Inst. Tokyo* **30**, 197–230.

Goodall, R. N. P., Norris, K. S., Harris, G., Oporto, J. A. and Castello, H. P. (1995b). Notes on the biology of the Burmeister's porpoise, *Phocoena spinipinnis*, off southern South America. *Rep. Int. Whal. Commn*, Spec. Issue No. 16, Cambridge, pp. 317–347.

Goodall, R. N. P., Schiavini, A. C. M. and Fermani, C. (1994). Net fisheries and net mortality of small cetaceans off Tierra del Fuego, Argentina. *Rep. Int. Whal. Commn*, Spec. Issue No. 15, Cambridge, pp. 295–304.

Goodall, R. N. P., Würsig, B., Würsig, M., Harris, G. and Norris, K. S. (1995a). Sightings of Burmeister's porpoise, *Phocoeana spinipinnis*, off southern South America. *Rep. Int. Whal. Commn*, Spec. Issue No. 16, Cambridge, pp. 297–316.

Guerra, C., Van Waerebeek, K., Portflitt, G. and Luna, G. (1987). The presence of cetaceans off the northern Chilean coast. *Estud. Oceanol.* **6**, 87–96.

Guiler, E. R., Burton, H. R. and Gates, N. J. (1987). On three odontocete skulls from Heard Island. *Sci. Rep. Whales Res. Inst. Tokyo* **38**, 117–124.

Huaman, P. and Reyes, J. C. (1986). *Xenobalanus globicipitis*, Steenstrup, 1851 (Thoracica, Coronulidae), cirripedo epizoo de la marsopa espinosa *Phocoena spinipinnis* Burmeister, 1865 (Cetacea, Phocoenidae). Report to the Octoavo Congreso Nacional de Biologia, 25–31 May, Arequipa, Peru (in Spanish).

International Whaling Commission. (1994). Report of the workshop on mortality of cetaceans in passive fishing nets and traps. *Rep. Int. Whal. Commn*, Spec. Issue No. 15, Cambridge, pp. 1–71.

Lescrauwaet, A. C. and Gibbons, J. (1994). Mortality of small cetaceans and the crab bait fishery in the Magallanes area of Chile since 1980. *Rep. Int. Whal. Commn*, Spec. Issue No. 15, Cambridge, pp. 485–494.

Loureiro, J. D. (1986). Varamiento de una marsopa espinosa (*Phocoena spinipinnis*); intentos para su recuperación en oceanario. *In* Abstracts of the Reun. Trab. Esp. Mamif. Acuat. Am. Sud, Vol. I, pp. 163–167 (in Spanish).

McKinnon, J. S. (1988). Feeding habits of two dolphins and a porpoise from the coastal waters of Peru. M.Sc. thesis, University of Guelph, 94 pp.

Oporto, J. A. and Brieva, L. M. (1994). Interacción entre la pesquería artesanal y pequeños cetáceos en la localidad de Quele (IX Región), Chile. *In* An. Reun. Trab. Esp. Mamif. Acuat. Vol. 4, pp. 187–205 (in Spanish).

Perrin, W. F., Donovan, G. P. and Barlow, J. (1994). Gillnets and cetaceans. *Rep. Int. Whal. Commn*, Spec. Issue No. 15, Cambridge.

Philippi, R. A. (1893). Los delfines de la punta austral de la América del Sur. *An. Mus. Nac. Chile*, Sec. I., *Zool.* **6**, 1–18 (in Spanish).

Philippi, R. A. (1896). Los cráneos de los delfines chilenos. *An. Mus. Nac. Chile*, Sec. 1, *Zool.* **12**, 1–19 (in Spanish).

Pinedo, M. C. (1989). Primeiro registro de *Phocoena spinipinnis* (Cetacea, Phocoenidae) para o litoral do Rio Grande do Sul, Brasil, com medidas osteológicas e análise do conteúdo estomacal. *Atlántica (Rio Grande)* **11**, 85–89 (in Portuguese).

Read, A. J. and Hohn, A. A. (1995). Life in the fast lane: the life history of harbor porpoise in the Gulf of Maine. *Mar. Mamm. Sci.* **11**, 423–440.

Reyes, J. C. and Van Waerebeek, K. (1995). Aspects of the biology of Burmeister's porpoise from Peru. *Rep. Int. Whal. Commn*, Spec. Issue No. 16, Cambridge, pp. 349–364.

Simões-Lopes, P. C. and Ximénez, A. (1989). *Phocoena spinipinnis* Burmeister 1865, na costa sul do Brasil (Cetacea-Phocoenidae). *Biotemas* **2**, 83–89 (in Portuguese).

True, F. W. (1903). On species of South American Delphinidae described by Dr R. A. Philippi in 1893 and 1896. *Proc. Biol. Soc. Washington* **16**, 133–144.

Van Bressem, M., Van Waerebeek, K., Reyes, J. C., Dekegel, D. and Pastoret, P. P. (1993). Evidence of poxvirus in dusky dolphin (*Lagenorhynchus obscurus*) and Burmeister's porpoise (*Phocoena spinipinnis*) from coastal Peru. *J. Wildl. Dis.* **29**, 109–113.

Würsig, B. and Würsig, M. (1980). Behavior and ecology of the dusky dolphin, *Lagenorhynchus obscurus*, in the South Atlantic. *Fish. Bull.* **77**, 871–890.

Würsig, M., Würsig, B. and Mermoz, J. F. (1977). Desplazamientos, comportamiento general y un varmiento de la marsopa espinosa, *Phocoena spinipinnis.*, en el Golfo San José (Chubut, Argentina). *Physis (Buenos Aires)* **36**, 71–79 (in Spanish).

Yañez, J., Canto, H. J. and Reyes, J. C. (1994). Cráneo fósil de *Phocoena spinipinnis* (Cetacea: Phocoenidae) hallado en Chile. *Not. Mens. Mus. Nac. Hist. Nat. Santiago* **324**, 25–29 (in Spanish).

16

Finless Porpoise
Neophocaena phocaenoides (G. Cuvier, 1829)

Toshio Kasuya

Genus and Species

Taxonomy

This species was described by G. Cuvier in 1829 under the name of *Delphinus phocaenoides* based on a skull allegedly from Cape of Good Hope. A new genus *Neomeris* was later established by Gray (1846). The name *Neomeris* was later found to have been preoccupied for a genus of polyps and therefore it was replaced by *Neophocaena* (Palmer, 1899). The type locality has been questioned, and the type specimens are now considered to have been brought most probably from India (Allen, 1923). There has been no indisputable South African record of the finless porpoise (Rice, 1977).

Interpretation of geographical variation of finless porpoises has been controversial. Using body colour, shape of head, extent of the dorsal denticulated area and osteology, Pilleri and Gihr (1975) and Pilleri and Chen (1980) classified them into three species, i.e. *Neophocaena phocaenoides* (G. Cuvier, 1829) in the Indian Ocean, *N. asiaeorientalis* (Pilleri and Gihr, 1972) from the Yangtze River and Chinese coast, and *N. sunameri* (Pilleri and Gihr, 1975) in Korean and

Handbook of Marine Mammals, Volume 6
ISBN 0-12-588506-7

Japanese waters. Van Bree (1973), accepting 1973 for the date of Pilleri and Gihr (1972), believed from osteological study that the difference between the Indian Ocean and Chinese populations is of a subspecific level, and that the difference between the Chinese and Japanese populations is insignificant. This view was later supported by Amano *et al.* (1992). Gao (1991) identified three subspecies of *Neophocaena p. phocaenoides* in the South China Sea and Indian Ocean, *N. p. asiaeorientalis* in the Yangtze River and *N. p. sunameri* in the Bohai, Yellow Sea (including the East China Sea) and Japanese waters based on the dorsal denticulated area and discriminant function analyses of osteology and external proportions. Although supporting this conclusion in principle, Wang (1992a,b) considered that those in the East China Sea are the same as the conspecifics in the Yangtze River. Additionally, there are suggestions on geographical variations in growth and reproduction within some of these local areas (see below).

Finless porpoises are distributed in a narrow band along the coast of tropical and temperate Asia including some rivers. Because of this almost linear distribution, the chance of gene exchange between distant populations appears small even though there are no geographical barriers or habitat gaps that inhibit free interbreeding between nearby areas. Therefore the finless porpoise may easily develop clinal variations; additionally the formation of morphologically distinct local stocks will be accelerated by some environmental gaps in the range. The existing information seems to be insufficient to establish more than one species in the genus *Neophocaena*.

The family Phocoenidae (Gray, 1825) includes the genera *Phocoena* and *Phocoenoides*, plus *Neophocaena phocaenoides* (G. Cuvier, 1829), which is the only extant member of the genus *Neophocaena* and the most primitive (Gaskin, 1982) or derived (Barnes, 1985) member of the family.

Common names

English names include finless porpoise, finless black porpoise, or Indian porpoise. Other names are tabi in Pakistan (Pilleri and Gihr, 1972), molagan among Tamil in south India (Owen, 1869), jiangchu (river pig) along the Yangtze (Chanjiang) River in China (Chen *et al.*, 1979), limbur in Indonesia (Tas'an and Leatherwood, 1984), and sunameri in Japan. Suzame is also in use in the Ise Bay, name or nameno-uwo in the eastern Inland Sea of Japan (ISJ), and zegon or zegondo in the western ISJ.

Distribution

Range

The westernmost range of the finless porpoise is in the Persian Gulf (Fig. 1) (Tomilin, 1967; Pilleri and Gihr, 1973–74; T. Preen, personal communication). They have been seen off Bahrain (S. H. Ridgway, personal communication) but are not found along the coasts of Oman (Baldwin and Salm, 1994) or in the Gulf of Adan off Djibouti (Robineau and Rose, 1984). In the south they range to the northern coast of Java and Strait of Sunda (Tas'an and Leatherwood, 1984), and the northern range is central Japan. They are also known from Pakistan, India, Ceylon, Bangladesh, Strait of Malacca, west and south coast of Borneo, Gulf of Siam, Vietnam (McKenna, 1992; Smith *et al.*, 1995), northern Palawan (Lorenzo Tan, 1995), China and Korea. The apparent lack of records along the coast of Burma may not indicate the discontinuity of distribution, because these areas have been the least studied for cetaceans. Finless porpoises appear to be absent from the African coast (see above), Sulaweshi, Halmahera and Timor (Tas'an and Leatherwood, 1984), the Philippines (Hammond and Leatherwood, 1984; T. Kataoka, personal communication), and the northern coast of Australia (Harwood *et al.*, 1984; H. Marsh, personal communication).

In China this species inhabits the entire coast from the Gulf of Tongking to the Korean border including the west coast of Taiwan (Yang, 1976; Wang, 1984). In summer, the finless porpoise inhabits the entire western half of the Yellow Sea, where water depth is less than 50 m (Miyashita *et al.*, 1995). Although the distribution in Korean waters is known only from Chollabukto (Zenra Hokudo) on the Yellow Sea coast (Howell, 1927), the distribution is possibly continuous from the Chinese border to the southern coast of the Korean Peninsula. The Japanese distribution has been confirmed from the west coast of Kyushu (Mizue *et al.*, 1965) to Hime (37° 17′N) on the west coast of the Toyama Bay, Sea of Japan (Anon., 1980; Yamada, 1989), and to the Sendai Bay bounded by the Oshika Peninsula (38° 20′N) on the Pacific coast (Ogawa, 1950; Nishiwaki, 1972; N. Kimura, personal communication).

The finless porpoise has not been confirmed along the coast of Tsushima and Iki Islands between Korea and Kyushu (Shirakihara, K. *et al.*, 1992; T. Kasuya, unpublished), suggesting infrequent intermingling between Korean and Japanese stocks. Off Okinawa, half digested finless porpoises were collected from a shark stomach in June (see below), but the distribution in the Ryukyu Islands and southern Kyushu has not been supported by Shirakihara, K. *et al.* (1992) and will require further confirmation. The species in the southern Kuril Islands suggested by Tomilin (1967) is questionable because the species has not been reported from the Hokkaido coast (Kasuya *et al.*, 1984), and the continental shelf of this northern area is inhabited by the more boreal common porpoise *Phocoena phocoena* (T. Kasuya, unpublished).

Habitat

The finless porpoise is a coastal, estuarine or riverine species (Fig. 1). In tropical Asia, it is found in the mangrove zones and penetrates into the Indus River for 60 km (Pilleri and Gihr, 1972, 1973–74; Pilleri and Pilleri, 1979), and into the Brahmaputra River for 40 km from the mouth (Lal Mohan, personal communication). In China it inhabits a 1670 km range of the Yangtze River (below Yichang) and the Dongting and Poyang Lakes (Chen *et al.*, 1979; Zhou *et al.*, 1979; Wang, 1984; Zhang *et al.*, 1993), and also is known to go up near Dangdong, about 20 km from the mouth of the Yalujiang River (Wang, 1984). Japanese finless porpoises in ISJ mostly live within 1 nautical mile (1.85 km) from the coast and their density over 3 nautical miles off the coast is only 1/10 of that within 1 nautical mile (Kasuya and Kureha, 1979).

Seasonal movements

Seasonal movements have been studied in detail in the ISJ, which is an area of about 14 300 square km, has a maximum depth of 98 m and over 700 islands in it, and is connected to the outer ocean by four narrow passes (Kasuya and

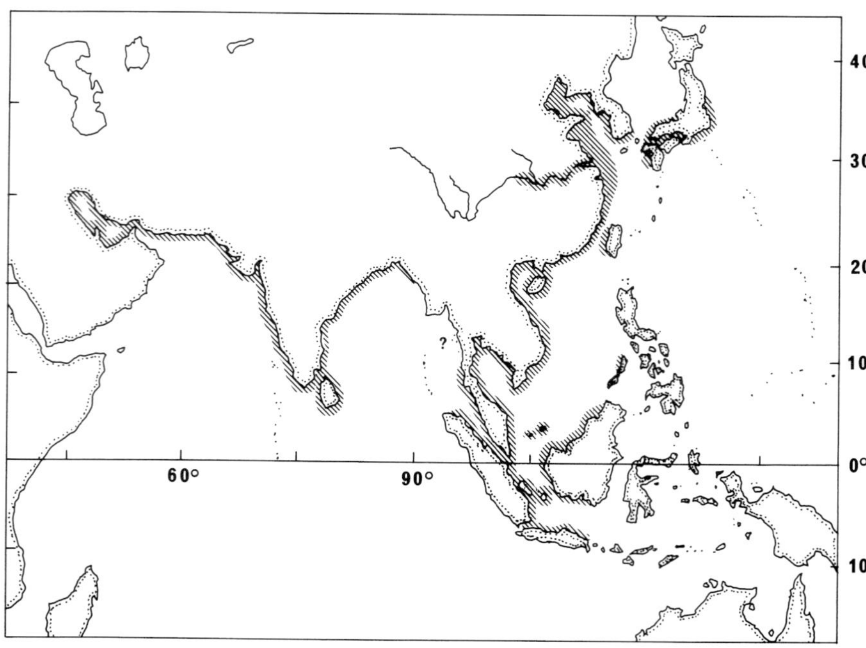

FIG. 1 Range of the distribution of finless porpoises (shaded area). The distribution of finless porpoises along the coast of Burma has not been studied.

Kureha, 1979). Some finless porpoises are found there yearly. In the ISJ, porpoises endure drastic seasonal changes in the surface water temperature between 6 °C (March) and 28 °C (September). However, density is lowest (40% of the peak season) in early winter, and it starts to increase in January, reaching its peak in April. Although the abundance in April has been estimated at about 4900 porpoises, the figure needs reevaluation using more recent sighting and evaluation methods. Migration appears to occur mainly through the two eastern passes (Kasuya and Kureha, 1979), but migration through a northwestern pass is not excluded (Shirakihara, M. *et al.*, 1992). The finless porpoise is known to occur yearly in the Ise and Mikawa Bays (Kataoka *et al.*, 1974, 1976) with a peak abundance of about 2000 individuals in April–June (Miyashita *et al.*, 1994) and in the Sendai Bay at its northern range (Shirakihara, K. *et al.*, 1992; Tobayama, T., personal communication). The distribution of finless porpoises off western Kyushu (Japanese southwestern range) also is yearly, apparently isolated from the rest of the ranges (Shirakihara *et al.*, 1993). The density is high in coastal waters less than 50 m deep in winter and spring seasons with about 3100 in the Ariake/Tachibana Bay and 200 in the Omura Bay (Shirakihara *et al.*, 1994; Yoshida *et al.*, 1997, 1998).

Along the Chinese coast, finless porpoises are present all year, but reported to have some seasonal density changes in Bohai and on the Yellow Sea coast (low in winter and high in summer/autumn) (Wang, 1984), where finless porpoises apparently move from shallow to deeper water in winter. Although Zhang *et al.* (1993) estimated the Yangtze population at about 2700 individuals, it is unclear if this represents abundance in winter, the highest density, or in the low density season of summer. Movement of finless porpoises between the Yangtze and the ocean has yet to be confirmed.

In the Indus Delta finless porpoise density is high in October–April; then they are believed to leave for the open sea following the seasonal migration of shrimps and prawns (Pilleri and Gihr, 1972; Pilleri and Pilleri, 1979). The extent of this offshore migration has not been examined and the abundance estimation needs refinement.

The above information suggests that finless porpoises are probably found yearly throughout their range, but that they show various degrees of seasonal movement and density change which are not well documented in most areas.

External Characteristics

Colour pattern

The newborn individual is black except for some light grey colour on the dorsal ridge area. The porpoise becomes lighter, probably after losing foetal epidermis by the age of one month. It takes 4–6 months to attain the light grey colour of the adult (M. Furuta, personal communication). The ventral surface is paler (Mizue *et al.*, 1965; Pilleri *et al.*, 1976). Indian Ocean finless porpoises are lighter in colour, especially on the ventral side (Pilleri and Chen, 1980). Body colour darkens during the *post mortem* time, as in other toothed whales.

Size and shape

Finless porpoises are slender with no dorsal fin. The single blowhole is located slightly to the left of the body midline and connected to a pair of nasal tracts and several nasal sacs (see below). There is no beak, and the forehead overhangs the upper lip (Fig. 2). Both sexes grow to 150 cm or more, but with some geographical variation. Table 1 shows external measurements among Japanese individuals (Mizue *et al.*, 1965) measuring over 155 cm, which are probably fully grown or close to it.

FIG. 2 Lateral (left) and ventral (right) aspects of a female finless porpoise from the Inland Sea of Japan.

TABLE 1 Range of some external measurements of four male finless porpoises off western Kyushu, Japan, measuring over 155 cm, expressed as a percentage of total length (compiled from data in Mizue *et al.*, 1965)

Measurement	Minimum (%)	Maximum (%)
1. From tip of rostrum to blowhole	6.8	7.3
2. From tip of rostrum to ear	9.4	10.5
3. From tip of rostrum to anterior insertion of flipper	15.6	17.1
4. From tip of rostrum to umbilicus	40.5	42.9
5. From anus to bottom of the tail notch	27.4	30.8
6. Total spread of tail flukes	26.8	29.6
7. Anterior insertion of flipper to tip	17.5	20.1

A dorsal ridge covered by thick denticulated epidermis, protrudes 2–3 cm above the surrounding mid-dorsum of the body (Howell, 1927). It was once considered to represent a vestige of dermal armature (Kukenthal, 1893) but this was questioned by Howell (1927). Pilleri and Chen (1979) suggested that it provides a secure surface on which mothers can carry their calves on their backs, but that it functions as a sensory organ (Liu, 1985) or organ of mutual contact is more plausible.

Weight

Body weight of six Pakistani (Pilleri and Gihr, 1972), 15 Chinese (Hsu *et al.*, 1973, excluding one anomaly) and 21 Japanese (Kasuya, T., unpublished), postnatal finless porpoises ranged between 5.7 kg (65.3 cm) and 71.8 kg (164 cm). Individual variation of body weight is considerable, and 10 individuals between 140 cm and 160 cm (mean = 146.9 cm) in body length weighed between 26.3 kg and 48.2 kg (mean = 38.8 kg). If logarithm of body weight is plotted against that of body length, all the 42 data come around a straight line, and show no geographical variation in the relationship. Thus the relationship between weight (W, kg) and body length (L, cm) is expressed by single least-squares regression:

$$W = 1.816 \times 10^{-4} \times L^{2.477}, \qquad 65.3 \leq L \leq 187$$

with a correlation coefficient of 0.95 between log W and log L. This is similar to that obtained by Shirakihara *et al.* (1993). An example of proportional weights of organs and tissues of a Japanese specimen is shown later in Table 4.

(A)

(B)

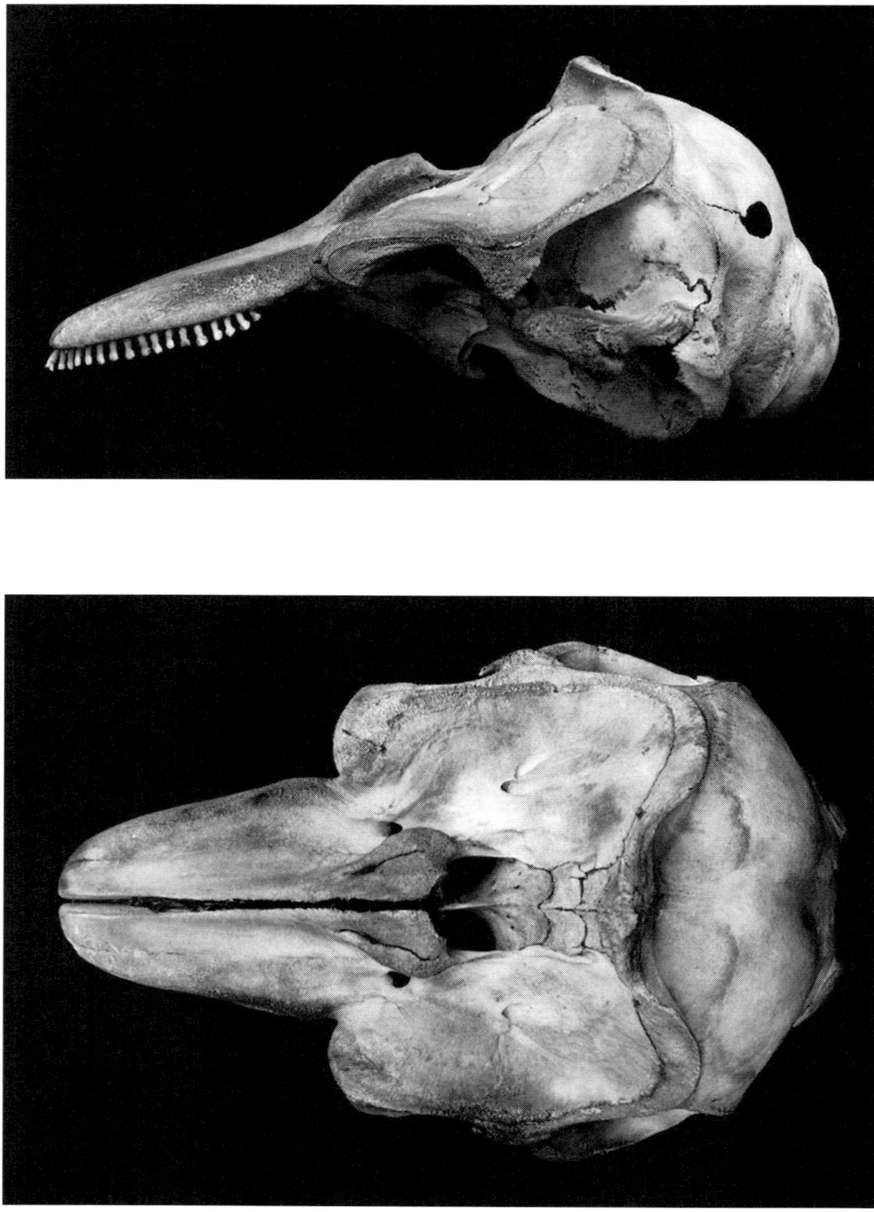

FIG. 3 Skull (A, B) of a male finless porpoise from the Inland Sea of Japan, 187 cm, 70.1 kg and aged at 14 tooth layers (National Science Museum, Tokyo, M.24957). Tooth crowns are worn on this individual.

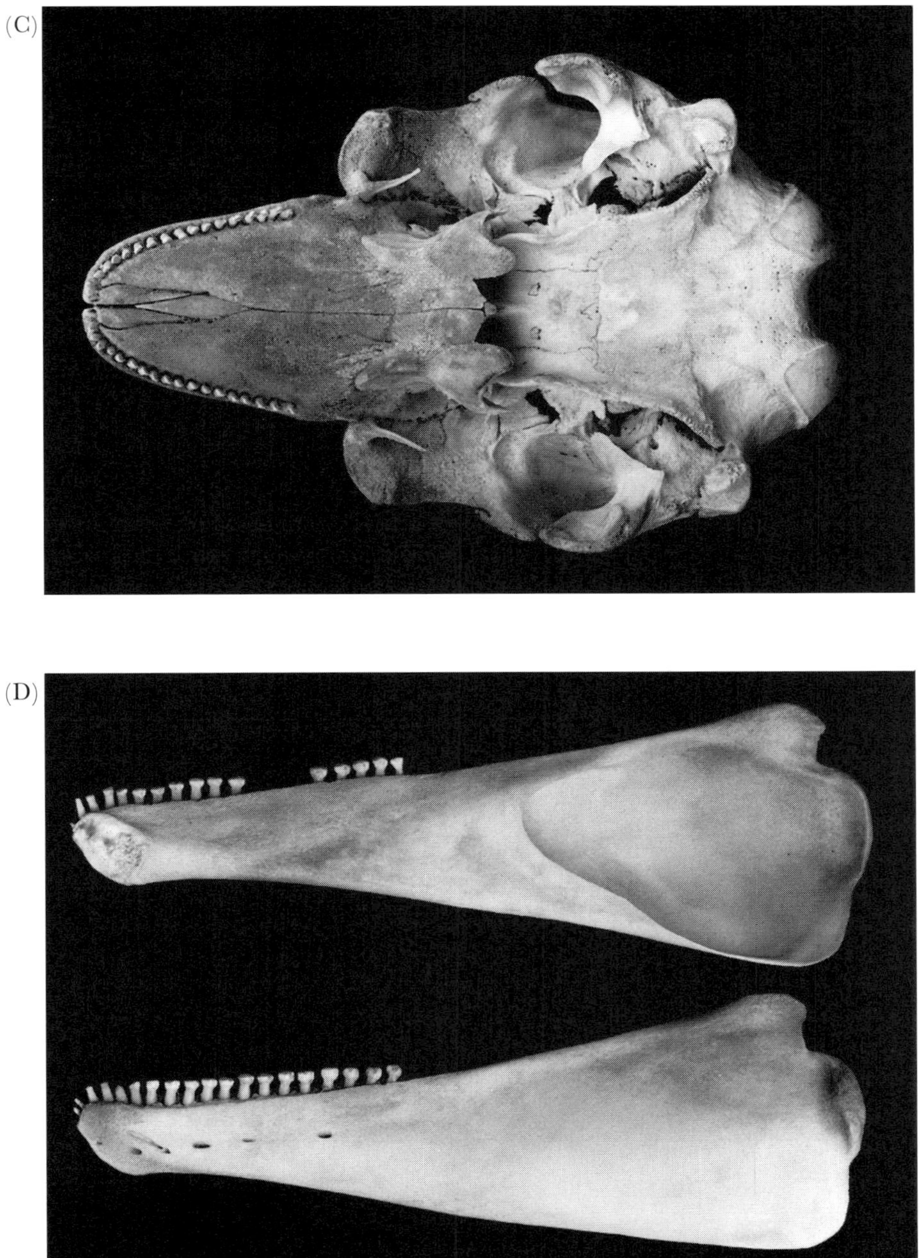

FIG. 3 *Continued* Skull (C) and mandibles (D) of a male finless porpoise.

Internal Anatomy

Skull

The condylobasal lengths range from 227 mm to 249 mm (14–17% of the body length ranging from 137 cm to 174 cm, $n = 19$) off Japan (Mizue *et al.*, 1965), or from 195 mm to 213 mm (14–15% of the body lengths ranging from 130 cm to 155 cm, $n = 5$) off Pakistan (Pilleri and Gihr, 1972). The rostrum is broad (Fig. 3) and not as pointed as that of other members of the family. The maxillae are slender and the lateral margins of the rostral portion are nearly parallel. Bony tubercles of the premaxillae are present in front of superior nares. The pterygoids are narrow and the laminae are widely spaced. The posterior process of the periotic bone is pointed and directed posteriorly. Dimensions of the skull are shown in Table 2.

Using geographical variation of skull morphology and DNA nucleotide sequence, Yoshida *et al.* (1995, 1997) distinguished five populations in major habitats of the species in Japan, i.e. Ariake/Tachibana Bays, Omura Bay, ISJ, Ise/Mikawa Bays, and the area from Tokyo to Senday Bay.

Finless porpoise teeth are about 15 mm long and 3 mm diameter at the neck, larger than those of *Phocoena* or *Phocoenoides* (Tomilin, 1967). The crown is laterally compressed and of spatulate shape except for two or three conical teeth in the anterior part of the tooth rows (Fig. 4). One or two pairs are in the premaxillaries (Tomilin, 1967).

The ranges of upper/lower tooth counts are 12–22/17–20 in the Indian Ocean (Pilleri and Gihr, 1972), 16–22/17–23 in the South China Sea (Gao, 1991), 15–21/14–22 in the Yangtze River (Gao, 1991), 17–21/15–22 in the East China

TABLE 2 Range of some skull measurements of finless porpoises expressed as a percentage of condylobasal length measuring over 230 mm, for eight males and three females from off western Kyushu, Japan (compiled from data in Mizue *et al.*, 1965)

Measurement	Minimum (%)	Maximum (%)
1. Width across zygomatic process of squamosals	59.7	65.1
2. Width of rostrum at base	29.0	31.6
3. Width of rostrum at the midlength	20.1	24.6
4. Length of rostrum	35.6	40.0
5. Length of mandible	69.2	72.8
6. Length of maxillary tooth row	29.7	31.9
7. Length of mandibular tooth row	39.8	44.0

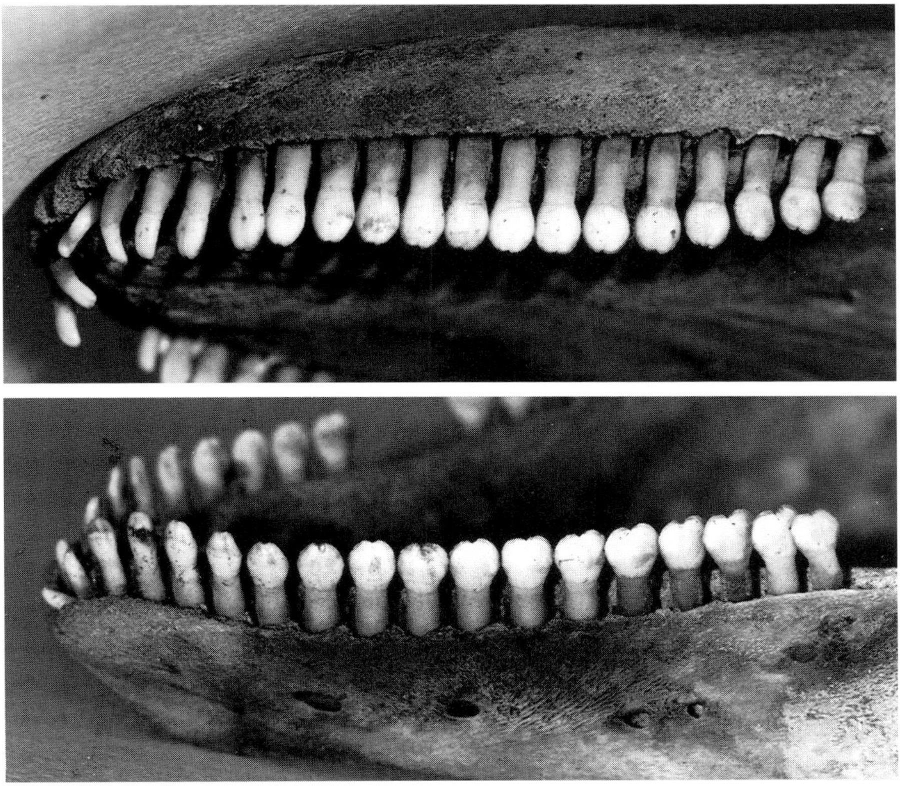

FIG. 4 Maxillary (upper) and mandibular (lower) teeth of a juvenile finless porpoise from the Inland Sea of Japan showing the overlapping teeth and compressed crown: female, 118 cm, 25.2 kg and aged at about one tooth layer (National Science Museum, Tokyo, M.24911).

Sea, Yellow Sea and Bohai (Gao, 1991), and 15–19/16–20 in Japan (Mizue *et al.*, 1965; Pilleri and Gihr, 1975; Kasuya, T., unpublished). There appears to be no significant geographical variation.

Postcranial skeleton

The vertebral formulae are C_7, $T_{12–14}$, $L_{10–13}$, $Ca_{26–33}$, total 58–64. However, Japanese finless porpoises have from 1 to 3 more vertebrae than Indian Ocean individuals, and the Chinese individuals cover both ranges (Table 3). The first three of the seven cervical vertebrae are fused together (with the exception of one individual having four fused cervicals reported by Mizue *et al.* (1965). This disagrees with Nishiwaki (1972), who stated that the first five cervicals were fused.

TABLE 3 Vertebral counts of finless porpoises

Localities	N	Cer.	Th.	Lum.	Ca.	Total
Pakistan and India[a]	7	7	12–14	12–13	26–30	59–61
South China Sea[b]	45	7	12–13	11–13	28–33	60–65
Changjiang River[b]	37	7	12–13	11–13	27–32	59–64
East China Sea/Bohai[c]	54	7	12–14	10–13	27–31	58–64
Japan, western Kyushu[d]	17	7	13–14	← 41–43	→	62–64
Japan, ISJ/Pacific[e]	4	7	13–14	11–12	29–32	61–64

[a] Lydekker (1909, cited in Allen, 1923); Pilleri and Gihr (1972).
[b] Gao (1991).
[c] Gao (1991) includes the Yellow Sea.
[d] Mizue *et al.* (1965)
[e] T. Kasuya (unpublished).

Out of 12–14 pairs of ribs, the anterior six to eight pairs are two headed (Mizue *et al.*, 1965; Pilleri and Gihr, 1972; Hsu *et al.*, 1973; Gao, 1991; T. Kasuya, unpublished). No geographical differences have been identified in the number of two-headed ribs. Japanese specimens are reported to have four pairs of sternal ribs while others usually have three (Pilleri and Gihr, 1975). Howell (1927) reported Chinese individuals having cervical ribs, which would be equivalent to bones of a few centimetres often found attached by tendon to the last cervical vertebrae of Japanese specimens (T. Kasuya, unpublished).

The manus consists of five carpal bones and five fingers (Kukenthal, 1893). The phalangeal formula is I_2, II_{5-8}, III_{5-7}, IV_{3-4}, V_2 (Kukenthal, 1893; Nishiwaki, 1972; Hsu *et al.*, 1973).

Organs, tissues and physiology

The finless porpoise has 9 or 10 nasal sacs (Gao and Zhou, 1989), (i.e. paired vestibular, nasofrontal premaxillary and vomeronasal sacs) and one or two accessory sacs. The vestibular sacs are well developed but the accessory sacs are degenerated and often present on only one side, features common among *Phocoenoides dalli* and *Phocoena phocoena* (Mead, 1975). Qian (1986) described hyaline cartilages from trachea to respiratory bronchiole and at the entrance of the alveolar duct, and the sphincter at the entrance of the alveolar duct.

According to Arvy and Pilleri (1972) the tongue is oval, measuring about 8 cm × 5 cm, and covered by keratinized epithelium. Salivary glands open into the bottom of minute depressions on the central and posterior area of the tongue. The tongue of the juvenile has numerous papillary projections on the lateral margin that degenerate with growth to disappear finally in the adult (Li *et al.*, 1984).

The oesophagus is about 22 cm long and 3.5 cm wide in the juvenile (Pilleri and Gihr, 1972). The inner wall is covered by keratinized stratified squamous epithelium and has no glands. Its lower end connects to the anterior end of the forestomach.

The stomach is divided into three or four compartments depending on definition (Pilleri and Gihr, 1972; Li *et al.*, 1984). The forestomach is the largest, covered, as is the oesophagus, by stratified squamous epithelium without glands. At the antero-dorsal end this compartment is connected to the antero-ventral end of the main stomach through an opening about 1 cm in diameter in dead specimens. The main stomach is a thick-walled large oval compartment characterized by longitudinal folds lined with reddish brown epithelium. Two kinds of gland open into the main stomach, the cardiac glands, which are limited to a small area adjacent to the entrance from the forestomach, and the fundic glands in the remainder of the main stomach. The postero-dorsal wall of this compartment is connected to the posterior left wall of the pyloric stomach through a connecting canal 3.5 cm long and about 1.8 cm in diameter. A constriction of about 1.2 cm divides the thin-walled pyloric compartment into two segments that have pyloric glands in the mucous membrane. The posterior segment opens to the duodenal ampulla through a passage about 2 mm in diameter (in dead specimens).

The duodenal ampulla receives secretions via bile and pancreatic ducts. The finless porpoise has no caecum, and the distinction between small and large intestines is grossly unclear. The length of intestines is about 7–11.2 times the body length (Qian *et al.*, 1985).

The liver weighs 0.87–1.96 kg or about 2.4% of body weight. The pancreas is at the left of the duodenal ampulla. The spleen is spherical, about 2–3 cm in diameter, connected to the wall of the forestomach (Pilleri and Gihr, 1972), and weighs 4.0–6.6 g in adults (T. Kasuya, unpublished). Several accessory spleens may be present in some individuals (Pilleri and Gihr, 1972). The kidney weight was 362–737 g (0.70–0.89% of body weight in four males and 1.07% in a female), and the heart weight was 252–308 g (0.37–0.61% in four males and 0.45% in a female) in Japanese specimens measuring 150–169 cm (T. Kasuya, unpublished).

Nervous system

Mean brain weight of four finless porpoises from Pakistan (130–155 cm in body length) was 466.3 g or 1.5% of the body weight, which is almost the same as the values of the harbour porpoise (Pilleri and Gihr, 1972) and of the Japanese finless porpoise (Table 4). Among the cranial nerves, the largest are auditory nerves, next are trigeminals followed by optic and facial nerves. Olfactory nerves are absent, as in other delphinoids (Pilleri and Gihr, 1972). Wu (1989) counted 44 pairs of spinal nerves on each of three finless porpoises examined.

The diameter of the eye ball is about 20 mm. As with many cetaceans, the distended pupil is U-shaped. The half-opened pupil is a horizontal slit of a cocoon shape with a constriction at the middle. The retina has rod cells but no typical cones, which probably have changed into cells having a cone-like nucleus, cone-

TABLE 4 Organ and tissue weights of a finless porpoise incidentally killed in a fish net on 4 December 1977 in the Ise Bay, Pacific coast of Japan

Body length and sex	144.0 cm	male
Maximum girth (at umbilicus)	86.0 cm	
Thickness of blubber		
mid-dorsal	3.4 cm	
mid-lateral	2.2 cm	
mid-ventral	2.0 cm	
Body weight of intact carcass	44.0 kg	100 %
1. Muscle	14.96 kg	34.0%
2. Blubber[a]	15.65 kg	35.6%
3. Bone[b]	5.30 kg	12.1%
4. Brain	0.661 kg	1.5%
5. Total viscera	5.500 kg	12.5%
lung		
right	0.470 kg	1.1%
left	0.400 kg	0.9%
heart	0.244 kg	0.6%
liver	0.937 kg	2.1%
spleen	0.0113 kg	0.0%
kidneys	0.252 kg	0.6%
pancreas	0.0727 kg	0.2%
intestine and stomachs	2.65 kg	6.0%
testis		
right	0.0129 kg	0.0%
left	0.0119 kg	0.0%
Sum of the above items 1–5	42.071 kg	95.6%

[a] With tail flukes.
[b] With skull (0.66 kg) and flippers.

like inner segment and rod-like outer segment (Chen *et al.*, 1980). Chen *et al.* (1980) and Gao and Zhou (1986) have reported that, compared with their oceanic conspecifics, finless porpoises in the Yangtze River have eyes situated more anteriorly, with a smaller eye slit, a more rounded eye ball, and smaller optical nerves. Some of these differences are claimed to be the result of degeneration of vision in the turbid water, but they are based on small but statistically significant differences of mean values. Evaluation of the claimed difference will require further information on the growth stages of the porpoise specimens measured.

The external auditory meatus opens about 5 cm behind the eye and leads to the tympanic membrane. Five pairs of air sinuses are adjacent to the middle ear (Liao, 1978).

Liu (1985) found many encapsulated nerve endings in the tubercles on the back of the finless porpoise, and claimed that the area serves as a sensory organ. Similar receptor endings are found in the dermis of the skin of the upper and lower jaws (Liu *et al.*, 1986a). Several vibrissae are present on the upper jaw of the late foetus (Chen *et al.*, 1982) and neonate (Furuta *et al.*, 1976), but are lost in the adult.

Karyotype

The karyotype of the finless porpoise was studied using cultured bone marrow cells of a male (Peng and Chen, 1985) and fibroblast cells of a female (Zhang *et al.*, 1989) from the Yangtze River. The diploid number is 44, including a large X chromosome and a small Y chromosome. Although the number is the same, the morphology of individual chromosomes is different from those of *Phocoena phocoena* and *Phocoenoides dalli* (Peng and Chen, 1985).

Life History and Population Dynamics

Life history studies of this species are hampered mainly by the difficulty of obtaining unbiased samples of carcasses or obtaining cues for individual identification of live animals. The difficulty also relates to the possibility of numerous local populations which may exhibit different life history characteristics. Our current knowledge of their life history is reviewed below for several geographical areas.

Growth and reproduction

In the Yangtze River body lengths of the smallest neonate and the largest foetus ranged from 60.0 cm (Howell, 1927) to 73.5 cm (Chen *et al.*, 1982). The latter authors suggest from their limited data that calves will be born at 68–74 cm in the Yangtze River.

In the ISJ/Pacific area (i.e. the Inland Sea and habitats off the Pacific coast of Japan) the smallest neonate and the largest foetus ranged between 65.3 cm (Kasuya and Kureha, 1979) and 79.0 cm (Furuta *et al.* 1989). Body lengths of stranded neonates of unknown age ranged from 65.3 cm to 97.6 cm (mean = 78.6 cm, $n = 14$) (Kasuya and Kureha, 1979), and neonates born in a Japanese aquarium ranged from 77.5 cm to 83.0 cm (mean = 80.3 cm, $n = 4$) (Kasuya *et al.*, 1986). These data suggest that the average neonatal length is about 80 cm or slightly less, which is probably greater than in the Yangtze River.

Two populations of finless porpoise off western Kyushu are isolated from other habitats off Japan. Shirakihara *et al.* (1993) reported that the range of neonatal body lengths was 71.5–84.0 cm (mean = 78.2 cm, $n = 12$). This is not significantly different from figures from the ISJ/Pacific area.

A conception peak has been estimated from testicular activity and foetal body lengths, and suggests geographical variation in the seasonality of reproduction. In Pakistan testes are active in November and December (see below) and females have small foetuses (9.3–15.1 cm) in November (Harrison and McBrearty, 1973–74), suggesting that some of their conceptions occur in autumn and winter.

In the Yangtze River, Chen *et al.* (1982) concluded from the inactive testes in winter and several foetal records that mating occurs in June and July. Zhang

(1992) fitted an exponential foetal growth $y = 1.4713^x$ (y is the foetal body length in cm and x is the foetal age in months) to predict the conception date of 14 foetuses and a neonate, but this does not match a generally accepted pattern of mammalian foetal growth where body length or cube root of body weight increases linearly with age (Perrin and Reilly, 1984). Fitting a linear foetal growth to his 15 data points gives a least-squares regression $Y = 0.2973X - 73.21$, $r = 0.9$ (Y is the body length in cm and X is the number of days from 1 January). If forced to pass through these data points this equation cuts the time axis to give a 118 day period with a peak in August/September. A correction for slower growth for the early foetal stage will place the mating peak 1–2 months earlier (i.e. June–August), which is close to the mating season reported by Chen *et al.* (1982).

Finless porpoises from the Ise Bay kept in a nearby aquarium copulated frequently in February–June, and one female in near-term pregnancy also mated frequently in February preceding her parturition in the following April of that year (Izawa and Kataoka, 1965). This is similar to reports from the Bohai coast, northern China, where finless porpoises are believed to breed in spring (Wang, 1984).

Shirakihara *et al.* (1993), assuming a gestation period of about 11 months (Kasuya *et al.*, 1986) and using their estimated parturition season, predicted a mating season in autumn to spring for the finless porpoise off western Kyushu, Japan. The peak will be in December/January. This is between the apparent mating season for the Yangtze River and the Pacific coast of Japan populations.

The above regression of foetal body lengths on data calculated for the Yangtze population predicts the 14 foetuses and one neonate to reach the neonatal body length of 70 cm (Chen *et al.*, 1982) in March–June, with a peak in April/May. If neonatal length is assumed at 80 cm as for Japanese populations, the parturition peak will be about one month later.

Kataoka and Motomura (1966) claimed that parturition is in April and May in the Ise Bay, Pacific coast of Japan, based on capture of three newborn calves and one aquarium birth. Kasuya and Kureha (1979) studied the incidence of mother/calf pairs in the ISJ. The density of mother/calf pairs was negligible in October to March but increased in April, preceded by the increase of porpoise density. The proportion of mother/calf pairs in the total sightings continued to increase until September, whereas the population density declined after May. They found also that strandings of neonates occurred in the ISJ/Pacific area mostly in April, with some in May and June. Finally, Kasuya and Kureha (1979) concluded that finless porpoises aggregate in the Inland Sea for breeding and parturition in spring, after which most of the mother/calf pairs and other individuals leave the area. Thus parturitions in the ISJ/Pacific area occur mostly in April, with some delayed ones in May and June.

Shirakihara *et al.* (1993) compared parturition season of Japanese finless porpoises between western Kyushu and the ISJ/Pacific area using two methods, i.e. (1) season of neonates and (2) prediction using foetal body lengths and a foetal growth rate of 0.277 cm per day (Kasuya, 1977). For the ISJ/Pacific area parturitions were estimated to occur in March–August with a peak in April, which supported earlier studies. However, the season for western Kyushu was estimated from August to April, with a distinct peak in November/December and another indistinct smaller peak in March.

The above information on the season of mating and parturition suggests that peaks of mating and parturition of finless porpoises are more diffused and probably later in the southern habitat. The northern populations probably adapted for calves to switch their major source of nutrition from milk to solid food during the most productive summer months by placing the parturitions as early as the newborn calves can endure the cold climate in the spring.

A gestation of about 11 months is estimated from a female that mated frequently in April and May and had a calf in the middle of the next April (Furuta *et al.*, 1977). Although these data contain an uncertainty on the length of receptive period before and after conception, the suggested gestation period is close to the 10.6–11.2 months estimated by Kasuya *et al.* (1986) using range of mean neonatal body length, 75–85 cm, and the interspecies relationship $\log Y = 0.4586 \log X + 0.1659$, between mean neonatal length (X in cm) and gestation length (Y in months) (Perrin *et al.*, 1977). One parturition observed in the Toba Aquarium was caudal presentation (Furuta *et al.*, 1976).

Zhang (1992) thought that finless porpoises start taking solid food by about 0.5 year of age based on a juvenile of 96 cm killed in winter in the Yangtze River and found with remains of solid food in the stomach. A calf born in the Toba Aquarium erupted teeth in 3 months, tried to take solid food at 4 months, and probably started to swallow food before 6 months. This individual was observed to suckle the mother until the age of 18 months, when the mother died and lactation was confirmed (M. Furuta, personal communication). Suckling will last long if a mother is kept with her calf and the mother does not conceive another calf. Kasuya and Kureha (1979) observed a decline of mother/calf pairs during October and November accompanied by an increase in solitary calves, and deduced that the majority of calves born in spring are weaned in the following October and November, although some calves may be nursed until the next mating season. This implies a suckling period of about 7 months with a possible range of 6–15 months.

Zhang (1992) reported on 16 females taken in the Dongting Lake in December, of which seven were sexually mature and five were pregnant. Kasuya and Kureha (1979) examined five adult females stranded or incidentally killed in the ISJ/Pacific area, of which one was pregnant (not lactating) with a near term foetus in April. Three were caught in June and had large follicles or a corpus luteum of ovulation and were neither pregnant nor lactating. Another female was pregnant and simultaneously lactating in November. These observations suggest that conceptions in two successive years are uncommon. About 11 months gestation and a uniformly short nursing period suggest that a two-year cycle will be usual (Kasuya and Kureha, 1979), but further study is needed to arrive at a firm conclusion because the sample composition reflects mortality which may be dependent on age and reproductive status.

Male growth and longevity

In the Indian Ocean, one 130 cm male caught in November off Pakistan had testes that weighed 78.3 g with epididymis. No free spermatozoa were present in the testis or epididymis, although spermatogonia and spermatids were dividing actively. Testes of a 150 cm male, killed in December weighed 603 g including epididymis. Spermatocytes were actively dividing, and some free spermatozoa were present in the tubules and epididymis. Harrison and McBrearty (1973–74) considered that the first animal was immature and in the process of entering its first active reproductive season, and that the second one probably was mature and killed during the breeding season. They concluded that male finless porpoises in the region will attain sexual maturity at 140–150 cm. The largest male known from the Indian Ocean is 150 cm and smaller than the largest female. This is probably due to the small sample size and does not necessarily indicate that females are larger.

In the Yangtze River, Chen *et al.* (1982) observed an increase in the combined weight of both testes and epididymides from below 35 g of immature males (104–155 cm, $n = 4$) to 277–441 g of mature males (156–164 cm, $n = 6$). Zhang (1992), using eight males (0.5–13 years, see below for aging) in the Yangtze River and 16 males (0.1–11 years) in the Yellow Sea/Bohai area, indicated the combined weight of both testes to increase from below 86.5 g (≤ 3 years) to about 200 g or over (≥ 4.5 years), and concluded that sexual maturity will occur in both areas at 4.5 years.

For the Yangtze population Zhang (1992) showed that body length increased from ≤ 121 cm of immature males ($n = 5$) to 152–168 cm of mature males ($n = 3$). If this information is combined with that of Chen *et al.* (1982), both mature and immature males occur in the body length range of 152–155 cm. The largest male measured in the Yangtze River was 168 cm.

Body lengths of 27 males from the South China Sea at about 23°N ranged from 118 cm to 168 cm (Gao, 1991).

In the Yellow Sea/Bohai area, body length was ≤ 140 cm for immature males ($n = 12$) and 147–201 cm for mature males ($n = 4$) (Zhang, 1992). This suggests sexual maturity at body lengths around 140–150 cm.

Shirakihara *et al.* (1993) examined sexual maturity in 50 males obtained from incidental takes and strandings off western Kyushu, 36 of which had no spermatozoa in the testis and were < 4.0 years from birth and ≤ 138 cm in body length (excluding one anomaly of a 128 cm male at 2.0–3.0 years of age that had a mature testis of 58 g) while the remaining 13 males had spermatozoa in every seminiferous tubule examined, and were ≥ 6.0 years and ≥ 139 cm in body length. The sample had a deficiency of males between 4 years and 6 years of age, but showed no distinct body length gap. The weight of a single testis and the mean diameter of the seminiferous tubules were in the ranges 1.7–39.2 g and 43–73 μm in mature males, respectively, while corresponding figures for the mature males were 58–863 g and 109–242 μm, respectively. The testis weight increased rapidly from 40 g to 150 g. These authors considered sexual maturity to occur in males at 4–5 years from birth and 135–140 cm in body length. The absence of pubertal males in the sample leaves uncertainty in the estimation. The authors believed that mean male growth was best expressed by $L = 165.5–157.2/(t + 1.683)$, where L is the body length (in cm) at t years from birth. The largest male in their sample was 175 cm and the oldest one was 23 years.

Two aquarium-reared males from the ISJ/Pacific area were thought to approach maturity at 160 cm and at an age of 4–5 years (Furuta, M., personal communication). The weight of a single testis without epididymis, or mean of both if available, ranged from 6 g to 886 g (114.5–169 cm in body length, $n = 10$), and showed a rapid increase from 13 g to 120 g at body lengths of 144–157 cm (T. Kasuya, unpublished). The smallest male with spermatozoa had a testis of 13 g and was measured at 144 cm, and the corresponding figures for the largest male without spermatozoa were 34 g and 155 cm, respectively. These data suggest that most animals are sexually mature at a body length of 145–155 cm, and at a single testis weight between 10 g and 100 g.

Annual deposition of growth layers in dentine and cementum has been compared between aquarium reared known-age individuals and specimens from the wild (Kasuya *et al.*, 1986).

The asymptotic body length of toothed whales may fluctuate with availability of resources (e.g. Kasuya, 1991), and the body length of the largest individual in a sample is to some degree a function of the sample size, i.e. a larger sample will have a higher probability of finding a greater body size. However, the body lengths of the largest males in the above studies suggest a significant geographical variation in their growth. It increases from the smallest in the Indian Ocean (150 cm) to the largest in the Yellow Sea/Bohai area (201 cm), and those in the Yangtze River (168 cm), South China Sea (168 cm), western Kyushu (175 cm) and ISJ/Pacific area (192 cm) are between the extremities. This seems to correlate with latitudes of the habitat, suggesting that the apparent clinal change might be due to differences in environmental temperature.

Female growth and longevity

Referring to the body length of six sexually mature females off India and Pakistan between 114 cm and 155 cm, Harrison and McBrearty (1973–74) considered that females attain sexual maturity at about 120 cm. They listed an additional pregnant female, 130 cm, from the Gulf of Siam.

From the Yangtze River, Chen *et al.* (1982) reported seven sexually mature females. The smallest was 130 cm in body length (corpora count not available), and the other three (136, 140 and 141 cm) were pregnant at their first ovulation. Zhang (1992) analysed 20 females from the Yangtze River. The body lengths increased from $\leq$ 126 cm for immature females ($\leq$ 3 years of age, $n = 10$) to 134–151 cm for mature ones (4.5 to $\geq$ 16 years, $n = 8$). These suggest that females in the Yangtze River attain sexual maturity at 3–4 years from birth and body lengths of 130–140 cm. The largest female obtained in the Yangtze River was 151 cm.

Body lengths of 19 females in the South China Sea at about 23°N ranged from 78 cm to 171 cm (Gao, 1991).

From the Yellow Sea/Bohai area, Zhang (1992) presented eight females. The oldest immature female was 4.0 years (no body length) and the youngest mature one was 6.0 years (156 cm), which suggests age at sexual maturity is 4–6 years. Females aged 4 years or over were greater than 141 cm ($n = 5$), suggesting a body length at sexual maturity of 140–155 cm. The largest female in the sample was 200 cm in body length.

Shirakihara *et al.* (1993) examined sexual maturity in 39 females from western Kyushu, Japan, of which 32 (< 5.0 years from birth and $\leq$ 134 cm in body length) were immature and seven ($\geq$ 7.0 years and $\geq$ 135 cm) were mature. There was a deficiency of 5.0–7.0 years old and 135–144 cm long females. The

TABLE 5 Various views and data on the growth of finless porpoises for major geographical areas. Precision is variable between data, and narrower range of some parameters does not necessarily indicate higher precision. For further details see text

Area	Indian Ocean	Yangtze	Yellow Sea/Bohai	ISJ/ Pacific	West. Kyushu
Average neonatal length	—	68–74	—	79	78
Largest known fetus (> 50 cm)	77	74	—	79	57
Smallest known neonate (< 100 cm)	98	60	—	65	72
Body length at maturation					
male	140–50	150–55	140–50	145–55	135–40
female	120	130–40	140–55	≤ 140	135–45
Average asymptotic body length					
male	—	—	—	160–69	166
female	—	—	—	160–69	158
Largest known individual					
male	150	168	201	192	175
female	155	151	200	180	165
Mating peak	fall/ winter	June/ Aug.	Spring	Spring	Dec./ Jan.
Parturition peak	—	?Apr./ May	—	April	Nov./ Dec.
Age at sexual maturity					
male	—	4.5	4.5	3–9	4–6
female	—	3–4	4–6	≤ 4	5–7

youngest mature female (which was 135 cm and aged between 7.0 years and 8.0 years) was neither pregnant nor lactating and had one corpus luteum and one corpus albicans in the ovaries. Another female of 151 cm and aged between 9.0 years and 10.0 years was pregnant with a 52 cm foetus and had one corpus luteum. Five females at 146–149 cm (9–24 years) had ovulated for 4–14 times. These data suggest that females of this population will usually attain sexual maturity between ages 5.0 years and 9.0 years and at body lengths of 135–145 cm, although the authors gave a slightly different interpretation (6.0–10.0 years from birth). The mean female growth curve was best approximated by $L = 157.7–118.1/(t + 1.624)$, where L indicates body length in cm at age t years from birth. The largest female was 165 cm and the oldest one was 23 years.

From the ISJ/Pacific area, Shirakihara *et al.* (1993) reported a female born in an aquarium that calved in her 3rd year of life. This indicates an ability to ovulate under favourable conditions at about 24 months of age. The body length of the largest individual among five immature ones was 118 cm (about 1 year old) and the smallest individual in eight sexually mature ones was 141 cm (20 years) (T. Kasuya, unpublished). The ages of the oldest immature and the youngest mature females were 1 year and 11 years, respectively, and the body lengths of four females of unknown maturity (4–10 years old) were larger than the smallest mature female (Fig. 5), suggesting that females in these waters will attain sexual maturity before 4 years of age and a body length of about 140 cm. The maximum

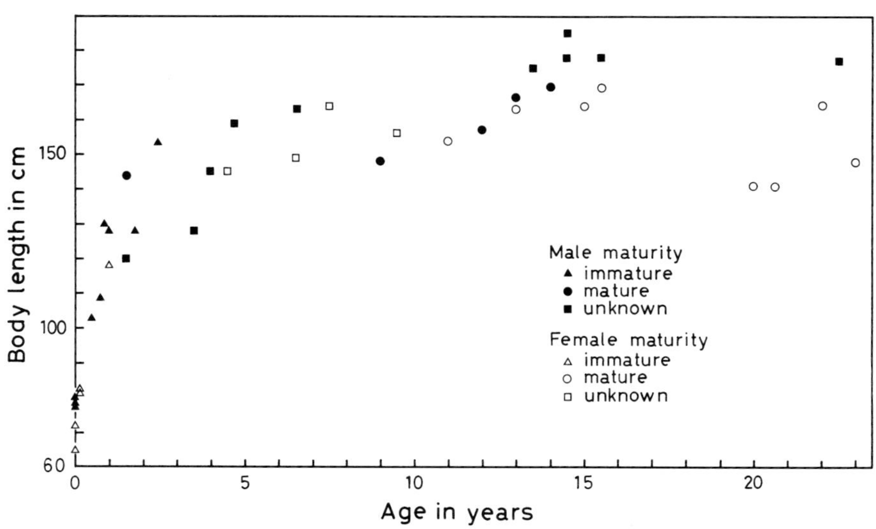

FIG. 5 Scatter plot of body length with age for a finless porpoise from the Inland Sea and Ise Bay (Japan). Age is estimated assuming annual deposition of tooth layers or based on direct observation of four males (0, 0.01, 2.4, and 4.7 years) and two females (0.05 and 0.08 year) born in the Toba Aquarium (Kasuya *et al.*, 1986). Body lengths are smaller in the more southern habitats such as the Indian Ocean.

number of corpora lutea and albicantia in the ovaries was 11 (no age), which was followed by 8 (20 years) and 6 (22 years).

The fusion of vertebral epiphyses with the centrum was examined on four sexually mature females in the ISJ/Pacific area (T. Kasuya, unpublished). One 154 cm (11 years old) female was physically immature, one 164 cm (15 years old) female was close to maturity having free epiphyses only on the 7th thoracic vertebra, and 141 cm (20 years old) and 162 cm (23 years old) females were physically mature. Thus, finless porpoises in these waters will attain physical maturity at wide ranges of body length over 140 cm and ages over 15 years. The average body length at physical maturity, lower bound of 160–169 cm, is presumed from Fig. 5 and a body length frequency of 37 females that had a peak of 10 individuals at 160–169 cm (Iwakawa, 1986; T. Kasuya and M. Furuta, unpublished). The largest female was 180 cm in the ISJ/Pacific area (Iwakawa, 1986; T. Kasuya, unpublished).

The maximum body length of the females reviewed above changes from the smallest in the Indian Ocean (155 cm) to the largest in the Yellow Sea/Bohai area (200 cm), with intermediate values in the Yangtze River (151 cm), South China Sea (171 cm), off western Kyushu (165 cm) and in the ISJ/Pacific area (180 cm). The order of the clinal change is similar to that observed in males. The larger body size increases the capacity for energy reserves, and probably is an adaptation to life in a colder and seasonally variable environment.

Behaviour

Social organization

In the Yangtze River, mothers were observed to carry the calves on their back. The calf clung to the mother's back with its flippers. This carrying behaviour was most common in April and May, the main calving season in the area (Chen *et al.*, 1979; Pilleri and Chen, 1979; Liu *et al.*, 1986b). This behaviour was not recorded during the extended sighting survey in the ISJ (Kasuya and Kureha, 1979), but possibly similar behaviour was observed with aquarium reared individuals from the Ise Bay (M. Furuta, personal communication). In the Toba Aquarium calves usually were carried by the mother while submerged as if the calf were riding the mother's "bow wave", or carried by the mother's trailing turbulence. The calf was usually apart from the mother by 5–10 cm, and positioned near the lumbar region and slightly off the midline of mother's body. Occasionally the calf touched the mother's body with its flipper or moved under the mother's body. In most of the cases when they breathed, the calf ascended first and surfaced slightly earlier than the mother, giving the appearance that the calf was being pushed up by the mother. However, it was uncommon that their bodies touched at surfacing. Further study is needed to confirm whether or not this behaviour is similar to that observed in the Yangtze River.

In the Indus delta, sightings were of one or two individuals, and groups exceeding four were not observed (Pilleri and Gihr, 1973–74). In the Yangtze River, school size was usually from two (Zhou *et al.*, 1979) or three to five individuals, with exceptional cases of up to 20 individuals (Chen *et al.*, 1979). Zhou *et al.* (1979) considered that a group of two was the basic unit.

In the ISJ, the school sizes ranged from 1 to 13 individuals (mean = 1.97, $n = 709$): 48% of them were solitary, 34% were groups of two animals, and schools of three individuals were 8%. The majority (87%) of single individuals were of adult size. Most of the schools of two individuals were represented by two adults (55%) or by a juvenile and an adult (37%), and those of three individuals by three adults (53%) or by two adults and one juvenile (40%). Solitary adults were rare in April–September and mother/calf pairs increased from March to September, reflecting the seasonality of reproduction.

These data suggest that the basic unit of a finless porpoise school is a mother/calf pair or two adults, and that schools of three or more individuals are aggregations of these units or of solitary individuals. Social structure seems to be undeveloped in the species, and the mother/calf pair is probably the only stable social unit.

Feeding

Finless porpoises are reported to eat fish and shrimps in the Yangtze River (Chen *et al.*, 1979; Wang, 1984), and fish, shrimps and squids in the Yellow Sea/Bohai area (Wang, 1984) and off Pakistan (Pilleri and Gihr, 1973–74). In Japanese waters they are known to eat fish, shrimps and squids (Mizue *et al.*, 1965; Kataoka *et al.*, 1976) or fish, shrimps, squids, cuttle fish and octopus (Shirakihara and Takemura, 1993). Finless porpoises are opportunistic feeders utilizing various kinds of available food items in the habitat. Seasonal changes in the food diet have not been studied.

The food consumption of two finless porpoises in the Toba Aquarium varied, as would be expected, with seasonal variations in water temperature. Kataoka *et al.* (1967) estimated average daily food consumption to be 5.2–5.8% of body weight, based on an assumed body weight of 60 kg. The food consumption estimates approximate those for other captive odontocetes such as *Lagenorhynchus obliquidens* and *Delphinapterus leucas* (Sergeant, 1969; Van Dyke and Ridgway, 1977).

Swimming and diving

The finless porpoise rolls slowly on the surface, exposing only part of the head and back. Jumping clear of the water or splashing is rare (Zhou *et al.*, 1979; Kasuya and Kureha, 1979). Although finless porpoises occasionally chase fast moving vessels and ride the stern wave, they have not been observed riding a ship's "bow wave" in the ISJ (Kasuya and Kureha, 1979). They have been reported to stay away from vessels in the Indian Ocean (Pilleri and Gihr, 1973–74). If they encounter moving vessels they often dive under the vessels (Zhou *et al.*, 1979; Kasuya and Kureha, 1979), which may result in an occasional collision with the vessel. An aquarium specimen from the Yangtze River has been reported to swim on its side near the bottom of the tank (Liu *et al.*, 1986b).

The dive time is usually around 20 sec with a range of 1–70 sec (Pilleri and Gihr, 1973–74; Chen *et al.*, 1979; Zhou *et al.*, 1979). The mean of 62 dives was 17.5 sec (Zhou *et al.*, 1979).

Aggregations appear to relate to the availability of food. They aggregate near the confluence of the Yangtze River with its tributaries (Chen *et al.*, 1979), or near points and straits in the ISJ (Kasuya and Kureha, 1979).

Sound production

Finless porpoises produce no whistles. Clicks contain dominant frequencies of about 128 kHz, presumably for echolocation, although the click may have low-frequency components of about 2 kHz (Mizue *et al.*, 1968; Pilleri *et al.*, 1980; Kamminga *et al.*, 1986, 1996). The frequency bandwidth is 15 kHz for longer pulses and 35 kHz for the shortest ones; the individual clicks last for 40 to 80 μsec and the number of clicks in a pulse train ranges from 15 to over 300, with a pulse repetition rate of 40–120 Hz (Kamminga *et al.*, 1986). The general characteristics of the clicks are very similar to those of *Phocoenoides dalli, Phocoena phocoena* and *Cephalorhynchus commersonii* (Kamminga *et al.*, 1986).

Predation

A great white shark, *Carcharodon carcharias*, taken by a local longline fishery in the East China Sea off Okinawa, was found with an almost complete finless porpoise of about 180 cm (bitten into three pieces) and a more digested small individual in the stomach (S. Uchida, personal communication).

Parasites and Disease

Ten parasites have been reported from finless porpoises (Neiland *et al.*, 1970; Dailey and Brownell, 1972; Petter and Pilleri, 1982; Tao, 1983): four trematodes *Campula folium, Nasitrema spathulatum, N. sunameri,* and *Orthosplanchus elongatus*; five nematodes *Crassicauda fuelleborni, Holocercus pingi, Stenurus auditivus, S. nanjingensis* and *Pharurus asiaeorientalis*; and one cestode *Dyphyllobothrium furhmanni*. Tapeworms are common in the intestine of finless porpoise in the Inland Sea of Japan (T. Kasuya, unpublished).

Chen *et al.* (1982) reported a female of 147 cm from the Yangtze River, which had a tumour (fibroma) of 2.57 g on the dorsal wall of the upper part of the vagina.

Live Maintenance

These porpoises have been kept and bred at Japanese aquaria. Their breeding, feeding, parturition, lactation and other life habits have been observed and documented as mentioned under various sections elsewhere in this chapter. More recently, animals from the Yangtze River have been kept in China.

Human Effects

Direct fisheries

Remains of finless porpoises are rare in Japanese archaeological sites (< 10 000 years BP) that produce large amounts of other temperate delphinoid remains (Kasuya *et al.*, 1985). There is some controversy about the usefulness of this species for human consumption. People in Ayukawa on the Oshika Peninsula at the northern limit of this species off the Pacific coast do not eat them, believing that they have a strong purgative effect. An experiment by Mr N. Kimura, then Director of Oshika Whale Museum, has shown that fat has to be carefully removed during cooking to avoid the effect (Kasuya *et al.*, 1985). However, along the west coast of Kyushu, Japan, finless porpoises were incidentally taken during autumn and winter seasons by trap nets, and some of them were sent to fish markets for human consumption (Mizue *et al.*, 1965). Only one of five persons who volunteered to eat cooked meat and blubber experienced a slight purgative effect (M. Shirakihara, personal communication).

By-catches

The type of incidental takes reported by Mizue *et al.* (1965) seem to have ceased due to discontinuation of the particular trap net fishery (Kasuya and Kureha, 1979). Shirakihara *et al.* (1993) recorded 114 specimens in their study, which were collected during 1985–1992 off the coasts of western and north-eastern Kyushu including part of the western ISJ: 84 of them were killed incidentally by fisheries, 25 were found dead on the beach or in the sea, and five were of unknown origin. Fisheries that killed the 84 porpoises were bottom gill net (58), surface gill net (17), trap net (7), trawl net (1) and drifting [discarded] net (1). The operation of such fishing gear is common in other parts of Japan and probably is killing finless porpoises in other Japanese habitats, although usually such catches remain unreported. Live captures have been reported from Ise Bay (Kasuya *et al.*, 1984).

Angling fishermen around the Awashima Island, ISJ, used the finless porpoise as an indicator of their target fish, *Chrysophrys major* (Kasuya and Kureha, 1979). Since around 1989 the species has been protected through regulation by the Prefecture Government of the hand harpoon fishery, followed by Fisheries Agency regulation of the catch throughout Japan.

Although finless porpoises are legally protected in China (Wang, 1984), there are some incidental catches by coastal and riverine fisheries (Zhang, 1986). Allen (1923) reported that Chinese took finless porpoises "by hanging a number of iron hooks from a stout cord, as a sort of barrage across part of the stream (Yangtze River)". Almost identical fishing gear, rolling hooks, are still in use for fishing and are causing a threat to baiji, *Lipotes vexillifer* (Lin *et al.*, 1985) and possibly to finless porpoises.

Around the Kerala coast finless porpoises are incidentally killed by monofilament gill net together with larger numbers of other dolphins (Lal Mohan, 1985). A considerable number of finless porpoises are incidentally caught in Pakistan in a net fishery of undescribed type, and the carcasses are discarded (Pilleri and Pilleri, 1979). Pilleri and Gihr (1973–74) reported on the incidence of this species in the drift nets off Phangnga, west of the Malay Peninsula. Incidental take in coastal gill net fisheries causes considerable damage to local stocks of finless porpoises. Jefferson and Curry (1994) presented a global review of gill net mortality.

Pollution

Blubber of finless porpoises from the ISJ/Pacific area contains DDT isomers and metabolites at a level of 12–132 ppm in females and of 84–132 ppm in males. The concentration of PCBs is 18–88 ppm in females and 64–96 ppm in males (O'Shea *et al.*, 1980). These are up to 10 times the concentration in the striped dolphin, *Stenella coeruleoalba*, off the Pacific coast of Japan analysed by the same authors, and are similar to the level in non-pregnant females of Baltic ringed seals with stenosis and occlusion of uteri reported by Helle *et al.* (1976). Although such pathology has not been reported in the finless porpoise, pollution is very possibly a threat to the species. Yang *et al.* (1988) analysed heavy metals, DDT and PCBs in some tissues of male finless porpoises from the lower Yangtze River. Iwata *et al.* (1994, 1995) reported high concentrations of toxic butyltins in this species from Japanese waters.

Acknowledgments

I am grateful to Dr R. L. Brownell, Jr, Mr M. Furuta, Dr G. Pilleri, Dr X. Zhang and an anonymous reviewer for their comments, and to Dr N. Miyazaki for his assistance in the photography of the skull.

References

Allen, G. M. (1923). The black finless porpoise, *Neomeris*. *Bull. Mus. Comp. Zool.* **65**, 233–256.

Amano, M., Miyazaki, N. and Kureha, K. (1992). A morphological comparison of skulls of the finless porpoise *Neophocaena phocaenoides* from the Indian Ocean, Yangtze River and Japanese waters. *J. Mamm. Soc. Japan* **17**, 59–69.

Anon. (1980). "Information relating to CITES". Fisheries Agency, Tokyo, 177 pp.

Arvy, L. and Pilleri, G. (1972). Comparison of the tongues of some odontocetes: *Pontoporia*, *Neomeris* and *Delphinus*. *Invest. Cetacea* **4**, 191–199.

Baldwin, R. and Salm, R. (1994). "Whales and Dolphins along the Coast of Oman". Published by R. Baldwin and distributed by Natural History Book Service, Totnes, 65 pp.

Barnes, L. G. (1985). Evolution, taxonomy and antitropical distribution of the porpoises (Phocoenidae, Mammalia). *Mar. Mamm. Sci.* **1**, 149–165.

van Bree, P. J. H. (1973). *Neophocaena phocaenoides asiaeorientalis* (Pilleri and Gihr, 1973), a synonym of the preoccupied name *Delphinus melas* Schlegel, 1841. (Notes on cetacea, delphinoidea VII). *Beaufortia* **21**, 17–24.

Chen, P., Liu, P., Liu, R. and Lin, K. (1979). Distribution, ecology and conservation of the dolphins of the middle reaches of Chanjiang (Yangtze) River (Wuhan-Yueyang). *Invest. Cetacea* **10**, 87–104.

Chen, P., Shao, Z. and Pilleri, G. (1980). Regression of optic system in the Changjiang (Yangtze) finless porpoise (*Neophocaena phocaenoides*) as a result of lack of light. *Invest. Cetacea* **15**, 115–120.

Chen, P., Liu, R. and Harrison, R. J. (1982). Reproduction and reproductive organs in *Neophocaena asiaeorientalis* from the Yangtze River. *Aquat. Mamm.* **9**, 9–16.

Dailey, M. D. and Brownell Jr, R. L. (1972). A checklist of marine mammal parasites. *In* "Mammals of the Sea, Biology and Medicine" (Ed. S. H. Ridgway) pp. 528–589. Charles C. Thomas, Springfield, IL.

Furuta, M., Tsukada, O., Kataoka, T. and Kitamura, S. (1976). On the birth of finless porpoise, *Neophocaena phocaenoides*, in captivity. *J. Jap. Assn Zool. Gardens Aquariums* **18**, 43–48 (in Japanese with English summary).

Furuta, M., Tsukada, O. and Kataoka, T. (1977). On the birth of finless porpoise *Neophocaena phocaenoides* in captivity. *In* "Report on the Rearing and Behavior of Finless Porpoise (*Neophocaena phocaenoides*) in Toba Aquarium", pp. 1–12. Toba Aquarium, Toba (in Japanese with English summary).

Furuta, M., Kataoka, T., Sekido, M., Yamamoto, K., Tsukada, O. and Yamashita, T. (1989). Growth of the finless porpoise, *Neophocaena phocaenoides* (G. Cuvier, 1829), from the Ise Bay, Central Japan. *Ann. Rep. Toba Aquarium* **1**, 89–102.

Gao, A. (1991). Morphological differences and genetic variations among the populations of *Neophocaena phocaenoides*. Doctoral thesis, Nanjin Normal College, 116 pp (in Chinese with English summary).

Gao, A. and Zhou, K. (1986). Anatomical and histological studies of the eyes of the finless porpoise, *Neophocaena phocaenoides*. *Acta Zool. Sinica* **32**, 248–254.

Gao, G. and Zhou, K. (1989). Anatomy of the nasal passage and associated structures of *Neophocaena phocaenoides*. *Acta Theriol. Sinica* **9**, 275–280 (in Chinese with English summary).

Gaskin, D. E. (1982). "The Ecology of Whales and Dolphins". Heinemann, London, 459 pp.

Gray, J. E. (1825). An outline of an attempt to the disposition of Mammalia into tribes and families, with a list of the genera apparently appertaining to each tribe. *Annals of Philosophy* **26**, 337–377 (cited in Barnes, 1985).

Gray, J. E. (1846). "The Zoology of the Voyages of H.M.S. Erebus and Terror", Vol. 1, "Mammalia". E. W. Janson, London, 53 pp.

Hammond, D. D. and Leatherwood, S. (1984). Cetaceans live-captured for Ocean Park, Hong Kong, April 1974–February 1983. *Rep. Int. Whal. Commn* **34**, 491–495.

Harrison, R. J. and McBrearty, D. A. (1973–74). Reproduction and gonads of the black finless porpoise, *Neophocaena phocaenoides*. *Invest. Cetacea* **5**, 225–230.

Harwood, M. B., McNamara, K. J., Anderson, G. R. V and Walter, D. G. (1984). Incidental catch of small cetaceans in a gillnet fishery in northern Australian waters. *Rep. Int. Whal. Commn* **34**, 555–559.

Helle, E., Olsson, M. and Jensen, S. (1976). PCB levels correlated with pathological changes in seal uteri. *Ambio* **5**, 261–263.

Howell, A. B. (1927). Contribution to the anatomy of the Chinese finless porpoise *Neomeris phocaenoides*. *Proc. U.S. Natl Mus.* **70**, 1–43.

Hsu, F., Kou, J., Lu, E. and Dong, Z. (1973). The external form and skeleton of *Neomeris phocaenoides*. *Acta Zool. Sinica* **19**, 104–111 (in Chinese with English summary).

Iwakawa, K. (1986). Strandings and incidental takes of cetaceans on the coast of Chita Peninsula. *Echolocation* **6**, 3 (in Japanese).

Iwata, H., Tanabe, S., Miyazaki, N. and Tatsukawa, R. (1994). Detection of butyltin compound residues in the blubber of marine mammals. *Mar. Pollut. Bull.* **28**, 607–612.

Iwata, H., Tanabe, S., Mizuno, T. and Tatsukawa, R. (1995). High accumulation of toxic butyltins in marine mammals from Japanese coastal waters. *Environ. Sci. Technol.* **29**, 2959–2962.

Izawa, K. and Kataoka, T. (1965). On the rearing and ecological observation of the finless black porpoise (*Neophocaena phocaenoides*). *J. Jap. Assn Zool. Gardens Aquariums* **7**, 56–62 (in Japanese with English summary).

Jefferson, T. A. and Curry, B. E. (1994). A global review of porpoise (Cetacea: Phocoenidae) mortality in gillnets. *Biol. Conserv.* **67**, 167–183.

Kamminga, C., Kataoka, T. and Engelsma, F. J. (1986). Investigations on cetacean sonar VII, underwater sounds of *Neophocaena phocaenoides* of the Japanese coastal population. *Aquat. Mamm.* **12**, 52–60.

Kamminga, C., Stuart, A. C. and Silber, G. K. (1996). Investigations on cetacean sonar: XI. Intrinsic comparison of the wave shapes of some members of the Phocoenidae family *Aquat. Mamm.* **22**, 45–55.

Kasuya, T. (1977). Age determination and growth of the Baird's beaked whale with a comment on the fetal growth rate. *Sci. Rep. Whales Res. Inst. Tokyo*, **29**, 1–20.

Kasuya, T. (1991). Density dependent growth in the North Pacific Sperm whales. *Mar. Mamm. Sci.* **7**, 230–257.

Kasuya, T., Kaneko, H. and Nishimoto, Y. (1985). Archaeology and related sciences, 7, zoology. *Archaeology* **11**, 91–95 (in Japanese).

Kasuya, T. and Kureha, K. (1979). The population of finless porpoise in the Inland Sea of Japan. *Sci. Rep. Whales Res. Inst. Tokyo* **31**, 1–44.

Kasuya, T., Tobayama, T. and Matsui, S. (1984). Review of the live capture of small cetaceans in Japan. *Rep. Int. Whal. Commn* **34**, 597–602.

Kasuya, T., Tobayama, T., Saiga, T. and Kataoka, T. (1986). Perinatal growth of delphinoids: information from aquarium reared bottlenose dolphins and finless porpoises. *Sci. Rep. Whales Res. Inst. Tokyo* **37**, 85–97.

Kataoka, T. and Motomura, Y. (1966). A trial on the hand rearing of an infant finless black porpoise, *Neophocaena phocaenoides*. *J. Jap. Assn Zool. Gardens Aquariums* **8**, 82–85 (in Japanese with English summary).

Kataoka, T., Motomura, Y., Kitamura, S. and Yamamoto, K. (1967). On the food quantity of the finless black porpoises (*Neophocaena phocaenoides*) in captivity. *J. Jap. Assn. Zool. Gardens Aquariums* **9**, 546–50 (in Japanese with English summary).

Kataoka, T., Furuta, M. and Kitamura, S. (1974). On the feeding to newly captured finless black porpoise, *Neophocaena phocaenoides*. *J. Jap. Assn Zool. Gardens Aquariums* **16**, 1–5 (in Japanese with English summary).

Kataoka, T., Kitamura, S., Sekido, M. and Yamamoto, K. (1976). On the feeding habit of finless porpoise (*Neophocaena phocaenoides*). *Mie Seibutsu* **24/25**, 29–36 (in Japanese with English summary).

Kukenthal, P. W. (1893). "Vergleichend-anatomische und entwicklungsgeschichtliche Untersuchungen an Walthieren", Part 2. Verlag von Gustav Fisher, Jena, 448 pp.

Lal Mohan, R. S. (1985). Observation of the by-catch of dolphins *Stenella longirostris, Tursiops aduncus, Sousa chinensis* and *Delphinus delphis tropicalis* in the gill nets off Culicut coast, India. Paper presented to the Symposium on Endangered Marine Animals in Marine Parks, Cochin, India, 12–16 January.

Li, Y., Quian, W., Shen, H. and Chen, Y. (1984). The digestive tract of the finless porpoise (*Neophocaena asiaerorientalis*) I. Tongue, oesophagus and stomach. *Acta Theriol. Sinica* **4**, 257–264 (in Chinese with English summary).

Liao, C. (1978). Anatomy of the auditory organ of the black finless porpoise, *Neophocaena phocaenoides*. *Acta Zool. Sinica* **24**, 278–288 (in Chinese with English summary).

Lin, K., Chen, P. and Hua, Y. (1985). Population size and conservation of *Lipotes vexillifer*. *Acta Ecol. Sinica* **5**, 77–85 (in Chinese with English summary, English translation is available in C. H. Perrin and W. F. Perrin (1986), US Administrative Report LJ-86-27, 14 pp).

Liu, R. (1985). The ultrastructure and function of the tubercle on the back of *Neophocaena phocaenoides* in the Changjiang River in China. *Acta Hydrobiol. Sinica* **9**, 209–212 (in Chinese with English summary).

Liu, R., Harrison, R. J. and Thurley, K. W. (1986a). Characteristics of the skin of *Neophocaena phocaenoides* from the Changjiang (Yangtse River), China. *In* "Research on Dolphins" (Eds M. M. Bryden and R. Harrison) pp. 23–31. Clarendon Press, Oxford.

Liu, R., Klinowska, M. and Harrison, R. J. (1986b). The behaviour of *Lipotes vexillifer* and *Neophocaena phocaenoides* in the Changjiang River and in captivity in China. *In* "Research on Dolphins" (Eds M. M. Bryden and R. Harrison) pp. 433–439. Clarendon Press, Oxford.

Lorenzo Tan, J. M. (1995). "A Field Guide to Whales and Dolphins in the Philippines", Bookmark, Inc., Manila, 125 pp.

Lydekker, R. (1909). On an Indian dolphin and porpoise. *Proc. Zool. Soc. Lond.* **2**, 802–808, plts 44, 45. (Cited in Allen, 1923.)

McKenna, V. (1992). "Into the Blue". The Aquarium Press, London, 144 pp.

Mead, J. G. (1975). Anatomy of the external nasal passage and facial complex in the Delphinidae (Mammalia: Cetacea). *Smithsonian Contrib. Zool.* **207**, 1–72.

Miyashita, T., Shimada, H., Taishaku, M. and Asai, Y. (1994). Density and its seasonal change of finless porpoise in the Ise/Mikawa Bays. Abstracts of the Autumn Meeting of the Japanese Society for the Science of Fisheries, p. 58 (in Japanese).

Miyashita, T., Wang, P., Cheng, J. H. and Yang, G. (1995). Report of the Japan/China joint whale sighting cruise in the Yellow Sea and the East China Sea in 1994 summer. Document IWC/SC/NP17, 12 pp.

Mizue, K., Yoshida, K. and Masaki, Y. (1965). Studies on the little toothed whales in the west sea area of Kyushu, XII. *Neomeris phocaenoides*, so called Japanese "sunameri", caught in the coast of Tachibana Bay, Nagasaki Pref. *Bull. Fac. Fish. Nagasaki Univ.* **18**, 7–29 (in Japanese with English summary).

Mizue, K., Takemura, A. and Nakasai, K. (1968). Studies of the little toothed whales in the west sea area of Kyushu, XV. Underwater sound of the Chinese finless porpoise caught in the Japanese coastal sea. *Bull. Fac. Fish. Nagasaki Univ.* **25**, 25–32.

Neiland, K. A., Rice, D. W. and Holden, B. L. (1970). Helminths of marine mammals, I. The genus *Nasitrema*, air sinus flukes of delphinid cetacea. *J. Parasitol.* **56**, 305–316.

Nishiwaki, M. (1972). General biology. *In* "Mammals of the Sea, Biology and Medicine" (Ed. S. H. Ridgway) pp. 3–204. Charles C. Thomas, Springfield, IL.

Ogawa, T. (1950). "Tales on Whales". Chuo-Koron, Tokyo (in Japanese).

O'Shea, T. J., Brownell Jr, R. L., Clark Jr, D. R., Walker, W. A., Gay, M. L. and Lamont, T. G. (1980). Organochlorine pollutants in small cetaceans from the Pacific and south Atlantic Oceans. *Pesticides Monitor. J.* **14**, 35–46.

Owen, F. R. S. (1869). On some Indian cetacea collected by Walter Elliot. *Trans. Zool. Soc. Lond.* **6**, 17–47.

Palmer, T. S. (1899). Notes on the three genera of dolphins. *Proc. Biol. Soc. Washington* **13**, 23–24.

Peng, X. and Chen, J. (1985). The study on karyotype of finless porpoise (*Neophocaena phocaenoides*). *Acta Theriol. Sinica* **5**, 161–165 (in Chinese with English summary).

Perrin, W. F., Holts, D. B. and Miller, R. B. (1977). Growth and reproduction of the eastern spinner dolphin, a geographical form of *Stenella longirostris* in the eastern tropical Pacific. *Fish. Bull.* **75**, 725–750.

Perrin, W. F. and Reilly, S. B. (1984). Reproductive parameters of dolphins and small whales of the family Delphinidae. *In* "Reproduction in Whales, Dolphins and Porpoises" (Eds W. F. Perrin, R. L. Brownell Jr and D. P. DeMaster) pp. 97–126. *Rep. Int. Whal. Commn*, Spec. Issue No. 6, Cambridge.

Petter, A. J. and Pilleri, G. (1982). *Pharurus asiaeorientalis* new species, metastrongylid nematode, parasite of *Neophocaena phocaenoides* (Phocoenidae, Cetacea). *Invest. Cetacea* **13**, 141–148.

Pilleri, G. and Chen, P. (1979). How the finless porpoise (*Neophocaena asiaeorientalis*) carries its calves on its back, and the function of the denticulated area of the skin, as observed in the Changjiang River, China. *Invest. Cetacea* **10**, 105–108.

Pilleri, G. and Chen, P. (1980). *Neophocaena phocaenoides* and *Neophocaena asiaoerientalis*: taxonomical differences. *Invest. Cetacea* **11**, 25–32.

Pilleri, G. and Gihr, M. (1972). Contribution to the knowledge of the cetaceans of Pakistan with particular reference to the genera *Neomeris, Sousa, Delphinus* and *Tursiops* and description of a new Chinese porpoise (*Neomeris asiaeorientalis*). *Invest. Cetacea* **4**, 107–162.

Pilleri, G. and Gihr, M. (1973–74). Contribution to the knowledge of the cetaceans of the Southwest and Monsoon Asia (Persian Gulf, Indus Delta, Malabar, Andaman Sea and Gulf of Siam). *Invest Cetacea* **5**, 95–149.

Pilleri, G. and Gihr, M. (1975). On the taxonomy and ecology of the finless black porpoise, *Neophocaena* (Cetacea, Delphinidae). *Mammalia* **39**, 657–673.

Pilleri, G. and Pilleri, O. (1979). Observations on the dolphins in the Indus Delta (*Sousa plumbea* and *Neophocaena phocaenoides*) in winter 1978–79. *Invest. Cetacea* **10**, 129–135.
Pilleri, G., Zbinden, K. and Gihr, M. (1976). The black finless porpoise (*Neophocaena phocaenoides* Cuvier, 1829) is not black *Invest. Cetacea* **7**, 161–164.
Pilleri, G., Zbinden, K. and Kraus, C. (1980). Characteristics of the sonar system of cetaceans with pterygoschisis directional properties of the sonar clicks of *Neophocaena phocaenoides* and *Phocoena phocoena* (Phocoenidae). *Invest. Cetacea* **11**, 157–188.
Qian, W. (1986). Study on the anatomy and histology of trachea and lung of *Neophocaena asiaeorientalis*. *Acta Theriol. Sinica* **6**, 184–189 (in Chinese with English summary).
Qian, W., Li, Y., Du, L. and Tang, Y. (1985). The digestive organs of the finless porpoise *Neophocaena asiaeorientalis*. II. Intestine, liver and pancreas. *Acta Theriol. Sinica* **5**, 1–9 (in Chinese with English summary).
Rice, D. W. (1977). A list of the marine mammals of the world. NOAA Technical Rept. NMFS SSRF-711, National Marine Fisheries Series, Seattle, 15 pp.
Robineau, D. and Rose, J. M. (1984). Les cetaces de Djibouti. Bilan des connaisance actuelles sur la faune cetologique de la mer Rouge et du golfe d'Aden. *Bull. Mus. Natl Hist. Nat. Paris*, 4th Ser., **6** (section A), 219–249.
Sergeant, D. E. (1969). Feeding rates of cetacea. *FiskDir. Skr.* Ser. *HavUnters.* **15**, 246–258.
Shirakihara, K., Yoshida, H., Shirakihara, M. and Takemura, A. (1992). A questionnaire survey on the distribution of the finless porpoise, *Neophocaena phocaenoides*, in Japanese waters. *Mar. Mamm. Sci.* **8**, 160–164.
Shirakihara, M. and Takemura, A. (1993). On the food of finless porpoise off the Nagasaki Prefecture. *In* Abstracts of the Spring Meeting of the Japanese Society for the Science of Fisheries (in Japanese).
Shirakihara, M., Yoshida, H., Shirakihara, K. and Takemura, A. (1992). Records of the finless porpoise, *Neophocaena phocaenoides*, in the waters adjacent to Kanmon Pass, Japan. *Mar. Mamm. Sci.* **8**, 82–85.
Shirakihara, M., Takemura, A. and Shirakihara, K. (1993). Age, growth and reproduction of the finless porpoise, *Neophocaena phocaenoides*, in the coastal waters of western Kyushu, Japan, *Mar. Mamm. Sci.* **9**, 392–406.
Shirakihara, M., Shirakihara, K. and Takemura, A. (1994). Distribution and seasonal density of the finless porpoise *Neophocaena phocaenoides* in the coastal waters of western Kyushu, Japan. *Fish. Sci.* **60**, 41–46.
Smith, B. D., Jefferson, T. A., Ho, D. T., Leatherwood, S., van Thuoc, C., Andersen, M. and Chiam, E. (1995). Marine mammals of Vietnam: a preliminary checklist. *Collect. Mar. Res. Works* **6**, 147–176.
Tao, J. (1983). A new species and a new Chinese record of Nematodes from porpoise *Neophocaena phocaenoides*. *Acta Zootaxon. Sinica* **8**, 350–353 (in Chinese with English summary).
Tas'an and Leatherwood, S. (1984). Cetaceans live-captured for Jaya Anchol Oceanarium, Djakarta, 1974–1982. *Rep. Int. Whal. Commn* **34**, 485–489.
Tomilin, A. G. (1967). "Mammals of the U.S.S.R. and Adjacent Countries", Vol. IX "Cetacea" (Ed. V. G. Heptner). Nauk USSR, Moscow (English translation by Israel Program for Scientific Translations, Jerusalem).
Van Dyke, D. and Ridgway, S. H. (1977). Diet for marine mammals. *In* "Handbook Series in Nutrition and Food" (Ed. M. Rechcigl) pp. 595–598. CRC Press, Boca Raton, FL.
Wang, P. (1984). Distribution, ecology and resource conservation of the finless porpoise in Chinese waters. *Trans. Liaoning Zool. Soc.* **5**, 106–110 (in Chinese with English summary)
Wang, P. (1992a). On the taxonomy of the finless porpoise in China. *Fish. Sci. China* **11**, 10–14 (in Chinese with English summary).

Wang, P. (1992b). The morphological characters and the problem of subspecies identification of the finless porpoise. *Fish. Sci. China* **11**, 4–9 (in Chinese with English summary).

Wu, B. (1989). The spinal cord of finless porpoise, *Neophocaena phocaenoides*. *Acta Theriol. Sinica* **9**, 16–23 (in Chinese with English summary).

Yamada, M. (1989). Cetaceans in the Sea of Japan. *IBP Rep. (Kamokawa, Japan)* **1**, 27–32 (in Japanese).

Yang, H. (1976). Studies on the whales, porpoises and dolphins of Taiwan. *Ann. Rep. Sci. Taiwan Mus.* **19**, 131–178 (in Chinese).

Yang, L., Yu, D. and Lu, P. (1988). Studies on the levels of some metals and organochlorine compounds in *Lipotes vexillifer* and *Neophocaena phocaenoides*. *Acta Theriol. Sinica* **8**, 122–127 (in Chinese with English summary)

Yoshida, H., Shirakihara, K., Kishino, H. and Shirakihara, M. (1997). A population size estimate of the finless porpoise, *Neophocaena phocaenoides*, from aerial sighting surveys in Ariake Sound and Tachibana Bay, Japan. *Res. Popul. Ecol.* **39**, 239–247.

Yoshida, H., Shirakihara, K., Kishino, H., Shirakihara, M. and Takemura, A. (1998). Finless porpoise abundance in Omura Bay, Japan: estimation from aerial sighting surveys. *J. Wildl. Mgmt* **62**, 286–291.

Yoshida, H., Shirakihara, K., Shirakihara, M. and Takemura, A. (1995). Geographical variation in the skull morphology of the finless porpoise *Neophocaena phocaenoides* in Japanese waters. *Fish. Sci.* **61**, 555–558.

Yoshida, H., Yoshioka, M., Chou, S. and Shirakihara, M. (1997). Study of geographical variation of mtDNA nucleotid sequence. *In* Abstract of the Spring Meeting of Japanese Society for the Science of Fisheries (in Japanese).

Zhang, X. (1986). Population biology of the finless porpoise. Thesis, Institute of Hydrobiology, Academia Sinica, Wuhan (in Chinese).

Zhang, X. (1992). Studies on the age determination, growth and reproduction of finless porpoise *Neophocaena phocaenoides*. *Acta Hydrobiol. Sinica* **16**, 289–298 (in Chinese with English summary).

Zhang, X., Cheng, Y. and Zhou, K. (1989). Banded karyotypes of the finless porpoise, *Neophocaena phocaenoides*, from the Yangtze River. *Acta Theriol. Sinica* **9**, 281–284 (in Chinese with English summary)

Zhang, X., Liu, R., Zhao, Q., Zhang, G., Wei, Z., Wang, Z. and Yang, J. (1993). The population of finless porpoise in the middle and lower reaches of the Changjiang River. *Acta Theriol. Sinica* **13**, 260–270 (in Chinese with English summary).

Zhou, K., Pilleri, G. and Li, Y. (1979). Observations on the baiji (*Lipotes vexillifer*) and the finless porpoise (*Neophocaena asiaeorientalis*) in the Changjiang (Yangtze) River between Nanjing and Taiyangzhou, with remarks on some physiological adaptations of the baiji to its environment. *Invest. Cetacea* **10**, 109–120.

17

Dall's Porpoise
Phocoenoides dalli (True, 1885)

Warren J. Houck and Thomas A. Jefferson

Genus and Species

Taxonomy

Dall's porpoise was originally described as *Phocoena dalli* by F. W. True (1885) based on a specimen collected in August 1873 by W. H. Dall west of Adak Island in the Aleutian Islands, Alaska. In 1911, R. C. Andrews described a new species of porpoise, *Phocoenoides truei*, proposed the new genus *Phocoenoides*, and placed the previously described species *Phocoena dalli* in this genus. The validity of the new genus is well established; however, the relationship between the two named forms in the genus, i.e. Dall's porpoise, *P. dalli*, and True's porpoise, *P. truei*, has remained controversial. They have been called species (Andrews, 1911), subspecies (Kuroda, 1954), and colour morphs of the same species, *P. dalli* (Houck, 1976). This problem has been discussed by Houck (1976), Kasuya (1978, 1982), Jefferson (1988), and others. Investigators since Andrews generally have agreed that the differences are below the species level, and most recent work supports the idea of colour morphs. In the remainder of this chapter, these named forms will be referred to as *dalli*-type and *truei*-type.

Barnes (1985) considered *Phocoenoides* to be more closely related to *Australophocaena* than to the other two genera in this family (*Phocoena* and

Handbook of Marine Mammals, Volume 6
ISBN 0-12-588506-7

Neophocaena). He proposed the new subfamily Phocoenoidinae to include these two extant genera, plus fossil forms. Recent molecular genetic evidence, however, does not support this affinity (Rosel *et al.*, 1995).

Common names

Following the nomenclature of Hershkovitz (1966), other common names include Dall porpoise, white-flank porpoise, spray porpoise, ishi-iruka (Japanese), rikuzen-iruka (Japanese), and belokrylaya morskaya svinya (Russian).

Distribution

Range

The distribution of Dall's porpoise, as shown in Fig. 1, is confined to the North Pacific Ocean and adjacent seas. In the eastern North Pacific, a beach-cast specimen (California Academy of Sciences #16237) is known from as far south as Scammon's Lagoon (approximately 28°N). Morejohn (1979) reported an additional specimen from the same locality. However, Dall's porpoises are seldom seen south of 32°N in the summer (Ridgway, 1966; Morejohn, 1979). The distribution extends northwards along the continental shelf and slope, through the

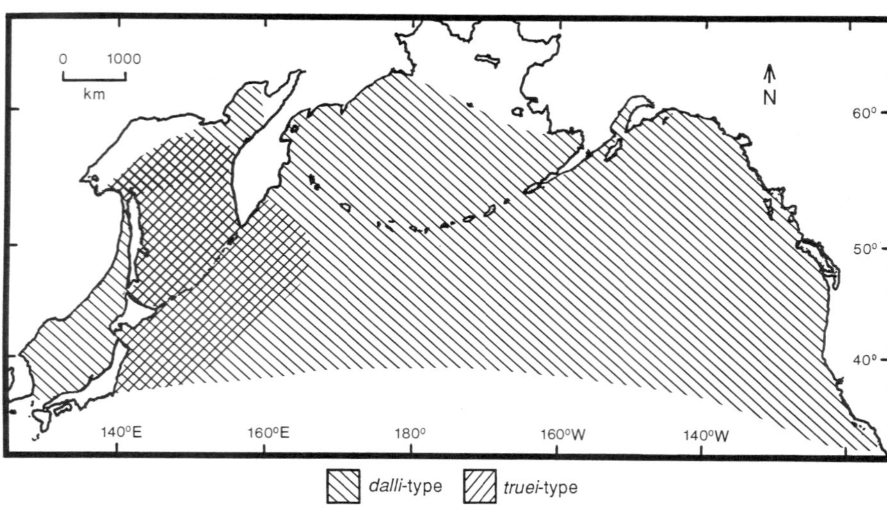

FIG. 1 Geographical distribution of *Phocoenoides*. This genus is confined to the colder waters of the North Pacific Ocean and adjacent seas.

Gulf of Alaska, throughout the Bering Sea (except for the shallow eastern flats) and, according to Sleptsov (1961), penetrates the Bering Strait into the Chukchi Sea on occasion. In the western Pacific, Kasuya (1978) reported the range to extend as far south as 35°N (central Honshu, Japan) in the winter. *Phocoenoides* also occurs in the Okhotsk Sea and Sea of Japan. In the central North Pacific, the *dalli*-type (Fig. 2) is known at least as far south as 39°N and over 1100 km from the nearest land (Jones *et al.*, 1987).

The distribution of the *truei*-type (Fig. 2) is much more restricted. It is generally found in the western Pacific, along the east coast of Japan and the Kurile Islands, from about 35°N in the south to 54°N in the north (Kasuya, 1978), and it is sympatric with the *dalli*-type in this area and in the Okhotsk Sea (Miyashita, 1991). *Truei*-types have been captured as far east as the western Aleutians (ca. 180°), and recently a single neonate having the *truei*-type pattern has been reported from the central California coast (Szczepaniak *et al.*, 1992). This record should not be taken to imply that the eastern Pacific is a normal part of the range of the *truei*-type. *Truei*-types are not known from the Sea of Japan, where the *dalli*-type is common. Within the area of overlap, both mixed

FIG. 2 Large adult male Dall's porpoises, showing the species' body shape, coloration, and two major colour types: *dalli*-type (top) and *truei*-type (bottom).

and unmixed schools are found. Miyazaki *et al.* (1984) reported on mixed schooling behaviour and the possibility of interbreeding between these types off Japan.

Habitat

Phocoenoides inhabits both the neritic and oceanic zones, although the major populations are oceanic. Porpoises of this genus rarely come close to shore, except in areas of relatively deep water. Sightings are common along the central and northern California coastlines (Sullivan and Houck, 1979; Jefferson, 1991), and they are common in the deep inshore waters of Washington, British Columbia, and Alaska (Leatherwood *et al.*, 1982). Dohl *et al.* (1983) reported only 28% of their sightings in water less than 180 m deep, and Pike and MacAskie (1969) had frequent sightings as far as 2400 km off the coast of British Columbia.

Dall's porpoise is a cold water species that inhabits water with temperatures of 3–24 °C (Kasuya, 1978; Miyashita and Kasuya, 1988), but generally avoids water warmer than about 17–18 °C (Norris and Prescott, 1961; Kasuya and Jones, 1984; Jones *et al.*, 1987). According to Leatherwood *et al.* (1982) and Ohsumi (1975), it is not found in the southern extremes of its range during the summer or warm water months.

Patterns in skull size of Dall's porpoises from different parts of the species' range correlate with primary productivity patterns, indicating that food availability affects growth patterns (Amano and Miyazaki, 1992).

FIG. 3 A newborn Dall's porpoise of the *dalli*-type (total length 101 cm). The absence of white "trim" or "frosting" on the dorsal fin and flukes is an age characteristic.

External Characteristics

Colour pattern

The colour pattern of adults is highly contrasting black and white (Fig. 2). The primary body colour is black; there is a large white patch on each flank, extending ventrally across the abdomen onto the other flank. In adults, the trailing edge of the flukes and the upper two-thirds or so of the dorsal fin are usually light grey to white (this is termed "trim" or "frosting"), and the extent of this lightening is highly variable. The lateroventral part of the tail stock often has a white or light grey pattern on it. The remainder of the head and body is black.

There are two major colour patterns, with many minor variations (Figs 2 and 3). The primary difference is in the extent of the lateral white blaze. In the *dalli*-type the blaze extends anteriorly at least to the level of the dorsal fin, but does not reach to the flippers. In the *truei*-type the lateral blaze extends to or beyond the flippers. In addition to these two commonly occurring types, completely black (Nishiwaki, 1966), white (Joyce *et al.*, 1982), and grey (Morejohn *et al.*, 1973; Hall, 1979) types have been reported. Also, some porpoises with brown rather than black backgrounds and some with the lateral blaze marked with black speckling are known. Kasuya (1978) listed 14 different minor pattern groups. Types apparently intermediate between *dalli*- and *truei*-types also have been found (Houck, 1976).

The colour pattern changes with age. Newborns are dark and light grey instead of black and white, with no trim (Fig. 3). As the animals age, coloration intensifies to black and white, and the trim develops and lightens (Morejohn, 1979; Kasuya, 1982; Jefferson, 1990). Morejohn *et al.* (1973) described sexual dimorphism in the colour patterns of the urogenital area; however, there is great individual variation that was not accounted for by these researchers (see Jefferson, 1990).

Size and shape

Dall's porpoises are born at a length of about 100 cm (Fig. 3), and the growth rate for both sexes is similar during the first two years, after which males grow somewhat faster (Kasuya, 1978; Newby, 1982). For most populations, the maximum size is about 210 cm for females and 220 cm for males, but males from the southern Okhotsk Sea grow to as large as 239 cm (R. L. Brownell Jr, personal communication). Mizue *et al.* (1966) gave a maximum weight of approximately 170 kg, but some Dall's may reach nearly 200 kg.

The body shape of Dall's porpoise is relatively stocky (Fig. 2). The head has no prominent beak, but is typically porpoise-like in that it forms a rapidly tapering cone. In adults, the body and caudal peduncle are deeper than in most other small cetaceans. The flippers are terminally rounded, and small in adults. The dorsal fin is triangular to slightly falcate and wide-based, often with a recurved tip. Previously unpublished external measurements for several individuals are presented in Table 1.

The head and flippers of newborn calves are relatively larger than in adults (Amano and Miyazaki, 1993; see Fig. 3). Sexual dimorphism in this species is moderate. Males grow longer and heavier than females (Kasuya, 1978; Newby, 1982). As is true for cetaceans in general, the distance between the anal and genital openings is much greater in males than in females. In adult males, the caudal peduncle is deeper than in females, and discernible sexual differences in dorsal fin shape, fluke shape, and post-anal hump depth also have been reported (Jefferson, 1990; Amano and Miyazaki, 1993).

Internal Anatomy

Skull

The skull of *Phocoenoides* (Fig. 4) is distinctive in several ways. It is large, relative to that of other phocoenids; an adult male may have a condylobasal length of 34 cm. The cranium may be as much as 1.7 times wider than its length. The rostrum is broad and short; its length usually is less than 1.5 times its basal breadth and only slightly more than 0.4 times the condylobasal length (Table 2). In profile, the skull of this species differs from those of both *Neophocaena* (finless porpoise) and *Phocoena* (harbour and Burmeister's porpoises) in that the posterior part of the maxillary bends dorsally abruptly to form the anterior face of the cranium. These "maxillary shields" make an angle of approximately 130° with the axis of the rostrum. In *Neophocaena* and *Phocoena* this angle is much flatter. In dorsal view the anterior surface of the cranium is greatly flattened. The suture between the nasal bones is only slightly more anterior than the lateral margins of the "maxillary shields". In older animals, the transverse ridge formed by the top of the frontals may slightly overhang the posterior edge of the maxillaries. In *Neophocaena* and *Phocoena* the anterior surface of the cranium is rounded.

The teeth of Dall's porpoises are smaller than those of *Phocoena* (some are only about 1 mm in diameter), and are not as clearly spatulate (although some of the teeth have crowns broader than the roots). There is a slight constriction between the crown and the root. These small teeth erupt only a short distance through the gums, which are rough and may help the teeth in grasping such slippery prey as squid (Morejohn, 1979). Miller (1929) described the gums of this species in detail. The tooth counts are highly variable, although most individuals have 21–28 teeth per tooth row (Table 2). In most cases, a greater number of teeth is found in the lower jaw. The teeth in the premaxillary region are rudimentary, and thus quite variable in number. Probably this variation is also partially a

TABLE 1 Unpublished external measurements from Dall's porpoises taken in Japanese waters.[a] All measurements are in cm. The numbers in parentheses are percentages of the total body length

	HSU #2694 ♂, *dalli*-type		HSU #2695 ♂, *truei*-type		HSU #2696 ♀, *dalli*-type		HSU #2697 ♀, *truei*-type		HSU #2698 ♂, *dalli*-type		HSU #2699 ♀, *truei*-type	
Total length (straight line)	185	(100)	178	(100)	187	(100)	173	(100)	180	(100)	187	(100)
Tip of upper jaw to centre of eye	18	(10)	18.5	(10)	19	(10)	18	(10)	19	(11)	18.5	(10)
Length of gape	9	(05)	9	(05)	9.5	(05)	9.5	(05)	10	(06)	10	(05)
Tip of upper jaw to anterior insertion of flipper	28	(15)	28.5	(16)	28.5	(15)	27.5	(16)	28	(16)	27	(14)
Tip of upper jaw to tip of dorsal fin	86	(46)	81	(46)	87	(47)	78	(45)	83	(46)	82	(44)
Flipper length (anterior insertion to tip)	21.5	(12)	21	(12)	22	(12)	19.5	(11)	22	(12)	23	(12)
Height of dorsal fin	15	(08)	18.5	(10)	16	(09)	13.5	(08)	15.5	(09)	18	(10)
Width of flukes	45.5	(25)	51	(29)	46	(25)	41	(24)	46	(26)	54	(29)

[a] HSU indicates the collection of Humboldt State University.

TABLE 2 Unpublished skull measurements from Dall's porpoises taken in Japanese waters.[a] All measurements are in cm. The numbers in parentheses are percentages of the total skull length

	HSU #2694 ♂, *dalli*-type	HSU #2695 ♂, *truei*-type	HSU #2696 ♀, *dalli*-type	HSU #2697 ♀, *truei*-type	HSU #2698 ♂, *dalli*-type	HSU #2699 ♀, *truei*-type
Total length (condylobasal)	32.9 (100)	32.6 (100)	31.3 (100)	33.1 (100)	34.2 (100)	32.0 (100)
Length of rostrum	13.2 (40)	12.8 (39)	12.4 (40)	14.6 (44)	14.0 (41)	13.1 (41)
Tip of rostrum to anterior nares	16.7 (51)	17.0 (52)	16.1 (51)	18.3 (55)	18.9 (55)	17.0 (53)
Width of rostrum at base	9.6 (29)	9.5 (29)	9.2 (29)	9.2 (28)	10.4 (30)	9.6 (30)
Width of rostrum at midpoint	5.6 (17)	5.1 (16)	6.0 (19)	5.3 (16)	6.4 (19)	5.8 (18)
Max. width of cranium (parietals)	17.3 (53)	18.8 (58)	16.2 (52)	17.0 (51)	17.5 (51)	17.5 (55)
Total length of mandible right side	25.1 (76)	25.0 (77)	24.1 (77)	25.8 (78)	26.8 (78)	25.3 (79)
Length of mandibular symphysis	4.2 (13)	3.9 (12)	3.4 (11)	4.2 (13)	3.6 (11)	3.6 (11)
Number of teeth right side:						
upper	31	26	22	24	24	24
lower	32	24	25	25	31	26

[a] HSU indicates the collection of Humboldt State University.

reflection of the fact that the premaxillary teeth are tiny and may be lost easily during skeletal preparation.

P. dioptrica (spectacled porpoise) has a higher braincase, a shorter and wider rostrum, and lower tooth counts (Hamilton, 1941; Brownell 1975; Barnes, 1985). It also has a shorter skull, with a maximum condylobasal length of 31.5 cm (Brownell, 1975).

Postcranial skeleton

Dall's porpoises have greatly elongated and slender dorsal and lateral processes on their vertebrae, and compressed centra (Miller, 1930). The total number of vertebrae in this genus is variable, ranging from 86 to 98, with a mean of 91.8–93.7 for different populations, and a mode of 96 (Walker and Sinclair, 1990; W. J. Houck, unpublished data). Although much of this variability is real, it must be noted that some low counts can be attributed to losses in skeletal preparation. The vertebral formula is C_7, $T_{15–18}$, $L_{24–27}$, $Ca_{44–49} = 92–98$ (Tomilin, 1957; Nishiwaki, 1963, 1972). The cervical vertebrae are fused into a unit; the sutures between the first six centra usually are indistinct; the separation between the centra of C6 and C7 is clear. The neural spine of C1, the atlas, is bifurcate. *Australophocaena* has fewer vertebrae (66–70) and shorter vertebral processes (Brownell, 1975; Barnes, 1985).

The usual number of ribs is 18 pairs, but 17 is not unusual. The ribs are also long and slender, compared with other species (Miller, 1930). The first 12 pairs are two-headed, while the remaining five or six pairs have single heads. The last rib is a "floating rib" and can be easily missed in preparation of the skeleton. A pair of cervical ribs, or a single small one (approximately 2 cm long) is not uncommon.

Organs, tissues and physiology

Relative to other small cetaceans, Dall's porpoise has a large skeletal muscle mass, relatively small brain, thick tracheal cartilage rings, large deeply folded vestibular sacs, and large adrenals and thyroid (Ridgway, 1966; Ridgway *et al.*, 1966; Pilleri and Gihr, 1970; Mead, 1975; Morejohn, 1979; Curry, 1992). Fatty pads lining the lungs are unique to this species and serve no known function (Ridgway, 1966).

Very little has been published on organ weights of *Phocoenoides*, although Miyazaki *et al.* (1986) presented a complete set of organ and body part weights for a stranded animal from Japan. The most significant studies have been the heart weight research of Ridgway and Johnston (1966). They found a mean heart weight of 1.31% of body weight in *Phocoenoides*, 0.85% in *Lagenorhynchus obliquidens* (Pacific white-sided dolphin), and 0.54% in *Tursiops truncatus* (bottlenose dolphin). This sequence reflects the postulated relative metabolic rates and activity levels of these three species, and more recent work supports this as well

(Ridgway and Kohin, 1995). McCormick (1969) never observed captive Dall's porpoises to engage in any activity resembling sleep, and although oxygen consumption rates have not been studied, they probably have one of the highest metabolic rates of all small cetaceans (see Ridgway, 1966; Ridgway and Johnston, 1966). Of these three species, *P. dalli* was found to have also the highest mean blood volume and packed-cell volume (haematocrit) (Ridgway and Johnston, 1966). For blood oxygen content, which is probably a better indicator of physiological capabilities, these same investigators found that *P. dalli* has a value nearly three times that of *Tursiops*.

Dall's porpoise is a cold water species, the Pacific white-sided dolphin is a more temperate species, and the bottlenose dolphin is a warm water species. While one would expect the thickness of the blubber layer to be greatest in the colder water species, Ridgway and Johnston (1966) found just the opposite, with mean blubber thicknesses of 1, 2, and 3 cm for these three species, respectively. This paradox seems best explained by the suggestion that Dall's porpoise has a higher metabolic rate, by which body temperature is maintained. The food consumption rates of captive animals of these three species support a higher metabolic rate for Dall's porpoise.

Recent studies have shown that the retina of the Dall's porpoise is similar to that in other species of cetaceans (Murayama *et al.*, 1995). Calculi have been found in the nasal diverticula of Dall's porpoises (Curry *et al.*, 1994).

Life History and Population Dynamics

Growth and reproduction

Almost all age data have been obtained by sectioning teeth longitudinally, then staining and counting the cementum layers (Kasuya, 1978; Perrin and Myrick, 1980). These cementum layers become difficult to interpret with increasing age, causing problems in aging older animals. Age data may be obtained also from length data for smaller individuals (Kasuya, 1978).

Reproductive parameters for various Dall's porpoise populations are summarized in Table 3. *Truei*-types attain sexual maturity at body lengths 12–17 cm greater than *dalli*-types of most populations (Kasuya and Shiraga, 1985), although *dalli*-types in the Sea of Japan/Okhotsk Sea reach maturity at lengths similar to those of the *truei*-type population (Amano and Kuramochi, 1992). Everywhere that this species has been studied, there appears to be a calving peak in the summer months (Okada and Hayashi, 1951; Kasuya, 1978; Newby, 1982; Jones *et al.*, 1987; Jefferson, 1989). A single neonate from northern California of 100.7 cm in length (Fig. 3) was found on 7 August with its teeth unerupted but with the umbilicus healed (W. J. Houck, unpublished data).

TABLE 3 Reproductive parameters estimated for various Dall's porpoise populations

Parameter	Population: Pacific coast of Japan (*truei*)[a]	Sea of Japan/ Okhotsk Sea[b]	NW North Pacific[c]	Bering Sea[c]
Length at birth (cm)	—	~ 100	97	—
Length at sexual maturity (cm)				
Male	195.7	192	175–180	180–185
Female	186.5	187	174–177	182
Age at sexual maturity (yrs)				
Male	7.9	—	3.6–4.0	4.0–5.0
Female	6.8	—	4.5	4.4
Calving season	Aug.–Sept.	May–June	mid June–late July	mid July–?
Length of gestation (months)	11.4	7–9	10.7	—
Length of lactation (yrs)	2	—	—	—
Calving interval (yrs)	3	—	1	—

[a] Data from Kasuya (1978).
[b] Data from Okada and Hayashi (1951) and Amano and Kuramochi (1992).
[c] Data from Jones *et al.* (1985, 1987, 1988).

The length of the lactation period is very difficult to estimate accurately. It is thought generally that lactation is much shorter than the 2 years estimated by Kasuya (1978), and probably less than 4 months (Loeb, 1972; Newby, 1982).

Mortality

The only known predator other than humans, is the killer whale (*Orcinus orca*). Killer whales have been observed preying on Dall's porpoises, and have been found with Dall's porpoise remains in their stomachs (Rice, 1968; Pike and MacAskie, 1969; Barr and Barr, 1972; Morejohn, 1979). Several authors have also suggested sharks as possible predators, and Newby (1982) illustrated a possible shark attack wound; however, because of the great speeds achieved by Dall's porpoises, sharks seem unlikely to be major predators (Morejohn, 1979).

Abundance

Several stocks have been recognized, based largely on geographic variation in morphology and colour patterns, parasite loads, and densities of mother/calf pairs (see Kasuya and Ogi, 1987; Walker, 1990; Walker and Sinclair, 1990;

(A)

(B)

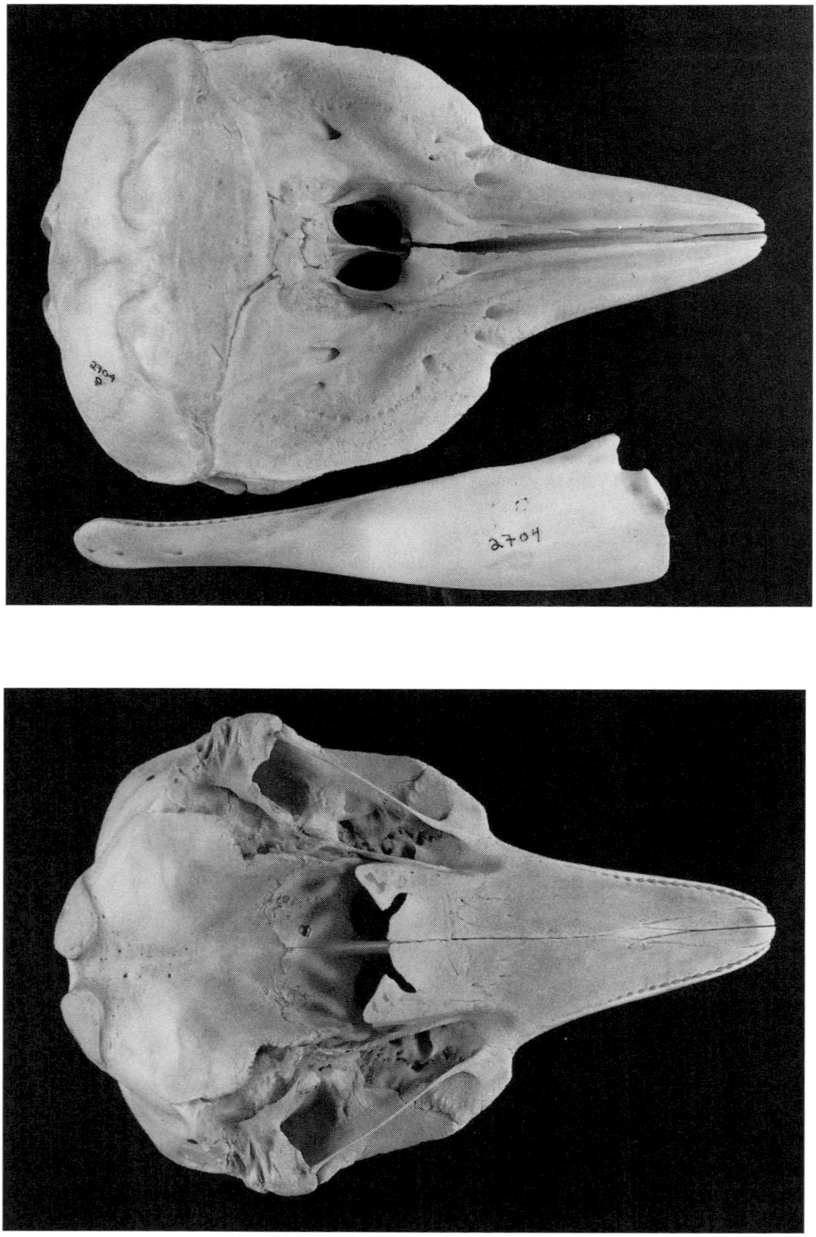

FIG. 4 Skull of *Phocoenoides dalli* (Humboldt State University #2704, female, condylobasal length 28.7 cm): (A) dorsal view of cranium and lateral view of mandible and (B) ventral view of cranium.

(C)

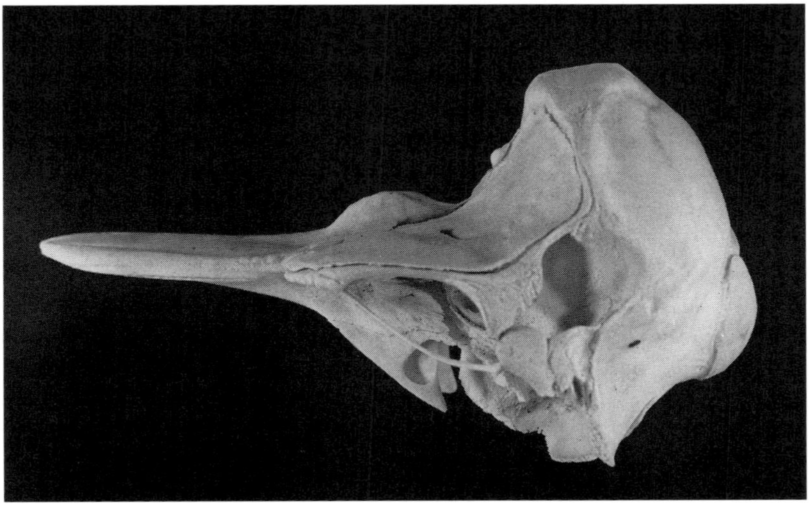

FIG. 4 *Continued* (C) Lateral view of cranium.

International Whaling Commission, 1991; Amano and Miyazaki, 1992). Genetic techniques also have been applied in recent years (Winans and Jones, 1988). Eight stocks (seven *dalli*-type and one *truei*-type) are currently recognized by the International Whaling Commission (1992).

The National Marine Fisheries Service, in cooperation with the government of Japan, has conducted sighting surveys since 1978 for the purpose of estimating the abundance of Dall's porpoise throughout much of its range. Estimates using line and strip transect methods have given a range of 1.4–2.8 million animals in the North Pacific and Bering Sea (Jones *et al.*, 1987). The most recent estimate is 1 186 000 (CV = 0.09) (Buckland *et al.*, 1993). Animal movement and other factors result in violations of some of the assumptions of the analyses, which bias population estimates.

Recently, several estimates for smaller portions of the species' range have been produced. Turnock *et al.* (1995) estimated that there are 141 800 in the western North Pacific. In the waters off the Pacific coast of Japan, Miyashita and Kasuya (1988) have estimated a total of 104 000 Dall's porpoises, more than half of them *truei*-types. Miyashita (1991) estimated the three stocks that inhabit the Sea of Okhotsk to number about 554 000 in total. Along the west coast of the United States there are estimated to be 2150 Dall's porpoises off coastal Washington and Oregon (Green *et al.*, 1992) and 78 400 off California (Barlow, 1995).

Behaviour

Social organization

Dall's porpoises are social animals, but group structure is thought to be fluid as in many other small cetaceans (Miller, 1990). Usually they are seen in groups of 2–12 animals; singles are relatively rare. Larger groups are known as well. They form aggregations of up to 20 or so for feeding and playing in Monterey Bay, California (Jefferson, 1991), and Pike and MacAskie (1969) recorded many large groups of over 30 in inshore waters of British Columbia. Scheffer (1950) reported a group of several thousand in the Gulf of Alaska, and Sullivan and Houck (1979) reported one of approximately 1000 off northern California. A group of about 3000 was seen passing through Stephens Passage, in southeastern Alaska (Jones *et al.*, 1987). Such large groups generally appear to be temporary aggregations of several smaller schools.

Usually Dall's porpoises do not arrange themselves in any discernible order; however, there are notable exceptions. Norris and Prescott (1961) described groups of 20–100 travelling in single file in southern California waters. So far, this behaviour has not been reported elsewhere, nor since, from southern California.

Swimming and diving

Dall's porpoises are extremely fast swimmers and may be the fastest of all the small cetaceans. Speeds as high as 32 km/hr (Ridgway and Johnston, 1966) and 55 km/hr (Leatherwood and Reeves, 1986) have been recorded or estimated, although such high speeds probably can be maintained only for short bursts. Law and Blake (1994) accurately measured maximum speed for wild Dall's porpoises at 22 km/hr.

Respiratory rates of wild individuals are difficult to obtain, except when they are bowriding. Jefferson (1987), while studying wild Dall's porpoises in British Columbia, did not record any dives in excess of 4 min. However, the sample was small and dives of over 5 min probably occur. Individual specimens are difficult to identify and track (see Miller, 1990; Jefferson, 1991). Ridgway (1972) found a mean respiratory rate of 3.2 per min in captivity, which is about the same as that for several other small cetaceans.

Dall's porpoises make a characteristically sloppy splash when surfacing to breathe while travelling quickly. This splashing has come to be known as "roostertailing" (Fig. 5). These distinctive splashes can be seen from long distances at sea and make the presence and identification of the animals known considerably before the porpoises themselves can be seen. When they decrease their speed and roll slowly at the surface, these splashes are not produced, and this type of surfacing is called "slow rolling". Slow rolling Dall's porpoises have been found to tra-

FIG. 5 Dall's porpoise at sea, showing the characteristic "roostertail" splash.

vel at about 5 km/hr (Jefferson, 1987). A third type of surfacing, recognized and named "fast surfacing" by Jefferson (1987), involves fast travel below the surface, but without the roostertail splash upon surfacing.

These porpoises rarely, if ever, leap from the water. Yocom (1946), in his account of Dall's porpoises west of San Francisco, mentioned leaping in several instances. It seems likely that he confused the splashes associated with roostertailing with those caused by leaping animals.

Dall's porpoises frequently will approach vessels at high speed and position themselves in front of and alongside the bow, although often they weave in and out of the bow wave with jerky movements (Fig. 6). Bowriding appears to be a play activity, most commonly undertaken by younger animals (Kasuya and Jones, 1984; Jefferson, 1991). They may swim in the normal position, on their sides as if to look at the observers on the bow, or occasionally upside down. This play may last anywhere from a few minutes to over half an hour. They terminate the "game" by veering off to the side at high speed or simply outracing the vessel. Certain vessels seem to be preferred over others; perhaps a combination of speed and bow configuration of some vessels produces pressure waves more to their liking. It is possible that sound may also be a factor.

Bowriding behaviour may have been learned from their association with larger cetaceans. Morejohn (1979) recorded associations with gray (*Eschrichtius*

FIG. 6 Bowriding Dall's porpoise on its side just below the surface.

robustus) and fin (*Balaenoptera physalus*) whales, and Dall's porpoises have been observed "bow riding" on the pressure waves produced by the heads of surfacing blue whales (*Balaenoptera musculus*) (Jefferson, 1991) and humpback whales (*Megaptera novaeangliae*) (T. A. Jefferson, unpublished observations). Associations with other smaller marine mammals are relatively uncommon.

Feeding

The diet of Dall's porpoise consists primarily of small fish and squid; size and abundance probably determine the prey (Table 4). Norris and Prescott (1961) found that prey items were all less than 25 cm in length. This is not surprising, as the teeth of this species are slender and incapable of shearing bites from larger prey.

Loeb (1972), in her study of Dall's in Monterey Bay, found that Pacific hake (*Merluccius*), juvenile rockfish (*Sebastes*), and market squid (*Loligo*) made up 85% of the food items over the year. Off Japan, Wilke and Nicholson (1958) found that squid accounted for 11% of the food items; lanternfish were over 70% of the remainder. The most common prey items in the Sea of Okhotsk were the Japanese pilchard (*Sardinops sagax*) and the squid *Berryteuthis magister* (Walker, 1996). Stomachs of Dall's porpoises from the northwestern North Pacific and Bering Sea contained mostly squid (Mizue *et al.*, 1966; Fiscus and Jones, 1990; Kuramochi *et al.*, 1993), but 33 species of epi- and mesopelagic fishes were also found (Crawford, 1981). Of the fish remains, 94% of the individuals

TABLE 4 Prey species recorded from stomach contents of Dall's porpoise

Scientific name	Common name	Source(s)
FISHES		
Merluccius productus	Pacific hake	Scheffer, 1953; Norris and Prescott, 1961; Loeb, 1972; Morejohn, 1979; Stroud *et al.*, 1981
Laemonema spp.	Hakes	Wilke *et al.*, 1953; Wilke and Nicholson, 1958; Jones, 1981
Trachurus symmetricus	Jack mackerel	Scheffer, 1953; Norris and Prescott, 1961
Scomber japonicus	Chub mackerel	Wilke *et al.*, 1953; Wilke and Nicholson, 1958; Kuzin and Belyaev, 1985
Clupea harengus	Pacific herring	Cowan, 1944; Poke and MacAskie, 1969; Loeb, 1972; Morejohn, 1979; Crawford, 1981
Engraulis mordax	Northern anchovy	Loeb, 1972; Morejohn, 1979; Crawford, 1981; Stroud *et al.*, 1981
Engraulis japonica	Anchovy	Wilke *et al.*, 1953
Cololabis saira	Pacific saury	Sleptsov, 1961; Stroud *et al.*, 1981; Kuzin and Belyaev, 1985
Sardinops sagax	Pacific sardine Japanese pilchard	Kuzin and Belyaev, 1986; Walker, 1996
Sebastes spp.	Rockfishes	Loeb, 1972; Morejohn, 1979
Mallotus villosus	Capelin	Scheffer, 1953; Stroud *et al.*, 1981
Thaleichthys pacificus	Eulachon	Stroud *et al.*, 1981
Tarletonbeania crenularis	Blue lanternfish	Loeb, 1972; Crawford, 1981
Protomyctophum thomsoni	Flashlightfish	Crawford, 1981
Myctophidae	Lanternfishes	Wilke *et al.*, 1953; Wilke and Nicholson, 1958; Loeb, 1972; Morejohn, 1979; Crawford, 1981
Citharichthys sordidus	Pacific sandab	Loeb, 1972; Morejohn, 1979
Oncorhynchus spp.[a]	Salmon	Mizue *et al.*, 1966
Spirinchus starski	Nightsmelt	Loeb, 1972; Morejohn, 1979
Bathylagus spp.	Deep sea smelts	Loeb, 1972; Morejohn, 1979; Crawford, 1981
Chilara taylori	Spotted cuskeel	Morejohn, 1979
Deprilus simillimus	Pacific butterfish	Morejohn, 1979
Microgadus proximus	Pacific tomcod	Jones, 1981
Zoarcidae	Eelpouts	Loeb, 1972; Morejohn, 1979
Macrouridae	Rattails	Loeb, 1972; Morejohn, 1979
Ammodytes hexapterus	Pacific sandlance	Crawford, 1981
Cottidae	Sculpins	Crawford, 1981
Scopelosaurus harryi	Paperbone	Crawford, 1981
Moridae	Moras	Crawford, 1981
Nansenia candida	Bluethroat argentine	Crawford, 1981

continued

TABLE 4 *Continued*

Scientific name	Common name	Source(s)
Pleurogrammus monopterygius	Atka mackerel	Crawford, 1981
Ichthyococcus sp.	Gonostomatid	Crawford, 1981
Stichaeidae	Pricklebacks	Crawford, 1981
Stomatiatiodae	Stomatiatiods	Crawford, 1981
Otophidium taylori	Cusk eel	Loeb, 1972
Peprilus simillimus	Pompano	Loeb, 1972; Morejohn, 1979
Anoploma fimbria	Sablefish	Loeb, 1972; Morejohn, 1979
Lampanyctus regalis	Pinpoint lampfish	Loeb, 1972; Morejohn, 1979
Liparis sp.	Snailfish	Loeb, 1972; Morejohn, 1979
Melamphaediae	Bigscales	Crawford, 1981
Opistoproctidae	Spookfishes	Crawford, 1981
Paralepididae	Barracudinas	Wilke *et al.*, 1953; Wilke and Nicholson, 1958; Crawford, 1981
Benthabella dentata	Pearleye	Crawford, 1981
Pleuronectidae	Righteye flounders	Stroud *et al.*, 1981
Theragra chalcogramma	Walleye pollock	Crawford, 1981; Walker 1996
Anguilliformes	Eels	Loeb, 1972
CEPHALOPODS		
Loligo opalescens	Market squid	Scheffer, 1953; Norris and Prescott, 1961; Ridgway, 1966; Loeb, 1972; Morejohn, 1979; Stroud *et al.*, 1981
Sepiolidae	Bobtail cuttlefishes	Morejohn, 1979
Ommastrephes sloani	Flying squid	Wilke *et al.*, 1953; Wilke and Nicholson, 1958; Sleptsov, 1961
Ommastrephidae	Flying squids	Morejohn, 1979
Onychoteuthis spp.	Hooked squids	Stroud *et al.*, 1981; Kuzin and Belyaev, 1985; Fiscus and Jones, 1990
Watasenia scintillans	Sparkling enope squid	Wilke *et al.*, 1953; Wilke and Nicholson, 1958
Octopoteuthis sp.	Octopus squid	Stroud *et al.*, 1981
Octopoteuthidae	Octopus squids	Morejohn, 1979
Gonatidae	Gonatid squids	Loeb, 1972; Morejohn, 1979; Stroud *et al.*, 1981; Fiscus and Jones, 1990
Gonatopsis spp.	? squid	Kuramochi *et al.*, 1993
Gonatus onyx	? squid	Kuramochi *et al.*, 1993
Berryteuthis spp.	? squid	Kuramochi *et al.*, 1993; Walker, 1996
Abraliopsis sp.	? squid	Loeb, 1972; Stroud *et al.*, 1981
Enoploteuthidae	Enope squids	Morejohn, 1979; Fiscus and Jones, 1990
Histioteuthis sp.	Jewell squid	Fiscus and Jones, 1990

continued

TABLE 4 *Continued*

Scientific name	Common name	Source(s)
Histioteuthidae	Jewell squids	Morejohn, 1979
Chiroteuthis sp.	? squid	Fiscus and Jones, 1990
Cranchiidae	Cranch squids	Stroud *et al.*, 1981; Fiscus and Jones, 1990
Bolitaenidae	Bolitaenid squids	Fiscus and Jones, 1990
Octopus sp.	Octopus	Loeb, 1972; Morejohn, 1979
OTHER INVERTEBRATES		
Euphausiidae[a]	Krill	Crawford, 1981
Decapoda[a]	Decapods	Wilke *et al.*, 1953; Morejohn, 1979
	Shrimp	Mizue and Yoshida, 1965

[a] Not considered a regular prey item.

were myctophids (lanternfish), with *Protomyctophum* accounting for 78% of all the fishes consumed.

The high percentage of deepwater and vertically migrating fishes in the diet of Dall's porpoises from many areas indicates that the porpoises were either feeding at night or at great depths, most likely both (see Morejohn, 1979; Crawford, 1981; Stroud *et al.*, 1981).

Movements

This species is present throughout much of its range year-round. This is true for at least central California (Morejohn, 1979; Jefferson, 1991), northern California (W. J. Houck, unpublished data), Puget Sound, Washington (Miller, 1990), and British Columbia (Pike and MacAskie, 1969). In these areas, waters remain cool (about 9–15 °C) year-round.

Norris and Prescott (1961) found Dall's porpoises in southern California waters only in the winter, generally when the water temperature was less than 15 °C. Inshore/offshore movements off southern California and British Columbia have been postulated (Brown and Norris, 1956; Pike and MacAskie, 1969). Although movements in the eastern Pacific also have a north/south component (Leatherwood *et al.*, 1982), there appear to be more distinct north/south movements in the western Pacific (Nishiwaki, 1967; Kasuya, 1978; Amano and Kuramochi, 1992).

These movements may be temperature-related or, as Morejohn (1979) postulated, food-dependent. Miyazaki *et al.* (1984) found that *truei*-type porpoises and mixed schools generally were found in warmer waters, while *dalli*-types were found in both warmer and colder waters.

Movement patterns of individual porpoises have not been studied well. Hall (1979) reported evidence of a summer home range in porpoises of this species from Prince William Sound, Alaska, and Miller (1989, 1990) reported resightings over various distances and periods of time of photoidentified Dall's porpoises in Puget Sound. One porpoise tagged in Prince William Sound was resighted 90 days later at a straight line distance of 821 km away (Hall, 1979).

Sound production

Only pulsed sounds, no pure-tone whistles, have been recorded from this species. The clicks produced by Dall's porpoises have been found to be similar to those known from other phocoenids and dolphins of the genus *Cephalorhynchus* (Evans *et al.*, 1988). They consist of high frequency (mostly 120–150 kHz), narrow-band single and double pulses (Awbrey *et al.*, 1979). The pulses are of short duration; single pulses are about 500 μs and double pulses are about 250 μs (Evans *et al.*, 1988). These sounds presumably are used for echolocation.

Parasites and Disease

Dall's porpoises suffer from a variety of parasitic infestations (Table 5). There has been some limited work on the diseases and pathology of this species (Ridgway, 1966; Dailey and Walker, 1978; Conlogue *et al.*, 1985; Cowan *et al.*, 1986). The trematode fluke *Nasitrema*, which infests and produces lesions in the respiratory system and brain, has been implicated as a possible causal factor in single strandings of Dall's porpoises in southern California (Dailey and Walker, 1978). Walker (1990) found certain parasites useful as stock indicators.

Live Maintenance

Dall's porpoises have been extremely difficult to maintain in captivity. Most attempts have ended with the premature death of the animal. Walker (1975) claimed this to be the most difficult to maintain of all the southern California cetaceans. Of eight porpoises live-captured, half died during capture or transport, two died within 60 days, one lived three months, and the remaining one survived slightly over 15 months. Walker (1975) reported that, although they had a high food intake, they never regained their original capture weight. The last specimen, and another reported by Ridgway (1966), are the only ones known to have survived longer than one year in captivity. Small numbers have also been live-captured in Japan for public display (Kasuya *et al.*, 1984).

TABLE 5 Parasites recorded from Dall's porpoise

Location	Species	Source(s)
Lungs	*Halocercus dalli*	Benson and Groody, 1942; Machida, 1974; Conlogue *et al.*, 1985
	Stenurus minor	Johnston and Ridgway, 1969
	Stenurus sp.	Walker, 1975
Nasal passages/ air sacs	*Nasitrema dalli*	Yamaguti, 1951; Dailey, 1971; Dailey and Walker, 1978; Cowan *et al.*, 1986
	Torynurus dalli	Dailey and Walker, 1978; Cowan *et al.*, 1986
	Pharurus dalli	Dailey, 1971; Kuramochi *et al.*, 1990
	Pharurus sp.	Walker, 1975
	Stenurus minor	Johnston and Ridgway, 1969; Dailey and Walker, 1978; Cowan *et al.*, 1986
	Stenurus truei	Kuramochi *et al.*, 1990
	Stenurus yamagutii	Kuramochi *et al.*, 1990
	Crassicauda sp.	Dailey and Walker, 1978; Cowan *et al.*, 1986
Inner ear	*Pharurus dalli*	Machida, 1974
	Pharurus sp.	Walker, 1975
	Stenurus truei	Machida, 1974
Brain	*Nasitrema* sp.	Cowan *et al.*, 1986
Liver/pancreas/ ducts	*Campula oblonga*	Yamaguti, 1951; Machida, 1974; Walker, 1975; Dailey and Walker, 1978; Conlogue *et al.*, 1985; Cowan *et al.*, 1986
Stomach	*Anisakis simplex*	Dailey, 1971
	Anisakis sp.	Machida, 1974
Urogenital blubber	*Crassicauda* sp.	Walker, 1975, 1990; Dailey and Walker, 1978; Conlogue *et al.*, 1985; Araki *et al.*, 1994
	Placentonema sp.	Ridgway, 1966
Mammaries	*Phyllobothrium* sp.	Dailey and Brownell, 1972; Walker, 1990
	Placentonema sp.	Ridgway, 1966
	Crassicauda sp.	Dailey and Walker, 1978
Intestines	*Bolbosoma* sp.	Machida, 1974
	Corynosoma sp.	Machida, 1974
External	*Neocyamus physeteris*	Leung, 1967

Human Effects

Directed fisheries

Along the coast of northern Japan, Dall's porpoises are taken for human consumption (Table 6). This fishery, in recent years, has been primarily a harpoon fishery, as shooting has been banned (see description of the fishery in Ohsumi, 1972; Miyazaki, 1983). Annual kills in the past were in the thousands, but in the late 1980s they increased to the tens of thousands. The 1988 catch of over 40 000

Dall's porpoises in this fishery (Kasuya, 1992) represented over one-third of the population known to migrate through the fishing area (Miyashita and Kasuya, 1988) and clearly was not sustainable. In recent years, the catch has been reduced somewhat, but still remains high (Table 6).

By-catches

Large numbers of Dall's porpoises are taken also in various gill net fisheries throughout much of their range, especially in the western Pacific and Bering Sea. The largest kills have been in the Japanese mothership and land-based salmon drift net fisheries (Table 6). Catches in the 1960s probably numbered between 10 000 and 20 000 animals per year (Mizue and Yoshida, 1965), but have declined in recent years, concomitant with reduced fishing effort. Since 1978, Japanese, Taiwanese, and Korean drift net fisheries for squid have developed in the area south of the mothership area. Although other species of small cetaceans are caught more frequently, a significant catch of Dall's porpoise is also known. Monitoring beginning in 1989 suggested that as many as 6000 Dall's porpoises may have been taken per year in the squid fisheries (International Whaling Commission, 1992). Although the effect of these takes on Dall's porpoise numbers is uncertain (Hobbs and Jones, 1993), there is evidence for a decline in abundance in the western North Pacific between 1984 and 1986 (Turnock and Buckland, 1995). A global moratorium on pelagic drift net

TABLE 6 Estimated incidental kill of Dall's porpoise in Japanese salmon drift net fisheries and reported direct kill in Japanese harpoon fishery

	Drift net fisheries		
Year	Mothership[a]	Land-based[a]	Harpoon fishery[b]
1981	2862	2936	9 767
1982	5903	6010	12 833
1983	4280	4429	12 776
1984	3355	3356	9 764
1985	3239	2979	10 378
1986	1719	1392	15 457
1987	1011	1229	25 871
1988	222	319	40 853
1989	36	282	29 048
1990	54	134	21 802
1991	—	—	17 634
1992	—	—	11 403
1993	—	—	14 318
1994	—	—	15 947

[a] Data from Jones (1990), Jones *et al.* (1987), and Anonymous (1991).
[b] Data from International Whaling Commission (1992, 1995, 1996), and Kasuya (1992).

fishing went into effect at the end of 1992, and should reduce or eliminate these kills. Extensive research in take reduction methods in the 1980s was largely ineffective (Hatakeyama *et al.*, 1994).

Relatively small numbers of Dall's porpoises are known to be taken also in gill net fisheries along the west coast of North America (Lennert *et al.*, 1994) and in drift nets for tuna and billfish in the central Pacific (Jones *et al.*, 1992). Dall's porpoises are taken occasionally in other types of fishing gear as well, such as in trawl nets along the west coast of the United States (see Perez and Loughlin, 1991).

Pollution

In recent years, some work has been done on the levels of pollutants in Dall's porpoise tissues (O'Shea *et al.*, 1980; Tanabe *et al.*, 1983; Subramanian *et al.*, 1986, 1987a). High levels of organochlorines have been found, but the effects of these substances remain largely unknown (Tanabe *et al.*, 1994, 1996; Jarman *et al.*, 1996). It has been discovered that females may transfer organochlorines to their offspring during gestation and especially through lactation (Subramanian *et al.*, 1988), and that testosterone levels in males may be reduced by high levels of PCBs and DDE (Subramanian *et al.*, 1987b). These findings suggest that current levels of contaminants in Dall's porpoise tissues may have detrimental effects on production and calf survival.

Acknowledgments

The senior author would like to thank Dr Toshio Kasuya, Dr Hideo Omura, and the late Dr Masaharu Nishiwaki for facilitating his studies of Dall's porpoise in Japan in 1963, 1964, and 1974. We thank two anonymous reviewers for their comments on the manuscript.

References

Amano, M. and Kuramochi, T. (1992). Segregative migration of Dall's porpoise (*Phocoenoides dalli*) in the Sea of Japan and Sea of Okhotsk. *Mar. Mamm. Sci.* **8**, 143–151.

Amano, M. and Miyazaki, N. (1992). Geographic variation and sexual dimorphism in the skull of Dall's porpoise, *Phocoenoides dalli*. *Mar. Mamm. Sci.* **8**, 240–261.

Amano, M. and Miyazaki, N. (1993). External morphology of Dall's porpoise (*Phocoenoides dalli*): growth and sexual dimorphism. *Can. J. Zool.* **71**, 1124–1130.

Andrews, R. C. (1911). A new porpoise from Japan. *Bull. Am. Mus. Nat. Hist.* **30**, 31–52.

Anonymous. (1991). Scientific review of the North Pacific high seas driftnet fisheries, Sidney, B.C., 11–14 June. Report for the United Nations.

Araki, J., Machida, M. and Kuramochi, T. (1994). Three species of *Crassicauda* (Nematoda, Spirurida) from cetaceans in Japanese and adjacent waters. *Bull. Nat. Sci. Mus.* **20A**, 59–65.

Awbrey, F. T., Norris, J. C., Hubbard, A. B. and Evans, W. E. (1979). The bioacoustics of the Dall porpoise-salmon driftnet interaction. *Hubbs/Sea World Res. Inst. Tech. Rep.* **79**, 1–41.

Barlow, J. (1995). The abundance of cetaceans in California waters. Part I: ship surveys in summer and fall of 1991. *Fish. Bull. U.S.* **93**, 1–14.

Barnes, L. G. (1985). Evolution, taxonomy and antitropical distributions of the porpoises (Phocoenidae, Mammalia). *Mar. Mamm. Sci.* **1**, 149–165.

Barr, N. and Barr, L. (1972). An observation of killer whale predation on a Dall porpoise. *Can Field-Naturalist* **86**, 170–171.

Benson, S. B. and Groody, T. C. (1942). Notes on the Dall porpoise (*Phocoenoides dalli*). *J. Mamm.* **23**, 41–51.

Brown, D. H. and Norris, K. S. (1956). Observations of captive and wild cetaceans. *J. Mamm.* **37**, 311–316.

Brownell Jr, R. L. (1975). "*Phocoena dioptrica*". Mammalian Species No. 66, 3 pp.

Buckland, S. T., Cattanach, K. L. and Hobbs, R. C. (1993). Abundance estimates of Pacific white-sided dolphin, northern right whale dolphin, Dall's porpoise and northern fur seal in the North Pacific, 1987–1990. *Int. N. Pacif. Fish. Commn Bull.* **53**, 387–407.

Conlogue, G. J., Ogden, J. A. and Foreyt, W. J. (1985). Parasites of the Dall's porpoise (*Phocoenoides dalli* True). *J. Wildl. Dis.* **21**, 160–166.

Cowan, D. F., Walker, W. A. and Brownell Jr, R. L. (1986). Pathology of small cetaceans stranded along the southern California beaches. *In* "Research on Dolphins" (Eds M. M. Bryden and R. Harrison) pp. 323–367. Clarendon Press, Oxford.

Cowan, I. M. (1944). The Dall porpoise, *Phocoenoides dalli* (True), of the northern Pacific Ocean. *J. Mamm.* **25**, 295–306.

Crawford, T. W. (1981). Vertebrate prey of *Phocoenoides dalli* (Dall's porpoise) associated with the Japanese high seas salmon fishery in the North Pacific Ocean. M.Sc. thesis, University of Washington, Seattle, WA.

Curry, B. E. (1992). Facial anatomy and potential function of facial structures for sound production in the harbor porpoise (*Phocoena phocoena*) and Dall's porpoise (*Phocoenoides dalli*). *Can. J. Zool.* **70**, 2103–2114.

Curry, B. E., Mead, J. G. and Purgue, A. P. (1994). The occurrence of calculi in the nasal diverticula of porpoises (Phocoenidae). *Mar. Mamm. Sci.* **10**, 81–86.

Dailey, M. D. (1971). Distribution of helminths in the Dall porpoise (*Phocoenoides dalli* True). *J. Parasitol.* **57**, 1348.

Dailey, M. D. and Brownell Jr, R. L. (1972). A checklist of marine mammal parasites. *In* "Mammals of the Sea: Biology and Medicine" (Ed. S. H. Ridgway) pp. 528–589. Charles C. Thomas, Springfield, IL.

Dailey, M. D. and Walker, W. A. (1978). Parasitism as a factor (?) in single strandings of southern California cetaceans. *J. Parasitol.* **64**, 593–596.

Dohl, T. P., Guess, R. G., Duman, M. L. and Helm, R. C. (1983). Cetaceans of central and northern California, 1980–1983: Status, abundance, and distribution. Report to Minerals Management Service, Contract #14-12-0001-29090, 284 pp.

Evans, W. E., Awbrey, F. T. and Hackbarth, H. (1988). High frequency pulses produced by free-ranging Commerson's dolphins (*Cephalorhynchus commersonii*) compared to those of phocoenids. *Rep. Int. Whal. Commn*, Spec. Issue No. 9, Cambridge, pp. 173–181.

Fiscus, C. H. and Jones, L. L. (1990). Cephalopods from the stomachs of Dall's porpoise (*Phocoenoides dalli*) from the northwestern Pacific and Bering Sea, 1978–1982.

International Whaling Commission Scientific Committee Report SC/42/SM7, 17 pp. (unpubl.).
Forney, K. A., Barlow, J. and Carretta, J. V. (1995). The abundance of cetaceans in California waters. Part II: Aerial surveys in winter and spring of 1991 and 1992. *Fish. Bull., U.S.* **93**, 15–26.
Green, G. A., Brueggeman, J. J., Grotefendt, R. A., Bowlby, C. E., Bonnell, M. L. and Balcomb III, K. C. (1992). Cetacean distribution and abundance off Oregon and Washington, 1989–1990. *In* "Oregon and Washington Marine Mammal and Seabird Surveys: Final Report" (Ed. J. J. Brueggeman) pp. 1/1–1/100. OCS Study MMS 91–0093 (unpublished contract report).
Hall, J. D. (1979). Aspects of the natural history of cetaceans of Prince William Sound, Alaska. Ph.D. thesis, University of California at Santa Cruz.
Hamilton, J. E. (1941). A rare porpoise of the South Atlantic, *Phocoena dioptrica* (Lahille, 1912). *Discov. Rep.* **21**, 227–234.
Hatakeyama, Y., Ishii, K., Akamatsu, T., Soeda, H., Shimamura, T. and Kojima, T. (1994). A review of studies on attempts to reduce the entanglement of the Dall's porpoise, *Phocoenoides dalli*, in the Japanese salmon gillnet fishery. *Rep. Int. Whal. Commn*, Spec. Issue No. 15, pp. 549–564.
Hershkovitz, P. (1966). Catalog of living whales. *Bull. U.S. Natl Mus.* **246**, 1–259.
Hobbs, R. C. and Jones, L. L. (1993). Impacts of high seas driftnet fisheries on marine mammal populations in the North Pacific. *Int. N. Pacif. Fish. Commn Bull.* **53**, 409–434.
Houck, W. J. (1976). The taxonomic status of the porpoise genus *Phocoenoides*. ACMRR/MM/SC/114. Report presented to the FAO Scientific Consultation on Marine Mammals, Bergen.
International Whaling Commission. (1991). Report of the subcommittee on small cetaceans. *Rep. Int. Whal. Commn* **41**, 1172–1190.
International Whaling Commission. (1992). Report of the subcommittee on small cetaceans. *Rep. Int. Whal. Commn* **42**, 178–234.
International Whaling Commission. (1995). Report of the subcommittee on small cetaceans. *Rep. Int. Whal. Commn* **45**, 165–186.
International Whaling Commission. (1996). Report of the subcommittee on small cetaceans. *Rep. Int. Whal. Commn*, **46**, 160–179.
Jarman, W. M., Nordstrom, R. J., Muir, D. C. G., Rosenberg, B., Simon, M. and Baird, R. W. (1996). Levels of organochlorine compounds, including PCDDS and PCDFS, in the blubber of cetaceans from the west coast of North America. *Mar. Pollut. Bull.* **32**, 426–436.
Jefferson, T. A. (1987). A study of the behavior of Dall's porpoise (*Phocoenoides dalli*) in the Johnstone Strait, British Columbia. *Can J. Zool.* **65**, 736–744.
Jefferson, T. A. (1988). "*Phocoenoides dalli*". Mammalian Species No. 319, 7 pp.
Jefferson, T. A. (1989). Calving seasonality of Dall's porpoise in the eastern North Pacific. *Mar. Mamm. Sci.* **5**, 196–200.
Jefferson, T. A. (1990). Sexual dimorphism and development of external features in Dall's porpoise *Phocoenoides dalli*. *Fish. Bull. U.S.* **88**, 119–132.
Jefferson, T. A. (1991). Observations on the distribution and behaviour of Dall's porpoise (*Phocoenoides dalli*) in Monterey Bay, California. *Aquat. Mamm.* **17**, 12–19.
Johnston, D. G. and Ridgway, S. H. (1969). Parasitism in some marine mammals. *J. Am. Vet. Med. Assn.* **155**, 1064–1072.
Jones, L. L. (1990). Incidental take of Dall's porpoise in high seas gillnet fisheries. International Whaling Commission Scientific Committee Report SC/42/SM12, 18 pp.
Jones, L. L., Rice, D. W. and Wolman, A. A. (1985). Reproductive biology of Dall's porpoise in the western North Pacific Ocean and Bering Sea. Abstract presented at the

Sixth Biennial Conference on the Biology of Marine Mammals, Vancouver, B.C., Canada.

Jones, L. L., Bouchet, G. C. and Turnock, B. J. (1987). Comprehensive report on the take, biology and status of Dall's porpoise. International North Pacific Fisheries Commission Report 3156, 78 pp.

Jones, L. L., Larntz, K., Garrott, R. and Hobbs, R. (1992). Incidental take of cetaceans by large-mesh driftnet fisheries in the North Pacific. International Whaling Commission Scientific Committee Report SC/44/SM26, 27 pp.

Jones, L. L., Gosho, M. E., Rice, D. W., Wolman, A. A. and Muto, M. M. (1988). Biology of male Dall's porpoise in the western North Pacific and Bering Sea. International North Pacific Fisheries Commission Report 3280, 25 pp.

Jones, R. E. (1981). Food habits of smaller marine mammals from northern California. *Proc. Calif. Acad. Sci.* **42**, 409–433.

Joyce, G. C., Rosapepe, J. V. and Ogasawara, J. (1982). White Dall's porpoise sighted in the North Pacific. *Fish. Bull. U.S.* **80**, 401–402.

Kasuya, T. (1978). The life history of Dall's porpoise with special reference to the stock off the Pacific coast of Japan. *Sci. Rep. Whales Res. Inst.* **30**, 1–63.

Kasuya, T. (1982). Preliminary report of the biology, catch and populations of *Phocoenoides* in the western North Pacific. *In* "Mammals of Seas", pp. 3–19. FAO Fisheries Ser. 5.

Kasuya, T. (1992). Examination of Japanese statistics for the Dall's porpoise hand harpoon fishery. *Rep. Int. Whal. Commn* **42**, 521–528.

Kasuya, T. and Jones, L. L. (1984) Behavior and segregation of the Dall's porpoise in the northwestern North Pacific Ocean. *Sci. Rep. Whales Res. Inst.* **35**, 107–128.

Kasuya, T. and Ogi, H. (1987) Distribution of mother–calf Dall's porpoise pairs as an indication of calving grounds and stock identity. *Sci. Rep. Whales Res. Inst.* **38**, 125–140.

Kasuya, T. and Shiraga, S. (1985). Growth of Dall's porpoise in the western North Pacific and suggested geographical growth differentiation. *Sci. Rep. Whales Res. Inst.* **36**, 139–152.

Kasuya, T., Tobayama, T. and Matsui, S. (1984). Review of the live-capture of small cetaceans in Japan. *Rep. Int. Whal. Commn* **34**, 597–602.

Kuramochi, T., Araki, J. and Machida, M. (1990). Pseudaliid nematodes from Dall's porpoise, *Phocoenoides dalli*. *Bull. Nat. Sci. Mus. Tokyo* **18A**, 97–103.

Kuramochi, T., Kubodera, T. and Miyazaki, N. (1993). Squids eaten by Dall's porpoises, *Phocoenoides dalli*, in the northwestern North Pacific and in the Bering Sea. *In* "Recent Advances in Fisheries Biology" (Eds T. Okutani, R. K. O'Dor and T. Kubodera) pp. 229–240. Tokai University Press, Tokyo.

Kuroda, N. (1954). On the affinity of the Dall's and True's porpoises. *Bull. Yamashina Ornithol. Res. Inst.* **5**, 44–46.

Kuzin, A. Y. and Belyaev, V. A. (1985). On the biology of the Dall's porpoise (*Phocoenoides dalli*, True 1885). International Whaling Commission Scientific Committee Report SC/36/SM1, 8 pp.

Law, T. C. and Blake, R. W. (1994). Swimming behaviors and speeds of wild Dall's porpoises (*Phocoenoides dalli*). *Mar. Mamm. Sci.* **10**, 208–213.

Leatherwood, S. and Reeves, R. R. (1986). Porpoises and dolphins. *In* "Marine Mammals of the Eastern North Pacific and Arctic Waters" (Ed. D. Haley) pp. 110–131. Pacific Search Press, Seattle, WA.

Leatherwood, S., Reeves, R. R., Perrin, W. F. and Evans, W. E. (1982). Whales, dolphins and porpoises of the eastern North Pacific and adjacent Arctic waters. NOAA Technical Report NMFS Circular 444, 245 pp.

Lennert, C., Kruse, S., Beeson, M. and Barlow, J. (1994). Estimates of incidental marine mammal bycatch in California gillnet fisheries for July through December, 1990. *Rep. Int. Whal. Commn*, Spec. Issue No. 15, pp. 449–463.

Leung, Y. (1967). An illustrated key to the species of whale-lice (Amphipoda, Cyamidae), ectoparasites of Cetacea, with a guide to the literature. *Crustaceana* **12**, 279–291.

Loeb, V. J. (1972). A study of the distribution and feeding habits of the Dall's porpoise in Monterey Bay, California. M.A. thesis, San Jose State University, CA.

Machida, M. (1974). Helminth parasites of the True's porpoise, *Phocoenoides truei*, Andrews. *Bull. Nat. Sci. Mus. Tokyo* **17**, 221–226.

McCormick, J. G. (1969). Relationship of sleep, respiration and anesthesia in the porpoise: a preliminary report. *Proc. Natl Acad. Sci. USA* **62**, 697–703.

Mead, J. G. (1975). Anatomy of the external nasal passages and facial complex in Delphinidae (Mammalia, Cetacea). *Smithsonian Contrib. Zool.* **207**, 1–72.

Miller, E. J. (1989). Distribution and behavior of Dall's porpoise (*Phocoenoides dalli*) in Puget Sound, Washington. M.Sc. thesis, University of Washington, Seattle, WA.

Miller, E. J. (1990). Photo-identification techniques applied to Dall's porpoise (*Phocoenoides dalli*) in Puget Sound, Washington. *Rep. Int. Whal. Commn*, Spec. Issue No. 12, pp. 429–437.

Miller Jr, G. S. (1929). The gums of the porpoise *Phocoenoides dalli* (True). *Proc. U.S. Natl Mus.* **74**, 1–4.

Miller Jr, G. S. (1930). A note on the skeletons of two Alaskan porpoises. *Smithsonian Misc. Collect.* **82**, 1–2.

Miyashita, T. (1991). Stocks and abundance of Dall's porpoise in the Okhotsk Sea and adjacent waters. International Whaling Commission Scientific Committee Report SC/43/SM7, 24 pp.

Miyashita, T. and Kasuya, T. (1988). Distribution and abundance of Dall's porpoises off Japan. *Sci. Rep. Whales Res. Inst.* **39**, 121–150.

Miyazaki, N. (1983). Catch statistics of small cetaceans taken in Japanese waters. *Rep. Int. Whal. Commn* **33**, 621–631.

Miyazaki, N., Jones, L. L. and Beach, R. (1984). Some observations on the schools of *dalli*- and *truei*-type Dall's porpoises in the northwestern North Pacific. *Sci. Rep. Whales Res. Inst.* **35**, 93–105.

Miyazaki, M., Nakamura, Y. and Amano, M. (1986). Two stranding records of Dall's porpoises on the coast of Joetsu, Niigata Prefecture. *Mem. Natl Sci. Mus.* **19**, 193–202.

Mizue, K. and Yoshida, K. (1965). On the porpoises caught by the salmon fishing gill-net in the Bering Sea and the North Pacific Ocean. *Bull. Faculty Fish. Nagasaki Univ.* **19**, 1–36.

Mizue, K., Yoshida, K. and Takemura, A. (1966). On the ecology of the Dall's porpoise in the Bering Sea and the North Pacific Ocean. *Bull. Faculty Fish. Nagasaki Univ.* **21**, 1–21.

Morejohn, G. V. (1979). The natural history of Dall's porpoise in the North Pacific Ocean. *In* "Behavior of Marine Animals", Vol. 3, "Cetaceans" (Eds H. E. Winn and B. L. Olla) pp. 45–83. Plenum Press, New York.

Morejohn, G. V., Loeb, V. and Baltz, D. M. (1973). Coloration and sexual dimorphism in the Dall's porpoise. *J. Mamm.* **54**, 977–982.

Murayama, T., Somiya, H., Aoki, I. and Ishii, T. (1995). Retinal ganglion cell size and distribution predict visual capabilities of Dall's porpoise. *Mar. Mamm. Sci.* **11**, 136–149.

Newby, T. C. (1982). Life history of the Dall porpoise (*Phocoenoides dalli*, True 1885) incidentally taken by the Japanese high seas salmon fishery in the northwest North Pacific and the western Bering Sea, 1978 to 1980. Ph.D. thesis, University of Washington, Seattle, WA.

Nishiwaki, M. (1963). Taxonomical consideration of genera of Delphinidae. *Sci. Rep. Whales Res. Inst.* **17**, 93–104.
Nishiwaki, M. (1966). A discussion of rarities among smaller cetaceans caught in Japanese waters. *In* "Whales, Dolphins and Porpoises" (Ed. K. S. Norris), pp. 192–204. University of California Press, Berkeley, CA.
Nishiwaki, M. (1967). Distribution and migration of marine mammals in the North Pacific area. *Bull. Ocean Res. Inst.* **1**, 1–64.
Nishiwaki, M. (1972). General biology. *In* "Mammals of the Sea: Biology and Medicine" (Ed. S. H. Ridgway) pp. 3–204. Charles C. Thomas, Springfield, IL.
Norris, K. S. and Prescott, J. H. (1961). Observations on Pacific cetaceans of Californian and Mexican waters. *Univ. Calif. Publ. Zool.* **63**, 291–402.
O'Shea, T. J., Brownell, R. L., Clark, D. R., Walker, W. A., Gay, M. L. and Lamont, T. G. (1980). Organochlorine pollutants in small cetaceans from the Pacific and South Atlantic Oceans, November 1968–June 1976. *Pest. Monit. J.* **14**, 35–46.
Ohsumi, S. (1972). Catch of marine mammals, mainly small cetaceans, by local fisheries along the coast of Japan. *Bull. Far Seas Fish. Res. Lab.* **7**, 137–166.
Ohsumi, S. (1975). Incidental catch of cetaceans with salmon gillnet. *J. Fish. Res. Bd Can.* **32**, 1229–1235.
Okada, Y. and Hayashi, R. (1951). Notes on the migration and foetus of *Phocoenoides dalli* (True) in Japan. *Rep. Fac. Fish. Praefect. Univ. Mie* **1**, 4–6.
Perez, M. A. and Loughlin, T.R. (1991). Incidental catch of marine mammals by foreign and joint venture trawl vessels in the U.S. EEZ of the North Pacific, 1973–88. NOAA Technical Report NMFS 104, 57 pp.
Perrin, W. F. and Myrick Jr, A. C. (1980). Report of the workshop. *Rep. Int. Whal. Commn*, Spec. Issue No. 3, pp. 1–50.
Pike, G. C. and MacAskie, I. B. (1969). Marine mammals of British Columbia. *Fish. Res. Bd Can. Bull.* **171**, 1–54.
Pilleri, G. and Gihr, M. (1970. The central nervous system of the mysticete and odontocete whales. *Invest. Cetacea* **2**, 89–144.
Rice, D. W. (1968). Stomach contents and feeding behavior of killer whales in the eastern North Pacific. *Norsk Hvalfangst-Tid.* **57**, 35–38.
Ridgway, S. H. (1966). Dall porpoise, *Phocoenoides dalli* (True): observations in captivity and at sea. *Norsk Hvalfangst-Tid.* **55**, 97–110.
Ridgway, S. H. (1972). Homeostasis in the aquatic environment. *In* "Mammals of the Sea: Biology and Medicine" (Ed. S. H. Ridgway) pp. 590–747. Charles C. Thomas, Springfield, IL.
Ridgway, S. H. and Johnston, D. G. (1966). Blood oxygen and ecology of porpoises of three genera. *Science* **151**, 456–458.
Ridgway, S. H., Flanigan, N. J. and McCormick, J. G. (1966). Brain-spinal cord ratios in porpoises: possible correlations with intelligence and ecology. *Psychon. Sci.* **6**, 491–492.
Ridgway, S. H. and Kohin, S. (1995). The relationship between heart mass and body mass for three cetacean genera: narrow allometry demonstrates interspecific differences. *Mar. Mamm. Sci.* **11**, 72–79.
Rosel, P. E., Haygood, M. G. and Perrin, W. F. (1995). Phylogenetic relationships among the true porpoises (Cetacea: Phocoenidae). *Molec. Phylogen. Evol.* **4**, 463–474.
Scheffer, V. B. (1950). Porpoises assembling in the North Pacific. *Murrelet* **31**, 16.
Scheffer, V. B. (1953). Measurements and stomach contents of eleven delphinids from the northeast Pacific. *Murrelet* **34**, 27–30.
Sleptsov, M. M. (1961). Observations of small cetaceans in the far east seas and north-west part of the Pacific Ocean. *Trud. Inst. Morfol. Zivotnix* **34**, 136–143.

Stroud, R. K., Fiscus, C. H. and Kajimura, H. (1981). Food of the Pacific white-sided dolphin, *Lagenorhynchus obliquidens*, Dall's porpoise, *Phocoenoides dalli*, and northern fur seal, *Callorhinus ursinus*, off California and Washington. *Fish. Bull. U.S.* **78**, 951–959.

Subramanian, A., Tanabe, S., Fujise, Y. and Tatsukawa, R. (1986). Organochlorine residues in Dall's and True's porpoises collected from northwestern Pacific and adjacent waters. *Mem. Natl Inst. Polar Res.*, Spec. Issue No. 44, pp. 167–173.

Subramanian, A., Tanabe, S. and Tatsukawa, R. (1987a). Age and size trends and male–female differences of PCBs and DDE in *dalli*-type Dall's porpoises, *Phocoenoides dalli* of the northwestern North Pacific. *Proc. Natl Inst. Polar Res. Symp. Polar Biol.* **1**, 205–216.

Subramanian, A., Tanabe, S., Tatsukawa, R., Saito, S. and Miyazaki, N. (1987b). Reduction in testosterone levels by PCBs and DDE in Dall's porpoises of northwestern North Pacific. *Mar. Poll. Bull.* **18**, 643–646.

Subramanian, A., Tanabe, S. and Tatsukawa, R. (1988). Use of organochlorines as chemical tracers in determining some reproductive parameters in *dalli*-type Dall's porpoise *Phocoenoides dalli*. *Mar. Environ. Res.* **25**, 161–174.

Sullivan, R. M. and Houck, W. J. (1979). Sightings and strandings of cetaceans from northern California. *J. Mamm.* **60**, 828–833.

Szczepaniak, I. D., Webber, M. A. and Jefferson, T. A. (1992). First record of a *truei*-type Dall's porpoise from the eastern North Pacific. *Mar. Mamm. Sci.* **8**, 425–428.

Tanabe, S., Mori, T., Tatsukawa, R. and Miyazaki, N. (1983). Global pollution of marine mammals by PCBs, DDTs and HCHs (BHCs). *Chemosphere* **12**, 1269–1275.

Tanabe, S., Iwata, H. and Tatsukawa, R. (1994). Global contamination by persistent organochlorines and their ecotoxicological impact on marine mammals. *Sci. Total Environ.* **154**, 163–177.

Tanabe, S., Kumaran, P., Iwata, H., Tatsukawa, R. and Miyazaki, N. (1996). Enantiomeric ratios of hexachlorocyclohexane in blubber of small cetaceans. *Mar. Pollut. Bull.* **32**, 27–31.

Tomilin, A. G. (1957). "Mammals of the U.S.S.R. and Adjacent Countries", Vol. IX, "Cetacea" (Ed. V. G. Heptner). Nauk USSR, Moscow (English translation, 1967, Israel Program for Scientific Translations, Jerusalem).

True, F. W. (1885). On a new species of porpoise, *Phocoena dalli*, from Alaska. *Proc. U.S. Natl Mus.* **8**, 95–98.

Turnock, B. J. and Buckland, S. T. (1995). Trends in abundance of Dall's porpoise in the western North Pacific, 1979 to 1989. *Rep. Int. Whal. Commn*, Spec. Issue No. 16, pp. 399–408.

Turnock, B. J., Buckland, S. T. and Boucher, G. C. (1995). Population abundance of Dall's porpoise (*Phocoenoides dalli*) in the western North Pacific Ocean. *Rep. Int. Whal. Commn*, Spec. Issue No. 16, pp. 381–398.

Walker, W. A. (1975). Review of the live-capture fishery for smaller cetaceans taken in southern California waters for public display, 1966–73. *J. Fish. Res. Bd Can.* **32**, 1197–1211.

Walker, W. A. (1990). Geographic variation of the parasites *Crassicauda* (Nematoda) and *Phyllobothrium* (Cestoda) in *Phocoenoides dalli* in the northern North Pacific, Bering and Okhotsk Sea. International Whaling Commission Scientific Committee Report SC/42/SM16, 20 pp.

Walker, W. A. (1996). Summer feeding habits of Dall's porpoise, *Phocoenoides dalli*, in the southern Sea of Okhotsk. *Mar. Mamm. Sci.* **12**, 167–181.

Walker, W. A. and Sinclair, E. H. (1990). Geographic variation of Dall's porpoise, *Phocoenoides dalli*, in the eastern North Pacific, Bering Sea, and southern Fisheries Conservation Zone of the western North Pacific Ocean. International Whaling Commission Scientific Committee Report SC/42/SM9, 29 pp.

Wilke, F. and Nicholson, A. J. (1958). Food of porpoises in waters off Japan. *J. Mamm.* **39**, 441–443.

Wilke, F., Taniwaki, T. and Kuroda, N. (1953). *Phocoenoides* and *Lagenorhynchus* in Japan, with notes on hunting. *J. Mamm.* **34**, 488–497.

Winans, G. A. and Jones, L. L. (1988). Electrophoretic variability in Dall's porpoise (*Phocoenoides dalli*) in the North Pacific Ocean and Bering Sea. *J. Mamm.* **69**, 14–21.

Yamaguti, S. (1951). Studies of helminth fauna of Japan. Part 45. Trematodes of marine mammals. *Acta Medidinae Okayama* **7**, 238–294.

Yocom, C. F. (1946). Notes on the Dall porpoise off California. *J. Mamm.* **27**, 364–368.

Index

D

E

OWN COPY. CARDED.